P9-AON-254

ENGINEERING HEAT TRANSFER

ENGINEERING HEAT TRANSFER

B. V. Karlekar
Professor of Mechanical Engineering
Rochester Institute of Technology

R. M. Desmond
Professor of Mechanical Engineering
Head of Department of Mechanical Engineering
Rochester Institute of Technology

WEST PUBLISHING COMPANY
St. Paul • New York • Los Angeles • San Francisco

Library of Congress Cataloging in Publication Data

Karlekar, Bhalchandra V 1939-
Engineering heat transfer.

Bibliography: p.
Includes index.

1. Heat–Transmission. I. Desmond, Robert M., joint author. II. Title.

TJ260.K28 621.4'022 76-25109

ISBN 0-8299-0054-3

2nd Reprint—1978

To the memory of

Shri Balasaheb Ranade

*

Preface

The present text is designed for a first-level undergraduate course in heat transfer. Material in the text is suitable for curricula in engineering as well as technology. It can be used for a single course given in a quarter or a semester system or as a two-course sequence in a quarter system.

During the last several years, educators have begun to concern themselves with the more down-to-earth aspects of engineering. With this in mind, this text has been written giving emphasis to the physical interpretation of problems without sacrificing mathematical rigor.

Our book deals with basic topics such as conduction, convection, and radiation, their application to heat exchangers, and numerical solutions to heat conduction problems. The philosophy in presenting the material is to discuss specific situations first and then to generalize. When discussing any topic, the reader is told beforehand the objective of the particular section and the ways or the steps involved in achieving the stated objective. We feel this to be a particularly essential aspect, since a student who reads the material for the first time is otherwise lost and does not appreciate the path laid out for him. Some of the material is introduced by describing simple physical experiments instead of merely stating mathematical models of the phenomena. Time and again, we have observed that the engineering student has difficulty in applying the calculus that he has learned in a calculus course to situations in engineering disciplines. Consequently, every attempt has been made to incorporate the methodology of calculus learned in a typical calculus course in dealing with the equations related to the heat transfer phenomena.

Chapter 1 gives an introduction to the general topic of heat transfer. Chapter 2 presents an in-depth discussion of one-dimensional heat conduction (including cylindrical and spherical systems). Systems with heat sources and variable thermal conductivity are also presented in Chapter 2. Chapter 3 discusses two-dimensional heat conduction with the emphasis on the conduction shape factor. An exact analytical solution is presented as an option in this chapter, and numerical methods are discussed later in the text. Chapter 4 contains material on transient heat conduction with emphasis on lumped parameter systems and chart solutions.

A separate chapter, devoted to numerical methods for heat conduction problems, shows by examples how problems, difficult to solve analytically, can be handled readily by numerical techniques. Although numerous analytical solutions are available to a host of problems, these usually relate to rather simple situations. If the boundary conditions are a trifle complex, the form of the analytical solution becomes too cumbersome to be of any practical use, and one is almost forced to go to numerical methods. A teacher interested in the first principles of heat transfer and numerical methods for finding approximate solutions to practical heat transfer problems in conduction should find Chapters 1 through 6 and Chapter 10 appropriate.

Chapter 6, *Thermal Radiation*, is different from that found in most heat transfer texts. It is divided into two parts; the first part introduces the minimum number of parameters to solve typical engineering heat transfer problems involving radiation. After the student has developed some feeling or confidence toward the subject matter, the details of the physics of radiation are then expounded in the second part.

Instead of discussing related fluid dynamics in Chapter 8, *Forced Convection*, a separate chapter (Chapter 7) is devoted to selected topics from fluid flow. A reader who is interested only in the heat transfer aspects may skip Chapter 7. Chapter 8 is devoted to forced convection presenting the integral method of solution and the differential formulation of the flat plate problem. A section on dimensional analysis is included in this chapter. Natural convection is discussed in Chapter 9. Every effort has been made to present the latest correlations for the Nusselt number.

Analysis of rectangular fins and double pipe heat exchangers form the subject matter of Chapter 10. The logarithmic mean temperature difference (LMTD) method and the number of transfer units (NTU) technique are discussed in this chapter. Chapter 11 introduces the topic of change of phase.

It is apparent that an increasing number of industries, states, and schools are going "metric." It has been anticipated that significant efforts for converting from the English engineering system of units to the Systeme Internationale d'Units, or the SI units, will be made in the next decade. Keeping such a trend in mind, tables or properties, sample problems, and homework problems are presented in both systems. Such a provision permits an instructor to use exclusively SI or English units or to intermix the two systems.

Although it is recognized that the conversion factor, g_c, is redundant, especially in the SI system, we feel that its inclusion in the text will assist the student in distinguishing between a pound-mass and a pound-force.

We wish to thank Professor E. M. Sparrow of the Mechanical Engineering Department of the University of Minnesota for his meticulous review of the manuscript during its preparation. His comments related to the chapters on radiation and forced convection were particularly helpful.

It is also a pleasure to acknowledge Mrs. Jackqueline Harrison, Mrs. Candy Giandana, Mrs. Antoinette Schmidt, Mr. Ewan Choroszylow, and Mr. Peter Michaels of the Rochester Institute of Technology for their efforts in assisting with the typing of the manuscript, the preparation of the figures, and the proofreading of the material.

B. V. Karlekar
R. M. Desmond

Nomenclature

ENGLISH SYMBOLS

Symbol	Description	English	SI
A	Surface area	ft^2	m^2
A_m	Profile area for a fin	ft^2	m^2
[A]	Coefficient matrix		
Bi	Biot number, (hL/k)	dimensionless	
{B}	Column vector		
C	Capacity rate	Btu/hr°F	W/°K
C_f	Skin friction or local drag coefficient	dimensionless	
$C_{f,\text{av}}$	Average drag coefficient	dimensionless	
c_p	Specific heat at constant pressure	Btu/lb_m °F	J/kg°K
c	Specific heat	Btu/lb_m °F	J/kg°K
D	Diameter	ft	m
D_H	Hydraulic diameter	ft	m
e	Emissive power	$Btu/hr\text{-}ft^2$	W/m^2
erf	Error function		
exp	Exponential function		
F_{i-j}	Shape factor of area i with respect to area j	dimensionless	

		English	SI
F	Force	lb_f	Newton
Fr	Froude number, u^2/gL	dimensionless	
f	Friction factor	dimensionless	
G	Irradiation	Btu/hr-ft^2	W/m^2
Gr	Grashof number, $g\beta\Delta TL_c^3/\nu^2$	dimensionless	
Gr*	Modified Grashof number, $g\beta L_c^4 q/(\nu^2 k)$	dimensionless	
g	Acceleration due to gravity	ft/sec^2	m/s^2
g_c	Gravitational constant	lb_m ft/lb_fsec^2	kg m/Ns^2
H	Distance separating two plates	ft	m
H	Height above a surface	ft	m
h	Convective heat transfer coefficient	Btu/hr-ft^2°F	W/m^2°K
h_{av}	Average coefficient of convective heat transfer	Btu/hr-ft^2°F	W/m^2°K
h_{fg}	Latent heat of vaporization	Btu/lb_m	J/kg
I	Electrical current	ampere	ampere
I.D.	Inside diameter	ft	m
i	Specific enthalpy	Btu/lb_m	J/kg
i	Intensity of radiation	Btu/hr-ft^2	W/m^2
i	Nodal coordinate in x coordinate direction		
i_b	Intensity of radiation leaving blackbody	Btu/hr-ft^2	W/m^2
J	Radiosity	Btu/hr-ft^2	W/m^2
j	Nodal coordinate in y coordinate direction		
k	Thermal conductivity	Btu/hr-ft°F	W/m°K
k_e	Effective thermal conductivity	Btu/hr-ft°F	W/m°K
L	Wall thickness	ft	m
LMTD	Logarithmic mean temperature difference	°F	°C
l	Depth	ft	m
M	Total number of heat flow lanes		
M	Total number of nodes in x coordinate direction		
m	Mass	lb_m	kg
$M\dot{o}m$	Rate of momentum flow	lb_m ft/sec^2	kg m/s^2
$\dot{m}$	Mass rate of flow	lb_m/sec	kg/s

Symbol	Description	English	SI
N	Total number of isotherms in a body for a flux plot		
N	Total number of nodes in y coordinate direction		
NTU	Number of transfer units, (UA/C)	dimensionless	
Nu	Nusselt number, (hL/k)	dimensionless	
n	Direction normal to a surface		
O.D.	Outside diameter	ft	m
P	Power	hp	watt
P	Perimeter	ft	m
Pe	Peclet number, $c_p \rho u_\infty L_c/k$	dimensionless	
Pr	Prandtl number, $\mu c_p/k$	dimensionless	
p	Pressure	lb_f/in^2	N/m^2
Q	Rate of heat flow	Btu/hr	W
q	Rate of heat flow per unit area or heat flux	Btu/hr-ft^2	W/m^2
$\dot{q}$	Rate of heat generation per unit volume per unit time	Btu/hr-ft^3	W/m^3
R	Electrical resistance	ohm	ohm
R	Thermal resistance	hr°F/Btu	°K/W
Ra	Rayleigh number, $g\beta\Delta TL_c^3/(\nu\alpha)$	dimensionless	
Re	Reynolds number, $\rho u_{av} L_c/\mu$	dimensionless	
R_f	Fouling factor	hr-ft^2°F/Btu	m^2°C/W
r	Radius	ft	m
S	Conduction shape factor	ft	m
S	Pitch	ft	m
St	Stanton number, $h/c_p \rho u_\infty$	dimensionless	
T	Temperature	°F, °R	°C, °K
{T}	Column vector of temperatures	°F	°C
t	Thickness	ft	m
U	Internal energy	Btu	J
U	Overall heat transfer coefficient	Btu/hr-ft^2°F	W/m^2°K
u	Velocity	ft/sec	m/s
u	Component of velocity in x coordinate direction	ft/sec	m/s
u^*	Friction velocity	ft/sec	m/s

		English	SI
u_{av}	Average velocity	ft/sec	m/s
u_∞	Free-stream velocity	ft/sec	m/s
V	Electrical voltage	volts	volt
V	Volume	ft^2	m^3
V	Velocity	ft/sec	m/s
v	Component of velocity in y coordinate direction	ft/sec	m/s
W	Rate of work done	ft lb_f/sec	J/s
w	Component of velocity in z coordinate direction	ft/sec	m/s
x	Coordinate direction	ft	m
y	Coordinate direction	ft	m
z	Coordinate direction	ft	m

GREEK SYMBOLS

		English	SI
α	Thermal diffusivity	ft^2/hr	m^2/s
α	Absorptivity	dimensionless	
β	Coefficient of thermal expansion	$°R^{-1}$	$°K^{-1}$
Δ	Difference, change		
δ	Thickness of a boundary layer	ft	m
δ	Width of an air gap	ft	m
ϵ	Surface emissivity	dimensionless	
ϵ	Height of roughness elements in a pipe	dimensionless	
ϵ	Heat exchanger effectiveness	dimensionless	
ϵ_H	Eddy diffusivity of heat transfer	lb_m/ft-hr	kg/ms
ϵ_M	Eddy diffusivity of momentum transfer	lb_m/ft-hr	kg/ms
η_f	Fin effectiveness	dimensionless	
θ	Angle	Radian	Radian
Θ	Dimensionless temperature	dimensionless	
λ	Wave length	micron	micron
μ	Viscosity	lb_m/ft-hr	kg/ms
ν	Kinematic viscosity	ft^2/hr	m^2/s
ρ	Density	lb_m/ft^3	kg/m^3
ρ	Reflectivity	dimensionless	

		English	SI
σ	Stefan–Boltzmann constant	Btu/hr-ft^2°R^4	W/m^2°K^4
σ	Vapor–liquid surface tension	lb_f/ft	N/m
τ	Time	hr	s
τ	Shear stress	lb_f/ft^2	N/m^2
τ	Transmissivity	dimensionless	
ϕ	Angle	radian	radian
ω	Solid angle	steradian	steradian

SUBSCRIPTS

av	average
B	burnout
b	blackbody
b	bulk conditions
b	bottom surface
cr	critical condition
c	geometric center
c	characteristic dimension
c	complementary solution to a differential equation
c	cold fluid
c	convection
cond	conduction
conv	convection
e	entrance region in a tube
eqv	equivalent
f	fin
g	gas
gen	generated
h	hot fluid
h	horizontal
i	interior conditions
i	nodal point
i	inlet conditions
in	inside condition
in	direction of flow
ins	insulation

l	logitudinal (S_1) direction
l	liquid
liq	liquid
M	conditions at Mth node
max	maximum
n	normal direction
o	exterior condition
o	centerline conditions in a tube at $r = 0$
o	outlet condition
out	exterior surface
out	direction of flow
p	pipe
p	particular solution to a differential equation
p	plate surface condition
rad	radiation
s	shear
s	surface condition
sat	saturation condition
surf	surface condition
t	thermal condition
t	transverse direction (S_t)
t	top surface
t	turbulent flow
tc	thermocouple
th	total (ΣR_{th})
v	vapor
v	vertical
vap	vapor phase
w	wall or surface condition
∞	free-stream condition

*

Contents

APPENDIXES 547

†

1

Introduction to Heat Transfer

1-1 INTRODUCTION

Heat is energy in transit due to a temperature difference. Heat transfer is the area of engineering that deals with the mechanisms responsible for transferring energy from one place to another when a temperature difference exists. In studying thermodynamics, the student is concerned with the conservation of energy and the direction in which energy may be transferred. The majority of time is spent studying equilibrium situations. In the study of heat transfer, both equilibrium and nonequilibrium processes are encountered. The science of heat transfer allows us to determine the time rate of energy transfer caused by a nonequilibrium of temperatures.

To demonstrate the difference between the study of thermodynamics and the study of heat transfer, let us consider the topic of thermodynamic cycles used for the generation of power.

In thermodynamics, we are concerned about the gross transfer of energy (heat or work or any other form) to or from a system. When the topic of power cycles is discussed in thermodynamics, we are interested in the heat energy going into the system, the work output of the system, and the resulting efficiency. No consideration is given either to the time or temperature difference required to bring about the transfer of heat energy, or to whether there is a uniform temperature within the thermodynamic system. On the other hand, the subject of heat transfer seeks to provide answers to such questions as

(1) How long does it take to transfer the heat energy?
(2) How much heat energy is transferred?
(3) What sort of temperature distribution exists in the system?

Answers to such questions render certain thermodynamic power cycles feasible (e.g., Rankine cycle) or impractical (e.g., Carnot cycle).

The subject of heat transfer has a great impact on all energy problems, covering a spectrum ranging from the routine task of heating or cooling buildings to the sophisticated problems associated with nuclear power generation. In the case of climate control for a building, heat balances must be made that equate heat addition from lights, electric motors, fossil-fuel fired engines, people, and solar energy entering windows with heat losses through walls, cracks, and doors. The heat transfer problems associated with a nuclear power plant are much more complex—a carefully controlled nuclear reaction must be monitored to release the correct amount of heat to turn water into steam to drive a steam turbine. The steam leaving the turbine is then condensed in a heat exchanger (condenser) requiring the use of cooling water, which is normally extracted from a river or a lake and which, when returned to its origin, may result in thermal pollution. There is extensive demand put upon heat transfer analyses to determine heat transfer rates during the steam generation and condensation processes and to estimate the quantity of heat that may be effectively dispersed into a given body of water without altering its biological balance.

During our everyday activities we see the results of heat transfer considerations. Let us follow a housewife through an average day. The items mentioned below all required heat transfer analysis in their design. She arises and immediately turns up the *furnace* thermostat to heat the house up to 72°F, goes into the kitchen using a *toaster*, *electric coffee pot*, and *electric fry pan* to prepare breakfast. During the morning, she washes and dries a load of clothes demanding hot water from her *hot water heater* and dry hot air from the *gas clothes drier*. After washing clothes, she vacuums her rug with her *electric vacuum cleaner*, polishes her wood floor with her *electric floor polisher*, and irons her clothes with her *electric steam iron*. For lunch she has corned beef hash cooked on her *gas stove* and watches her favorite soap opera on her *television*. In the afternoon, she drives her *car* to the high school to pick up her children. After supper, she washes her dishes in the *electric dish washer*, listens to her *radio* and before going to bed turns on her *electric air conditioner*.

The practicing engineer encounters numerous heat transfer problems during his daily activities. As examples, the chemical engineer must concern himself with heat transfer rates in various chemical processing operations; the electrical engineer must design his electrical motors so that they do not overheat, and he must worry about proper sizing of electrical transmission wires to prevent excess loss of power during transmission due to Joulean dissipation; the civil and structural engineer must be careful to prevent the creation of thermal stresses in concrete structures since heat is generated during the curing (drying) of concrete resulting in differential expansion of the structural components; metallurgical and ceramic engineers must control temperatures accurately during heat treatment of various metals and ceramics to achieve the desired properties of the heat-treated material; the biomedical engineer is often interested in the effects of temperature level on living organisms; and the mechanical engineer is concerned with heat transfer rates when designing the heating systems for buildings, developing new power plants, improving thermodynamic cycle efficiencies, and working on thermal pollution problems. In addition, modern processes such as xerography demand extensive knowledge of heat conduction, convection, and radiation. Consequently, we see that the engineering science of heat transfer has broad applications in technology and is not limited in scope to one or two isolated areas.

Although heat transfer has been studied for a number of years, its popularity in engineering curricula was brought about as a result of the space effort. The many problems related to generating the power necessary to put man in space and of shielding the space capsule upon re-entry into the earth's atmosphere made it necessary to embark upon a thorough study of the mechanisms of heat transfer.

With the energy shortage currently upon us, it becomes even more important to study heat transfer so that we can utilize our energy reserves more efficiently. By improved methods of energy transport, by new designs that decrease heat losses, by more efficient generation of power, and by improved usage of power, we can draw upon our limited energy resources in an economical manner.

The following sections of this chapter will introduce the mechanisms present when heat is transferred by conduction, convection, and radiation. We will give examples of combined heat transfer processes in which two or more of these mechanisms may be working simultaneously, and we will introduce a dual system of units. The two systems of units to be introduced will be the *English Engineering System,* which is the system that has been used most commonly in engineering practice in the United States and Great Britain, and the *SI* (*le Système International d' Unités*) system of units, which is expected to come into use in this country in the next decade.

1-2 CONDUCTION

The early development of heat conduction is largely due to the efforts of the French mathematician, Fourier (1822), who first proposed the law that is known today as Fourier's Law of Heat Conduction. Fourier's Law is a generalization of empirical information. It predicts how heat is conducted through a medium from a region of high temperature to a region of low temperature. Let us consider an oven, hot on the inside and cool on the outside. The rate of heat transferred, Q, from the inside of the oven to the outside of the oven, is directly proportional to the surface area of the wall, A, which is normal to the direction of heat flow, directly proportional to the temperature difference across the wall, $(T_{in} - T_{out})$, and inversely proportional to the wall thickness, L. Therefore

$$Q \propto \frac{A(T_{in} - T_{out})}{L} \quad \text{or} \quad Q = \frac{kA(T_{in} - T_{out})}{L} \tag{1-1}$$

where k is the proportionality constant and is called the *thermal conductivity* of the wall. It is a physical property of the material in question. It should be pointed out that in order to maintain the temperatures at T_{in} and T_{out}, a quantity of heat, Q, as given by equation (1-1) must be supplied to that face of the wall that is at temperature T_{in}. It could be due to an electric or gas heater located inside the oven.

If equation (1-1) is carried to the differential form, we have

$$Q = -kA\frac{dT}{dx} \tag{1-2}$$

Observe that the quantity (dT/dx) is the change of temperature with respect to the increase in the x coordinate. Since we want the amount of heat, Q, flowing in the positive x direction to be a positive quantity, and since we know from the second law of thermodynamics that heat flows in the direction of decreasing temperature (i.e., dT is negative), we must have a negative sign in equation (1-2). If we consider, for example, a copper rod with its left-hand end placed in a furnace as shown in Figure 1-1, we know from experience that as we move away from the furnace, the temperature of the rod decreases. The quantity, $T(x)$, is the temperature of the rod, which is a function of the distance x from the wall. We know also that heat is flowing from the furnace to the rod and ultimately to the air in the room. Hence, since dT is negative for increasing values of x, there must be a negative sign in equation (1-2) for Q to be a positive heat flow.

Equation (1-2) becomes, in effect, Fourier's Law of Heat Conduction. Writing equation (1-2) as

$$Q = -kA\frac{\partial T}{\partial n} \tag{1-3}$$

implies that $(\partial T/\partial n)$ is the rate of change of temperature in the direction normal to the area, A, regardless of the orientation of the coordinate system chosen. It is commonly referred to as the temperature gradient. Figure 1-2 shows the relationship between the various quantities in equation (1-3).

Let us take a closer look at the thermal conductivity from a physical point of view. From Fourier's equation, equation (1-3), we note that its magnitude tells us

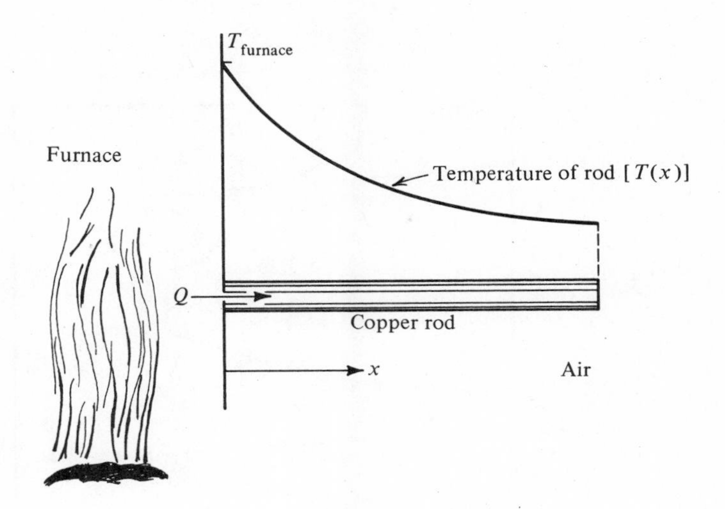

Figure 1-1 Heat flow in a copper rod.

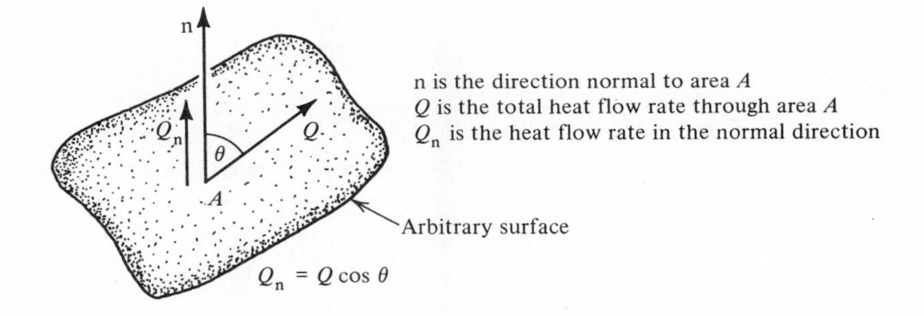

Figure 1-2 Fourier's Law of heat conduction.

how well a material transports energy by conduction. The mechanism for conduction and the mechanism for the transport of electrical current are both highly dependent upon the flow of free electrons. Consequently, materials that are good electrical conductors are good thermal conductors.

A secondary mechanism for thermal conduction in solids is associated with lattice vibrations. This effect explains why, as temperature increases, the thermal conductivity of good electrical conductors usually decreases (see Figure 1-3). The lattice vibrations impede the motion of the free electrons, thus causing that component of the thermal conductivity to decrease faster than the increased component of the thermal conductivity due to greater lattice vibrations. For this reason, we find that the thermal conductivity of pure metals tends to decrease with increasing temperature, while the thermal conductivity of alloys and thermal insulators, which have a few free electrons and depend mainly upon lattice vibrations to conduct heat, tends to increase with increasing temperature (see Figures 1-4 and 1-5).

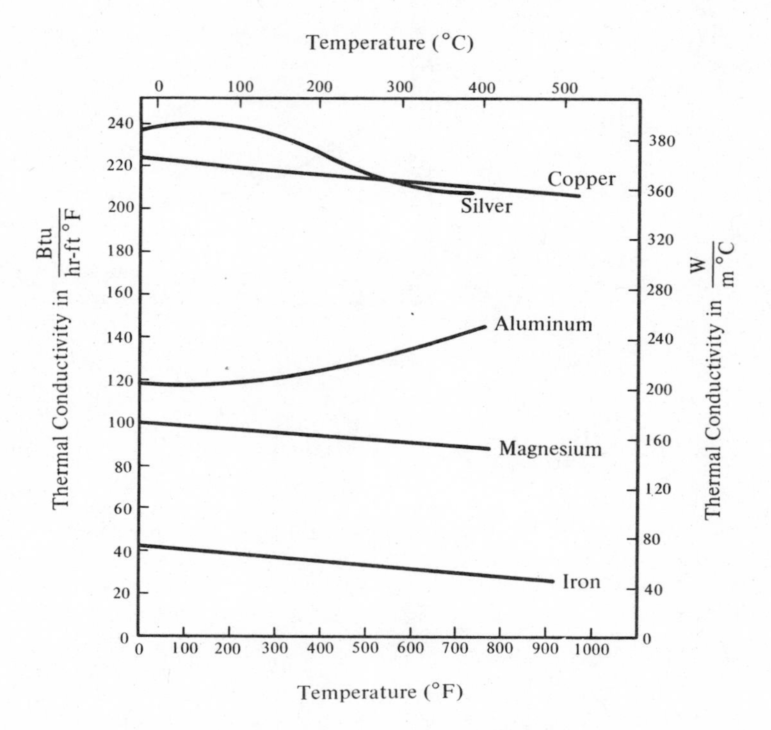

Figure 1-3 Thermal conductivity of metals.

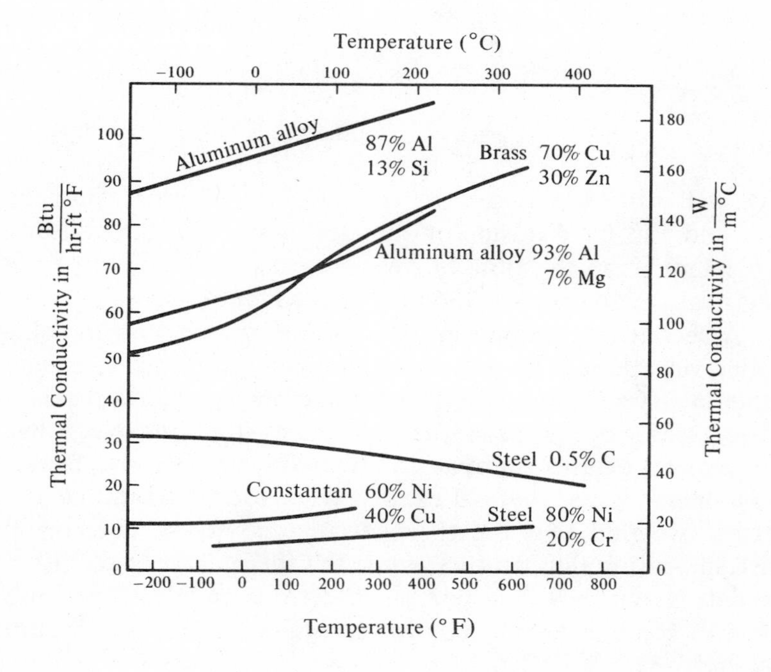

Figure 1-4 Thermal conductivity of alloys.

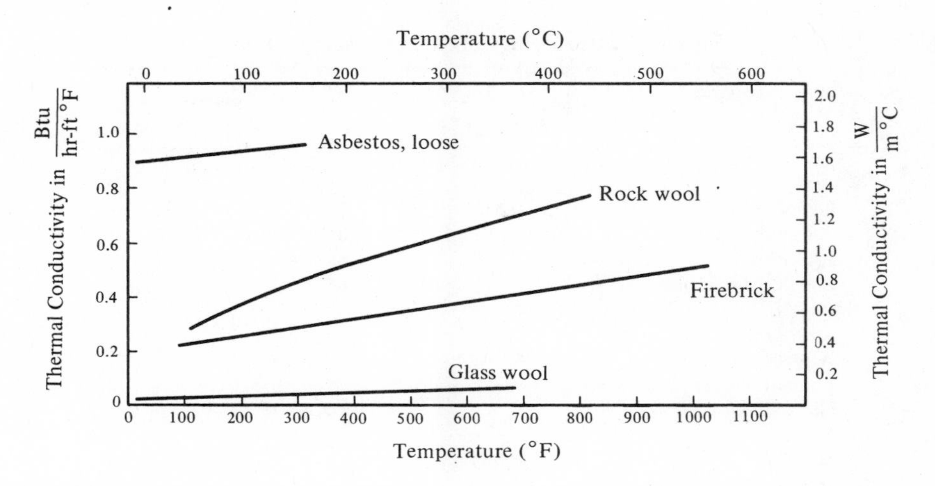

Figure 1-5 Thermal conductivity of insulating materials.

In the case of gases, free electrons are of no significance, and their thermal conductivity is determined by the exchange of energy that takes place during molecular collisions. From kinetic theory, the kinetic energy of an ideal gas is found to be proportional to the square root of the absolute temperature, and consequently, the thermal conductivity of an ideal gas is found to be proportional to the square root of the absolute temperature. The case of liquids is not so easily explained. Representative values of thermal conductivities of different materials are given in Table 1-1. A more detailed list of thermal conductivity values is given in the Appendix.

We should note that materials may have different values of thermal conductivity in the different coordinate directions. As an example, wood exhibits one value of thermal conductivity parallel to its grain and a different value of thermal conductivity normal to its grain. For most engineering calculations, it is assumed that the thermal conductivity is constant and is not directionally dependent. If a material possesses constant properties that are the same in all directions at a point, it is said to be isotropic.

1-3 ANALOGY BETWEEN HEAT CONDUCTION, ELECTRICAL, AND HYDRAULIC SYSTEMS

If an electrical resistor having a constant value of resistance, R, has a voltage difference, V, impressed across it, we know from experience that an electrical current, I, will flow through that resistor. The magnitude of the current will vary linearly with the magnitude of the voltage difference impressed across the resistor. An analogous situation is encountered in the case of the low velocity flow of water flowing from a faucet. In a half-inch pipe, if the water flows at a velocity of

TABLE 1-1 Representative Values of Thermal Conductivity

Material	*k* Btu/hr-ft °F	*T* °F	*k* W/m °C	*T* °C
Metals:				
Copper (pure)	223	68	386	20
Aluminum (pure)	118	68	204	20
Iron (pure)	42	68	72.7	20
Steel (.5% carbon)	31	68	53.6	20
Nonmetals:				
Structural				
Asphalt	.43–.44	68–132	.74–.76	20–55
Cement, cinder	.44	75	.76	24
Glass, window	.45	68	.78	20
Stone				
Marble	1.2–1.7	–	2.08–2.94	–
Wood				
Balsa	.032	86	.055	30
White Pine	.065	86	.112	30
Oak	.096	86	.166	30
Insulating Material:				
Asbestos (sheets)	.096	124	.166	51
Cork (ground)	.025	90	.043	32
Wood shavings	.034	75	.059	24
Saturated Liquids:				
Ammonia, NH_3	.301	68	.521	20
Carbon Dioxide, CO_2	.0504	68	.087	20
Engine Oil	.081	140	.140	60
Gases at Atmos. Pressure:				
Air	.0104	–100	.018	–74
	.0157	100	.027	38
Helium	.0536	–200	.093	–130
	.0977	200	.169	93
Hydrogen	.0567	–190	.098	–123
	.145	350	.251	175
Oxygen	.00790	–190	.0137	–123
	.02212	350	.0383	175

0.6 ft/sec or less, the flow will be laminar* and the analogy will hold. The pressure difference between the upstream of the faucet and the downstream of the faucet is the driving force analogous to the potential difference in the electrical circuit. The extent to which the faucet is opened represents the reciprocal of resistance. The flow rate of water from the faucet is then analogous to the current flow in the electrical circuit, and the magnitude of that flow is directly proportional to the pressure differential upstream and downstream of the faucet. Extending this analogy to the case of heat flow through a wall, we are able to make the following comparisons. The temperature difference across the wall represents the potential difference or driving force, Q represents the current flow, and (L/kA) in equation (1-1) represents the thermal resistance. These ideas are summarized in Figure 1-6.

*See Chapter 7.

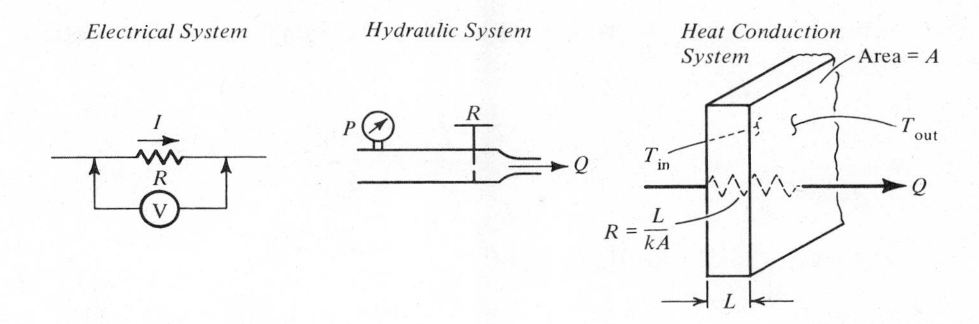

Figure 1-6 Electrical, hydraulic, and thermal analog.

1-4 COMPOSITE WALLS

Consider Figure 1-7(a). There is one wall composed of material "a," and we can write

$$Q = \frac{k_a A_a (T_{in} - T_{out})}{L_a}$$

where k_a refers to the thermal conductivity of the wall, A_a refers to the area of the wall normal to heat flow, and L_a refers to the thickness of the wall. T_{in} and T_{out} are the temperatures of those faces where heat is entering and leaving, respectively.

For Figure 1-7(b), there are two walls in series with each other, a situation which is similar to an electrical circuit consisting of two resistors and a battery

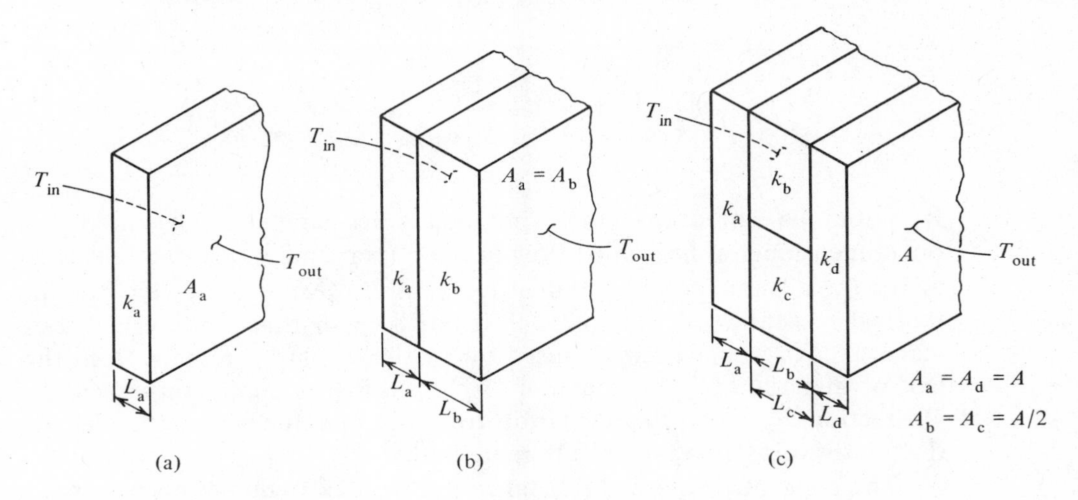

Figure 1-7 Sample composite walls.

across them; the total resistance through the composite wall is given by Figure 1-6 to be

$$\frac{L_a}{k_a A_a} + \frac{L_b}{k_b A_b}$$

and gives a total heat flow, Q, of

$$Q = \frac{T_{in} - T_{out}}{(L_a/k_a A_a) + (L_b/k_b A_b)}$$

where the denominator is referred to as the total thermal resistance. In general, we write

$$Q = \frac{\Delta T_{overall}}{\Sigma R_{th}} \tag{1-4}$$

where $\Delta T_{overall}$ is the overall temperature difference and ΣR_{th} is the sum of the individual thermal resistances.

In Figure 1-7(c), heat flows through material "a" and then has a choice of two parallel paths through materials "b" and "c," finally passing through material "d." For this case $\Delta T_{overall} = T_{in} - T_{out}$ with the ΣR_{th} determined from the set of resistances shown in Figure 1-8. We write the heat flow rate as

$$Q = \frac{T_{in} - T_{out}}{\dfrac{L_a}{k_a A_a} + \left[\dfrac{1}{L_b/(k_b A_b)} + \dfrac{1}{L_c/(k_c A_c)}\right]^{-1} + \dfrac{L_d}{k_d A_d}}$$

It should be noted that the combined series-parallel arrangement is not truly one-dimensional as heat must flow in the y (perpendicular direction) when leaving material "a" and entering either material "b" or "c." As an example, if the thermal resistance of material "c" is much larger than the thermal resistance of material "b," some energy must flow in the upward direction from the bottom half of material "a" into material "b," resulting in a temperature variation in the y direction. The resulting deviation from the one-dimensional model depends on the relative resistances of the alternate paths.

The concept of thermal resistance will be used in numerous places throughout the text when dealing with various modes of heat transfer.

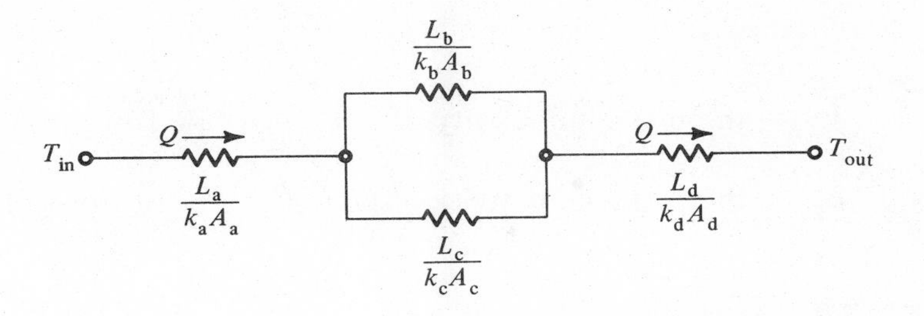

Figure 1-8 Electrical analog for composite wall of Figure 1-7c.

Sample Problem 1-1 A wall of a house measures 8 ft by 20 ft, has no windows, and consists of ¼-inch thick oak paneling and two inches of white pine. The inside temperature of the wall is 70°F and the outside temperature of the wall is 10°F. Determine the heat loss through the wall in Btu/hr.

Solution:

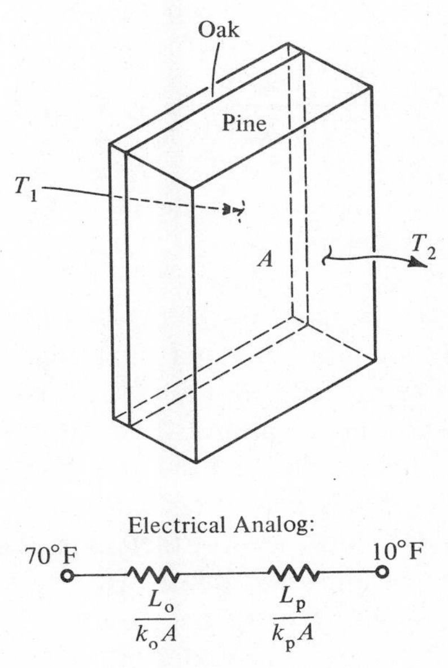

$$Q = \frac{\Delta T_{\text{overall}}}{\Sigma R_{\text{th}}}$$

$$L_o = \text{thickness of oak} = (1/4)/12 = 1/48 \text{ ft}$$

$$L_p = \text{thickness of pine} = 2/12 = 1/6 \text{ ft}$$

$$A = 8 \times 20 = 160 \text{ ft}^2$$

From Table 1-1

k_o = thermal conductivity of oak = 0.096 Btu/hr-ft°F

k_p = thermal conductivity of pine = 0.065 Btu/hr-ft°F

$$\Sigma R_{th} = \frac{L_o}{k_o A} + \frac{L_p}{k_p A}$$

$$= \frac{1/48}{(0.096)(160)} + \frac{1/6}{(0.065)(160)}$$

$$= 1.356 \times 10^{-3} + 16.03 \times 10^{-3}$$

$$= 17.39 \times 10^{-3} \text{ hr°F/Btu}$$

$$\Delta T_{overall} = 70 - 10 = 60°\text{F}$$

$$Q = \frac{60}{17.39 \times 10^{-3}} = 3450 \text{ Btu/hr}$$

1-5 CONVECTION

Convection heat transfer problems are considerably more difficult than those encountered in conduction, and analytical solutions are frequently impossible. This difficulty arises from the fact that the basic mechanism for convection is a combination of conduction and fluid motion. Convection occurs whenever a surface comes in contact with a fluid at a temperature that is different from its own. For example, consider a hot vertical wall in contact with a cold fluid. As time passes, the fluid in intimate contact with the wall is heated by conduction, causing the fluid to become less dense. Due to the difference in density, a buoyancy force results, causing the lighter fluid to rise and to be replaced by cooler fluid with this process being repeated continually. Since the motion of the fluid is set up by natural forces, this type of convection is called natural or free convection. Other examples of natural convection are the mechanisms associated with the flow of heat from household radiators (they radiate little, but convect considerably) and the flow of cigarette smoke in a still room.

If the wall discussed above were the wall of a room in a home and a fan were turned on and directed toward the wall, the motion of the fluid would be caused by an external source and forced convection would result. Another example of forced convection is the car radiator. Once again, there is little radiation heat transfer from a car "radiator," but rather the mechanism for heat flow is forced convection. Should the velocity of the air from a fan hitting a wall be very small, say 0.5 ft/sec, then the overall motion of the air would be due in part to both the fan and buoyancy force, and mixed convection would result.

In practice, engineers use the following expression to determine convection heat transfer rates.

$$Q = hA(T_s - T_\infty) \tag{1-5}$$

Referring to Figure 1-9, we have:

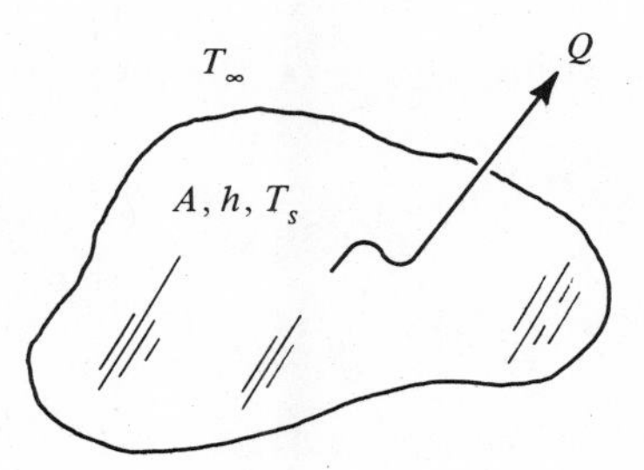

Figure 1-9 Convection heat transfer from a surface.

Q = heat transferred from surface to the surrounding fluid, Btu/hr, W

A = area of the surface, ft^2, m^2

T_s = temperature of the surface, °F, °C

T_∞ = temperature of the surrounding fluid, °F, °C. The subscript ∞ is used to imply that part of the fluid that is sufficiently away from the surface that is to be unaffected by the heat transfer process.

h = convective heat transfer coefficient, Btu/hr-ft^2 °F, W/m^2 °C

This expression does not explain the mechanism of convective heat transfer but rather defines the *convective heat transfer coefficient*. Consequently, much effort is expended in determining h, which is a complicated function of geometry, fluid flow, and fluid properties. The convective heat transfer coefficient will be discussed in greater detail in Chapter 8. Table 1-2 gives estimates for convective heat transfer coefficients under different conditions.

The concept of convective resistance can be introduced in a manner similar to the one for conduction through a wall. Starting with the general equation for convective heat transfer

$$Q = hA(T_s - T_\infty)$$

TABLE 1-2 Representative Values of the Convective Heat Transfer Coefficient

Condition	*h* *Btu/hr ft² °F*	*h* *W/m² °C*
Air, free convection	1–3	5–15
Air or superheated steam forced convection	3–50	15–300
Oil, forced convection	10–300	50–1700
Water, forced convection	50–2,000	300–12,000
Water, boiling	500–10,000	3000–55,000
Steam, condensing	1,000–20,000	5500–100,000

we note that the current flow is Q, the driving force is $(T_s - T_\infty)$, and since

$$Q = \frac{\Delta T_{\text{overall}}}{\Sigma R_{\text{th}}}$$

the thermal resistance for convection must be equal to $(1/hA)$. (See Figure 1-10.)

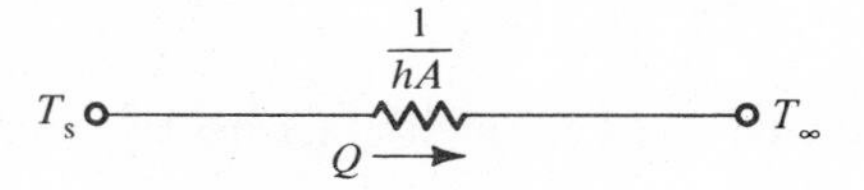

Figure 1-10 Convective resistance for electrical analog.

Consider a wall with convective heat transfer taking place on the right-hand face and the left-hand face being maintained at T_{in} as shown in Figure 1-11. The electrical analogy is shown in Figure 1-12.

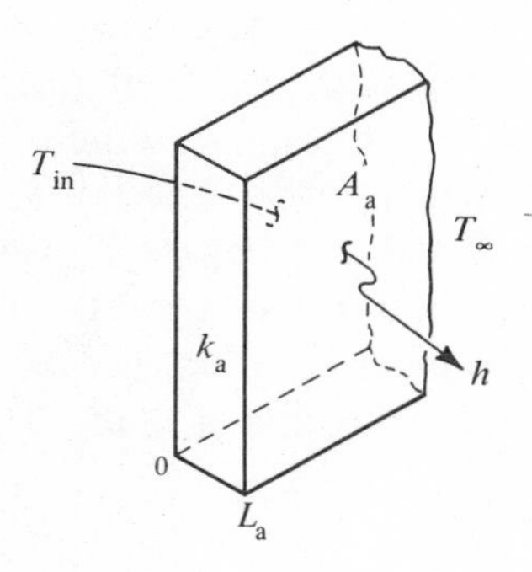

Figure 1-11 Plane wall with convective heat transfer at its right-hand surface.

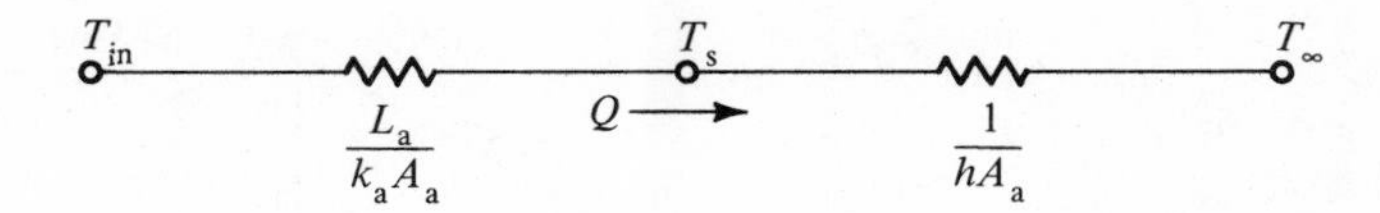

Figure 1-12 Electrical analog for plane wall with convective heat transfer at its right-hand surface.

The resulting expression for the heat flow, Q, is

$$Q = \frac{T_{in} - T_\infty}{\left[\frac{L_a}{k_a A_a} + \frac{1}{hA_a}\right]}$$

In general, note that whereas the conduction resistance, (L/kA), is dependent on the thickness, L, of the wall, no such parameter appears in the convective resistance $(1/hA)$. The latter may be regarded as a resistance at the surface. The former is a property of the body, while the latter is controlled by the physical laws governing convection.

Sample Problem 1-2 Consider the same wall as the one analyzed in Sample Problem 1-1. This time let the inside *air* temperature be 70°F and the outside *air* temperature be 10°F. Referring to Table 1-2, we will estimate the natural convective heat transfer coefficient on the inside of the wall to be 2 Btu/hr-ft^2°F. For maximum heat loss, we will assume a windy day and take a value of the forced convective heat transfer coefficient on the outer wall to be 25 Btu/hr-ft^2°F. Under these conditions, estimate the heat loss through the wall in Btu/hr.

Solution:

$$Q = \frac{\Delta T_{overall}}{\Sigma R_{th}}$$

h_o = value of h at oak-air interface

h_p = value of h at pine-air interface

$$\frac{1}{h_o A} = \frac{1}{(2)(160)} = 3.12 \times 10^{-3} \frac{\text{hr°F}}{\text{Btu}}$$

$$\frac{1}{h_p A} = \frac{1}{(25)(160)} = 0.25 \times 10^{-3} \frac{\text{hr°F}}{\text{Btu}}$$

$$\Sigma R_{th} = (17.39 + 3.12 + 0.25) \times 10^{-3} = 20.76 \times 10^{-3} \frac{\text{hr°F}}{\text{Btu}}$$

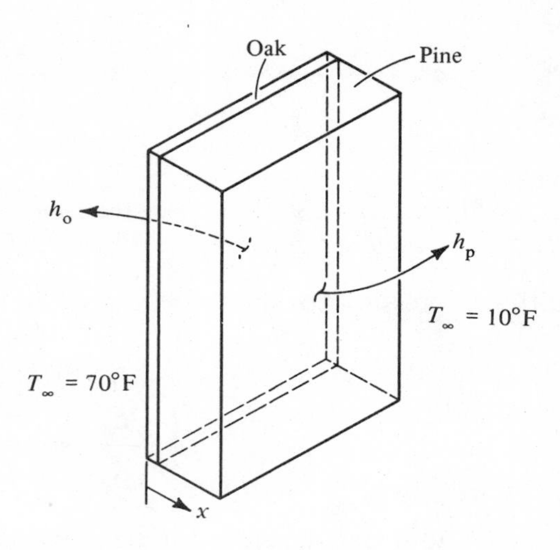

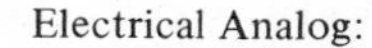

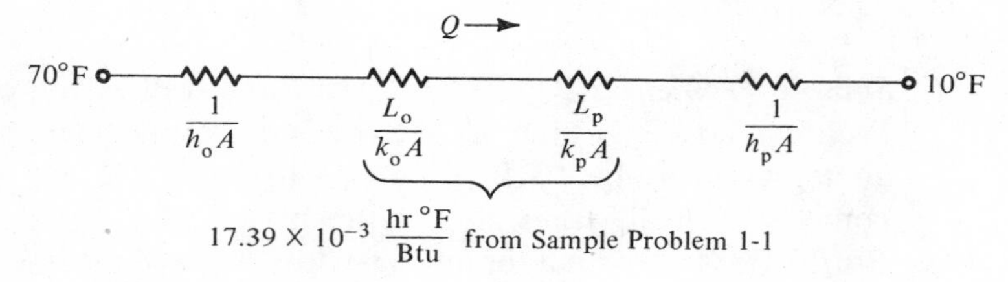

Hence

$$Q = \frac{(70 - 10)}{20.8 \times 10^{-3}} = 2890 \frac{\text{Btu}}{\text{hr}}$$

We note from the solution that consideration of the convective heat transfer coefficient on the inside and the outside surfaces adds significant thermal resistance, cutting the total heat loss through the wall from 3450 Btu/hr to 2890 Btu/hr or nearly 20 percent.

1-6 RADIATION

Unlike convective heat transfer, a transmitting medium is not required for surfaces to exchange heat by radiation. This is true since thermal radiation is electromagnetic radiation that is emitted in the wavelength band between 0.1 and 100 microns (1 micron equals 10^{-6} meters) solely as a result of the temperature of a surface. Therefore, it possesses the same properties as x-rays, visible light, and radio waves, the distinguishing feature being its wavelength band. Visible radiation occurs between wavelengths of 0.35 and 0.75 microns, x-rays occur between wavelengths of 10^{-5} and 2×10^{-2} microns, and radio waves possess wavelengths greater than 10^4 microns. Liquids, solids, and some gases (especially water vapor and hydrocarbons) emit thermal radiation as a result of their temperatures. An ideal emitter, called a *blackbody*, emits thermal radiation according to the *Stefan-Boltzmann* equation

$$e_b = \sigma T^4 \tag{1-6}$$

where e_b is the *emissive power* of the blackbody and is the energy emitted per unit surface area and time; σ is the Stefan-Boltzmann constant; and T is the temperature in degrees absolute. The two absolute temperature scales used are the Rankine and Kelvin scales. We should note that

$$T(°R) = T(°F) + 460°$$

and

$$T(°K) = T(°C) + 273° \tag{1-7}$$

where $T(°F)$ and $T(°C)$ are the temperatures in degrees Fahrenheit and Celsius, respectively. Values of the Stefan-Boltzmann constant in the British and the SI systems are:

$$\sigma = 0.1713 \times 10^{-8} \text{ Btu/hr-ft}^2\ °\text{R}^4$$

$$\sigma = 5.668 \times 10^{-8} \text{ W/m}^2\ °\text{K}^4$$

Nonideal surfaces radiate according to the equation

$$e = \epsilon e_b \tag{1-8}$$

where ϵ is the *emissivity* of the surface and ranges from 0 for an ideal reflector to 1.0 for a blackbody.

To calculate the radiant energy gained or lost by a surface, we define a quantity, F_{1-2}, which is called the *shape factor* and is the fraction of energy leaving surface 1 and headed toward surface 2.

Consider surface 1 to be completely enclosed by surface 2; then the rate of net radiant energy loss from surface 1 may be calculated from

$$Q = A_1 \epsilon_1 (e_{b1} - e_{b2}) \tag{1-9}$$

Equation (1-9) can be used if surface 2 is black or if the area of surface 2 is much greater than the area of surface 1. The quantity, ϵ_1, is the emissivity of body 1. This equation will be derived later in Chapter 6.

Sample Problem 1-3 A black surface is positioned in a vacuum jar so that it absorbs incident solar radiant energy at the rate of 950 W/m^2. If the surface conducts no heat to its surroundings, determine its equilibrium temperature.

Solution: Since there is no conduction and because the surface is in a vacuum, there can be no convection; consequently, the only mechanism available for the transfer of heat is radiation.

At equilibrium, the radiant energy absorbed equals the radiant energy emitted, or

$$q = 950 \text{ W/m}^2 = e_b = \sigma T^4$$

Therefore,

$$T = \left(\frac{950}{\sigma}\right)^{1/4}$$

Knowing that

$$\sigma = 5.668 \times 10^{-8} \text{ W/m}^2 \text{ °K}^4$$

we obtain

$$T = 360\text{°K} = 87\text{°C}$$

Consider a blade of grass in an open field. The upper surface of the grass is exposed or "sees" the night sky. On a clear night, the effective temperature of the night sky may be as low as –40°F. The upper surface of the grass will lose heat by radiation to the sky, and, if it is in thermal equilibrium, it must gain heat from the air by convection, since practically no heat is conducted to it from the ground through the grass itself. The temperature of the ground and the grass are very nearly the same, resulting in a very small radiant energy exchange between them, which may be neglected.
Therefore

$$Q_{rad} = Q_{conv}$$

We know that for heat to be convected to the grass

$$Q_{conv} = hA_{surface}(T_{air} - T_{grass})$$

For Q_{conv} to be a positive quantity, T_{grass} should be lower than T_{air}, and consequently the surface temperature is artifically depressed. We also note for a given Q, that the smaller h is, the greater the temperature difference between

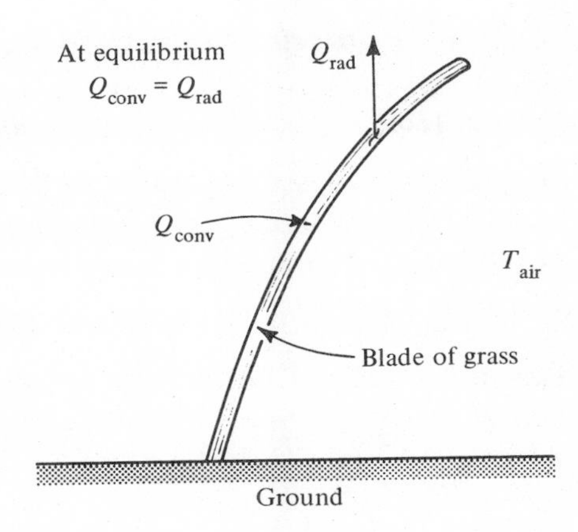

Figure 1-13 Formation of white frost on a blade of grass.

the grass surface and the air becomes. Since h increases with increasing air velocity, on a still night h will be small and T_{grass} may drop to the ice point. When this happens, frost may occur, although T_{air} may be 35°F to 45°F. (See Figure 1-13.) Such a frost is referred to as a "white" frost and can be damaging to certain fruits and vegetables. In states where oranges and apples grow, farmers frequently use smug pots to place an umbrella of smoke over the fruit trees in order to prevent the fruit from being exposed to the cold night sky. In other cases, farmers will use huge fans powered by airplane engines to increase the air velocity and, consequently, increase the value of h so that $(T_{air} - T_{grass})$ becomes small for a given Q and thereby prevents the formation of frost. For orchards on hillsides, farmers frequently cut down trees and brush so that the air will circulate over the fruit trees by natural convection, thereby increasing the value of h.

In certain cases, conduction may also play a part in the formation of white frost. An example is the formation of the frost on the roof of a house. If the ceiling of the house is well insulated, the same process takes place as takes place in an open field, but if the ceiling is poorly insulated, heat may be conducted to the outside and prevent frost. In travelling past a group of homes on the morning of a white frost, one can determine which houses are well insulated by the presence or absence of frost on the various rooftops.

1-7 SYSTEM OF UNITS

At the present time, many countries are in the process of converting from the English Engineering System of units to the SI system of units. With this in mind, both sets of units will be used in this text.

The units and dimensions encountered in this text, and a summary of conversion factors that should prove useful in solving problems are given in the Appendix.

PROBLEMS (English Engineering System of Units)

Refer to Table 1-1 for thermal conductivity values.

1-1. The walls of an apple storage room are made of cork. If one of the walls is 10 ft by 20 ft, what will be the heat loss through that wall on a cold winter day when its inside and outside surface temperatures are 34°F and −10°F, respectively. The wall is 8 inches thick.

1-2. Rework problem 1-1 for the same surface temperatures, but let the cork portion of the wall be sandwiched between 2-inch thick pine paneling on both sides.

1-3. Determine the temperature at both pine-cork interfaces in problem 1-2.

1-4. The walls of an apple storage room are 8 inches thick and are made of cork. If one of the walls is 10 ft by 20 ft, what will be the heat loss through that wall on a cold winter day when the **inside** and **outside air temperatures** are 34°F and −10°F, respectively. The natural convection heat transfer coefficient on the inside surface is 2 Btu/hr-ft^2°F, and the forced convection heat transfer coefficient on the outside surface is 20 Btu/hr-ft^2°F.

1-5. Rework problem 1-4 for the same environmental conditions, but let the cork portion be sandwiched between 2-inch thick pine paneling on both sides.

1-6. Determine the temperatures at both pine-air and pine-cork interfaces in problem 1-5.

1-7. Consider the white frost problem. Recall that on a still night, the heat lost by the dew on the grass by radiation is equal to the energy gained by convection. Let us consider a patch of dew on the top of the grass for which all the radiant energy leaving it goes to outer space. Further, if we assume the dew to be a black body and outer space to be a black body at −40°F, at what air temperature will the dew temperature drop to 32°F if the convective heat transfer coefficient between the dew and the air is 5 Btu/hr-ft^2°F?

1-8. A piece of aluminum is placed in a vacuum jar. The incident radiant energy from the sun is 300 Btu/hr-ft^2. The aluminum absorbs 10% of the incident solar energy. If, in the steady state, 50% of the absorbed energy is conducted to the surroundings and the remaining 50% is re-radiated to space, estimate the temperature of the aluminum if its emissivity is 0.05.

1-9. A metal plate is placed on a driveway and receives 300 Btu/hr-ft^2 of incident radiant energy from the sun. The plate absorbs 80% of the incident solar energy and has an emissivity of 0.05. Consider the lower surface of the plate to be thermally insulated from the driveway. If the air temperature is 60°F and the natural convection heat transfer coefficient between the plate's surfaces and the surrounding air is 2 Btu/hr-ft^2°R, estimate the temperature of the plate.

PROBLEMS (SI System of Units)

Refer to Table 1-1 for thermal conductivity values.

1-1. A glass window 60 cm by 30 cm is 16 mm thick. If its **inside** and **outside surface temperatures** are 20°C and –20°C, respectively, determine the conduction heat loss through the window.

1-2. A thermally insulated glass window 60 cm by 30 cm is made of two 8-mm thick pieces of glass sandwiching an 8-mm thick air space. Determine the conduction heat loss through the window for the same surface temperatures as problem 1-1.

1-3. Determine the temperature at both internal glass-air interfaces in problem 1-2. Neglect all convective heat transfer.

1-4. A glass window 60 cm by 30 cm is 16 mm thick. If the **inside** and **outside air temperatures** are 20°C and –20°C, respectively, determine the conduction heat loss through the window. The natural convection heat transfer coefficient on the inside surface is 10 $W/m^2 °C$, and the forced convection heat transfer coefficient on the outside surface is 100 $W/m^2 °C$.

1-5. A thermally insulated glass window 60 cm by 30 cm is made of two 8-mm thick pieces of glass sandwiching an 8-mm thick air space. Determine the conduction heat loss through the window if the inside air temperature is 20°C, the outside air temperature is –20°C, the natural convection heat transfer coefficient on the inside surface is 10 $W/m^2 °C$, and the forced convection heat transfer coefficient on the outside surface is 100 $W/m^2 °C$.

1-6. Determine the surface temperatures for problem 1-5.

1-7. Consider the white frost problem. Recall that on a still night, the heat lost by the dew on the grass by radiation is equal to the energy gained by convection. Let us consider a patch of dew on the top of the grass for which all the radiant energy leaving it goes to outer space. Further, if we assume the dew to be a black body and outer space to be a black body at –40°C, at what air temperature will the dew temperature drop to 0°C? The convective heat transfer coefficient between the dew and the air is 30 $W/m^2 °C$.

1-8. A piece of aluminum is placed in a vacuum jar. The incident radiant energy from the sun is 950 W/m^2. The aluminum absorbs 10% of the incident solar energy. If, in the steady state, 50% of the absorbed energy is conducted to the surroundings and the remaining 50% is re-radiated to space, estimate the temperature of the aluminum if its emissivity is 0.05.

1-9. A metal plate is placed on a driveway and receives 950 W/m^2 of incident radiant energy from the sun. The plate absorbs 80% of the incident solar energy and has an emissivity of 0.05. Consider the lower surface of the plate to be thermally insulated from the driveway. If the air temperature is 20°C and the natural convection heat transfer coefficient between the plate's surface and the surrounding air is 10 $W/m^2 °C$, estimate the temperature of the plate.

REFERENCES

[1] Holman, J.P., *Heat Transfer*, 3rd ed., McGraw Hill Book Company, 1972.

[2] Hsu, S. T., *Engineering Heat Transfer,* D. VanNostrand Company, Inc., 1963.

[3] Kreith, F., *Principles of Heat Transfer,* 3rd ed., Intext Educational Publishers, New York, 1973.

[4] Reddick, H. W., and Miller, F. H., *Advanced Mathematics for Engineers,* John Wiley & Sons, 1955.

[5] Rohsenow, W. M., and Hartnett, J. P., *Handbook of Heat Transfer,* McGraw Hill Book Company, 1973.

[6] Sears, F. W., and Zemansky, M. W., *University Physics,* 4th ed., Addison Wesley Publishing Company, 1970.

2

Steady-State One-Dimensional Heat Conduction

2-1 INTRODUCTION

There are two quantities of special interest in the study of heat conduction problems: They are the *heat flow rate* and the *temperature distribution.* Heat flow rates tell us about the energy demands on a given system while the temperature distribution may be required to properly design the system from a materials point of view. In any event, if the temperature distribution is known, heat fluxes may be determined from Fourier's Law. In thermodynamics, the work and heat flow are often determined from First Law considerations. Likewise, in heat conduction, temperature distributions are determined from First Law considerations.

In this chapter, we discuss several one-dimensional problems for which temperature distributions and heat flow rates are calculated utilizing First Law considerations and Fourier's Law. After this material is mastered, the general three-dimensional heat conduction equation is presented and shown to give the same results as those previously obtained. By presenting the general equation after the discussion of several specific problems, it is hoped that the student will gain an insight into the physical interpretation of the terms that comprise the general three-dimensional heat conduction equation.

2-2 HEAT CONDUCTION THROUGH A PLANE WALL

The first problem we will consider is the plane wall problem (Figure 2-1). The plane wall is considered to be made out of a constant thermal conductivity material and to extend to infinity in the y and z directions. It is important to note that the thermal conductivity is constant and is not a function of either location or temperature. The heat conducted through the wall of a room where negligible energy is lost through the edges of the wall might be modeled as a plane wall. For such a problem, the temperature is only a function of x; consequently it is said to be a one-dimensional problem. That is, the dependent variable is the temperature, and the lone independent variable is the location x in the wall.

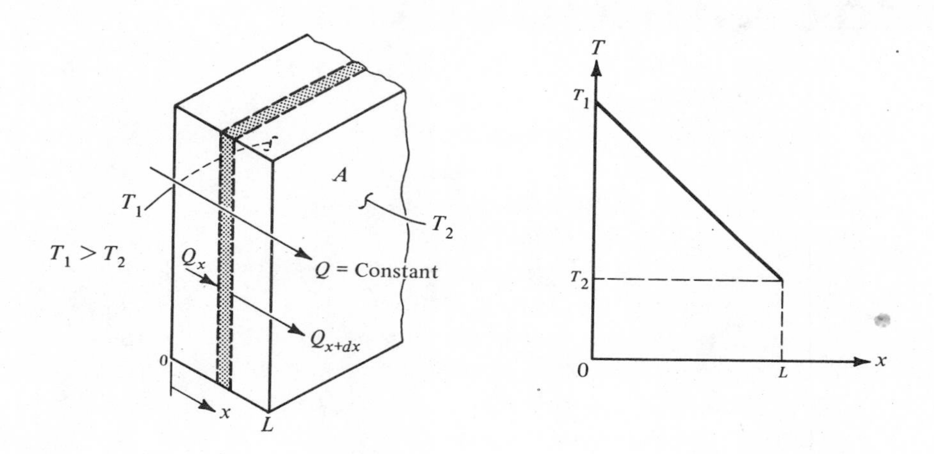

Figure 2-1 Heat conduction through a plane wall.

The governing differential equation can be obtained by making an energy balance on a small volume element of the wall having a thickness, dx, and a cross-sectional area, A.

Let Q_x be the heat conducted into the volume element at $x = x$, and let Q_{x+dx} be the heat conducted out of the volume element at $x = x + dx$. For steady-state conditions, the temperature cannot be a function of time. Hence, the volume element will not experience a change in its internal energy. Since the temperature is assumed to vary only with x, there will be no conduction in the y or z direction (i.e., the temperature gradients in these directions are zero). Assuming that there is no internal heat generation, which occurs when electrical current flows through a conductor, the quantities Q_x and Q_{x+dx} must be equal.

$$Q_x = Q_{x+dx}$$

From Fourier's Law, equation (1-3)

$$Q_x = -kA\frac{\partial T}{\partial x} \tag{2-1}$$

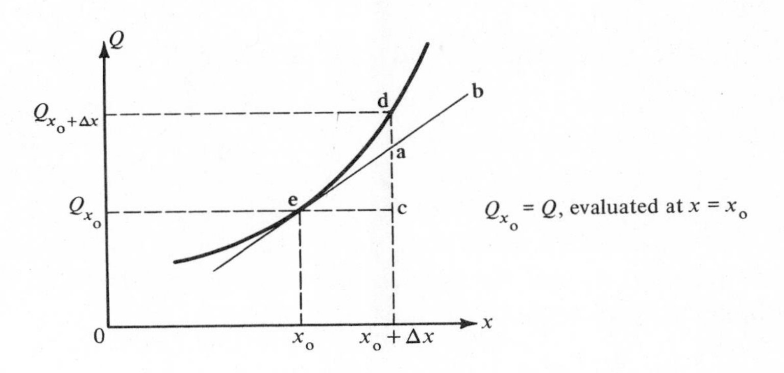

Figure 2-2 Determination of Q_{x+dx}.

To determine Q_{x+dx}, consider the variation of Q with x as shown in Figure 2-2. The following analysis applies to the general three-dimensional case, and the result will also permit us to determine Q_{y+dy} and Q_{z+dz}. The curved line shows an arbitrary variation in Q_x for specific values of y and z (say y_o and z_o) as x changes, while the line **eab** is drawn tangent to the curve at **e**. Point **e** represents the value of Q_{x_o} at x_o and point **d** represents the value of $Q_{x_o+\Delta x}$ at $x_o + \Delta x$. We wish to express $Q_{x_o+\Delta x}$ in terms of Q_{x_o} and its derivatives at x_o.

Let us determine the value of Q at point **a**. First, line **eab** represents the quantity $(\partial/\partial x)(Q_x)$, or, in other words, it represents the slope of the Q versus x curve, evaluated at $x = x_o$. This slope is equal to $\mathbf{ac}/\Delta x$, which tells us that the segment **ac** is equal to $[(\partial/\partial x)(Q_x)](\Delta x)$. Now, if Δx is made sufficiently small, both points **a** and **d** move toward point **e**, and then point **a** approaches point **d**. Thus, in such a limiting process, Δx becomes dx and $Q_{x_o+\Delta x}$ becomes Q_{x_o+dx}. Therefore, dropping the subscript o in x_o, we write *

$$Q_{x+dx} = Q_x + \frac{\partial}{\partial x}(Q_x)dx \tag{2-2}$$

*The same form is obtained when $Q_{x+\Delta x}$ is expanded in a Taylor's series about x_o and terms of second order and higher are neglected as $\Delta x \to dx$.

$$Q_{x_o+\Delta x} = Q_{x_o} + (\partial Q_x/\partial x)_{x=x_o}\Delta x + (\partial^2 Q_x/\partial x^2)_{x=x_o}[(\Delta x)^2/2!] + (\partial^3 Q_x/\partial x^3)_{x=x_o}[(\Delta x)^3/3!] + \ldots$$

or

$$Q_{x_o+dx} = Q_x + \partial/\partial x[(Q_x)dx]$$

or

$$\frac{d}{dx}\left(kA\frac{dT}{dx}\right)dx = 0$$

Since k and A are constants and since dx cannot be zero

$$\frac{d^2T}{dx^2} = 0 \tag{2-3}$$

For the case of a plane wall, the temperature is just a function of one independent variable, x, and the partial derivatives in equations (2-1) and (2-2) become ordinary derivatives; so we can write

$$Q_x = Q_{x+dx} = Q_x + \frac{d}{dx}(Q_x)dx$$

$$-kA\frac{dT}{dx} = -kA\frac{dT}{dx} + \frac{d}{dx}\left(-kA\frac{dT}{dx}\right)dx$$

Equation (2-3) is a second-order differential equation, which indicates that two boundary conditions are necessary for its solution. They are:

$$\text{at } x = 0, \quad T = T_1$$
$$x = L, \quad T = T_2$$

Integrating equation (2-3) once yields

$$\frac{dT}{dx} = C_1$$

where C_1 is a constant of integration. Integrating twice gives

$$T = C_1x + C_2$$

where C_2 is another constant of integration. At $x = 0$, $T = T_1$ so that $C_2 = T_1$, and the temperature is given by

$$T = C_1x + T_1$$

Also, at $x = L$, $T = T_2$ so that

$$T_2 = C_1 L + T_1$$

which gives

$$C_1 = \frac{T_2 - T_1}{L}$$

resulting in

$$T = (T_2 - T_1)\frac{x}{L} + T_1 \tag{2-4}$$

Equation (2-4) is the temperature distribution in the plane wall. It demonstrates to us that the temperature distribution is a linear function of x.

Now that we know how the temperature varies with x, we can determine the heat flux through the wall since Fourier's Law states that $Q = -kA(dT/dx)$.

Proceeding with equation (2-4)

$$T = (T_2 - T_1)\frac{x}{L} + T_1$$

we obtain

$$\frac{dT}{dx} = \frac{T_2 - T_1}{L}$$

and

$$Q = \frac{-kA(T_2 - T_1)}{L}$$

or

$$Q = \frac{kA(T_1 - T_2)}{L} \tag{2-4a}$$

which is the same equation as presented in Section 1-2. It is this quantity of heat, Q, that must be supplied to the left face of the wall so that the temperature difference ($T_1 - T_2$) is maintained. Note that Q is not a function of x, which is also reinforced by the electrical analogy

$\xrightarrow{Q}$

T_1 ——— L/kA ——— T_2

We do not expect the current to change in its value as it travels through the resistor in the steady state.

An alternate approach is to first find the heat flow and then the temperature distribution since we have steady-state conditions and Q is constant. To do that we start with Fourier's equation for the plane wall and proceed as follows:

$$Q = -kA\frac{dT}{dx}$$

Integration gives

$$Q\int_0^L dx = -kA\int_{T_1}^{T_2} dT \qquad (k \text{ and } A \text{ are constants})$$

$$QL = -kA(T_2 - T_1)$$

or

$$Q = \frac{kA(T_1 - T_2)}{L}$$

which is the same as equation (2-4a).

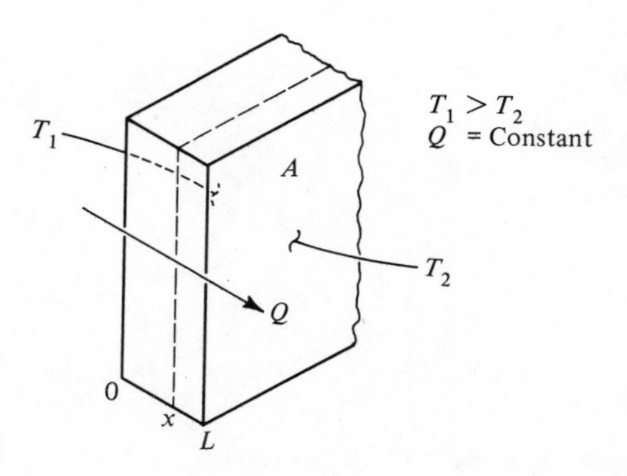

Figure 2-3 Heat flow through a plane wall.

Since the heat flow is the same through all cross sections of the wall (Figure 2-3), we can write, for a portion of the wall between $x = 0$ and $x = x$,

$$Q = \frac{kA(T_1 - T_x)}{x}$$

If Q is substituted from equation (2-4a), there follows, after rearrangement

$$T = (T_2 - T_1)\frac{x}{L} + T_1$$

which is the same as equation (2-4).

Sample Problem 2-1 A household oven is modeled to be a hollow, rectangular box having inside dimensions of 46 cm × 61 cm × 76 cm and outside dimensions of 51 cm × 66 cm × 81 cm. If the heat losses through the corners and edges are ignored, the inside wall temperature is 204°C, the outside wall temperature is 38°C, and the wall material is asbestos, estimate the power input in watts necessary to maintain this steady-state condition.

Solution:

$$Q = \frac{kA(T_1 - T_2)}{L} \quad \text{for each wall}$$

From Table 1-1, $k = 0.166$ W/m°C.

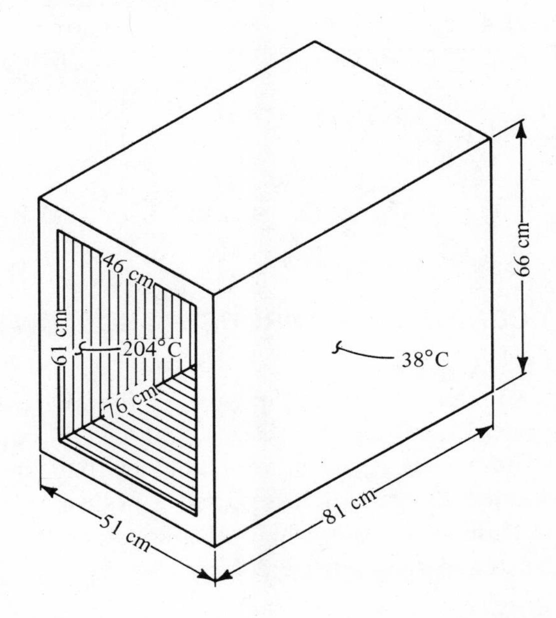

Ignoring corners and edges there are:

(A) 2 walls each having dimensions 46 cm × 61 cm × 2.5 cm
(B) 2 walls each having dimensions 61 cm × 76 cm × 2.5 cm
(C) 2 walls each having dimensions 46 cm × 76 cm × 2.5 cm

For A above

$$Q_A = \frac{2kA(T_1 - T_2)}{L} = \frac{(2)(0.166)\dfrac{(46)(61)}{10^4}(204 - 38)}{2.5/100}$$

$$Q_A = 618 \text{ W}$$

For B above

$$Q_B = \frac{2kA(T_1 - T_2)}{L} = \frac{(2)(0.166)\dfrac{(61)(76)}{10^4}(204 - 38)}{2.5/100}$$

$$Q_B = 1022 \text{ W}$$

For C above

$$Q_C = \frac{2kA(T_1 - T_2)}{L} = \frac{(2)(0.166)\dfrac{(46)(76)}{10^4}(204 - 38)}{(2.5/100)}$$

$$Q_C = 771 \text{ W}$$

$Q_{\text{total}} = Q_A + Q_B + Q_C = 2411$ W or 2.41 kW to maintain steady state

2-3 RADIAL HEAT CONDUCTION THROUGH A HOLLOW SPHERE

In addition to the plane wall problem, there are two other straightforward steady-state one-dimensional problems that we will consider: They are the case of a hollow cylinder that is either very long so that end losses are negligible or its ends are insulated to prevent losses, and a hollow sphere. In both problems, the inside and outside surfaces are held at a constant temperature. Consider first the hollow sphere as shown in Figure 2-4.

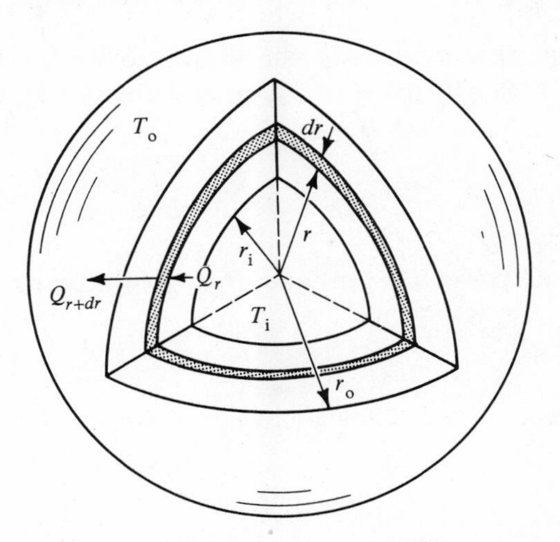

Figure 2-4 Heat conduction through a hollow sphere.

We approach this problem by making an energy balance on a differential volume element in order to determine the appropriate differential equation. Noting that the thermal conductivity is constant, that steady-state conditions exist, and that there are no heat sources, we write the following energy balance:

$$Q_r = Q_{r+dr} \tag{2-5}$$

where

Q_r = heat conducted into a sphereical shell at $r = r$

Q_{r+dr} = heat conducted out of a spherical shell at $r = r + dr$

$$= Q_r + \frac{d}{dr}(Q_r)dr$$

We are able to write the ordinary derivative $(d/dr)(Q_r)$ since the temperature is a function of r alone (i.e., there is only one independent variable, r).

The quantity, Q_r, is given by

$$Q_r = -kA_r \frac{dT}{dr}$$

where

$$A_r = 4\pi r^2$$

Observe that the area A_r in the above equation is not a constant but rather a function of r. Substituting for Q_r in equation (2-5), we obtain

$$-kA_r \frac{dT}{dr} = -kA_r \frac{dT}{dr} + \frac{d}{dr}\left(-kA_r \frac{dT}{dr}\right) dr$$

Now, k is a nonzero constant and dr cannot be zero.

Therefore

$$\frac{d}{dr}\left(A_r \frac{dT}{dr}\right) = 0$$

Substituting for A_r gives

$$\frac{d}{dr}\left(4\pi r^2 \frac{dT}{dr}\right) = 0$$

or

$$\frac{d}{dr}\left(r^2 \frac{dT}{dr}\right) = 0 \qquad (2\text{-}6)$$

Equation (2-6) is the appropriate differential equation for the case of a hollow sphere.

The boundary conditions associated with this problem are

$$\text{at } r = r_i, \quad T = T_i \quad \text{(1st boundary condition)}$$

$$r = r_o, \quad T = T_o \quad \text{(2nd boundary condition)}$$

Integrating once gives

$$r^2 \frac{dT}{dr} = C_1$$

and separating variables, we have

$$dT = C_1 \frac{dr}{r^2}$$

Integrating a second time leads to

$$T = -C_1 \frac{1}{r} + C_2$$

Let $C_3 = -C_1$. Thus

$$T = C_3 \frac{1}{r} + C_2 \tag{2-6a}$$

Applying the 1st boundary condition we get

$$T_i = C_3 \frac{1}{r_i} + C_2$$

Applying the 2nd boundary condition, we obtain

$$T_o = C_3 \frac{1}{r_o} + C_2$$

Solving the two equations for C_2 and C_3 and substituting the resulting expressions in equation (2-6a) yields

$$T(r) = \frac{r_o}{r}\left(\frac{r - r_i}{r_o - r_i}\right)(T_o - T_i) + T_i \tag{2-7}$$

Knowing $Q = -kA_r\,(dT/dr)$, we can then show that

$$Q = \frac{4\pi r_o r_i k(T_i - T_o)}{r_o - r_i} \tag{2-7a}$$

A simpler approach to this problem would be to start by utilizing Fourier's Equation as follows:

$$Q = -kA_r \frac{dT}{dr}$$

Substituting for A_r, we obtain

$$Q = -4\pi k r^2 \frac{dT}{dr}$$

Since Q is constant (steady state), one can readily integrate to yield

$$Q \int_{r_i}^{r_o} \frac{dr}{r^2} = -4\pi k \int_{T_i}^{T_o} dT$$

to obtain

$$-Q\left(\frac{1}{r_o} - \frac{1}{r_i}\right) = -4\pi k(T_o - T_i)$$

or

$$-Q\left(\frac{r_i - r_o}{r_o r_i}\right) = -4\pi k(T_o - T_i)$$

and

$$Q = \frac{4\pi r_o r_i k\,(T_i - T_o)}{r_o - r_i}$$

The above equation is the same as equation (2-7a), which gives the rate of heat flow through the hollow sphere.

Now using equation (2-7a) and equating the rate of heat flow for a portion of the sphere between $r = r_i$ and $r = r$ and the rate for the whole sphere equation (2-7) can be verified.

Sample Problem 2-2 A hollow sphere of pure iron contains a liquid chemical mixture which releases 10^5 Btu/hr. If the inside diameter of the sphere is ½ ft, the outside diameter of the sphere is 1 ft, steady state conditions prevail, and the outside surface temperature of the sphere is 100°F determine the temperature at a location 1 inch from the outer surface of the sphere.

Solution: To determine the temperature at r = 5 inches, it is necessary to find T_i, the temperature at the inner surface of the sphere.
Referring to equation (2-7a)

$$Q = \frac{4\pi r_o r_i k(T_i - T_o)}{r_o - r_i}$$

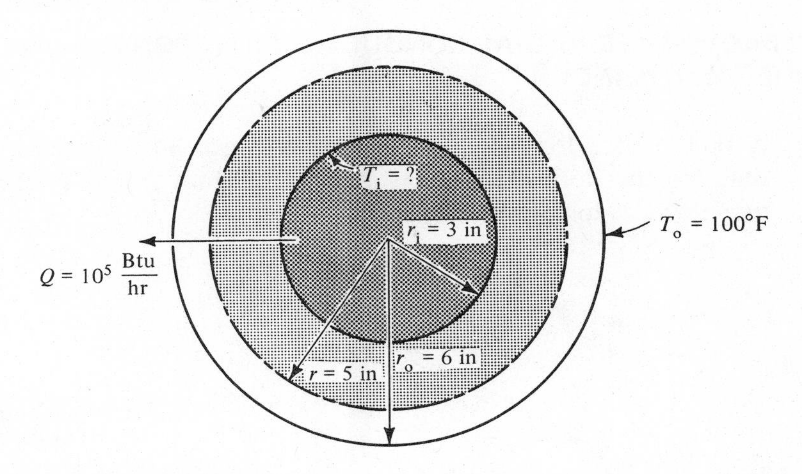

or

$$T_i = \frac{Q(r_o - r_i)}{4\pi r_o r_i k} + T_o$$

Also

$$k = 42 \text{ Btu/hr-ft °F from Table 1-1.}$$

Substituting the appropriate values of Q, r_o, r_i, k, and T_o, we obtain

$$T_i = \frac{10^5\,(6/12 - 3/12)}{(12.56)\,(6/12)\,(3/12)\,(42)} + 100$$

or

$$T_i = 379 + 100 = 479°\text{F}$$

Equation (2-7) is

$$T(r) = \frac{r_o}{r}\left(\frac{r - r_i}{r_o - r_i}\right)(T_o - T_i) + T_i$$

and

$$T(5'') = \frac{6/12}{5/12}\left(\frac{5/12 - 3/12}{6/12 - 3/12}\right)(100 - 479) + 479$$

$$T(5'') = 176°\text{F}$$

2-4 STEADY-STATE RADIAL CONDUCTION IN A LONG HOLLOW CYLINDER

A sketch of a long hollow cylinder that can be analyzed in a fashion similar to that for the hollow sphere is shown in Figure 2-5. A steam pipe can usually be modeled as a long hollow cylinder.

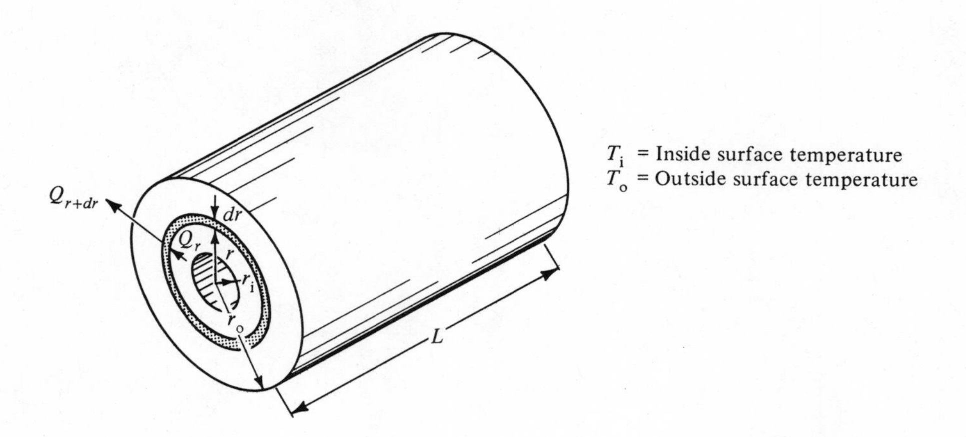

Figure 2-5 Radial heat conduction through a hollow cylinder.

What follows is an abbreviated analysis because of the similarity between this problem and the one presented for the hollow sphere. Noting that the thermal conductivity is constant, that steady state conditions exist, and that there are no heat sources, we write the following energy balance:

$$Q_r = Q_{r+dr} \tag{2-8}$$

where

$$Q_r = \text{heat conducted into a cylindrical shell at } r = r$$

$$Q_{r+dr} = \text{heat conducted out of a cylindrical shell at } r = r + dr$$

$$= Q_r + \frac{d}{dr}(Q_r)dr$$

$$Q_r = -kA_r\frac{dT}{dr}$$

$$A_r = 2\pi rL$$

Proceeding as we did for the hollow sphere, we arrive at the following set of equations:

$$\frac{d}{dr}\left(r\frac{dT}{dr}\right) = 0 \tag{2-9}$$

at

$$r = r_i, \quad T = T_i$$

$$r = r_o, \quad T = T_o$$

The solution to this problem is

$$T(r) = T_i - (T_i - T_o)\frac{\ln(r/r_i)}{\ln(r_o/r_i)} \tag{2-10}$$

To calculate the heat flux for the hollow cylinder, we start with Fourier's equation:

$$Q = -kA_r\frac{dT}{dr} \qquad A_r = 2\pi rL$$

Separating the variables as before, we have

$$Q\int_{r_i}^{r_o}\frac{dr}{r} = -2\pi kL\int_{T_i}^{T_o} dT$$

Integration yields

$$Q\ln\left(\frac{r_o}{r_i}\right) = -2\pi kL(T_o - T_i)$$

or

$$Q = \frac{2\pi kL(T_i - T_o)}{\ln\left(\frac{r_o}{r_i}\right)} \tag{2-11}$$

Sample Problem 2-3 To determine the thermal conductivity of gases, a hollow tube with a heating wire concentric to the walls of the tube is often used. In essence, the gas between the wire and the wall is a hollow cylinder, and the electric current passing through the wire acts as a heat source. Using the data given below, determine the thermal conductivity of the gas in the cell.

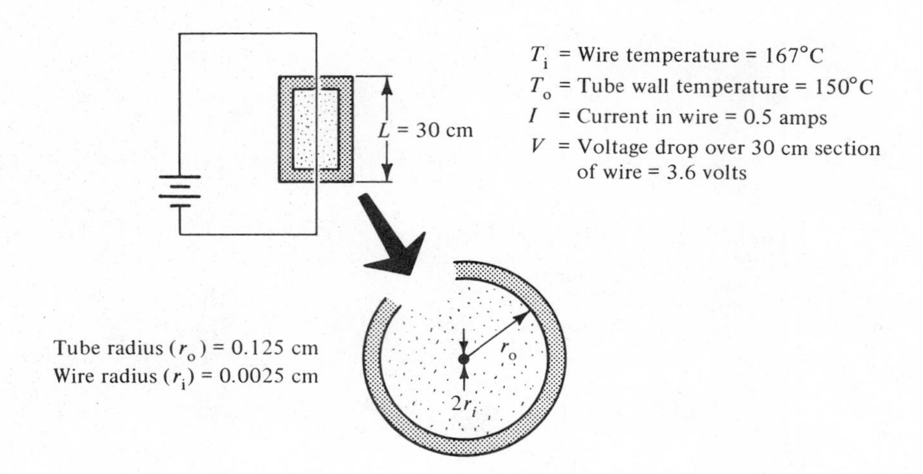

Solution: The power dissipated in the wire = IV.
This is equal to the energy conducted through the gas.

$$Q = IV = (0.5 \text{ amps})(3.6 \text{ volts}) = 1.8 \text{ W}$$

Equation (2-11) gives

$$k = \frac{Q \ln\left(\frac{r_o}{r_i}\right)}{2\pi L(T_i - T_o)}$$

Substituting for Q, r_o, r_i, L, T_i, and T_o, we obtain

$$k = \frac{(1.8)\ln\left(\frac{0.125}{0.0025}\right)}{(6.28)(0.3)(167 - 150)}$$

or

$$k = 0.22 \text{ W/m°C}$$

Referring to Table 1-1, we would suspect that the gas being tested might be hydrogen.

In the preceding sections, we calculated steady state temperature distributions and heat flow rates for a plane wall, a hollow sphere, and a long hollow cylinder. Two approaches were taken in obtaining the desired results. They were:

(1) Using a small volume element, an energy balance was made to determine the differential equation describing temperature as a function of position. The differential equation was solved and the boundary conditions were used to evaluate the constants in the solution. Fourier's Law of heat conduction was then applied to find the heat flow rate.

(2) The heat flow rate was determined directly by integration of Fourier's Equation assuming constant conductivity, and one-dimensional, steady-state heat flow. Next it was noted that the heat transfer rate is the same at any cross section, so that the relation between heat transfer and temperature difference can be applied to any portion of the body. By applying this relation both to a part of the body and to the entire body, the quantity, Q, is eliminated and the temperature distribution is obtained.

2-5 SUMMARY OF THERMAL RESISTANCES

The concept of thermal resistance was introduced in Chapter 1. A summary of the thermal resistances for the three geometries discussed in this chapter is presented here in Table 2-1 for convenience.

TABLE 2-1 Thermal Resistance

Geometry	*Equation for Heat Flow*	*Thermal Resistance*
Plane Wall	$Q = \dfrac{kA(T_1 - T_2)}{L}$	$\dfrac{L}{kA}$
Long Hollow Cylinder	$Q = \dfrac{2\pi kL(T_i - T_o)}{\ln(r_o/r_i)}$	$\dfrac{\ln(r_o/r_i)}{2\pi kL}$
Hollow Sphere	$Q = \dfrac{4\pi r_o r_i k(T_i - T_o)}{r_o - r_i}$	$\dfrac{r_o - r_i}{4\pi r_o r_i k}$
Convective Surface	$Q = hA(T_s - T_\infty)$	$\dfrac{1}{hA}$

Fundamental Equation $Q = \dfrac{\Delta T_{\text{overall}}}{\Sigma R_{\text{th}}}$

2-6 CRITICAL RADIUS OF INSULATION

Suppose we have a steam pipe that we wish to insulate to prevent loss of energy and to prevent people from burning themselves. If the steam is not superheated, some steam will be condensing on the inside of the pipe. The entire inside surface of the pipe will be at a constant temperature $T_{p,\,in}$, approximately equal to the saturation temperature, T_{sat}, corresponding to the pressure of the steam, since the convective resistance under such conditions is negligibly small. We have

$$T_{p,\,in} \simeq T_{sat}$$

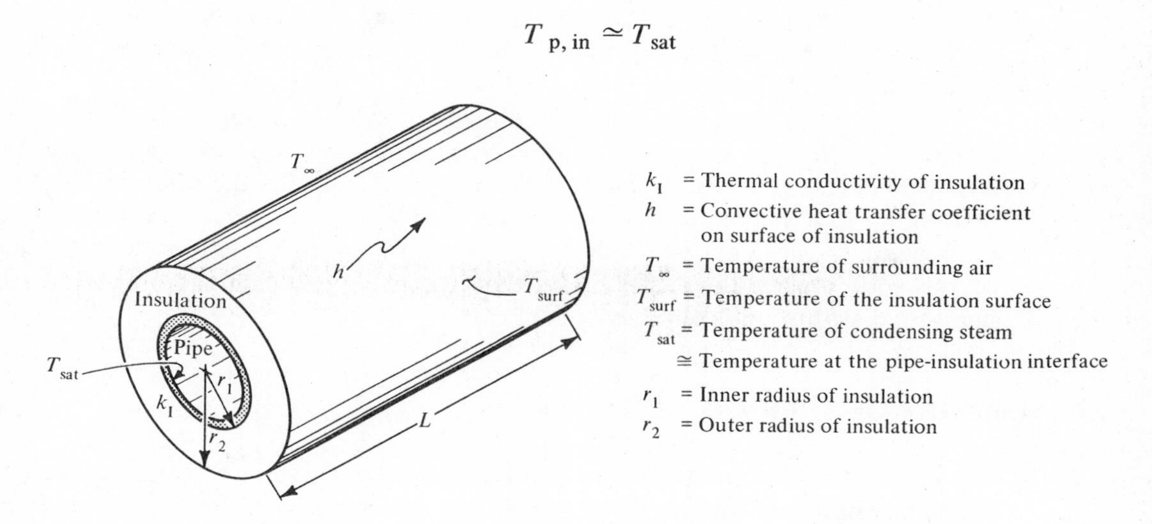

Figure 2-6 Steam pipe used to illustrate the critical radius of insulation.

In a qualitative sense, the temperature drop across the pipe wall is small compared with the drop across the insulation. Referring to Figure 2-6, we note that

$$Q = Q_{\text{pipe wall}} = -k_p A_p \left(\frac{dT}{dr}\right)_{\text{pipe}}$$

$$Q = Q_{\text{insulation}} = -k_I A_I \left(\frac{dT}{dr}\right)_{\text{ins}}$$

We also observe that the thermal conductivity of the pipe material is several orders of magnitude greater than that of the insulation while the ratio of the mean area of the insulation to the mean area of the pipe for radial heat conduction is only in the order of two. Since the same Q is transferred through the pipe as through the insulation under steady-state conditions, it follows that

$$\left(\frac{dT}{dr}\right)_{\text{Ins}} \gg \left(\frac{dT}{dr}\right)_{\text{pipe}}$$

which means that the temperature drop across the pipe wall will be very small. In fact, it will be considered negligible, and the temperature on the inside surface of the insulation will be taken to be T_{sat}.

Figure 2-7 shows an electrical analog constructed for this simplified problem.

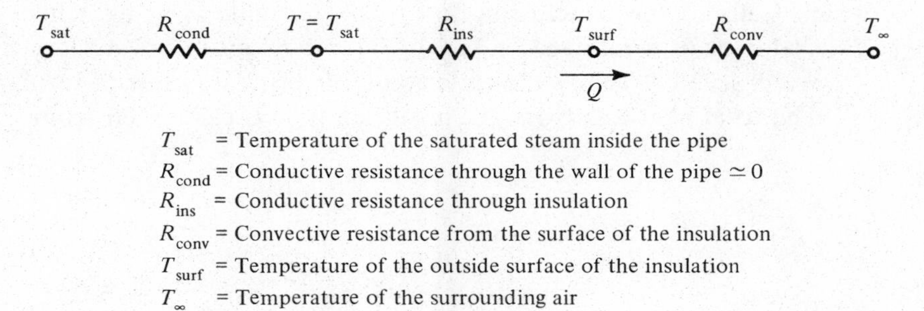

Figure 2-7 Electrical Analog for Heat Flow through an Insulated Steam Pipe

Note that the same quantity of heat flows through all resistors in Figure 2-7, so that Q can be determined by dividing the temperature difference across any resistor or set of resistors by the appropriate resistances, i.e.,

$$Q = \frac{T_{sat} - T_\infty}{R_{cond} + R_{ins} + R_{conv}} \tag{2-12a}$$

or

$$Q = \frac{T_{surf} - T_\infty}{R_{conv}} \tag{2-12b}$$

or

$$Q = \frac{T_{sat} - T_{surf}}{R_{ins}} \tag{2-12c}$$

Consider T_{sat}, and T_∞ (temperature of the room) to remain constant. Let r_i and r_o be the inner and outer radii of the insulation. Then as insulation is added, r_o increases and R_{ins} will also increase, since

$$R_{ins} = \frac{\ln(r_o/r_i)}{2\pi kL}$$

However, since R_{conv} is equal to $(1/h2\pi r_o L)$, the convective resistance will decrease as r_o increases. It is possible that R_{conv} may decrease faster than R_{ins} increases, causing an increase in Q, as revealed by equation (2-12a). We also know that if an infinite amount of insulation were added, Q would approach zero, which leads to the conclusion that there is a value of r_o for which Q is maximum. This value or r_o is referred to as r_{cr}, the critical radius of insulation.

We proceed as follows to determine the critical radius of insulation.

The total heat loss from the insulated pipe is calculated from

$$Q = \frac{\Delta T_{overall}}{\Sigma R_{th}}$$

where

$$\Delta T_{overall} = T_{sat} - T_\infty$$

and from Figure 2-7

$$\Sigma R_{th} = R_{cond} + R_{ins} + R_{conv}$$

where $R_{cond} \simeq 0$. Referring to Table 2-1, we see that

$$R_{ins} = \frac{1}{2\pi kL} \ln\left(\frac{r_o}{r_i}\right)$$

and

$$R_{conv} = \frac{1}{hA} = \frac{1}{h2\pi r_o L}$$

Thus

$$Q = \frac{T_{sat} - T_\infty}{\dfrac{1}{2\pi kL} \ln\left(\dfrac{r_o}{r_i}\right) + \dfrac{1}{2\pi r_o hL}}$$

To determine the value of r_o for which Q is a maximum, we find the value of r_o for which $(dQ/dr_o) = 0$. Then, substituting this value of r_o into (d^2Q/dr_o^2), we are able to verify whether we have found the conditions for a maximum.

$$\frac{dQ}{dr_o} = \frac{0 - (T_{sat} - T_\infty)\left[\dfrac{1}{2\pi kLr_o} - \dfrac{1}{2\pi hLr_o^2}\right]}{\left[\dfrac{1}{2\pi kL}\ln\left(\dfrac{r_o}{r_i}\right) + \dfrac{1}{2\pi hLr_o}\right]^2} = 0$$

If the solution is to be other than trivial, the denominator cannot become infinitely large nor can $(T_{sat} - T_\infty)$ be zero. Therefore

$$\frac{1}{2\pi kLr_o} - \frac{1}{2\pi hLr_o^2} = 0$$

and

$$r_o = \frac{k}{h} = r_{cr} \tag{2-13}$$

where r_{cr} = the critical radius of insulation.

The same result is obtained if ΣR_{th} is minimized by varying r_o. Substitution of $r_o = (k/h)$ into (d^2Q/dr_o^2) results in a negative quantity verifying that

$$r_{cr} = \frac{k}{h}$$

is the value of r_o for which the heat loss is a maximum.

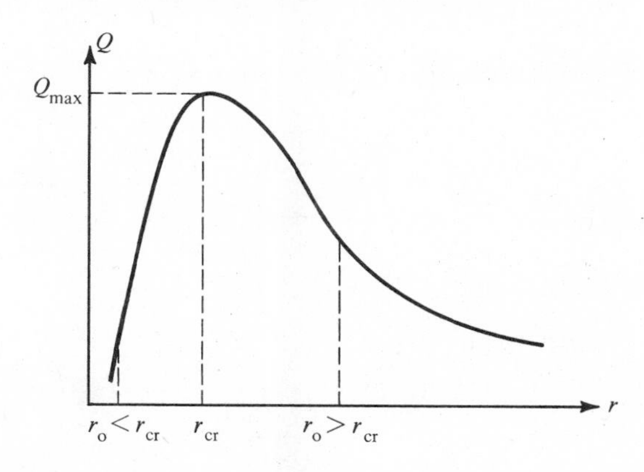

Figure 2-8 Qualitative relationship between r_o and Q.

The effect of increasing the radius r_o is shown in Figure 2-8. It demonstrates that if r_i is less than r_{cr} and insulation is added to the pipe, heat losses will increase and go through a maximum at r_{cr} and then decrease. However, if r_i is greater than r_{cr} and if insulation is added, the heat loss will continually decrease.

The important fact to remember is that as insulation is added, heat losses may actually increase. In some cases, this fact leads to the use of thin layers of insulation to protect people from injury, since additional insulation is costly from a materials cost standpoint and leads to additional energy loss.

Sample Problem 2-4 Wet steam at 325°F passes through a 3 inch O.D. pipe which is insulated with asbestos. The convective heat transfer coefficient between the outer surface of the asbestos and the surrounding 70°F air is 0.5 Btu/hr-ft^2 °F. Determine the critical radius of insulation. For this value of r_o, calculate the heat loss per foot of pipe, and the outer surface temperature.

Solution:

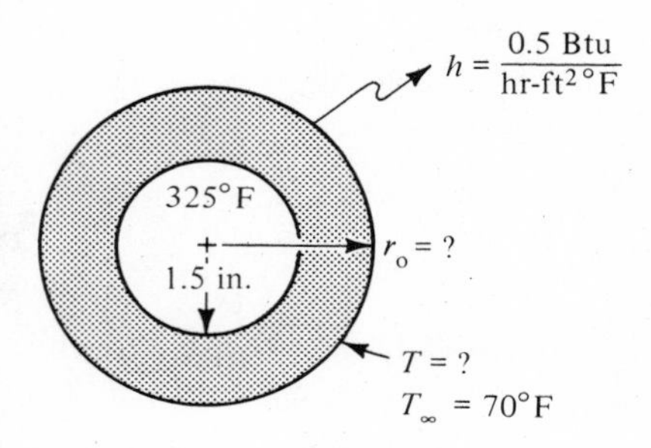

(1) Determination of r_{cr}:

$$r_{cr} = \frac{k}{h} = \frac{0.096 \dfrac{\text{Btu}}{\text{hr-ft°F}}}{0.5 \dfrac{\text{Btu}}{\text{hr-ft}^2\text{°F}}} = 0.192 \text{ ft} = 2.3 \text{ inches}$$

(2) Determination of Q/L when $r_o = r_{cr}$:

$$Q = \frac{\Delta T_{overall}}{\Sigma R_{th}} = \frac{T_i - T_\infty}{R_{ins} + R_{conv}}$$

Noting Table 2-1,

$$Q = \frac{T_i - T_\infty}{\dfrac{1}{2\pi kL} \ln\left(\dfrac{r_{cr}}{r_i}\right) + \dfrac{1}{2\pi r_{cr} hL}}$$

where

$$r_o = r_{cr} = 0.192 \text{ ft}, r_i = 1.5'' = 0.125 \text{ ft}$$
$$h = 0.5 \text{ Btu/hr-ft}^2\ ^\circ\text{F}, k = 0.096 \text{ Btu/hr-ft}^\circ\text{F}$$
$$T_i = 325^\circ\text{F}, T_\infty = 70^\circ\text{F}$$

Substitution gives

$$\frac{Q}{L} = \frac{325 - 70}{\dfrac{1}{(6.28)(0.096)} \ln\left(\dfrac{0.192}{0.125}\right) + \dfrac{1}{(6.28)(0.192)(0.5)}}$$

or

$$\frac{Q}{L} = 107.5 \text{ Btu/hr-ft}$$

We should note that values of r_o other than r_{cr} will yield smaller values of Q/L.

(3) Determination of T_{surf}
An electrical analog for the problem is:

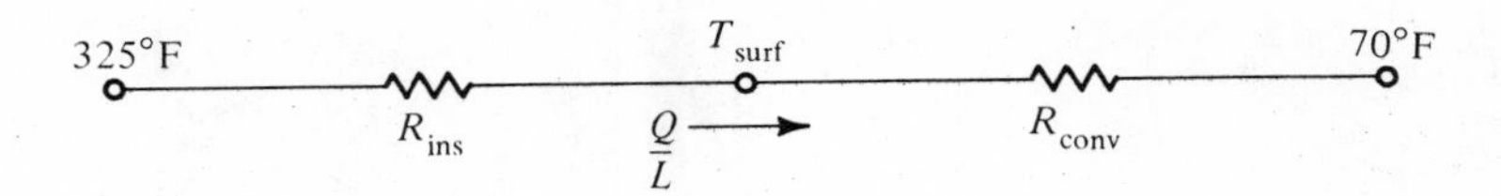

From part (2):

325°F — 0.712 — T_{surf} — 1.666 — 70°F, $\frac{Q}{L}$ →

$$T_{surf} = 70^\circ\text{F} + \left(\frac{1.66}{2.372}\right)(325 - 70) = 248^\circ\text{F}$$

This result tells us that more insulation should be added to prevent human injury. Since more insulation also decreases Q/L, less energy will be wasted during operation of the steam pipe, and the only added costs will be the initial investment in insulation.

2-7 HEAT SOURCE PROBLEMS

Many problems encountered in heat transfer require an analysis that takes into account the generation or absorption of heat within a body. Such problems are encountered in materials through which electrical current flows, in nuclear reactors, in the chemical processing industry, and in combustion processes. Also, thermal stresses are set up in concrete when it dries or "cures" as heat is generated in the curing process causing temperature differences to occur in the structure.

In this section we will consider a plane wall, a long solid cylinder, and a solid sphere with uniform heat sources present. The heat source will be called $\dot{q}$ and will be considered to be uniformly distributed throughout the material and to be constant in time. It will have the units of energy/time-volume. In all cases, it will be assumed that the material has constant conductivity, the heat flow is one-dimensional, and that steady-state conditions exist.

Consider a slab of copper submerged in a constant temperature bath at temperature T_∞. Let an electrical current pass through the slab causing a uniform heat generation, $\dot{q}$, per unit time and volume. The convective heat transfer coefficient on each face of the slab is the same, resulting in a temperature T_w on both faces.

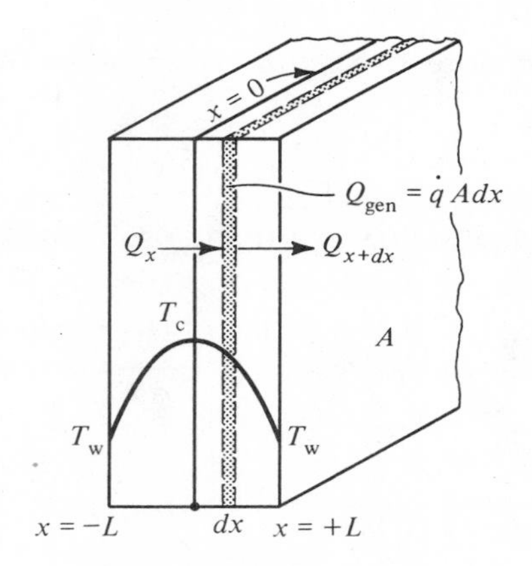

Figure 2-9 Plane wall with uniform heat generation.

To solve for the temperature distribution in the slab, we need to know the appropriate differential equation. This is accomplished by making an energy balance on a slab of thickness, dx, and cross-sectional area, A, as shown in Figure 2-9. The resulting energy balance equation is

$$Q_x + Q_{gen} = Q_{x+dx}$$

where Q_x and Q_{x+dx} are as previously described in Section 2-2 and Q_{gen} is the heat generated per unit time in the slab of thickness, dx, and cross-sectional area, A, and represents an energy increase in the volume element. The quantity Q_{gen} is distinct from the heat conducted into or out of the volume element. It depends only on $\dot{q}$ and the volume of the element. Thus

$$Q_{gen} = \dot{q}Adx$$

Consequently, we have

$$-kA\frac{dT}{dx} + \dot{q}Adx = -kA\frac{dT}{dx} - kA\frac{d^2T}{dx^2}dx$$

which yields

$$\frac{d^2T}{dx^2} + \frac{\dot{q}}{k} = 0 \tag{2-14}$$

Since this is a second-order equation, two boundary conditions are required for a solution. One possible condition is

$$\text{at } x = L, \quad T = T_w \tag{2-14a}$$

where T_w is known. Also, because $\dot{q}$ is uniform throughout the wall material and since $T = T_w$ at both $x = +L$ and at $x = -L$, we expect the temperature distribution to be symmetrical about the center plane of the wall.

From the physics of the problem, if steady-state conditions are to prevail, all the heat generated within the wall must be convected away to the surrounding fluid. Note that the temperature on each face is T_w. Now as one proceeds toward the center of the wall from each face, the temperature must continually increase so that the generated heat may be conducted to the surfaces so that it may be convected away. Consequently, the maximum temperature must occur at the wall centerline with one half of the total heat generated in the wall flowing to each face. Mathematically, this means that

$$\text{at } x = 0, \quad \frac{dT}{dx} = 0 \tag{2-14b}$$

and

$$+kA\frac{dT}{dx}\bigg|_{x=-L} = \frac{1}{2}(\dot{q}2LA) = -kA\frac{dT}{dx}\bigg|_{x=+L} \tag{2-14c}$$

Of the above two possible conditions, we will use equation (2-14b) as the second boundary condition. This means that

$$\text{at } x = 0, \quad \frac{dT}{dx} = 0$$

Referring back to equation (2-14), we can separate variables and integrate once to obtain

$$\frac{dT}{dx} = -\frac{\dot{q}x}{k} + C_1$$

Separating variables again and integrating yields

$$T = -\frac{\dot{q}x^2}{2k} + C_1 x + C_2$$

Applying the second boundary condition,

$$\text{at } x = 0, \quad \frac{dT}{dx} = 0$$

results in

$$0 = -\frac{\dot{q}\,(0)}{k} + C_1$$

or

$$C_1 = 0$$

Thus

$$T = -\frac{\dot{q}x^2}{2k} + C_2$$

Now applying the first boundary condition

$$\text{at } x = L, \quad T = T_w$$

gives

$$T_w = -\frac{\dot{q}L^2}{2k} + C_2$$

or

$$C_2 = T_w + \frac{\dot{q}L^2}{2k}$$

This results in the following temperature distribution

$$T - T_w = \frac{\dot{q}}{2k}(L^2 - x^2) \tag{2-15}$$

Now, the temperature along the centerline, T_c, can be determined by letting $x = 0$ in equation (2-15). This results in

$$T_c = T_w + \frac{\dot{q}L^2}{2k} \tag{2-16}$$

It can be shown that T_c is the maximum temperature in the wall.

Next, let us consider a solid sphere with a uniformly distributed heat source. It is made of a constant conductivity material, and its surface is maintained at a constant temperature T_w.

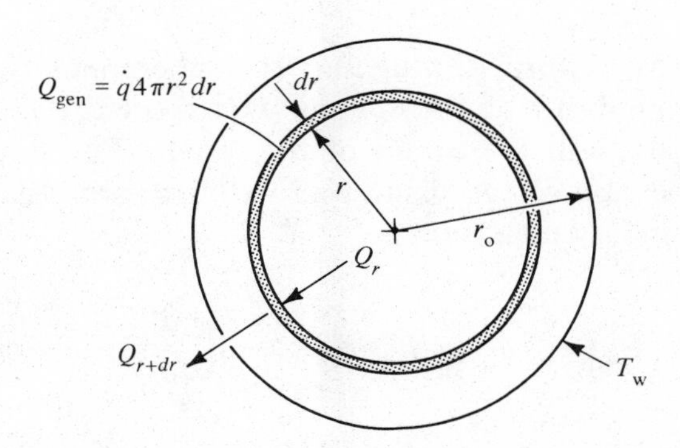

Figure 2-10 Solid sphere with a uniform heat source.

As before, we proceed to formulate the appropriate differential equation for this problem. This equation may be obtained by making an energy balance on a spherical shell of thickness, dr, as shown in Figure 2-10. This approach results in the following:

$$Q_r + Q_{gen} = Q_{r+dr}$$

where Q_r and Q_{r+dr} are as previously described in Section 2-3. Q_{gen} is the heat generated per unit time in the spherical shell of thickness, dr, and cross-sectional area, $4\pi r^2$, and represents an energy increase in the volume element.

$$Q_{gen} = \dot{q} 4\pi r^2 dr$$

Consequently, we have

$$-k4\pi r^2 \frac{dT}{dr} + \dot{q}4\pi r^2 dr = -k4\pi r^2 \frac{dT}{dr} - k4\pi \frac{d}{dr}\left(r^2 \frac{dT}{dr}\right)dr$$

giving

$$\frac{d}{dr}\left(r^2 \frac{dT}{dr}\right) + \frac{r^2 \dot{q}}{k} = 0 \tag{2-17}$$

Since this is a second-order differential equation, two boundary conditions are required for a solution. One condition is

$$\text{at } r = r_o, \quad T = T_w \quad (T_w \text{ is known}) \tag{2-17a}$$

Because $\dot{q}$ is uniform throughout the sphere and T_w is constant over the entire surface (boundary) of the sphere, we expect the temperature distribution to be symmetrical about the center of the sphere. The rationale is the same as that put forth in the case of the plane wall with uniform heat generation. This means our second boundary condition is

$$\text{at } r = 0, \quad \frac{dT}{dr} = 0 \tag{2-17b}$$

We can separate variables in equation (2-17) and integrate once to give

$$r^2 \frac{dT}{dr} = -\frac{\dot{q}r^3}{3k} + C_1$$

Separating variables again and integrating yields

$$T = -\frac{\dot{q}r^2}{6k} - \frac{C_1}{r} + C_2$$

Applying the second boundary condition, equation (2-17b), results in

$$0 = -(0) - \frac{C_1}{0}$$

In order that the above equation be satisfied, it is necessary to have

$$C_1 = 0$$

yielding

$$T = -\frac{\dot{q}r^2}{6k} + C_2$$

Now, applying the first boundary condition, equation (2-17a)

$$T_w = -\frac{\dot{q}r_o^2}{6k} + C_2$$

or

$$C_2 = T_w + \frac{\dot{q}r_o^2}{6k}$$

This results in the following temperature distribution:

$$T - T_w = \frac{\dot{q}}{6k}(r_o^2 - r^2) \tag{2-18}$$

The temperature, T_c, at the center of the sphere ($r = 0$) can be determined by letting $r = 0$ in the above equation. We thus have

$$T_c = T_w + \frac{\dot{q}r_o^2}{6k} \tag{2-19}$$

The term T_c can be shown to be the maximum temperature in the sphere.

The last problem that we wish to discuss in this section is that of uniform heat generation in a long solid cylinder having negligible heat loss out of its end faces (see Figure 2-11). The thermal conductivity of the material is assumed to be constant. The outside surface of the cylinder is maintained at a known temperature, T_w.

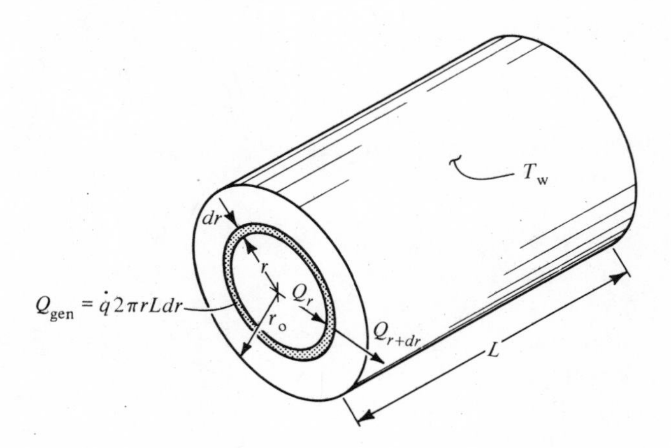

Figure 2-11 Solid cylinder with a uniform heat source.

As with the plane wall and sphere problems, it is necessary to determine the differential equation that describes the temperature distribution. This is done by making an energy balance on a cylindrical shell of thickness, dr. The resulting differential equation is

$$\frac{1}{r}\frac{d}{dr}\left(r\frac{dT}{dr}\right) + \frac{\dot{q}}{k} = 0 \tag{2-20}$$

The development of this equation is left to the reader since its derivation runs parallel to the derivation of equation (2-17). It should be noted that whereas the area of the spherical shell was $4\pi r^2$, the area of the cylindrical shell is $2\pi rL$.

The two boundary conditions used to solve equation (2-20) for $T(r)$ are

$$\text{at } r = 0, \quad \frac{dT}{dr} = 0 \quad \text{(by symmetry)} \tag{2-21}$$

$$\text{at } r = r_o, \quad T = T_w \tag{2-21a}$$

The desired solution to equation (2-20), subject to the above boundary conditions, is

$$T - T_w = \frac{\dot{q}}{4k}(r_o^2 - r^2) \tag{2-22}$$

and

$$T_c = T_w + \frac{\dot{q} r_o^2}{4k} \tag{2-23}$$

where T_c is the maximum temperature in the cylinder and occurs at the center of the cylinder.

In some problems T_w may not be known, but rather $\dot{q}$, h, and T_∞ will be known. In order to solve for the temperature distribution in the three cases discussed in this section, T_w must be determined. This is done by noting that in the steady state, all the heat generated in the solid must be convected away to the surrounding fluid. If this were not so, there would be a buildup of energy in the solid resulting in an increase of the internal energy of the material that would, in turn, necessitate a change in temperature with respect to time. This would conflict with our assumption that steady-state conditions exist. Consequently, for the three geometries discussed we may determine T_w as follows. In general

$$\dot{q}V = hA_{\text{surface}}(T_w - T_\infty)$$

where V is the volume of the entire body and A_{surface} is the total surface area of the body convecting heat to the surrounding fluid at T_∞.

(1) Plane wall of thickness 2L

Total heat generated in the wall = $\dot{q}(A)(2L)$
Heat convected from wall to surrounding fluid = $h2A(T_w - T_\infty)$ and

$$\dot{q}A2L = h2A(T_w - T_\infty)$$

or

$$T_w = \frac{\dot{q}L}{h} + T_\infty \tag{2-24}$$

(2) Sphere of radius r_o

Total energy generated in the sphere = $\dot{q}(4/3)\pi r_o^3$

Heat convected from sphere to surrounding fluid = $h4\pi r_o^2(T_w - T_\infty)$
and

$$\dot{q}(4/3)\pi r_o^3 = h4\pi r_o^2 (T_w - T_\infty)$$

or

$$T_w = \frac{\dot{q}r_o}{3h} + T_\infty \tag{2-25}$$

(3) Cylinder of length L and radius $r_o (L >> r_o)$
Total heat generated in the long solid cylinder = $\dot{q}\pi r_o^2 L$
Heat convected from cylinder to surrounding fluid = $h2\pi r_o L(T_w - T_\infty)$
and

$$\dot{q}\pi r_o^2 L = h2\pi r_o L(T_w - T_\infty)$$

or

$$T_w = \frac{\dot{q}r_o}{2h} + T_\infty \tag{2-26}$$

One last comment relative to the one-dimensional internal heat generation problems discussed above is appropriate. That is, for both the sphere and the cylinder, the maximum temperature will always occur at the center of symmetry if the heat source is uniform and if the convective heat transfer coefficient is constant over the entire surface. However, for the plane wall with a uniform heat source, the maximum temperature occurs at the center plane only if the convective heat transfer coefficients and the ambient temperatures for both faces are equal. If they are not equal, then the problem is solved by noting that the heat conducted to each face is convected away to the fluid in contact with that face. That is, for the left hand face

$$+\left(kA\frac{dT}{dx}\right)_{x=-L} = hA(T_{w1} - T_{\infty 1})$$

where

T_{w1} is the wall temperature on the left-hand face

$T_{\infty 1}$ is the ambient fluid temperature that is exposed to the left-hand face and for the right-hand face

$$-\left(kA\frac{dT}{dx}\right)_{x=+L} = hA(T_{w2} - T_{\infty 2})$$

where

T_{w2} is the wall temperature on the right-hand face

$T_{\infty 2}$ is the ambient fluid temperature that is exposed to the right-hand face

Sample Problem 2-5 An electrical transmission line made of one-inch diameter annealed copper cable carries 200 amps and has a resistance of 0.0104 ohms per foot of length. On a breezy day the convective heat transfer coefficient between the cable and the air is 20 Btu/hr-ft^2 °F. Determine the surface and centerline temperatures of the cable. Consider one foot of length and the air temperature to be 70°F.

Solution:

$$\dot{q}\,(\text{volume}) = I^2R \qquad r_o = \frac{1}{2}\text{ in} = \frac{1}{24}\text{ ft}$$

$$\dot{q}\pi r_o^2 L = I^2R \qquad L = 1\text{ ft}$$

$$\dot{q} = \frac{I^2R}{\pi r_o^2 L}$$

$$\dot{q} = \frac{(4)(10^4)(0.0104)}{(3.14)(1/24)^2(1)}$$

$$\dot{q} = 7.63 \times 10^4\ \frac{\text{Watt}}{\text{ft}^3}$$

$$\dot{q} = 2.23 \times 10^4\ \frac{\text{Btu}}{\text{hr-ft}^3}$$

From equation (2-26)

$$T_w = \frac{\dot{q}r_o}{2h} + T_\infty$$

and

$$T_w = \frac{(2.23)(10^4)(1/24)}{(2)(20)} + 70$$

or

$$T_w = 23 + 70 = 93°F$$

From equation (2-23)

$$T_c = T_w + \frac{\dot{q}r_o^2}{4k} = 93 + \frac{(2.23)(10^4)(1/24)^2}{(4)(223)}$$

$$T_c = 93 + 0.04 = 93.04°F$$

2-8 VARIABLE THERMAL CONDUCTIVITY

For all the problems we have solved so far we have assumed that the thermal conductivity of the conducting medium is constant. However, if we examine Figures 1-3, 1-4, and 1-5, we note that thermal conductivities do change with temperature. In many cases, the dependence is approximately linear. Consequently, it is often possible to describe the thermal conductivity by an equation of the form

$$k = k_o[1 + \beta(T - T_o)] \tag{2-27}$$

where

k_o is the value of the thermal conductivity at temperature T_o

T_o is the reference temperature

T is the temperature at which the conductivity is to be calculated

β is a constant and is positive if k increases with T, or is negative if k decreases with T. Its value is usually small.

Let us determine the heat flow in a plane wall made from a material for which the thermal conductivity varies linearly with temperature. We assume, as before, steady-state conditions without heat sources and begin with Fourier's Law.

$$Q = -kA\frac{dT}{dx}$$

Note that

$$\text{at } x = 0, \quad T = T_1$$

$$\text{at } x = L, \quad T = T_2$$

$$k = k_o[1 + \beta(T - T_o)]$$

and A and Q are constants.
Substituting for k and separating variables, we obtain

$$Q\int_0^L dx = -k_oA\int_{T_1}^{T_2} [1 + \beta(T - T_o)]dT$$

Integration gives

$$QL = -k_oA\left[(T_2 - T_1) - \frac{\beta}{2}(T_2^2 - T_1^2) - \beta T_o(T_2 - T_1)\right]$$

and

$$Q = \frac{k_oA}{L}\left[(1 - \beta T_o)(T_1 - T_2) + \frac{\beta}{2}(T_1^2 - T_2^2)\right] \tag{2-28}$$

We see that if the thermal conductivity were constant and had a value equal to k_o, the constant β would be zero and equation (2-28) would reduce to

$$Q = \frac{k_oA}{L}(T_1 - T_2)$$

The above equation is the same as equation (2-4) developed in Section 2-2 for steady-state conduction in a plane wall of constant thermal conductivity.

2-9 GENERAL THREE-DIMENSIONAL HEAT CONDUCTION EQUATION

Having discussed several specific examples of steady-state one-dimensional heat conduction, it is now time to present the general three-dimensional heat conduction equation. We will show that when it is properly interpreted, it will yield the same results as previously obtained.

To derive the general three-dimensional heat conduction equation we will consider a small cubical element in Cartesian coordinates having sides dx, dy, and

dz parallel to the x, y, and z axes respectively as illustrated in Figure 2-12. Utilizing the first law of thermodynamics (conservation of energy), the following energy balance may be written for the element:

$$\left\{\begin{array}{l}\text{Heat conducted in at}\\ \text{faces } x = x,\ y = y,\\ z = z, \text{ per unit time}\end{array}\right\} + \left\{\begin{array}{l}\text{Internal heat}\\ \text{generated}\\ \text{per unit time}\end{array}\right\} = \left\{\begin{array}{l}\text{Heat conducted out at faces}\\ x = x + dx, y = y + dy,\\ z = z + dz, \text{ per unit time}\end{array}\right\} +$$

$$\left\{\begin{array}{l}\text{Change in internal}\\ \text{energy per unit time}\end{array}\right\} + \left\{\begin{array}{l}\text{Work done by volume}\\ \text{element per unit time}\end{array}\right\} \tag{2-29}$$

Since the expansion of solids due to temperature changes is extremely small, the last term on the right-hand side of the above equation is negligible, and it is, therefore, dropped in the following development.

Let

Q_x = heat conducted in per unit time at $x = x$, face **ABCD**

Q_y = heat conducted in per unit time at $y = y$, face **DD′C′C**

Q_z = heat conducted in per unit time at $z = z$, face **BCC′B′**

Q_{x+dx} = heat conducted out per unit time at $x = x + dx$, face **A′B′C′D′**

Q_{y+dy} = heat conducted out per unit time at $y = y + dy$, face **AA′B′B**

Q_{z+dz} = heat conducted out per unit time at $z = z + dz$, face **AA′D′D**

$\dot{q}$ = the internal heat generation per unit time and per unit volume

k = thermal conductivity of material

ρ = density of material

c = specific heat of material

τ = time

From Fourier's Law [equation (1-3)]

$$Q_x = -k_x A_x \frac{\partial T}{\partial x}$$

where

k_x is the thermal conductivity in the x direction

A_x is the area normal to the x direction

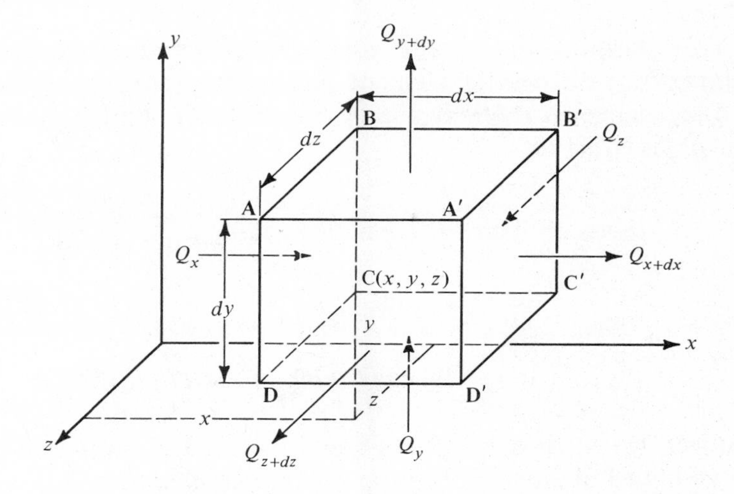

Figure 2-12 Volume element used for the derivation of the general three-dimensional heat conduction equation.

$\frac{\partial T}{\partial x}$ is the temperature gradient in the x direction

or

$$Q_x = -k_x \frac{\partial T}{\partial x} dy\, dz$$

Similarly

$$Q_y = -k_y \frac{\partial T}{\partial y} dx\, dz$$

$$Q_z = -k_z \frac{\partial T}{\partial z} dx\, dy$$

Also

$$Q_{x+dx} = -k_x \frac{\partial T}{\partial x} dy\, dz - \frac{\partial}{\partial x}\left(k_x \frac{\partial T}{\partial x}\right) dx\, dy\, dz$$

$$Q_{y+dy} = -k_y \frac{\partial T}{\partial y} dx\, dz - \frac{\partial}{\partial y}\left(k_y \frac{\partial T}{\partial y}\right) dx\, dy\, dz$$

$$Q_{z+dz} = -k_z \frac{\partial T}{\partial z} dx\, dy - \frac{\partial}{\partial z}\left(k_z \frac{\partial T}{\partial z}\right) dx\, dy\, dz$$

To determine the energy generated per unit time, $\dot{q}$ is multiplied by the volume of the differential element, giving the quantity $\dot{q}\ (dxdydz)$.

The change in internal energy for the differential element over a period of time, $d\tau$, is equal to

$$\text{(mass of element) (specific heat)} \begin{pmatrix}\text{change in the temperature} \\ \text{of the element in time } d\tau\end{pmatrix}$$

or

$$(\rho dxdydz)\,(c)\,dT = (\rho c dT)\,(dxdydz)$$

However, for equation (2-29) we need the change of internal energy per unit time. It is obtained by dividing the above expression by $d\tau$ and is written as

$$[\rho c\,(\partial T/\partial \tau)\,dxdydz]$$

Substitution of the quantities developed above into equation (2-29) and simplification results in the general three-dimensional heat conduction equation in Cartesian coordinates.

$$\frac{\partial}{\partial x}\left(k_x \frac{\partial T}{\partial x}\right) + \frac{\partial}{\partial y}\left(k_y \frac{\partial T}{\partial y}\right) + \frac{\partial}{\partial z}\left(k_z \frac{\partial T}{\partial z}\right) + \dot{q} = \rho c \frac{\partial T}{\partial \tau} \tag{2-30}$$

In this form, the general three-dimensional heat conduction equation may be used to solve problems where the thermal conductivity is a function of position and/or temperature. However, for many engineering problems, materials are often considered to possess a constant thermal conductivity.

For the case of constant conductivity materials, the general three-dimensional heat conduction equation becomes

$$\frac{\partial^2 T}{\partial x^2} + \frac{\partial^2 T}{\partial y^2} + \frac{\partial^2 T}{\partial z^2} + \frac{\dot{q}}{k} = \frac{1}{k/\rho c}\,\frac{\partial T}{\partial \tau} \tag{2-31}$$

The quantity $k/\rho c$ is called the thermal diffusivity, α, of the material and tells us how fast heat propagates or diffuses through a material. Even though the thermal conductivity of metals is many times greater than that of gases, the density of gases is sufficiently small that heat diffuses through gases at approximately the same rate as it does through metals. Rewriting equation (2-31) using the thermal diffusivity, we have

$$\frac{\partial^2 T}{\partial x^2} + \frac{\partial^2 T}{\partial y^2} + \frac{\partial^2 T}{\partial z^2} + \frac{\dot{q}}{k} = \frac{1}{\alpha}\,\frac{\partial T}{\partial \tau} \tag{2-32}$$

For steady-state problems, the temperature at any given point in the body does not change as time changes. Therefore, for steady-state heat flow $(\partial T/\partial \tau) = 0$ and equation (2-32) becomes

$$\frac{\partial^2 T}{\partial x^2} + \frac{\partial^2 T}{\partial y^2} + \frac{\partial^2 T}{\partial z^2} + \frac{\dot{q}}{k} = 0 \quad \textit{Poisson equation} \tag{2-32a}$$

For steady-state heat flow without any generation or release of heat energy within the body (i.e., in the absence of heat sources), equation (2-32) becomes,

$$\frac{\partial^2 T}{\partial x^2} + \frac{\partial^2 T}{\partial y^2} + \frac{\partial^2 T}{\partial z^2} = 0 \quad \textit{Laplace equation} \tag{2-32b}$$

For no heat sources but unsteady-state conditions, equation (2-32) becomes,

$$\frac{\partial^2 T}{\partial x^2} + \frac{\partial^2 T}{\partial y^2} + \frac{\partial^2 T}{\partial z^2} = \frac{1}{\alpha}\frac{\partial T}{\partial \tau} \quad \textit{Heat or Diffusion equation} \tag{2-32c}$$

If the general three-dimensional heat conduction equation is written in cylindrical coordinates for a material with constant thermal conductivity, we have

$$\frac{\partial^2 T}{\partial r^2} + \frac{1}{r}\left(\frac{\partial T}{\partial r}\right) + \frac{1}{r^2}\left(\frac{\partial^2 T}{\partial \phi^2}\right) + \frac{\partial^2 T}{\partial z^2} + \frac{\dot{q}}{k} = \frac{1}{\alpha}\frac{\partial T}{\partial \tau} \tag{2-33}$$

where r, ϕ, and z are shown in Figure 2-13.

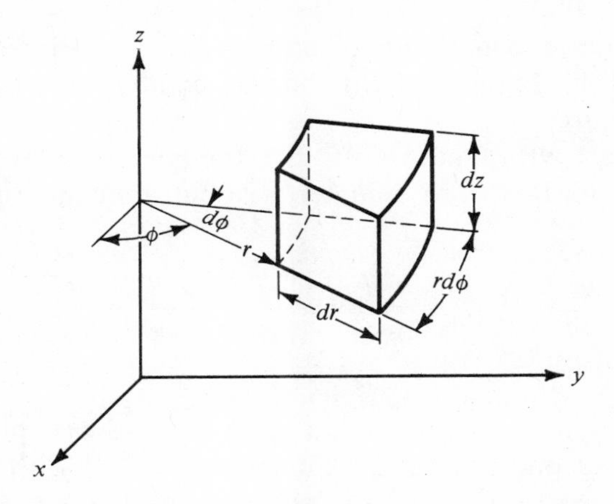

Figure 2-13 Elemental volume in cylindrical coordinates for three-dimensional heat conduction equation.

If the general three-dimensional heat conduction equation is written in spherical coordinates for a constant conductivity material, we have

$$\frac{1}{r^2}\frac{\partial}{\partial r}\left(r^2\frac{\partial T}{\partial r}\right)+\frac{1}{r^2\sin\theta}\frac{\partial}{\partial\theta}\left(\sin\theta\frac{\partial T}{\partial\theta}\right)+\frac{1}{r^2\sin^2\theta}\frac{\partial^2 T}{\partial\phi^2}+\frac{\dot{q}}{k}=\frac{1}{\alpha}\frac{\partial T}{\partial\tau} \quad (2\text{-}34)$$

where r, θ, and ϕ are shown in Figure 2-14.

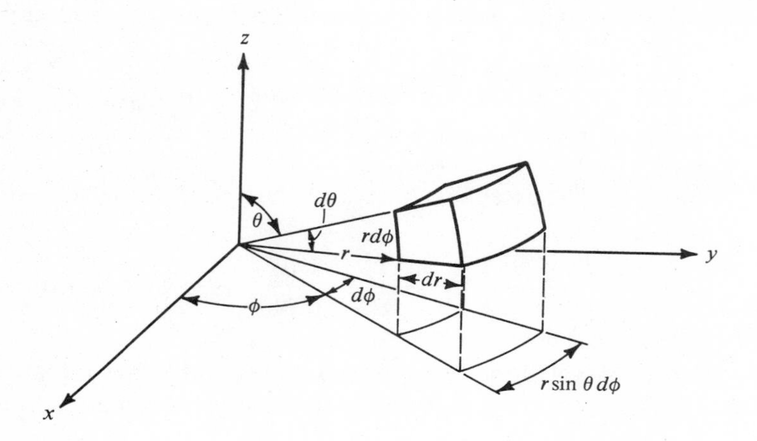

Figure 2-14 Elemental volume in spherical coordinates for three-dimensional heat conduction equation.

To demonstrate the application of the general three-dimensional heat conduction equation, let us reduce it for two of the problems that we considered earlier in order to show that it gives the same differential equation as we obtained earlier.

Consider the problem of steady-state one-dimensional heat conduction in a plane wall having a constant thermal conductivity (see Figure 2-1) and possessing no heat sources.

We start with equation (2-32), the general three-dimensional equation in rectangular coordinates for a material of constant thermal conductivity.

$$\frac{\partial^2 T}{\partial x^2}+\frac{\partial^2 T}{\partial y^2}+\frac{\partial^2 T}{\partial z^2}+\frac{\dot{q}}{k}=\frac{1}{\alpha}\frac{\partial T}{\partial\tau}$$

Note that $q = 0$ since no heat sources are present. Also $(\partial T/\partial\tau) = 0$ as steady-state conditions exist. Furthermore $(\partial^2 T/\partial y^2) = (\partial^2 T/\partial z^2) = 0$, since there is no temperature variation in the y or z direction. Once we eliminate the terms that do not apply to our specific problem, we are left with the following equation:

$$\frac{\partial^2 T}{\partial x^2} = 0$$

But since there is only one independent variable, x, this becomes the ordinary differential equation

$$\frac{d^2 T}{dx^2} = 0 \qquad \text{[same as equation (2-3) obtained earlier]}$$

When this is coupled with the boundary conditions

$$\text{at } x = 0, \quad T = T_1$$

$$\text{at } x = L, \quad T = T_2$$

we obtain the same results as we did in Section 2-2. That is

$$T = (T_2 - T_1)\frac{x}{L} + T_1 \tag{2-4}$$

Also, using equation (2-4) and substituting into Fourier's equation, we find the heat flux to be

$$Q = \frac{kA\,(T_1 - T_2)}{L} \tag{2-4a}$$

Now, consider the problem of steady-state one-dimensional heat conduction in a solid cylinder possessing a constant thermal conductivity and a uniformly distributed heat source, $\dot{q}$, of constant intensity (see Figure 2-11).

We start with equation (2-33), the general three-dimensional equation in cylindrical coordinates for a material of constant thermal conductivity.

$$\frac{\partial^2 T}{\partial r^2} + \frac{1}{r}\frac{\partial T}{\partial r} + \frac{1}{r^2}\frac{\partial^2 T}{\partial \phi^2} + \frac{\partial^2 T}{\partial z^2} + \frac{\dot{q}}{k} = \frac{1}{\alpha}\frac{\partial T}{\partial \tau}$$

Note that $(\partial T/\partial \tau) = 0$ as steady state prevails. Also $(\partial^2 T/\partial \phi^2) = (\partial^2 T/\partial z^2) = 0$, since there is no temperature variation in the ϕ or z direction.

Once we eliminate the terms that do not apply to our specific problem, the general equation reduces to

$$\frac{\partial^2 T}{\partial r^2} + \frac{1}{r}\frac{\partial T}{\partial r} + \frac{\dot{q}}{k} = 0$$

But since there is only one independent variable, r, this becomes the ordinary differential equation

$$\frac{d^2 T}{dr^2} + \frac{1}{r}\frac{dT}{dr} + \frac{\dot{q}}{k} = 0$$

Multiplying by r gives

$$r\frac{d^2 T}{dr^2} + \frac{dT}{dr} + \frac{\dot{q}r}{k} = 0$$

Noting that the quantity $[r\,(d^2T/dr^2) + dT/dr]$ is the exact differential $d/dr\ [(r)\,dT/dr)]$ we may write the differential equation as

$$\frac{d}{dr}\left(r\frac{dT}{dr}\right) = -\frac{\dot{q}r}{k} \qquad \text{[same as equation (2-20) obtained earlier]}$$

If the two boundary conditions

$$\text{at } r = r_o, \quad T = T_w$$

$$\text{at } r = 0, \quad \frac{dT}{dr} = 0$$

are applied to the above equation, the same results as those presented in Section 2-7 will be obtained.

That is

$$T - T_w = \frac{\dot{q}}{4k}(r_o^2 - r^2) \tag{2-22}$$

and

$$T_c = T_w + \frac{\dot{q}r_o^2}{4k} \tag{2-23}$$

where T_c is the maximum temperature in the cylinder, which occurs along its axis.

Once the terms comprising the general three-dimensional heat conduction equation are understood, it is no longer necessary to set up an energy balance on a volume element for each conduction problem in order to determine the appropriate differential equation. Rather, the general equation may be reduced for a specific situation by elimination of the terms that are identically equal to zero.

PROBLEMS (English Engineering System of Units)

2-1. A furnace wall is made of 9 inches of fire brick ($k = 0.05$ Btu/hr-ft°F) and 3 inches of asbestos cement ($k = 1.05$ Btu/hr-ft°F). The temperature at the outside surface of the fire brick is 800°F and at the outside surface of the asbestos cement is 80°F. Determine the temperature at the interface of the two materials.

2-2. Determine the rate of heat flow through a 20 ft^2 section of an oak wall, 4 inches thick, whose inside and outside surfaces are at 68°F and −2°F, respectively.

2-3. Rework problem 2-2 noting that in this problem the 68°F and −2°F are the temperatures of the inside and outside ambient air. Take the value of the convective heat transfer coefficient on the inside and outside walls to be 3 Btu/hr-ft°F and 20 Btu/hr-ft^2°F respectively. Also determine the new inside and outside wall temperatures.

2-4. The wall of an apple storage room is made of cinder blocks and cork. The cinder blocks are 6 inches thick and have an effective thermal conductivity of 0.6 Btu/hr-ft°F. If the heat loss is not to exceed 20 Btu/hr in a 10 ft^2 section of the wall, what thickness of cork must be used if the inside wall temperature is 34°F and the outside wall temperature is −10°F? The thermal conductivity of cork may be taken as 0.025 Btu/hr-ft°F.

2-5. Hot water flows through a steel pipe 2.0-inch I.D. and 2.5-inch O.D. The average temperature of the water is 200°F and that of the outside ambient air is 60°F. The convective heat transfer coefficient between the water and the inside pipe surface is 200 Btu/hr-ft^2°F and that between the ambient air and the outside pipe surface is 5 Btu/hr-ft^2°F. Take the thermal conductivity of steel to be 30 Btu/hr-ft°F. Determine the rate of heat loss per linear foot of pipe.

2-6. A hollow cylinder with 4-inch I.D. and 12-inch O.D. has an inner surface temperature of 500°F and an outer surface temperature of 300°F. Determine the temperature halfway between the inner and outer surfaces.

2-7. In problem 2-6, the thermal conductivity of the cylinder material is 25 Btu/hr-ft°F. Determine the rate of heat flow through the cylinder per linear foot.

2-8. A hollow sphere with 4-inch I.D. and 12-inch O.D. has an inner surface temperature of 500°F and an outer surface temperature of 300°F. Determine the temperature one-third way between the inner and outer surfaces.

2-9. In problem 2-8, the thermal conductivity of the sphere material is 25 Btu/hr-ft°F. Determine the rate of heat flow through the sphere.

2-10. In an experiment for determining the thermal conductivity of a given metal, a specimen 1 inch in diameter and 6 inches long is maintained at 212°F at one end and at 32°F at the other end. If the cylindrical surface is completely insulated and electrical measurements show a heat flow of 4 watts, determine the thermal conductivity of the specimen material.

2-11. A steel pipe with a 4.0-inch I.D. and a 0.25-inch wall thickness is covered with 4 inches of high-temperature insulation and 2 inches of low-temperature insulation. The temperature on the inside surface of the pipe is 450°F, and the temperature on the outside surface of the low-temperature insulation is 75°F. Determine the heat flow rate and the temperatures at the interface of the steel and high-temperature insulation and at the interface of the high- and low-temperature insulations. Treat the problem as one-dimensional heat conduction in concentric cylinders.

Given:

Material	*Thermal Conductivity* (Btu/hr-ft °F)
steel	30
high-temperature insulation	0.07
low-temperature insulation	0.05

2-12. Rework problem 2-11 noting that in this problem the 450°F and the 75°F are the temperatures of the steam in the pipe and of the air in contact with the low-temperature insulation, respectively. Take the value of the convective heat transfer coefficient between the steam and the inner pipe surface to be 75 Btu/hr-ft^2°F and the value of the convective heat transfer coefficient between the low-temperature insulation and the air to be 10 Btu/hr-ft^2°F. In addition to the quantities asked for in problem 2-11, also determine the inside and outside temperatures of the steel pipe.

2-13. Determine the temperature distribution in a plane wall for which the thermal conductivity varies according to $k = k_o e^{-x/L}$, where k_o is a constant, and L is the wall thickness. *Assume:* $T = T_1$ at $x = 0$ and $T = T_2$ at $x = L$.

2-14. A steel steam pipe 4.0-inch I.D. and 4.5 inch O.D. is covered with insulation having a thermal conductivity of 0.75 Btu/hr-ft°F. If the convective heat transfer coefficient between the insulation surface and the surrounding air is 1.5 Btu/hr-ft^2°F, determine the critical radius of insulation.

2-15. For problem 2-14, let the steam temperature be 400°F and the ambient air temperature be 80°F. Permit the radius of insulation, r_2, to vary, and make a plot of r_2 versus heat lost. Use the following values for r_2: 3 inch, 4 inch, 5 inch, 6 inch, 7 inch, 8 inch, 9 inch, 10 inch, and 20 inch. Hence, determine the critical radius of insulation.

2-16. Consider a hollow metal sphere whose outside temperature remains constant at temperature T_s independent of the amount of insulting material applied to it. If the thermal conductivity of the insulating material is k, the ambient air temperature is T_∞, the radius of insulation is r_2, and the convective heat transfer coefficient between the air and the insulation is h; find an expression for the critical radius of insulation for the spherical geometry.

2-17. A plane wall 4 inches thick generates heat at the rate of 10^4 Btu/hr-ft^3 when an electric current is passed through it. The convective heat transfer coefficient between each face of the wall and the ambient air is 10 Btu/hr-ft^2°F. Determine:

(a) the surface temperature
(b) the maximum temperature in the wall

Assume the ambient air temperature to be 70 °F, and the thermal conductivity of the wall material to be 10 Btu/hr-ft °F.

2-18. A plane wall 4 inches thick generates heat at the rate of 10^3 Btu/hr-ft^3 when an electric current is passed through it. One face of the wall is insulated and the other face is exposed to 70 °F air. If the convective heat transfer coefficient between the air and the exposed surface of the wall is 10 Btu/hr-ft^2 °F, determine the maximum temperature in the wall. The thermal conductivity of the wall material is 5 Btu/hr-ft °R.

2-19. Determine the heat flux (rate of heat flow through a unit area) through a wall for which the thermal conductivity varies according to

$$k = a + bT$$

where a and b are constants. Assume $T = T_1$ at $x = 0$, and $T = T_2$ at $x = L$.

2-20. The shielding wall for a nuclear reactor is exposed to gamma ray radiation, which results in heat generation according to the following equation:

$$\dot{q} = \dot{q}_o e^{-ax}$$

where a and $\dot{q}_o$ are constants. Using this relationship for $\dot{q}$, determine the temperature distribution in the shield if the wall is L units thick and its left- and right-hand faces are maintained at temperatures T_1 and T_2, respectively.

2-21. Determine the steady-state temperature distribution in a plane wall containing a uniformily distributed heat source, $\dot{q}$, with the left-hand face maintained at temperature T_1 and the right-hand face maintained at temperature, T_2. The thickness of the wall is $2L$.

2-22. An electrical transmission wire made of a one-inch diameter annealed copper wire carries 200 amps and has a resistance of 0.4×10^{-4} ohm per cm length. If the surface temperature is 350°F and the ambient air temperature is 50°F, determine the heat transfer coefficient between the wire surface and the ambient air and the maximum temperature in the wire. Assume $k = 100$ Btu/hr-ft°F.

2-23. A thick steel pipe is cut in half and insulated along its upper and lower surfaces as indicated in E Figure 2-23.

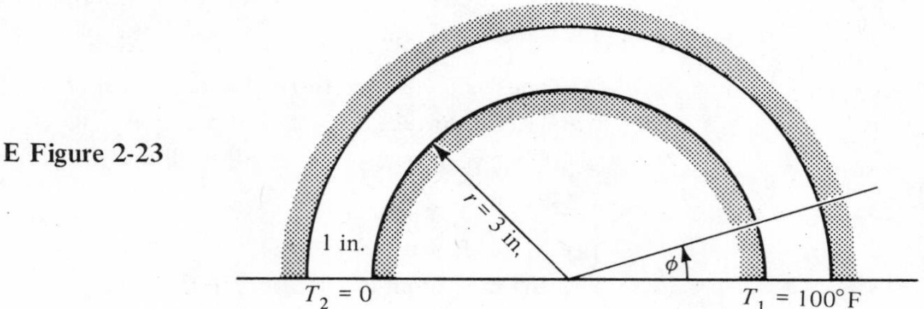

E Figure 2-23

(a) Starting with the general three-dimensional heat conduction equation in cylindrical coordinates [equation (2-33)], determine the differential equation that, when solved, will yield the temperature as a function of the angle, ϕ.

(b) Determine $T(\phi)$ from your answer to part (a).

(c) Determine the heat flow if $k = 30$ Btu/hr-ft °F.

Note: Assume that the pipe extends indefinitely into the paper.

PROBLEMS (SI System of Units)

2-1. The wall of a house consists of 15 cm cinder blocks ($k = 1$ W/m°C) covered by 6 mm pine ($k = 0.11$ W/m°C) paneling on the inside. The temperature on the outside surface of the cinder blocks is 1°C and the inside surface of the paneling is 25°C. Determine the heat flow per square meter of wall area and the temperature at the cinder block/pine interface.

2-2. Determine the heat flow through a 5 m^2 section of a pine wall 5 cm thick whose inside and outside surfaces are at 22°C and 2°C, respectively.

2-3. Rework problem 2-2 noting that in this problem the 22°C and 2°C are the temperatures of the inside and outside ambient air. Take the value of the convective heat transfer coefficient on the inside and outside walls to be 12 W/m^2°C and 100 W/m^2°C, respectively. Also determine the new inside and outside wall temperatures.

2-4. An insulating wall is composed of 15 cm of a material having a thermal conductivity of 1 W/m°C and an unknown thickness of cork (k = 0.045 W/m°C). If the inside wall temperature is 22°C and the outside wall temperature is −10°C, determine the amount of cork needed to keep the heat loss to 8 W/m^2.

2-5. Hot water flows through a steel pipe 5 cm I.D. and 6.5 cm O.D. The average temperature of the water is 95°C and that of the outside ambient air is 15°C. The convective heat transfer coefficient between the water and the inside surface is 1000 W/m^2°C and that between the ambient air and the outside pipe surface is 25 W/m^2°C. Take the thermal conductivity of steel to be 50 W/m°C. Determine the heat loss per linear meter of pipe.

2-6. A hollow cylinder 10 cm I.D. and 20 cm O.D. has an inner surface temperature of 300°C and an outer surface temperature of 100°C. Determine the temperature halfway between the inner and outer surfaces.

2-7. In problem 2-6, the thermal conductivity of the cylinder material is 50 W/m°C. Determine the heat flow through the cylinder per linear meter.

2-8. A hollow sphere 10 cm I.D. and 30 cm O.D. has an inner surface temperature of 300°C and an outer surface temperature of 100°C. Determine the temperature one-fourth way between the inner and outer surfaces.

2-9. In problem 2-8, the thermal conductivity of the sphere material is 50 W/m°C. Determine the heat flow through the sphere.

2-10. In an experiment for determining the thermal conductivity of a given metal, a specimen 2.5 cm in diameter and 15 cm long is heated to 100°C at one end and cooled to 0°C at the other end. If the cylindrical surface is completely insulated and electrical measurements show a heat flow of 3 watts, determine the thermal conductivity of the specimen material.

2-11. A steel steam pipe 10 cm I.D. and 6 mm wall thickness is covered with 10 cm of high-temperature insulation and 5 cm of low-temperature insulation. The temperature on the inside surface of the pipe is 300°C and the temperature on the outside surface of the low-temperature insulation is 25°C. Determine the heat flow rate and the temperatures at the interface of the steel and high-temperature insulation and at the interface of the high- and low-temperature insulation. Use the electrical analogy for one-dimensional heat conduction in concentric cylinders.

Given:

Material	*Thermal Conductivity* (W/m°C)
steel	50
high-temperature insulation	0.40
low-temperature insulation	0.20

2-12. Rework problem 2-11 noting that in this problem the 300°C and the 25°C are the temperatures of the steam in the pipe and of the air in contact with the low-temperature insulation, respectively. Take the value of the convective heat transfer coefficient between the steam and the inner pipe surface to be 85 W/m^2 °C and the value of the convective heat transfer coefficient between the low-temperature insulation and the air to be 40 W/m^2 °C. In addition to the quantities asked for in problem 2-11, also determine the inside and outside surface temperatures of the steel pipe.

2-13. Determine the temperature distribution in a plane wall for which the thermal conductivity varies according to $k = k_o e^{-x/L}$, where k_o is a constant, and L is the wall thickness. *Assume:* $T = T_1$ at $x = 0$ and $T = T_2$ at $x = L$.

2-14. A steel steam pipe 10 cm I.D. and 11 cm O.D. is covered with insulation having a thermal conductivity of 1 W/m°C. If the convective heat transfer coefficient between the insulation surface and the surrounding air is 8 W/m^2 °C, determine the critical radius of insulation.

2-15. For problem 2-14, let the steam temperature be 200°C and the ambient air temperature be 20°C. Permit the radius of insulation, r_2, to vary, and make a plot of r_2 versus heat lost, noting the critical radius of insulation. Use the following values for r_2: 6 cm, 8 cm, 10 cm, 12 cm, 13 cm, 15 cm, and 20 cm.

2-16. Consider a metal sphere whose outside temperature remains constant at temperature T_s independent of the amount of insulating material applied to it. If the thermal conductivity of the insulating material is k, the ambient air temperature is T_∞, the radius of insulation is r_2, and the convective heat transfer coefficient between the air and the insulation is h, find the expression for the critical radius of insulation for the spherical geometry.

2-17. A plane wall 10 cm thick generates heat at the rate of 4 × 10^4 W/m^3 when an electric current is passed through it. The convective heat transfer coefficient between each face of the wall and the ambient air is 50 W/m^2 °C. Determine:

(a) the surface temperature
(b) the maximum temperature in the wall

Assume the ambient air temperature to be 20°C and the thermal conductivity of the wall material to be 15 W/m°C.

2-18. A plane wall 10 cm thick generates heat at the rate of 30,000 W/m^3 when an electric current is passed through it. One face of the wall is insulated, and the outer face is exposed to 25°C air. If the convective heat transfer coefficient between the air and the exposed surface of the wall is 50 W/m^2 °C, determine the maximum temperature in the wall. The thermal conductivity of the wall material is 3 W/m°C.

2-19. Determine the heat flux (the rate of heat flow per unit area) through a wall for which the thermal conductivity varies according to

$$k = a + bT$$

where a and b are constants.

2-20. The shielding wall for a nuclear reactor is exposed to gamma ray radiation, which results in heat generation according to the following equation:

$$\dot{q} = \dot{q}_o e^{-ax}$$

where $\dot{q}_o$ and a are constants. Using this relationship for $\dot{q}$, determine the temperature distribution in the shield if the wall is L units thick and its left- and right-hand faces are maintained at temperatures T_1 and T_2, respectively.

2-21. Determine the steady-state temperature distribution in a plane wall containing a uniformly distributed heat source, $\dot{q}$, with the left-hand face maintained at temperature, T_1, and the right-hand face maintained at temperature, T_2. The thickness of the wall is $2L$.

2-22. An electrical transmission line made of a 25 mm diameter annealed copper wire carries 200 amps and has a resistance of 0.4×10^{-4} ohm per cm length. If the surface temperature is 200°C and the ambient air temperature is 10°C, determine the heat transfer coefficient between the wire surface and the ambient air and the maximum temperature in the wire. Assume $k = 150$ W/m°K.

2-23. A thick steel pipe is cut in half and insulated along its upper and lower surfaces as indicated in SI Figure 2-23.

SI Figure 2-23

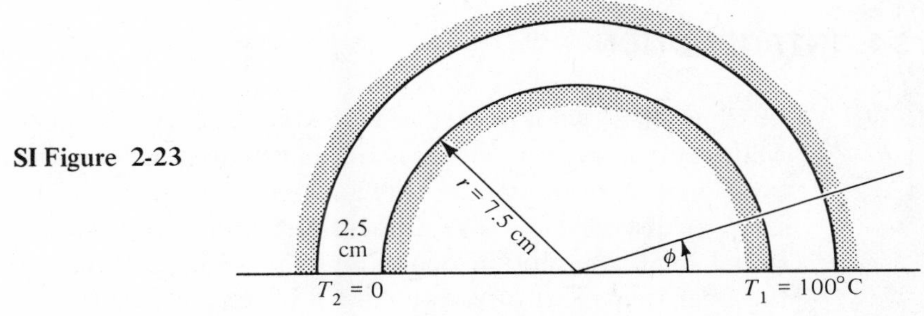

(a) Starting with the general three-dimensional heat conduction equation in cylindrical coordinate [equation (2-33)], determine the differential equation that, when solved, will yield the temperature as a function of the angle, ϕ.

(b) Determine $T(\phi)$ from your answer to part (a).

(c) Determine the heat flow if $k = 50$ W/m°C.

Note: Assume that the pipe extends indefinitely into the paper.

3

Two-Dimensional Steady-State Conduction

3-1 INTRODUCTION

In Chapter 2, we analyzed steady-state one-dimensional heat conduction in plane walls, cylinders, and spheres. In many engineering applications, temperatures in a body may vary in two or three coordinate directions, and it thus becomes necessary to discuss two- and three-dimensional heat conduction. Such multidimensional heat conduction occurs in the block of an internal combustion engine, in the heat treatment of various metal parts, and inside of any composite body made of materials possessing different thermal conductivities. For the present, we will limit our discussion to two-dimensional steady-state conduction (viz., temperature does not vary with time) without heat sources.

Recall once again that the prime objective of any heat transfer analysis is the determination of the temperature distribution and the heat flow within and at the boundary of a given body. In solving two-dimensional problems, analytical, graphical, or numerical techniques are usually used. The technique employed is dependent upon the relative complexity of the situation. Analytical and graphical methods are normally utilized on the simpler problems, and numerical techniques are used on the more complex problems.

Perhaps the most practical and useful tool used to solve multidimensional problems is finite-difference techniques. We will now discuss two-dimensional steady-state conduction utilizing the approximate methods of solution, including the conduction shape factor and analog methods. As an option for those interested in the analytical approach, the method of separation of variables is presented at the end of the chapter. Analysis of three-dimensional problems will be covered in Chapter 5, entitled Numerical Methods in Heat Conduction.

To begin our discussion of steady-state two-dimensional heat conduction in a constant conductivity material without heat sources, we write the general three-dimensional heat conduction equation as

$$\frac{\partial^2 T}{\partial x^2} + \frac{\partial^2 T}{\partial y^2} + \frac{\partial^2 T}{\partial z^2} + \frac{\dot{q}}{k} = \frac{1}{\alpha}\frac{\partial T}{\partial \tau} \tag{2-4}$$

For steady state conditions, $\partial T/\partial \tau = 0$.

For no heat source, $\dot{q}/k = 0$.

For two-dimensional conduction, $\partial^2 T/\partial z^2 = 0$.

The differential equation for two-dimensional steady-state conduction in a material with constant conductivity and without heat sources becomes

$$\frac{\partial^2 T}{\partial x^2} + \frac{\partial^2 T}{\partial y^2} = 0 \tag{3-1}$$

This same equation can be developed by beginning with an energy balance.

Since we have a partial differential equation containing the second partial derivatives of the dependent variable, T, with respect to x and y, we must have four boundary conditions in order to determine a solution to the problem. These boundary conditions must specify the temperature or its derivatives at two values of x and y, respectively. An example illustrating the analytical solution to a two-dimensional problem is given in Section 3-4. A similar problem will be solved in Chapter 5 using the finite difference technique.

The solution to equation (3-1) will yield the temperature as a function of x and y. Then, using Fourier's heat conduction equation, the local heat flow in the x and y directions may be determined.

For two-dimensional problems the situation arises where the direction of heat flow is neither in the x direction nor in the y direction. Figure 3-1 shows an arbitrary isotherm located somewhere within a given body.

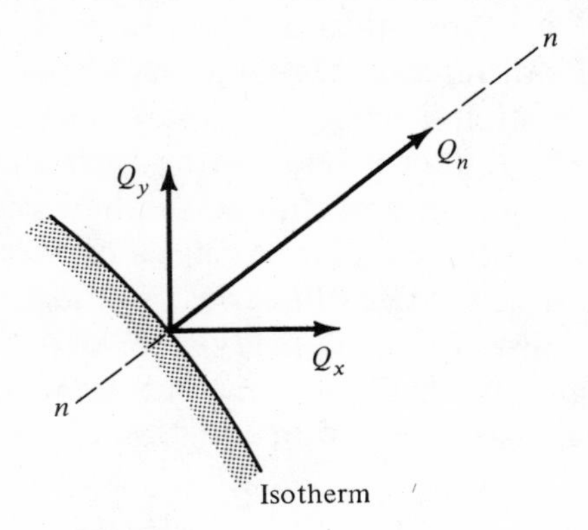

Figure 3-1 Two-dimensional heat flow.

We know from Fourier's Law that heat flows in a direction normal to lines of constant temperature, hence

$$Q_n = -kA_n \frac{\partial T}{\partial n}$$

where

$\frac{\partial T}{\partial n}$ = the temperature gradient in the n direction

A_n = the area normal to the n direction

Q_n = the total heat flow through A_n

By vector calculus it can be shown that

$$Q_n = Q_x + Q_y$$

where

$$Q_x = -kA_x \frac{\partial T}{\partial x}$$

$$Q_y = -kA_y \frac{\partial T}{\partial y}$$

The terms A_x and A_y are the x and y projections of A_n, and $\partial T/\partial x$ and $\partial T/\partial y$ are the temperature gradients in the x and y directions.

3-2 GRAPHICAL ANALYSIS OF TWO-DIMENSIONAL HEAT CONDUCTION

Due to the irregular geometries associated with specific problems and due to the imposition of certain boundary conditions, it is often very difficult or impossible to achieve an analytical solution to many problems. Oftentimes an approximate solution may be arrived at through graphical means. This is true especially if the boundaries of the body in question are isothermal.

In reality, to obtain a graphical solution, the problem solver needs some insights that are gained only through extensive exposure to heat conduction problems. The work of laying out a solution is somewhat of an "art form," and the beginning student should not expect immediate results from this approach.

To generate a graphical solution, a network of curvilinear squares is created by drawing isotherms and heat flow lines according to the following guidelines:

(1) Heat flow lines are always drawn perpendicular to isotherms and to isothermal boundaries and bisect the angle at a corner where two isothermal boundaries come together.
(2) The isotherms run perpendicular to insulated surfaces.
(3) The diagonals of a curvilinear square intersect at right angles.
(4) All sides of a curvilinear square are of approximately the same length although one curvilinear square may be larger or smaller than another.

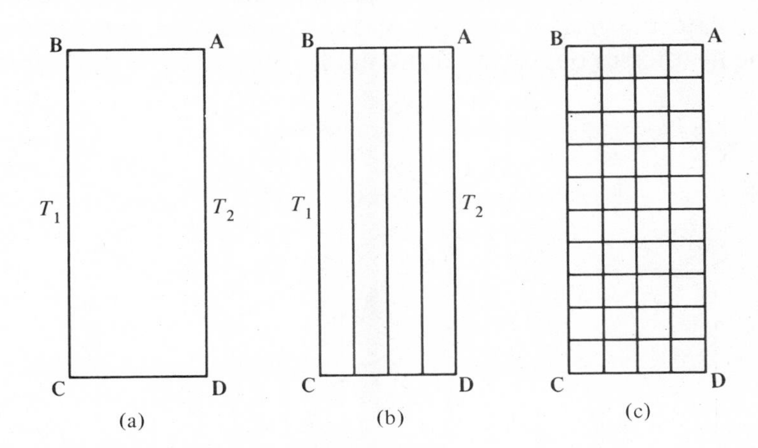

Figure 3-2 A network of curvilinear squares for a plane wall.

Let us apply the guidelines for drawing curvilinear squares to the plane wall problem. Figure 3-2(a) shows a plane wall with face **BC** at temperature, T_1, and face **AD** at temperature T_2. We observe that **BC** and **AD** are isotherms and **BA** and **CD** are heat flow lines. In Figure 3-2(b), the heat flow line **BA** is divided into four equal parts. Next three isotherms are drawn connecting **BA** and **CD**. The process of generating the desired network of curvilinear squares as shown in Figure 3-2(c) is completed using the guidelines outlined above. In this case the squares have straight sides, but we note, however, that if the wall is bent into a semicircle, the squares distort as shown in Figure 3-3. In fact the isotherms nearest T_1 are more closely spaced since the same heat flows across an area **A-A′** as across **B-B′**, and since area **B-B′** is larger, the temperature gradient across **A-A′** must be greater for the same Q. This sort of reasoning must be employed in sketching any graphical solution.

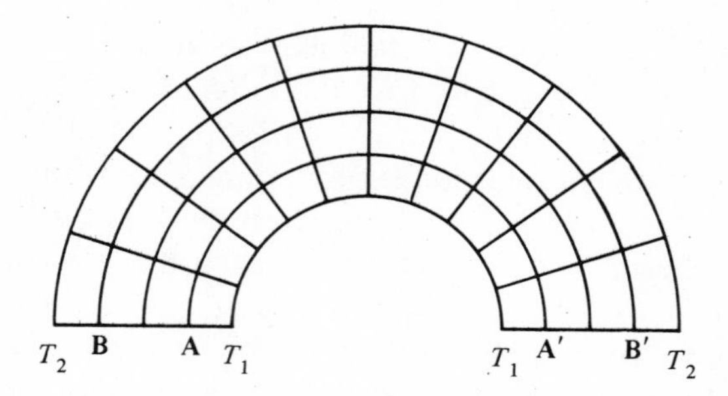

Figure 3-3 A network of curvilinear squares for a plane wall bent into a semicircle.

We will now consider a tubular furnace as illustrated in Figure 3-4. The inside wall and the outside wall of the furnace are at uniform temperatures T_1 and T_2, respectively. We wish to determine the heat losses through the furnace wall in order to properly size the heating elements. In the analysis that follows, we will calculate the heat losses out of the top, bottom, and both sides of the furnace. The heat losses out of the front and back will not be calculated.

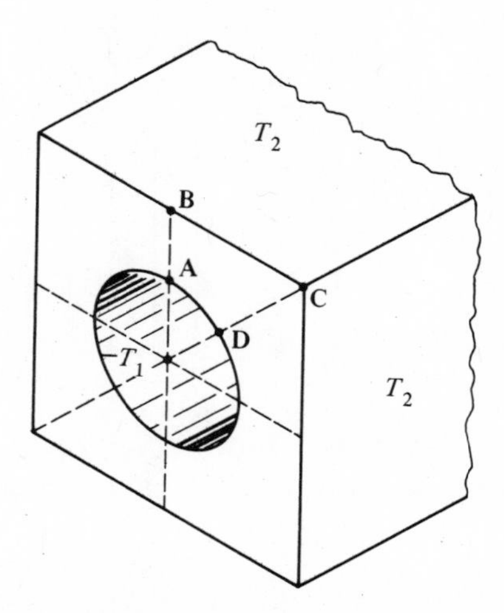

Figure 3-4 Tubular furnace.

We observe that due to the circular cross section of the heating volume and the square cross section of the furnace walls, the heat flow is symmetric. The total heat loss through the furnace walls will be eight times that lost through section **ABCD**, shown in Figure 3-4. In performing a graphical solution, the basic principle employed is that heat flows in the direction of decreasing temperatures and perpendicular to lines of constant temperature or isotherms.

We proceed by drawing section **ABCD** as shown in Figure 3-5. Note that the isotherms for temperatures T_1 and T_2 are the lines **AD** and **BC** on the inside and the outside of the furnace wall, and the heat flow lines are the lines **BA** and **CD**. Next, we divide the heat flow line **BA** into an arbitrary number of equal parts, N. Through each point on line **BA** we now draw one isotherm meeting the line **CD**. For the most part, these isotherms will be only approximately parallel to the boundary isotherms. Next, a series of heat flow lines are drawn perpendicular to the isotherms in such a way as to create a network of curvilinear squares. In setting up the network of curvilinear squares, we have applied the guidelines discussed earlier. We observe that near corner **C** it is difficult to follow the guidelines rigorously.

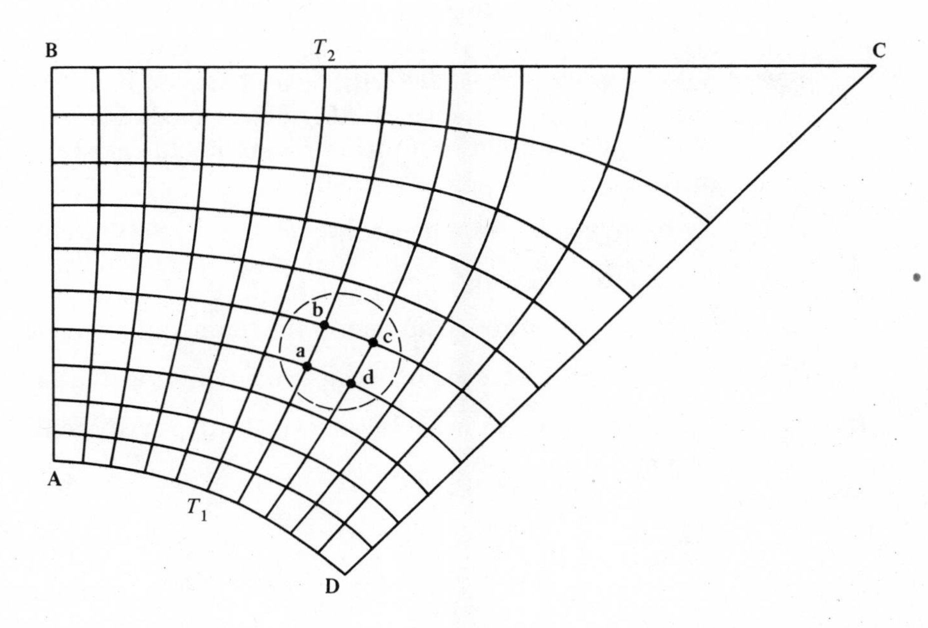

Figure 3-5 Section **ABCD** of Figure 3-4 used in graphical solution.

3-2.1 Analysis of Heat Flow Through Network of Curvilinear Squares

The question is now asked: What good has been done by creating the network of curvilinear squares? To respond, let us examine a typical curvilinear square for the tubular furnace problem (Figure 3-6). We observe that

$$N = \text{number of } \Delta T\text{s across body}$$
$$= \text{number of isotherms} - 1$$

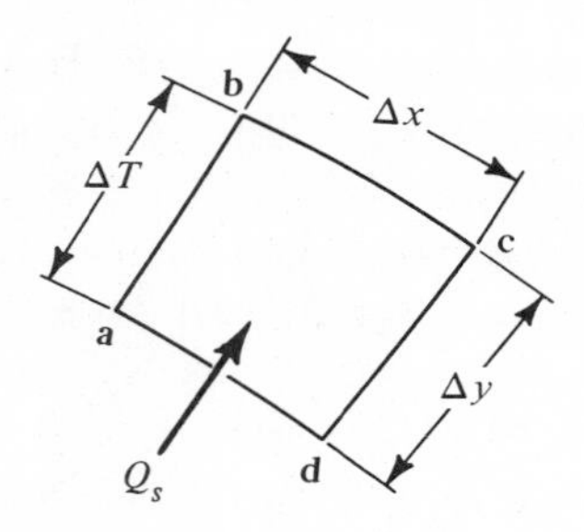

Figure 3-6 Typical curvilinear square.

Recall that the number of isotherms was chosen at the beginning of the graphical solution.

M = number of heat flow lanes (the space between two adjacent heat flow lines is one heat flow lane). M is determined by simply counting the heat flow lanes that result from generating the network of curvilinear squares.

$\Delta x \approx \Delta y$ for any given curvilinear square

k = thermal conductivity

Q_s = heat flow rate through **ad** due to the temperature gradient, $(\Delta T/\Delta y)$

ΔT = temperature difference between two adjacent isotherms

$(\Delta x)\ (l)$ = area through which heat flows; (l) is the depth of the furnace wall

Now

$$Q_s = \frac{k \Delta x l \Delta T}{\Delta y}$$

but $\Delta x = \Delta y$, so that $Q_s = kl\Delta T$. Since there are N increments of ΔT

$$\Delta T = \frac{T_1 - T_2}{N} \quad \text{and} \quad Q_s = \frac{lk\,(T_1 - T_2)}{N}$$

The quantity Q_s is the same through each curvilinear square, which means that Q_s flows through each heat flow lane. If there are M heat flow lanes, then the total heat flow Q, through **ABCD**, is given by

$$Q = MQ_s = \frac{Ml}{N} k (T_1 - T_2)$$

The number of ΔTs, N, is chosen arbitrarily, and the number of heat flow lanes, M, is determined by drawing heat flow lines in accordance with the guidelines mentioned earlier. The ratio (Ml/N) will be the same for any value of N chosen. For our furnace problem, if we choose $N = 9$, then from the network of curvilinear squares, $M \approx 10$, and letting $l = 1$, we have

$$Q = 1.11k (T_1 - T_2)$$

for one octant of the furnace wall.

The quantity (Ml/N) is often referred to as the conduction shape factor, S. It has been tabulated by graphical techniques, from analytical solutions, and from the results of electrical analogs for various geometries. It is given in Table 3-1 for some of the more frequently encountered configurations. If S is known, Q may be calculated from

$$Q = kS\Delta T \tag{3-2}$$

where ΔT is the overall temperature difference between two isothermal boundaries.

Sample Problem 3-1 A laboratory furnace 30 cm by 30 cm by 30 cm inside dimensions is constructed of firebrick having a thermal conductivity of 0.80 W/m°C. If the wall is 15 cm thick and the inside and outside wall temperatures are 540°C and 40°C, respectively, determine the heat lost through the walls. If 540°C is the maximum temperature at which the furnace is to be operated, estimate to the nearest kilowatt the size of the heating element needed.

Solution:

$$Q = kS\Delta T$$

For one wall: (Table 3-1) $S = \frac{A}{L} = \frac{(0.3)(0.3)}{0.15} = 0.6$

TABLE 3-1 Conduction Shape Factors†

Physical system	*Schematic*	*Shape factor*	*Restrictions*
Plane wall	A L	$\frac{A}{L}$	One-dimensional heat flow
Conduction through the edge section of two walls–inner and outer surface temperatures uniform	L Δx L	$0.54L$	Inside dimension must be greater than $(1/5)\Delta x$
Conduction through the corner section of three homogeneous walls–inner and outer surface temperatures uniform	Δx Δx Δx	$0.15\Delta x$	$\Delta x \ll L$ L is the length of the wall

Isothermal cylinder of radius r buried in semi-infinite medium having isothermal surface	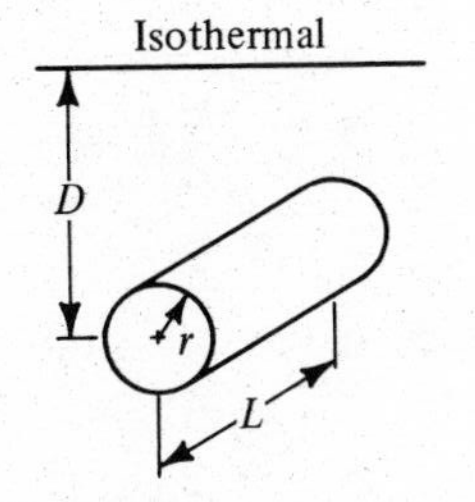	$\dfrac{2\pi L}{\cosh^{-1}(D/r)}$	$L \gg r$
		$\dfrac{2\pi L}{\ln(2D/r)}$	$L \gg r$ $D > 3r$
		$\dfrac{2\pi L}{\ln(L/r)\left[1 - \dfrac{\ln(L/2D)}{\ln(L/r)}\right]}$	$D \gg r$ $L \gg D$
Conduction between two isothermal cylinders buried in infinite medium	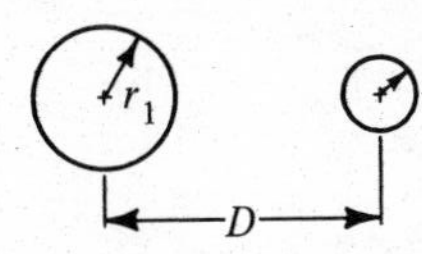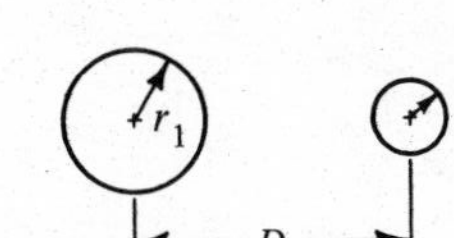	$\dfrac{2\pi L}{\cosh^{-1}\left(\dfrac{D^2 - r_1^2 - r_2^2}{2r_1r_2}\right)}$	$L \gg r$ $L \gg D$
Isothermal sphere of radius r buried in semi-infinite medium having isothermal surface	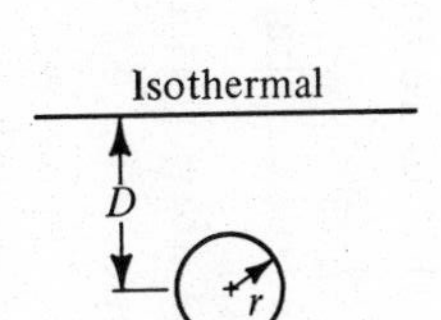	$\dfrac{4\pi r}{1 - r/2D}$	
Isothermal cylinder of radius r placed in semi-infinite medium as shown	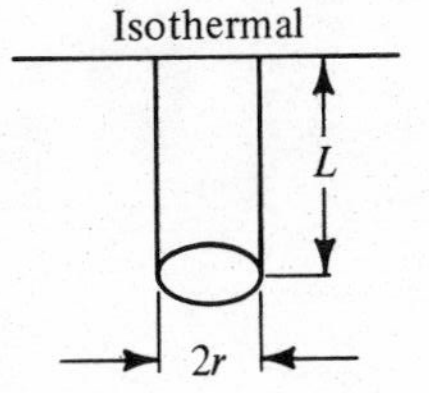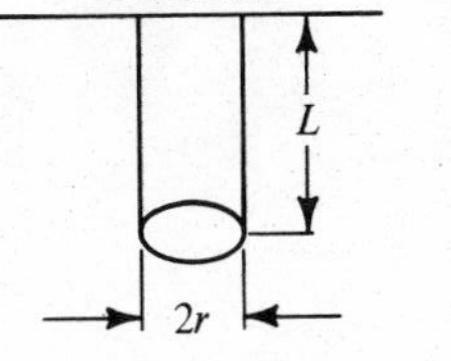	$\dfrac{2\pi L}{\ln(2L/r)}$	$L \gg 2r$

†Summarized from references 1, 2, and 3.

For one edge: $S = 0.54L = (0.54)(0.3) = 0.162$
(Table 3-1)

For one corner: $S = 0.15\Delta x = (0.15)(0.15) = 0.0225$
(Table 3-1)

Since there are six walls, twelve edges, and eight corners, the total shape factor is:

$$S = (6)(0.6) + (12)(0.162) + (8)(0.0225) = 5.724$$

and

$$Q = kS\Delta T$$

$$Q = (0.8)(5.724)(540 - 40) = 2290 \text{ watt or } 2.29 \text{ kW}$$

Hence, if 540°C were the maximum temperature at which the furnace were to be operated, a 3-kW heating element would suffice.

Sample Problem 3-2 A one-foot diameter steam pipe buried below the earth's surface carries 272°F steam to a building for heating purposes. The centerline of the pipe is five feet below the earth's surface. The soil in which the pipe resides has an average thermal conductivity of 0.22 Btu/hr-ft°F. Using a flux plot, determine the heat loss per linear foot of pipe if the surface temperature of the soil is 32°F. Check your result against that obtained by using the appropriate shape factor in Table 3-1.

Solution: To draw our flux plot, we will assume that the temperature drop across the pipe wall is small and that the temperature on the outside of the pipe is 272°F. Next, we note that the overall temperature difference for this problem is 272 − 32 = 240°F, and we will arbitrarily select eight temperature increments of 30°F each. The resulting flux plot is sketched below. Only one-half of the heat-flow field is shown because of the symmetry of the problem.
We note that there are 8 ΔTs and a total of 18 heat flow lanes. Therefore

$$S = \frac{18}{8} = 2.25$$

and

$$Q = Sk\Delta T_{\text{overall}}$$

$$Q = (2.25)(0.22)(272 - 32) = 118.8 \text{ Btu/hr}$$

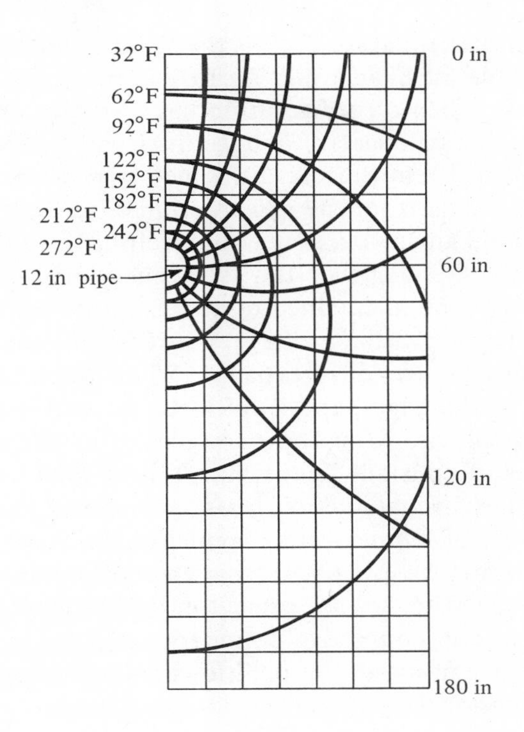

From Table 3-1

$$S = \frac{2\pi}{\cosh^{-1}(10/1)} = \frac{6.28}{3.0} = 2.09$$

and the heat loss per foot of pipe is

$$Q = (2.09)(0.22)(272 - 32) = 110.4 \text{ Btu/hr}$$

The difference in the two answers is 8.4 Btu/hr or about 7.5%.

3-3 ELECTRICAL ANALOGY FOR TWO-DIMENSIONAL CONDUCTION

If one considers the steady-state conduction of an electrical current through a constant electrical conductivity material, the differential equation that describes electrical potential as a function of position is

$$\frac{\partial^2 E}{\partial x^2} + \frac{\partial^2 E}{\partial y^2} = 0 \tag{3-3}$$

where E is the electrical potential.

Note that equation (3-3) is the same as equation (3-1) except that the dependent variable is E and not T. This is the Laplace equation in two dimensions. Therefore, when a configuration has a voltage difference impressed across it, the resulting loci of constant voltage lines will be analogous to the constant temperature lines if a temperature difference were to be impressed across it instead. This becomes the basis for the construction of an electrical analog.

To build an analog, a sheet of electrically conducting paper of high resistance* is cut into the shape of the two-dimensional system to be studied. The edges, which are to be maintained at a constant temperature, are modeled by maintaining them at a constant voltage, whereas a thermally insulated surface corresponds to the plain edge of the paper. When operating, the appropriate voltages are applied to the edges, and a voltmeter is used with a probe to determine lines of constant voltage. After these are marked on the paper (recognizing them to represent lines of constant temperature), heat flow lines are drawn to form a network of curvilinear squares. Next, from a count of the number of heat flow lanes and the number of temperature increments, the shape factor, S, is determined, and the heat flow may be calculated from equation (3-2).

Through the use of capacitors and added resistances, both unsteady-state conditions and convective boundaries may be analyzed. References 5, 6, 7, and 8 should be consulted for additional information on this technique for solving two-dimensional heat conduction problems.

3-4 ANALYTICAL SOLUTION FOR TWO-DIMENSIONAL HEAT CONDUCTION**

In order to obtain an analytical solution for two-dimensional heat conduction problems, it is necessary to introduce the concept of a Fourier series expansion of some function, say $f(x)$. During the solution of a two-dimensional heat conduction problem, a point is reached where sine and cosine terms appear on the right-hand side of an equal sign and $f(x)$ appears on the left-hand side of the same equal sign. At this point, it becomes necessary to expand $f(x)$ in a Fourier series in order to determine unknown coefficients.

A function that is sectionally continuous, single-valued, finite, and possesses a finite number of maxima and minima over a given interval is said to be a *piecewise regular* function. If $f(x)$ is piecewise regular over an interval $(-L, L)$, then it may be expanded in a sine and cosine series of the form

$$f(x) = a_o + \sum_{n=1}^{\infty} \left[a_n \cos\left(\frac{n\pi x}{L}\right) + b_n \sin\left(\frac{n\pi x}{L}\right) \right]$$

*Commercially called *Teledeltos* paper.

**Coverage of this material is optional in a first course in heat transfer.

where

$$\left.\begin{aligned} a_o &= \frac{1}{2L}\int_{-L}^{L} f(x)\,dx \\ a_n &= \frac{1}{L}\int_{-L}^{L} f(x)\cos\left(\frac{n\pi x}{L}\right)dx \\ b_n &= \frac{1}{L}\int_{-L}^{L} f(x)\sin\left(\frac{n\pi x}{L}\right)dx \end{aligned}\right\} \quad (3\text{-}4)$$

The above is the general Fourier series expansion of any piecewise regular function, $f(x)$. Since temperatures are well-behaved functions, they can normally be expanded in terms of a Fourier series.

Next, let us examine the form of the Fourier series expansion when $f(x)$ is an even or an odd function. An even function is one for which $f(-x) = f(+x)$ as illustrated in Figure 3-7(a). An odd function is one for which $f(-x) = -f(x)$ as illustrated in Figure 3-7(b).

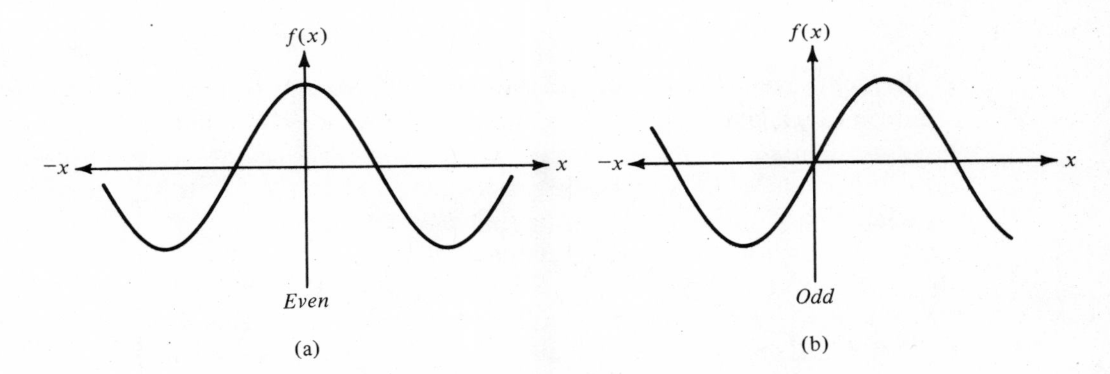

Figure 3-7 Even and odd functions.

Consequently sin x is an odd function and cos x is an even function. Noting that

(even function) × (even function) = even function

(odd function) × (odd function) = even function

(odd function) × (even function) = odd function

and that

$$\int_{-L}^{L} f(x)\,dx = 0, \qquad \text{when } f(x) \text{ is an odd function}$$

$$\int_{-L}^{L} f(x)\,dx = 2\int_{0}^{L} f(x), \qquad \text{when } f(x) \text{ is an even function}$$

we can rewrite the Fourier series expansion for the cases when $f(x)$ is odd and even.

If $f(x)$ is odd (Fourier sine series expansion), it follows from equation (3-4) that a_o and a_n must vanish, and

$$b_n = \frac{2}{L}\int_{0}^{L} f(x) \sin\left(\frac{n\pi x}{L}\right) dx$$

Hence

$$f(x) = \sum_{n=1}^{\infty} b_n \sin\left(\frac{n\pi x}{L}\right) \tag{3-5}$$

If $f(x)$ is even (Fourier cosine series expansion), the b_n terms vanish in view of equation (3-4), giving

$$a_o = \frac{1}{L}\int_{0}^{L} f(x)\,dx$$

and

$$a_n = \frac{2}{L}\int_{0}^{L} f(x) \cos\left(\frac{n\pi x}{L}\right) dx$$

Hence

$$f(x) = a_o + \sum_{n=1}^{\infty} a_n \cos\left(\frac{n\pi x}{L}\right) \tag{3-6}$$

Now, we are ready to analyze a two-dimensional heat conduction problem. Consider a long metal rod of rectangular cross section as shown in Figure 3-8. This might well be a metal billet that has undergone a heat treatment.

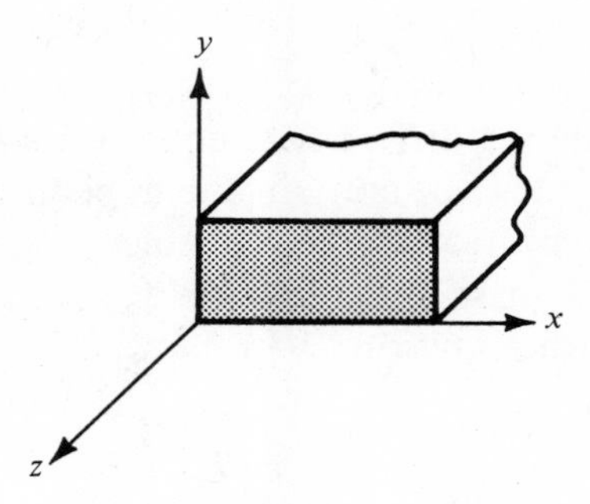

Figure 3-8 Long metal rod of rectangular cross section.

At any cross section (z = constant), the temperature is just a function of x and y. That is

$$T = T(x, y)$$

Consequently, such a problem is analyzed by solving the two-dimensional Laplace equation (3-1). Figure 3-9 shows one set of boundary conditions for this problem.

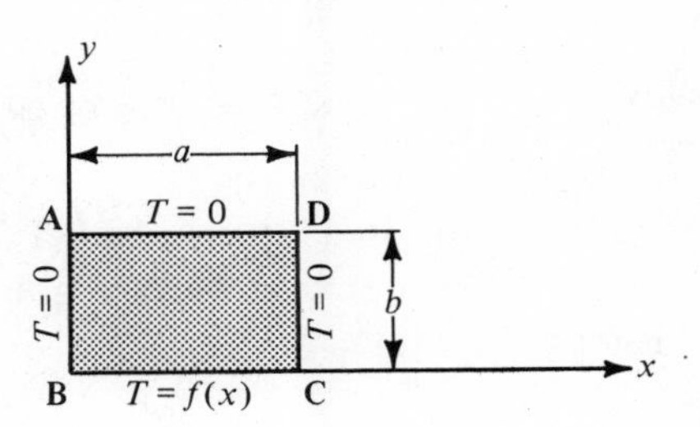

Figure 3-9 Two-dimensional heat conduction problem.

The value of $T = 0$ indicated on three surfaces (**BA**, **AD**, and **DC**) is merely a reference temperature. This is done only for convenience. It will be explained later how this requirement can be lifted.

The problem at hand has three homogeneous and one nonhomogeneous boundary conditions. To be homogeneous, a boundary condition must contain the dependent variable or its derivative in all of its nonzero terms (see Reference 8). The boundary condition $T = f(x)$ along the lower surface is nonhomogeneous since $f(x)$ does not contain the dependent variable, T. This same restriction is placed on a differential equation to determine its homogeneity; for example

$$\frac{\partial^2 T}{\partial x^2} + \frac{\partial^2 T}{\partial y^2} = 5$$

is not a homogeneous differential equation since the term on the right-hand side of the equal sign does not contain the dependent variable, T.

To obtain a solution to our problem, we write the appropriate differential equation that, when solved, will yield $T(x, y)$ and satisfy the boundary conditions. We start with equation (3-1)

$$\frac{\partial^2 T}{\partial x^2} + \frac{\partial^2 T}{\partial y^2} = 0 \tag{3-7}$$

From Figure 3-9, the boundary conditions are

$$\text{at } x = 0, T = 0 \tag{3-7a}$$

$$x = a, T = 0 \tag{3-7b}$$

$$y = b, T = 0 \tag{3-7c}$$

$$y = 0, T = f(x) \tag{3-7d}$$

Since T is a function of x and y only, we will propose a solution of the form

$$T = X(x)\,Y(y)$$

where $X(x)$ is a function of x alone, and $Y(y)$ is a function of y alone. The justification for such a form is that we are able to satisfy equations (3-7), as we will now show. First note that

$$\frac{\partial^2 T}{\partial x^2} = \left[\frac{d^2 X(x)}{dx^2}\right] Y(y)$$

and

$$\frac{\partial^2 T}{\partial y^2} = X(x)\left[\frac{d^2 Y(y)}{dy^2}\right]$$

Substituting the forgoing into equation (3-7) yields

$$\left[\frac{d^2 X(x)}{dx^2}\right] Y(y) + X(x)\left[\frac{d^2 Y(y)}{dy^2}\right] = 0$$

Now divide by $T(x, y) = X(x)Y(y)$, giving

$$\underbrace{\frac{1}{X(x)}\left[\frac{d^2 X(x)}{dx^2}\right]}_{\text{depends only on } x} + \underbrace{\frac{1}{Y(y)}\left[\frac{d^2 Y(y)}{dy^2}\right]}_{\text{depends only on } y} = 0$$

and

$$\frac{1}{X(x)}\left[\frac{d^2 X(x)}{dx^2}\right] = \frac{-1}{Y(y)}\left[\frac{d^2 Y(y)}{dy^2}\right] \tag{3-8}$$

The terms on either side of equation (3-8) must be equal to some constant. It is essential to determine what values the constant may take on in order to find a proper solution to the problem. If the constant should be zero, the solution to equation (3-8) will yield a linear relationship for $X(x)$ and $Y(y)$ that will not satisfy certain boundary conditions. Similarly, if the constant should be positive, the solution to equation (3-8) will not satisfy the prescribed boundary conditions. Therefore we are required to select the value of the separation constant to be $(-\lambda^2)$. The minus sign is selected so that the function that is to be expanded in a Fourier series, $f(x)$, will be in terms of sine and cosine when equation (3-8) is solved. Such a situation will allow us to satisfy all the prescribed boundary conditions.

From equation (3-8), with $-\lambda^2$ introduced as the separation constant

$$\frac{d^2 X(x)}{dx^2} + \lambda^2 X(x) = 0$$

which gives

$$X(x) = \mathbf{A} \sin(\lambda x) + \mathbf{B} \cos(\lambda x)$$

Also

$$\frac{d^2 Y(y)}{dy^2} - \lambda^2 Y(y) = 0$$

and the solution to the above equation is

$$Y(y) = c_1 e^{\lambda y} + c_2 e^{-\lambda y}$$

It can also be written in the following form

$$Y(y) = \mathbf{C} \sinh(\lambda y) + \mathbf{D} \cosh(\lambda y)$$

Since $T(x, y) = X(x)Y(y)$, we write

$$T = [\mathbf{A} \sin(\lambda x) + \mathbf{B} \cos(\lambda x)]\,[\mathbf{C} \sinh(\lambda y) + \mathbf{D} \cosh(\lambda y)]$$

We are now ready to apply the boundary conditions, equations (3-7a) to (3-7d). Applying equation (3-7a) gives

$$0 = [\mathbf{A} \underbrace{\sin(0)}_{=0} + \mathbf{B} \underbrace{\cos(0)}_{=1}]\,\underbrace{[\mathbf{C} \sinh(\lambda y) + \mathbf{D} \cosh(\lambda y)]}_{\text{cannot be zero or we have a trivial solution}}$$

Therefore, $\mathbf{B} = 0$. Applying equation (3-7b), we obtain

$$0 = \mathbf{A} \sin(\lambda a)\,\underbrace{[\mathbf{C} \sinh(\lambda y) + \mathbf{D} \cosh(\lambda y)]}_{\text{once again} \neq 0}$$

Also $\mathbf{A} \neq 0$ or we have a trivial solution, therefore

$$\sin(\lambda a) \equiv 0$$

This is true when $\lambda a = \pi, 2\pi, 3\pi$, etc. Or

$$\lambda = \frac{n\pi}{a}, \quad n = 1, 2, 3, \ldots.$$

The values of λ (the λ_ns) that satisfy the boundary condition are called *eigenvalues* or *characteristic values.*

We now write

$$T(x, y) = \left[\mathbf{A} \sin\left(\frac{n\pi x}{a}\right)\right]\left[\mathbf{C} \sinh\left(\frac{n\pi y}{a}\right) + \mathbf{D} \cosh\left(\frac{n\pi y}{a}\right)\right]$$

For compactness, let $(\mathbf{A})(\mathbf{C}) = \mathbf{E}$ and $(\mathbf{A})(\mathbf{D}) = \mathbf{F}$

Applying equation (3-7c), there results

$$0 = \underbrace{\left(\sin \frac{n\pi x}{a}\right)}_{\neq 0} \underbrace{\left[\mathbf{E} \sinh \left(\frac{n\pi b}{a}\right) + \mathbf{F} \cosh \left(\frac{n\pi b}{a}\right)\right]}_{\text{must be identically zero}}$$

Therefore

$$\mathbf{E} = \frac{-\mathbf{F} \cosh \left(\frac{n\pi b}{a}\right)}{\sinh \left(\frac{n\pi b}{a}\right)}$$

and

$$T = \mathbf{F} \sin \left(\frac{n\pi x}{a}\right) \left\{ \frac{-\cosh \left(\frac{n\pi b}{a}\right)}{\sinh \left(\frac{n\pi b}{a}\right)} \sinh \left(\frac{n\pi y}{a}\right) + \cosh \left(\frac{n\pi y}{a}\right) \right\}$$

or

$$T = \mathbf{F} \sin \left(\frac{n\pi x}{a}\right) \left\{ \frac{-\cosh \left(\frac{n\pi b}{a}\right) \sinh \left(\frac{n\pi y}{a}\right) + \cosh \left(\frac{n\pi y}{a}\right) \sinh \left(\frac{n\pi b}{a}\right)}{\sinh \left(\frac{n\pi b}{a}\right)} \right\}$$

But since

$$\sinh \left[\frac{n\pi}{a} (b - y)\right] = \cosh \left(\frac{n\pi y}{b}\right) \sinh \left(\frac{n\pi b}{a}\right) - \cosh \left(\frac{n\pi b}{a}\right) \sinh \left(\frac{n\pi y}{a}\right)$$

we can write

$$T = \frac{\mathbf{F}}{\sinh \left(\frac{n\pi b}{a}\right)} \sin \left(\frac{n\pi x}{a}\right) \left\{ \sinh \left[\frac{n\pi}{a} (b - y)\right] \right\}$$

Also, if we let

$$G_n = \frac{\mathbf{F}}{\sinh\left(\frac{n\pi b}{a}\right)} \qquad n = 1, 2, 3, \ldots\ldots$$

we obtain

$$T_n = G_n \sin\left(\frac{n\pi x}{a}\right) \sinh\left[\frac{n\pi}{a}(b - y)\right]$$

Recall that the governing differential equation and the boundary conditions are linear, therefore

$$\sum_{n=1}^{\infty} T_n$$

is also a solution to our problem. In fact, it is this solution for which the boundary condition (3-7d), when applied, will yield the coefficient G_n.

Thus

$$T = \sum_{n=1}^{\infty} T_n = \sum_{n=1}^{\infty} G_n \sin\left(\frac{n\pi x}{a}\right) \sinh\left[\frac{n\pi}{a}(b - y)\right]$$

satisfies the differential equation and the first three boundary conditions. We must ask ourselves if it is possible that $T(x, y)$ reduces to $f(x)$ if $y = 0$.

Applying equation (3-7d), we have

$$f(x) \stackrel{?}{=} \sum_{n=1}^{\infty} G_n \sin\left(\frac{n\pi x}{a}\right) \sinh\left(\frac{n\pi b}{a}\right)$$

We have inserted the question mark at this point since we have *assumed* a solution of the form $T(x, y) = X(x)Y(y)$. On the other hand, such an equality may exist according to the theorems of the Fourier series, which state that $f(x)$ can be expanded into a series, provided that $f(x)$ is well-behaved. Since $f(x)$ represents a temperature variation, it may be assumed to meet the necessary conditions.

Noting that $G_n \sinh (n\pi b/a) = \text{constant} = K_n$, we rewrite the foregoing equation as

$$f(x) \stackrel{?}{=} \sum_{n=1}^{\infty} K_n \sin\left(\frac{n\pi x}{a}\right)$$

We are home free if we recognize K_n to be the Fourier sine series coefficient of equation (3-5)

$$K_n = b_n = \frac{2}{a}\int_0^a f(x) \sin\left(\frac{n\pi x}{a}\right) dx$$

The final form of the solution is

$$T(x, y) = \frac{2}{a}\sum_{n=1}^{\infty}\left[\frac{1}{\sinh [(n\pi b)/a]}\int_0^a f(x) \sin\left(\frac{n\pi x}{a}\right) dx\right] \cdot \sin\left(\frac{n\pi x}{a}\right)\left\{\sinh\left[\left(\frac{n\pi}{a}\right)(b - y)\right]\right\} \tag{3-9}$$

We now have the solution to a simple two-dimensional problem with only one nonhomogeneous boundary condition. If we were confronted with a more difficult problem having four nonhomogeneous boundary conditions, we would solve the problem by superposition as follows.

If the temperature were not zero on all of the faces, a solution would be reached by solving four subproblems and adding their solutions. The subproblems would be such that in subproblem number 1, all temperatures would be zero except on the surface where $y = 0$ (face **BC** in Figure 3-9); in subproblem number 2, all temperatures would be zero except on the surface where $y = b$ (face **AD**); in subproblem number 3, all temperatures would be zero except on the surface where $x = 0$ (face **AB**); and in subproblem number 4, all temperatures would be zero except on the surface where $x = a$ (face **DC**). This is called the principle of superposition and is a result of the linearity of the differential equation and of the boundary conditions describing the temperature as a function of position. Had there been three boundaries with nonzero temperatures, then there would have been three subproblems, etc. The reader is referred to Reference 9 for a detailed discussion of such problems.

Sample Problem 3-3 illustrates the application of the superposition principle.

Sample Problem 3-3 Consider the following two-dimensional problem:

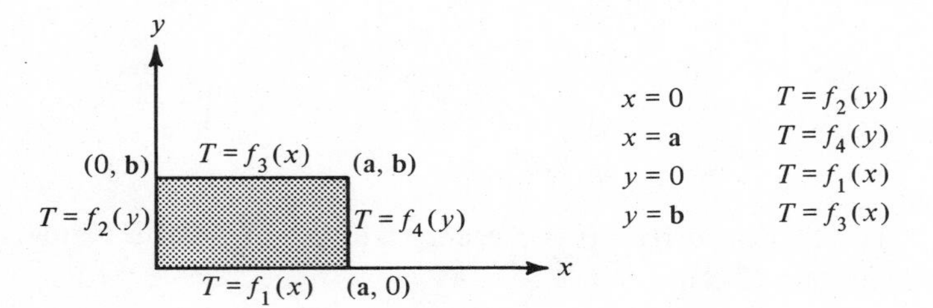

Outline of solution:

The problem can be broken down into four subproblems as follows:

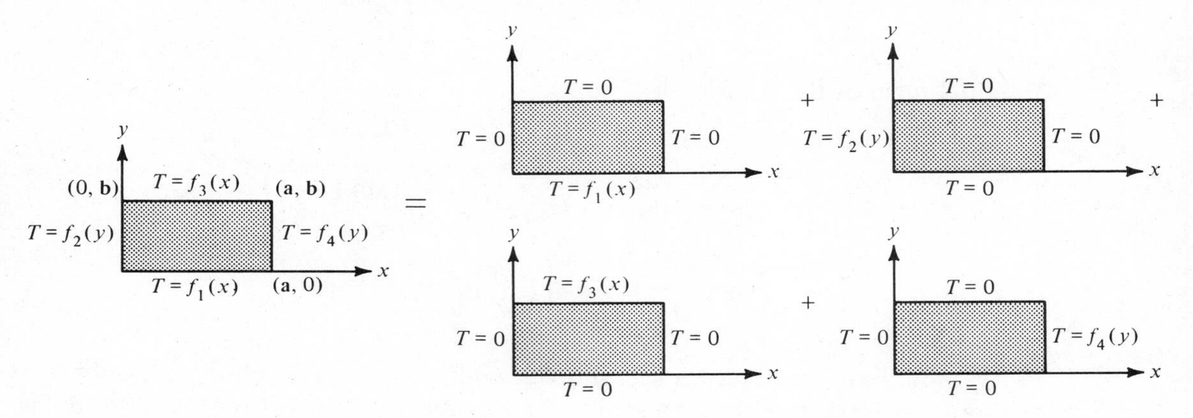

The equality in the above sketch holds since the governing differential equation and the boundary conditions are linear. That is

$$T(x, y) = T_1(x, y) + T_2(x, y) + T_3(x, y) + T_4(x, y)$$

where the terms on the right side are solutions to the subproblems in the above sketch.

It follows that

$$\nabla^2 T = \nabla^2 T_1 + \nabla^2 T_2 + \nabla^2 T_3 + \nabla^2 T_4 \; *$$

and we note that the boundary conditions of the main problem equal the sum of the boundary conditions of the subproblems.

$$* \; \nabla^2 T = \frac{\partial^2 T}{\partial x^2} + \frac{\partial^2 T}{\partial y^2}$$

Equation (3-9) is the solution to the subproblem for temperature T_1 (x, y). To obtain the solution for

T_2 (x, y) interchange x and y, a and b in equation (3-9)

T_3 (x, y) replace $(b - y)$ by y in equation (3-9)

T_4 (x, y) interchange x and y, a and b in solution for T_3 (x, y).

At this point, many texts discuss numerical solutions to two- and three-dimensional heat conduction problems. In this text, numerical solutions for steady-state heat conduction problems are discussed in Chapter 5 along with numerical solutions for transient problems.

PROBLEMS (English Engineering System of Units)

3-1. The cross section of a furnace wall is shown in EFigure 3-1. Using the graphical approach, determine the shape factor, S, and the resulting heat flow for the conditions shown.

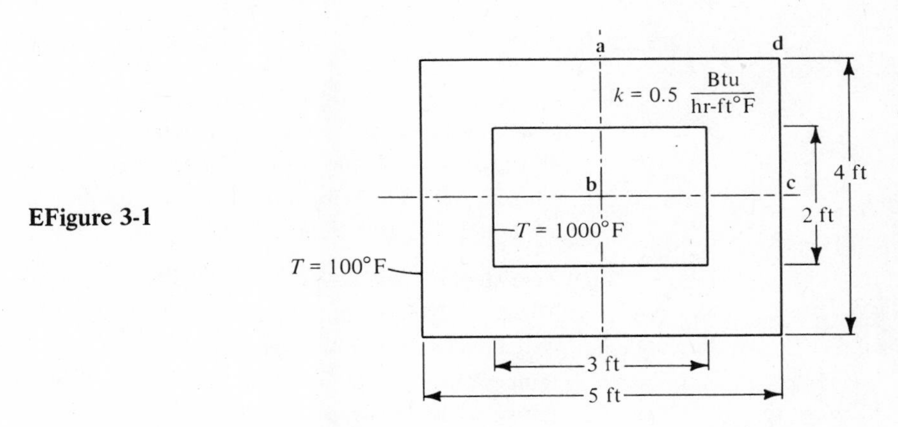

EFigure 3-1

Hint: Due to symmetry, the heat flow will be four times that through section **abcd.**

3-2. A sphere 3 feet in O.D. containing a radioactive material is buried such that its uppermost point is 5 feet below the earth's surface. If the outside surface of the sphere is at 800°F and the thermal conductivity of the soil is 0.2 Btu/hr-ft°F, determine the heat lost by the sphere. The surface temperature of the soil is 60°F.

3-3. A 1-inch O.D. heating rod is eccentrically embedded in a 4-inch O.D. cylinder as shown in EFigure 3-3. For the conditions shown, make a flux plot to determine the heat flow from the heating rod.

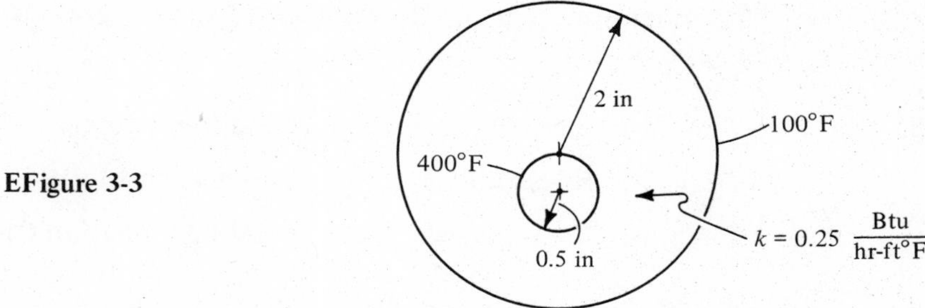

EFigure 3-3

3-4. Determine the heat flow per unit depth of the object shown in EFigure 3-4 using a flux plot.

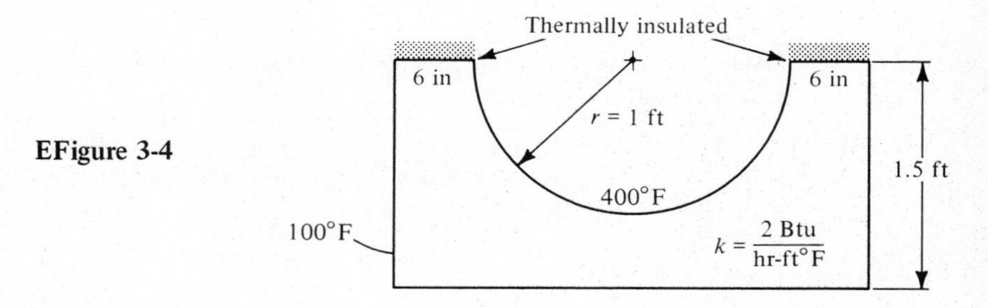

EFigure 3-4

3-5. A small laboratory furnace having inside dimensions of 12″ × 18″ × 18″ and a wall thickness of 4″ is made of a material having a thermal conductivity of 0.4 Btu/hr-ft°F. With the help of Table 3-1, determine the heat lost through the furnace wall. The temperatures on the inside and outside of the furnace walls are 500°F and 100°F, respectively.

3-6. Two steam pipes are buried parallel to each other deep in the earth's surface. The first pipe has an O.D. of 4 inches and an outside surface temperature of 350°F, while the second pipe has an O.D. of 6 inches and an outside surface temperature of 500°F. With the aid of Table 3-1, estimate the heat flow between the two pipes if the thermal conductivity of the soil is taken to be 0.3 Btu/hr-ft°F and the distance between the centerlines of the two pipes is 18 inches.

3-7. A ten-foot long 6-inch O.D. pipe is buried 30 inches below the surface of the soil having an average thermal conductivity of 0.25 Btu/hr-ft°F. The temperature at the surface of the soil is 35°F, and the temperature on the surface of the cylinder is 225°F. Determine the heat lost per foot of pipe by:

a. Using the graphical method (heat flux plot).
b. Using the appropriate shape factor given in Table 3-1.

3-8. Determine the heat flow per unit depth of the object shown in EFigure 3-8 using a flux plot.

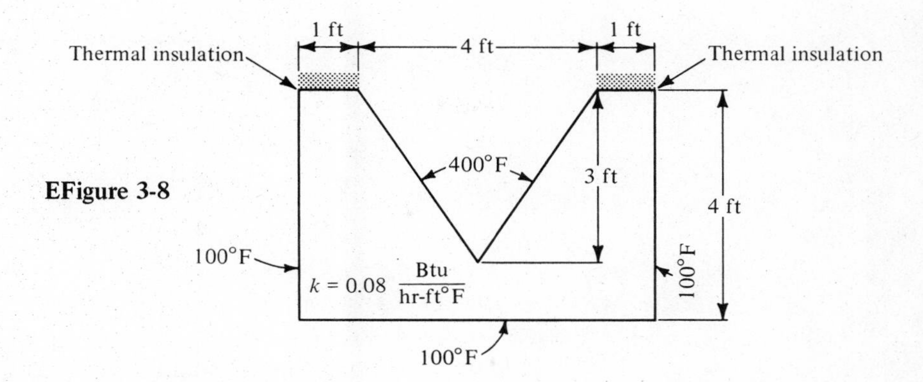

EFigure 3-8

3-9. Determine $T(x, y)$ for the two-dimensional problem in EFigure 3-9. T_o is a constant.

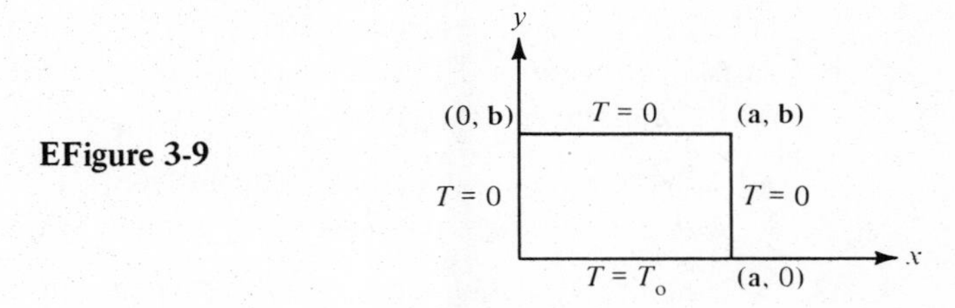

EFigure 3-9

3-10. In problem 3-9, if **a** = **b**, determine T (**a**/2, **a**/2).

3-11. In problem 3-9, let there be a convective boundary on the surface $y = b$; the heat transfer coefficient, h, and the ambient temperature, T_∞, are assumed to be constant. For this condition determine $T(x, y)$. See EFigure 3-11.

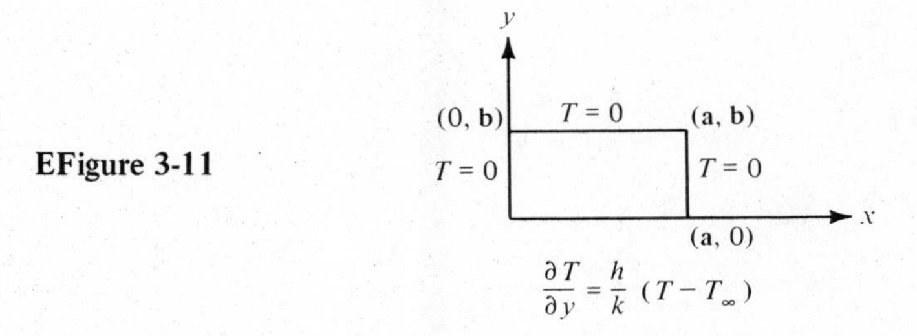

EFigure 3-11

PROBLEMS (SI System of Units)

3-1. The cross section of a furnace wall is shown below. Using the graphical approach, determine the shape factor, S, and the resulting heat flow for the conditions shown in SIFigure 3-1.

SIFigure 3-1

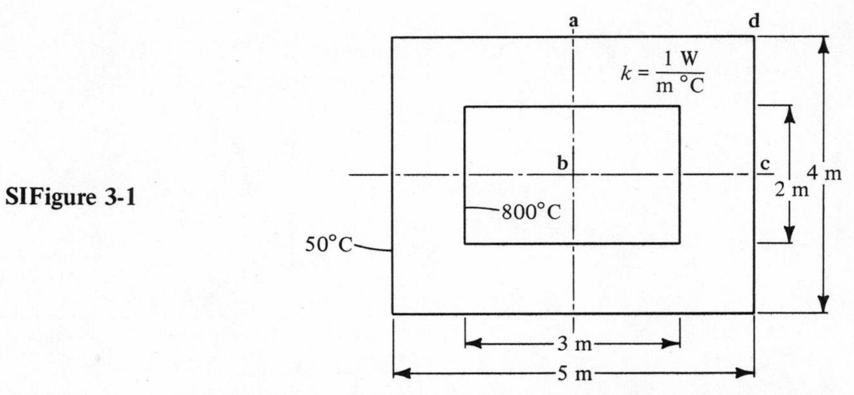

Hint: Due to symmetry, the heat flow will be four times that through section **abcd**.

3-2. A 4-meter long 15-cm O.D. pipe is buried 75 cm below the surface of the soil having an average thermal conductivity of 0.6 W/m°C. The temperature at the surface of the soil is 0°C, and the temperature on the surface of the cylinder is 110°C. Determine the heat lost per meter of pipe by:

a. Using the graphical method (heat flux plot).
b. Using the appropriate shape factor given in Table 3-1.

3-3. A 25-mm O.D. heating rod is eccentrically embedded in a 10-cm O.D. cylinder as shown in SIFigure 3-3. For the conditions shown, make a flux plot to determine the heat flow from the heating rod.

SIFigure 3-3

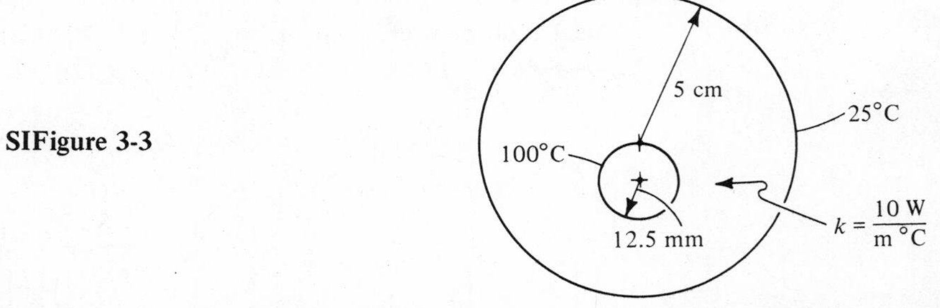

3-4. Determine the heat flow per unit depth of the object shown in SIFigure 3-4 using a flux plot.

SIFigure 3-4

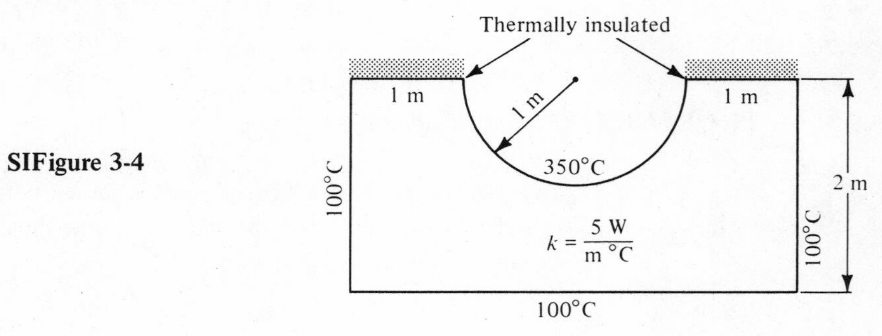

3-5. A small laboratory furnace having inside dimensions of 30 cm X 45 cm X 45 cm and a wall thickness of 10 cm is made of a material having a thermal conductivity of 1 watt/m°C. With the help of Table 3-1, determine the heat lost through the furnace walls. The temperatures on the inside and outside of the furnace wall are 300°C and 50°C, respectively.

3-6. Two steam pipes are buried parallel to each other deep in the earth's surface. The first pipe has an O.D. of 10 cm and an outside surface temperature of 175°C, while the second pipe has an O.D. of 15 cm and an outside surface temperature of 325°C. With the aid of Table 3-1, estimate the heat flow between the two pipes if the thermal conductivity of the soil is taken to be 0.75 W/m°C, and the distance between the centerlines of the two pipes is 45 cm.

3-7. A sphere 1 meter in O.D. containing a radioactive material is buried such that its uppermost point is 2 meters below the earth's surface. If the outside surface of the sphere is at 425°C and the thermal conductivity of the soil is 1 W/m°C, determine the heat lost by the sphere. The surface temperature of the soil is 25°C.

3-8. Determine the heat flow per unit depth of the object shown in SIFigure 3-8 using a flux plot.

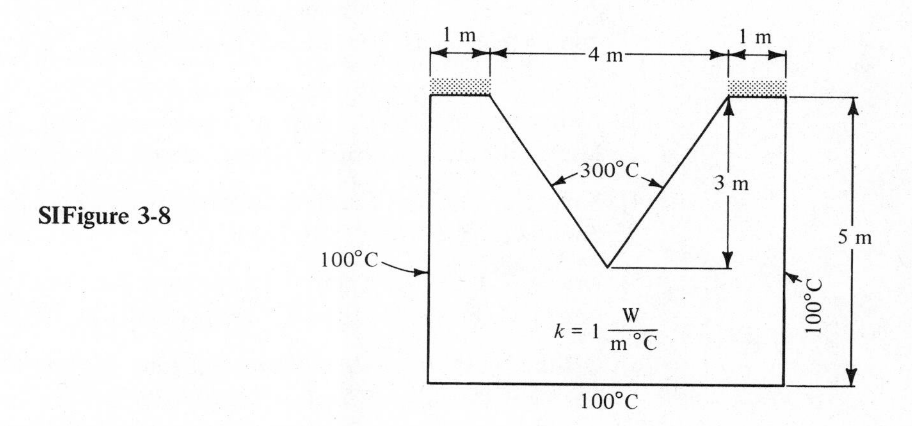

SIFigure 3-8

3-9. Determine $T(x, y)$ for the two-dimensional problem in SIFigure 3-9. T_o is a constant.

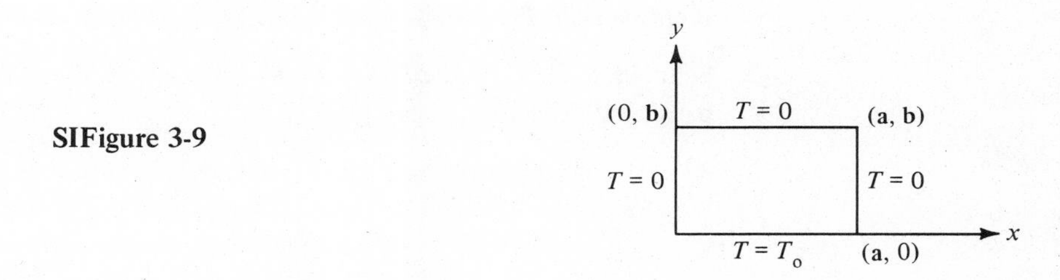

SIFigure 3-9

3-10. In problem 3-9, if **a** = **b**, determine $T(\mathbf{a}/2, \mathbf{a}/2)$.

3-11. In problem 3-9, let there be a convective boundary on the surface $y = 0$. For this condition, determine $T(x, y)$. The heat transfer coefficient h and the ambient fluid temperature, T_∞, are assumed to be constant. See SIFigure 3-11.

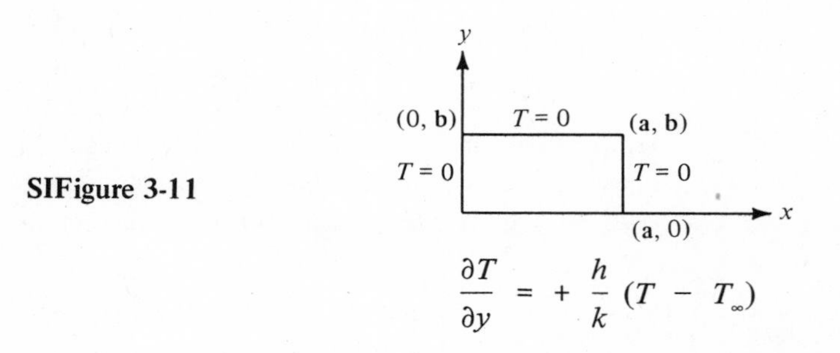

SIFigure 3-11

REFERENCES

[1] Langmuir, I., E. Q. Adams, and F. A. Meikle, "Flow of Heat Through Furnace Walls," *Trans. Am. Electrochem. Soc.,* **24**, (1913), p. 53.

[2] Rudenberg, R., 'Die Ausbrertung der Luft-und Erdfelder um Hochapannungaleitungen besonders bei Erd-und Kurzschlerssen," *Electrotech. Z.,* **46**, (1945), p. 1342.

[3] Andrews, R. V., "Solving Conductive Heat Transfer Problems with Electrical-Analogue Shape Factors," *Chem. Eng. Progr.,* **5**, No. *2,* (1955), p. 67.

[4] Kayan, C. F., "An Electrical Geometrical Analogue for Complex Heat Flow," *Trans. ASME,* **67**, (1945), p. 713.

[5] Kayan, C. F., "Heat Transfer Temperature Patterns of a Multicomponent Structure by Comparative Methods," *Trans. ASME,* **71**, (1949).

[6] Oziaik, N. M., *Boundary Value Problems of Heat Conduction,* International Textbook Company, Scranton, Pa., (1968).

[7] Instructions for Analog Field Plotters, Catalogues 112L152G1 and G2, General Electric Company, Schenectady, N. Y.

[8] Miller, K. S., *Partial Differential Equations in Engineering Problems,* Prentice-Hall, Inc., Englewood Cliffs, N. J., (1961).

[9] Schneider, P. J., *Conduction Heat Transfer,* Addison-Wesley Publishing Company, Inc., Reading, Mass., (1955).

4

Transient Heat Conduction

4-1 INTRODUCTION

In our discussion of heat conduction, we have thus far been restricted to situations where the temperature varies only with space coordinates. However, in many engineering problems, the temperature may also vary with time. In fact, whenever the boundary temperatures imposed on a system are changed an unsteady-state situation ensues, the temperature being a function of time as well as position. If a steady state is ultimately attained, it is the limit of the transient temperature distribution for large values of time. The time necessary for steady state to be reached is dependent upon the problem at hand. The period of time during which the temperature varies with time is often referred to as the transient period, i.e., that period of time that is required to attain steady-state conditons. We run across many transient heat conduction problems in our daily activities. Some examples are the cooling of food, the freezing of ice cubes, the heating of an electric iron, and the heating up of an automobile engine.

Let us consider the problem of transient conduction in a wall initially at temperature, T_2, which has its left-hand face raised to temperature, T_1, at time $\tau = 0$ while the right-hand face is continued to be maintained at temperature T_2.

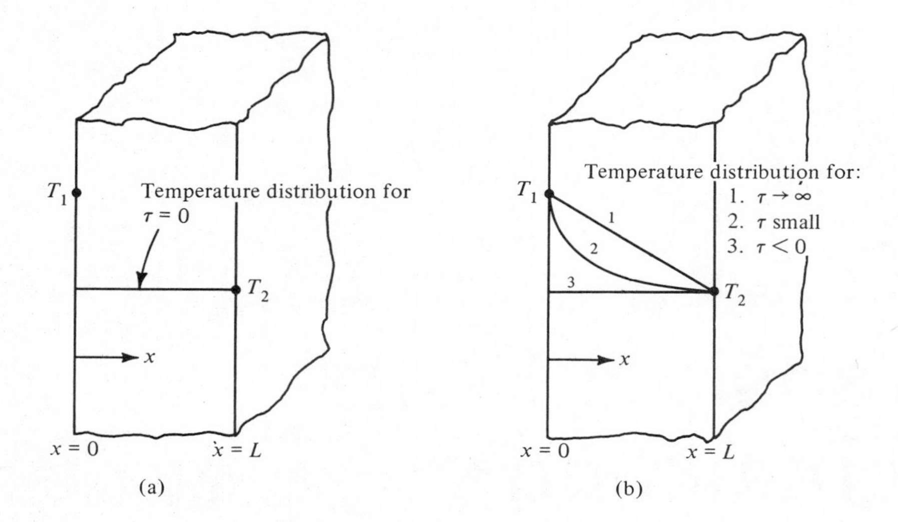

Figure 4-1 Plane wall undergoing transient heat conduction.

For $\tau = 0$, the temperature distribution is shown in Figure 4-1(a) and as τ increases from 0 to a large value ($\tau \to \infty$) the shape of the temperature distribution changes to that of the well-known straight-line distribution for steady-state one-dimensional heat conduction shown in Figure 4-1(b). The statement **as time tends to infinity** is not to be interpreted to mean thousands of hours have to pass before reaching steady state. The transient period can be anywhere from a few seconds to several hours depending upon the specific problem.

In many problems of engineering interest, it is necessary to know how long it will take to change the temperature by a specified amount at a given point within a body if the thermal conditions at the body surface are altered abruptly. In this chapter, we will be looking at ways to determine temperature distributions for transient heat conduction.

For one-dimensional transient heat conduction problems, the complexity of the solution is comparable to that presented in Chapter 3 for steady-state two-dimensional heat conduction. Two- and three-dimensional transient heat conduction problems are even more involved. The most powerful and commonly used approach is the finite difference method, which will be presented in Chapter 5. In this chapter, our emphasis will be placed on chart solutions. A chart solution is simply a plot of an analytical solution for a variety of parameter values. In discussing a semi-infinite body, a body for which there is always at a given time, an internal point that is unaffected by the alteration of thermal conditions at its boundaries, the *heat balance integral* will be used. This should prove helpful, as it will introduce the basic concepts presented in treating the integrated boundary layer equations used in convective heat transfer analyses.

In addition to transient heat flow, we will also discuss periodic heat flow. Problems of this type exist when temperatures within a system are forced to vary on a regular periodic basis. An example of this situation is the cyclic temperature fluctuation in the wall of a laboratory furnace whose heating element is turned on and off at regular intervals. In this case the temperature on the inside surface undergoes regular fluctuations between given limits depending on the thermostat setting, which results in regular temperature fluctuations within the wall. Another example is the variation of the temperature of the surface of the earth during a twenty-four hour period.

4-2 SYSTEMS WITH NEGLIGIBLE INTERNAL RESISTANCE

The class of transient problems that lend themselves most readily to analysis are those with negligible internal resistance to the flow of heat. In such problems, the convective resistance at the system boundary is very large when compared to the internal resistance due to conduction. In essence, the solid behaves as though it has an infinite thermal conductivity in that the temperature throughout the entire solid is always uniform and varies only with time. In reality, such a situation is never precisely attained since all materials have a finite thermal conductivity and, as heat is added or removed, temperature gradients must exist as demonstrated by Fourier's law of heat conduction, $Q = -kA(dT/dx)$. However, when the convective resistance at the boundary of the solid is large compared to the internal resistance due to conduction, the main part of the spatial temperature variation occurs across the system boundary, with only small variation in the internal temperature. Figure 4-2 illustrates this point.

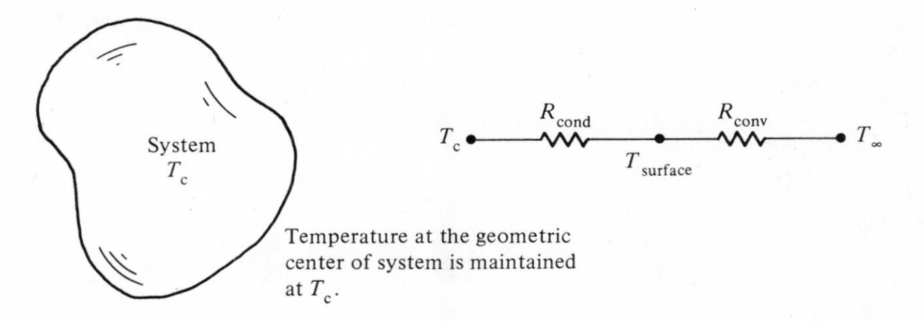

Figure 4-2 Steady state system with large external convective resistance and small internal conductive resistance.

Referring to Figure 4-2, we let R_{conv} be 999 times R_{cond}, a case where the external convective resistance is much greater than the internal conductive resistance. We note that for a given instant of time if T_∞ were 1000°F and T_c were 0°F, $T_{surface}$ would be 1°F, since

$$T_{surface} = \left(\frac{R_{cond}}{R_{conv} + R_{cond}}\right)(1000 - 0) = 1°F$$

and that the maximum temperature difference in the solid at that instant would be 1°F. Hence, if we say that the temperature in the solid is independent of location within the body, our statement would be accurate to within 1°F.

If we now let T_∞ vary with time, we find that, for all practical purposes, the temperature at any given time is a constant throughout the body varying only by a fraction of a degree. This is the basic concept behind the analysis of systems with negligible internal resistance. Such an approach is sometimes called the *lumped parameter* approach. To reiterate: If the external convective resistance is large compared to the internal conductive resistance, the temperature within a body can be treated as uniform throughout and, therefore, will be considered to vary only with time.

The systems to be analyzed will be limited to those that are at some initial temperature, T_o, throughout and that are placed in a new environment at a constant temperature, T_∞. Also, the convective heat transfer coefficient between the solid and the new surroundings is h, and the ratio of external resistance to internal resistance of the system is large.

A dimensionless number called the Biot number is introduced to determine the validity of the lumped parameter approach. The Biot number is the ratio of internal resistance to external resistance. Consequently, a small Biot number means a low value of the internal resistance in relation to the external resistance, thereby satisfying the prerequisite for a system of negligible internal resistance. To write the Biot number we note

(1) Internal resistance $\propto L_c/k$ where L_c is a characteristic length and is equal to the volume of the solid divided by its surface area, and k is the thermal conductivity of the solid.
(2) External resistance $\propto 1/h$ where h is the convective heat transfer coefficient at the boundary of the solid.

Therefore

$$\text{Bi} \propto \frac{\text{internal resistance}}{\text{external resistance}} \propto \frac{hL_c}{k}$$

Taking the constant of proportionality as unity, we have

$$\text{Bi} = \frac{hL_c}{k} \tag{4-1}$$

It has been found for simple geometric shapes such as plates, cylinders, and spheres, that if $\text{Bi} < 0.1$, the error introduced by assuming the temperature to be spatially uniform at any given time is less than 5 percent. Some commonly encountered characteristic lengths are listed in Table 4-1.

TABLE 4-1 Characteristic Length to be Used in Biot Number

Geometry	L_c – *Characteristic Length*
Plane wall–thickness $2L$	L
Long cylinder–radius r_o	$\frac{r_o}{2}$
Sphere–radius r_o	$\frac{r_o}{3}$
Cube–side a	$\frac{a}{6}$

Let us consider a steel ball to be used as a ball bearing, which undergoes a heat treating process. It comes out of a furnace at some initial uniform temperature, T_o, and is quenched in a vat of oil that is maintained at a temperature equal to T_∞. The convective heat transfer coefficient between the steel ball and the oil is h, the surface area of the ball is A, its volume is V, and its thermal conductivity is k. When the Biot number is calculated, it is found to be less than 0.1, hence the concept of negligible internal resistance may be applied.

To determine the differential equation that describes the temperature of the steel ball as a function of time, we write an energy balance on the steel ball.

$$\left\{\begin{matrix}\text{Rate of heat loss by} \\ \text{convection to oil}\end{matrix}\right\} = \left\{\begin{matrix}\text{Rate of decrease in internal} \\ \text{energy of the steel ball}\end{matrix}\right\}$$

Letting T be the temperature of the steel ball at time, τ, we write

$$hA\,(T - T_\infty) = -\rho c V \frac{dT}{d\tau} \tag{4-2}$$

where

ρ = density of ball material lb_m/ft^3, kg/m^3

c = specific heat of ball material, $Btu/lb_m\,°F$, $J/kg°C$

All other terms are as previously defined, and all thermal properties are assumed to be constant. The negative sign appears on the right-hand side since the internal energy is decreasing with time if the ball loses heat.

To solve equation (4-2), rewrite it as

$$\frac{dT}{d\tau} + \left(\frac{hA}{\rho c V}\right)T = \left(\frac{hA}{\rho c V}\right)T_\infty \tag{4-3}$$

Equation (4-3) is a nonhomogeneous ordinary differential equation with constant coefficients describing T as a function of τ with the term on the right-hand side representing the nonhomogeneity (it does not contain the dependent variable, T). To solve, we set the right-hand side equal to zero and use the standard operator technique to determine the complementary or transient solution, T_c. Then, noting that the nonhomogeneous part of the differential equation is a constant, we assume $T = K$ (constant) and substitute back into equation (4-3) to determine the particular or steady state solution, T_p. After determining T_c and T_p, they are added together to determine the general solution to equation (4-3). At this point, and **only at this point**, the initial condition that

$$\text{at } \tau = 0, T = T_o \tag{4-3a}$$

may be applied to determine the unknown constant. Proceeding as indicated

$$\frac{dT}{d\tau} + \left(\frac{hA}{\rho c V}\right)T = 0$$

characteristic equation: $P + \dfrac{hA}{\rho c V} = 0$

characteristic root: $P = -\dfrac{hA}{\rho c V}$

complimentary solution: $T_c = Ce^{-\left(\frac{hA}{\rho c V}\right)\tau}$

where C is a constant of integration.

Next we assume $T_p = K$ (constant) and substitute in equation (4-3) to obtain

$$\cancelto{0}{\frac{dK}{d\tau}} + \frac{hA}{\rho c V} K = \frac{hA}{\rho c V} T_\infty$$

which gives

$$K = T_\infty$$

Thus the particular solution, T_p, is given by $T_p = T_\infty$.

The complete solution may now be stated as

$$T = T_c + T_p = Ce^{-\left(\frac{hA}{\rho c V}\right)\tau} + T_\infty$$

or

$$T - T_\infty = Ce^{-\left(\frac{hA}{\rho c V}\right)\tau}$$

Application of the initial condition, at $\tau = 0$, $T = T_o$, gives

$$T_o - T_\infty = Ce^o$$

or

$$C = T_o - T_\infty$$

and

$$\frac{T - T_\infty}{T_o - T_\infty} = e^{-\left(\frac{hA}{\rho c V}\right)\tau} \quad \text{if Bi} < 0.1 \tag{4-4}$$

Equation (4-4) gives the temperature as a function of time for a body initially at temperature, T_o, which is placed in a convective environment at temperature, T_∞, for values of Biot number less than 0.1.

The quantity $(\rho c V/hA)$ in equation (4-4) has the units of time and is often referred to as the time constant of the system. At time, τ, equal to one time constant

$$\frac{T - T_\infty}{T_o - T_\infty} = e^{-1}$$

or

$$\frac{T - T_\infty}{T_o - T_\infty} = 0.368$$

This means that at the end of a time period equal to one time constant, the difference in the temperature of the body and that of the ambient fluid would be

36.8 percent of the initial temperature difference; or in other words, the temperature difference would be reduced by 63.2 percent. For all practical purposes, a system is said to have achieved steady state after a time of 4 time constants has elapsed; i.e.,

$$\frac{T - T_\infty}{T_o - T_\infty} = e^{-4} = 0.018$$

Sample Problem 4-1 An aluminum sphere weighing 7 Kg and initially at a temperature of 260°C is suddenly immersed in a fluid at 10°C. If $h = 50$ W/m^2°C, determine the time required to cool the aluminum to 90°C.

Solution:

For Aluminum:

$\rho = 2707$ kg/m^3

$c = 900$J/kg°C

$k = 204$ W/m°C

$$\text{Bi} = \frac{hL_c}{k}$$

$$V = \frac{4}{3}\pi r_o^3 \qquad L_c = \frac{r_o}{3}$$

$$V = \frac{\text{mass}}{\rho} = 7/2707 = 2.58 \times 10^{-3}\,\text{m}^3$$

$$h = 50\ \text{W/m}^2{}^\circ\text{C}$$

$$r_o = (3V/4\pi)^{1/3} = 0.085\ \text{m}$$

$$L_c = \frac{r_o}{3} = 0.028\ \text{m}$$

Therefore

$$\text{Bi} = \frac{hr_o}{3k} = \frac{(50)(0.085)}{(3)(204)} \ll 0.1$$

and the concept of negligible internal resistance applies. It follows that

$$\frac{T - T_\infty}{T_o - T_\infty} = e^{-\left(\frac{hA}{\rho c V}\right)\tau}$$

$$T = 90^\circ\text{C}$$

$$T_o = 260^\circ\text{C}$$

$$T_\infty = 10^\circ\text{C}$$

$$\frac{hA}{\rho c V} = \frac{3h}{\rho c r_o} = 7.2 \times 10^{-4}/\text{sec}$$

$$\frac{80}{250} = e^{-7.2\times10^{-4}\tau}$$

$$3.125 = e^{7.2\times10^{-4}\tau}$$

Therefore

$$\tau = 1580 \text{ sec} = 26.3 \text{ minutes}$$

In some problems, the change in internal energy of the system is important. During a time increment $d\tau$ the change in internal energy is

$$dU = \rho c V dT$$

Also, differentiating equation (4-4) gives

$$dT = (T_o - T_\infty)\left(-\frac{hA}{\rho c V}\right) e^{-\left(\frac{hA}{\rho c V}\right)} d\tau$$

The quantity of heat given off by the system, in accordance with the First Law of Thermodynamics, is simply dU since the work term is negligibly small. The total quantity of heat given off, during a time interval, (0 to τ), equals $(U_0 - U_\tau)$ where U_0 and U_τ represent the internal energies of the system at time, $\tau = 0$ and $\tau = \tau$, respectively. Hence

$$U_0 - U_\tau = \int_0^\tau - dU$$

or

$$U_0 - U_\tau = \int_0^\tau - \rho c V \, dT$$

$$U_0 - U_\tau = \int_0^\tau - \rho c V (T_o - T_\infty)\left(\frac{-hA}{\rho c V}\right) e^{-\left(\frac{hA}{\rho c V}\right)\tau} d\tau$$

Simplification and integration gives

$$U_0 - U_\tau = \rho c V (T_o - T_\infty)\left[e^{-\left(\frac{hA}{\rho c V}\right)\tau} - 1\right] \tag{4-5}$$

Systems comprised of several different materials may also be treated by the negligible internal resistance concept, but the algebra becomes quite complicated. Such systems are encountered when composite materials are subjected to environmental changes. References 1 and 2 treat a two-capacity system and are suggested for those wishing to pursue this concept.

It is also possible to consider systems in which the surrounding temperature varies periodically with time. Such systems exist during many chemical processing operations and in daily environmental cycles. The governing differential equation is the same as equation (4-2). The only difference in the solution is in that portion which results from the variation in T_∞ with time, so that the particular solution differs from the one previously obtained. If T_∞ varies sinusoidally with time, T_p would be assumed to have the form: $T_p = A \sin(\omega\tau) + B \cos(\omega\tau)$ where A and B are constants and ω is the frequency of the sine wave. If T_∞ increases linearly with time, then T_p would be assumed to have the form: $T_p = A\tau + B$ where A and B are constants. After obtaining the particular solution due to T_∞, it would be added to the complementary solution so that the unknown constants of integration could be determined by applying the initial condition.

4-3 TRANSIENT HEAT CONDUCTION IN PLANE WALLS, CYLINDERS, AND SPHERES WITH CONVECTIVE BOUNDARY CONDITIONS

In this section we will consider heat flows and temperature distributions in plane walls, cylinders, and spheres, for which the internal resistance is not negligible. To determine these quantities for such geometries when the Biot number is greater than 0.1, numerical, analytical, or chart solutions must be used. Only analytical and chart solutions will be considered here. Numerical solutions will be discussed in Chapter 5.

4-3.1 Plane Wall—Chart Solutions

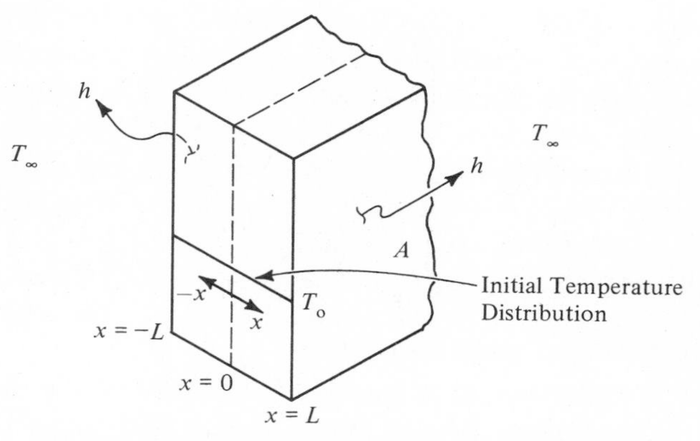

Figure 4-3 Plane wall with convective boundary.

We begin our discussion by considering chart solutions for a wall of thickness $2L$. A practical example of such a problem would be commercial freezing of vegetables or meats in the form of slabs. Typically, the prediction of the temperature at the midplane of the slab as a function of time is an important consideration in insuring proper processing of the food. Figure 4-3 shows a plane wall with convective boundary conditions.

Figure 4-4 (called Heisler charts after their original creator, see Reference 3) gives the results of the analytical solution in the form of charts. It shows a plot of

$$\left[\frac{T(0,\tau) - T_\infty}{T_o - T_\infty}\right] \text{ versus } \left(\frac{\alpha\tau}{L_c^2}\right)$$

with the reciprocal of the Biot number as a parameter, where

$T(0,\tau)$ = the centerline temperature at time, τ

T_∞ = the surrounding fluid temperature (constant)

T_o = initial temperature in the wall (constant)

$\dfrac{T(0,\tau) - T_\infty}{T_o - T_\infty}$ = dimensionless temperature ratio

$\dfrac{\alpha\tau}{L_c^2}$ = Fourier modulus (dimensionless time)

α = thermal diffusivity

τ = time

$L_c = L$ = one-half wall thickness

Bi = Biot number as previously defined in equation (4-1)

As indicated in Chapter 2, α is the thermal diffusivity (equal to $k/\rho c$) and is a measure of how rapidly heat diffuses through a material. It is interesting to note that although metals have much larger thermal conductivities than gases, their densities are very large, and the rate of diffusion through metals and gases are about the same. This is so because gases have low values of thermal conductivity and density. Thus, for a similar geometry, the time for heat to diffuse through a gas or a metal will be approximately the same.

The Fourier modulus is a dimensionless time containing the thermal diffusivity, α, time, τ and the characteristic length, L_c. For a given body, it varies linearly with time.

The temperature at some location other than at the centerline can be determined from Figure 4-5. In Figure 4-5 a plot of $[(T(x,\tau) - T_\infty)/(T(0,\tau) - T_\infty)]$ is made against $(1/\text{Bi})$ with (x/L) as a parameter. Remember that x is measured

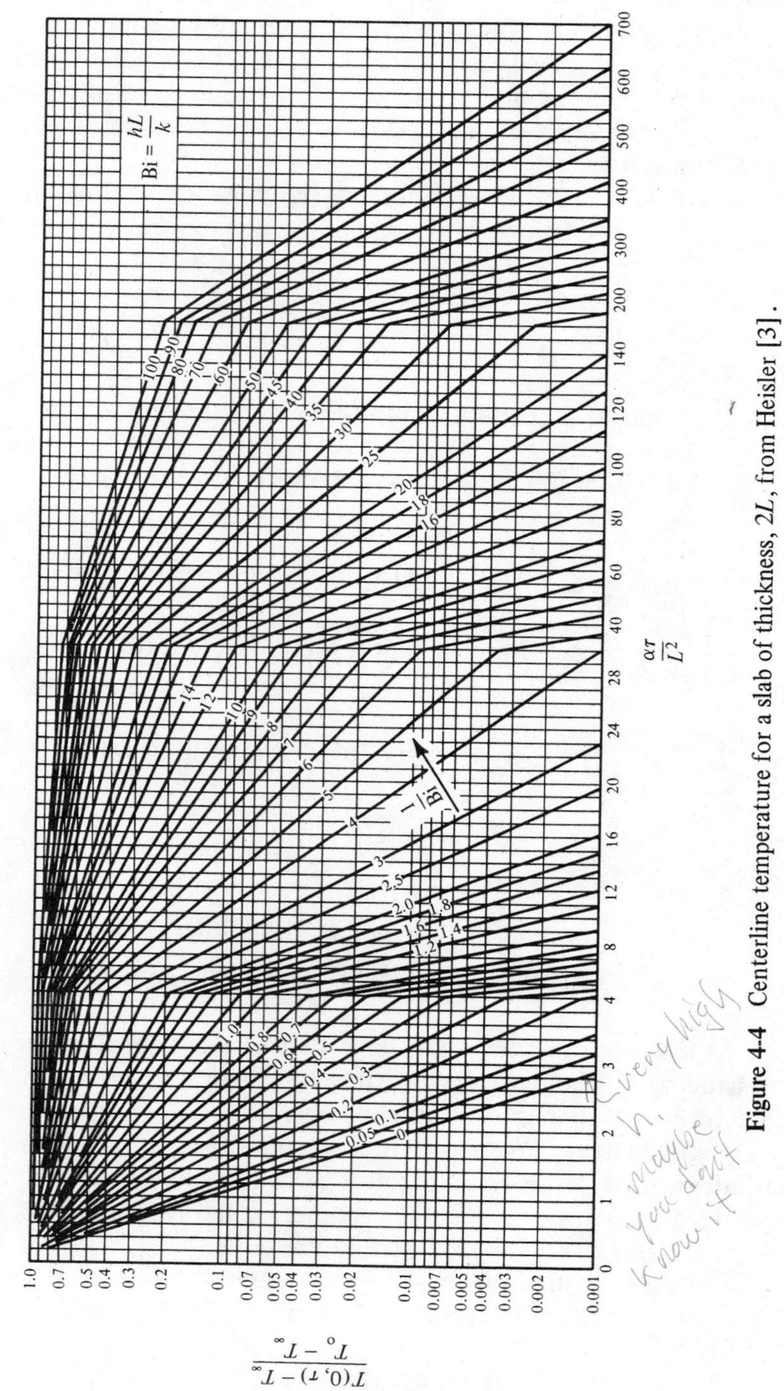

Figure 4-4 Centerline temperature for a slab of thickness, $2L$, from Heisler [3].

from the centerline toward the surface, and that due to the symmetry of the problem, it does not matter in which direction one proceeds from the center. The value of the temperature will be the same for two points on either side of the centerline if the points are equidistant from the centerline. This is true only if h and T_∞ are the same for both the right-hand and the left-hand faces of the wall. For Figure 4-5, we have

$T(x, \tau)$ = temperature at x at time, τ

$\dfrac{x}{L}$ = dimensionless position, x being measured from the centerline toward the surface

All other parameters are as previously defined.

When the slab at temperature, T_o, is placed in the fluid at temperature, T_∞, it possesses a potential at that moment for taking on or releasing a quantity of heat, U_o. This quantity of heat equals the change in internal energy corresponding to a temperature change from T_o to T_∞. That is

$$U_o = \rho c V (T_o - T_\infty) \tag{4-6}$$

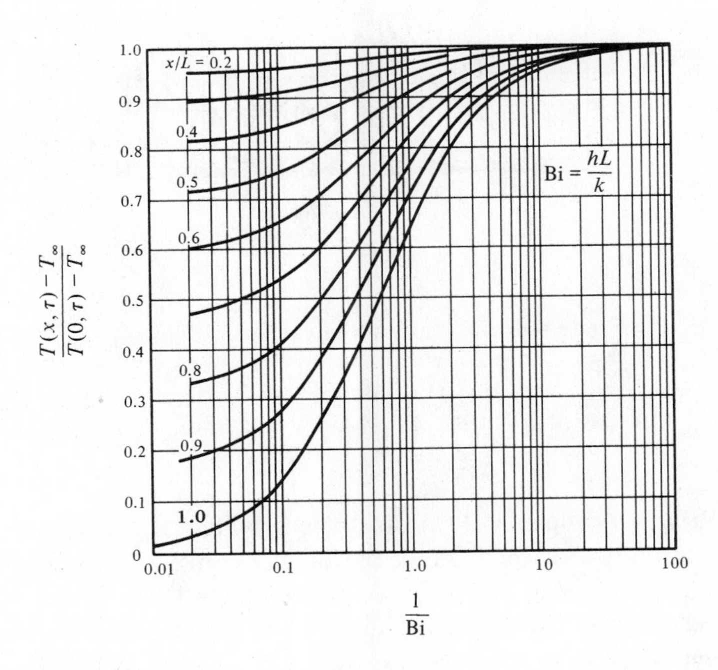

Figure 4-5 Temperature as a function of centerline temperature in a slab of thickness, $2L$, from Heisler [3].

The heat lost or gained during some time τ may be determined through the use of Figure 4-6 (Reference 4). In Figure 4-6, (U/U_o) is plotted against $(h^2\alpha\tau/k^2)$ with the Biot number as a parameter, where U is the heat lost or gained during time, τ.

If one face of the slab is insulated, the temperature distribution can be found by letting $x = 0$ at the insulated surface and letting $x = L$ at the surface exposed to the convective environment. This can be done because the chart solution corresponds to the case where at $x = 0$, $dT/dx = 0$, an at $x = L$, there is convection to the environment. The only difference is that the quantity, L, is now the entire wall thickness of the insulated wall. Thus, Figures 4-4 and 4-5 represent solutions to two different types of problems.

Let us look at a sample problem where the Biot number is approximately 0.2, which is where the chart solution first becomes usable. We will then compare the solution to that obtained by assuming neglibible internal resistance to see if a discrepancy exists between these two solutions.

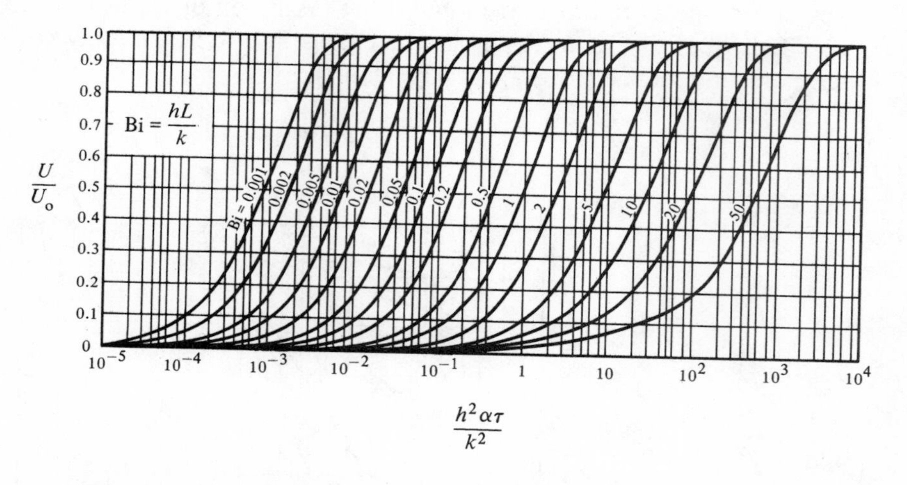

Figure 4-6 Dimensionless heat loss U/U_o of a slab of thickness, $2L$, with time, from *Fundamentals of Heat Transfer* by H. Grober, S. Erk, and H. Grigull. Copyright 1961 by McGraw-Hill Book Company. Used with permission.

Sample Problem 4-2 A steel sheet 10 feet by 10 feet square and 1 inch thick is initially at 1100°F and is placed in a convective environment at 100°F. If the convective heat transfer coefficient is 144 Btu/hr-ft^2 °F, determine the temperature at the centerline and at 1/8 inch from the surface using both the negligible internal resistance concept and the charts for a cooling time of 1.8 minutes. Compare the results. The thermal conductivity of the steel is 30 Btu/hr-ft°F, the thermal diffusivity is 0.6 ft^2/hr, the specific heat is 0.1 Btu/lb$_m$ °F, and the density is 500 lb$_m$/ft^3.

Solution: Since

$$L = \frac{1}{24}\text{ ft}$$

$$\text{Bi} = \frac{hL_c}{k} = \frac{(144)(1/24)}{30} = 0.2$$

Method of Charts: For $\tau = 1.8$ min $= 0.03$ hr., and $\alpha = 0.6$ ft^2/hr, the Fourier modulus is

$$\frac{\alpha\tau}{L_c^2} = \frac{(0.6)\,(0.03)}{(1/24)^2} = 10.3$$

The reciprocal of the Biot number is

$$\frac{1}{\text{Bi}} = \frac{1}{0.2} = 5$$

From Figure 4-4, the centerline temperature is given by

$$\frac{T(0,\tau) - T_\infty}{T_o - T_\infty} = 0.15$$

or

$$T\,(0,\tau) = (0.15)\,(1100 - 100) + 100$$
$$T\,(0,\tau) = 250°\text{F}$$

From Figure 4-5, at $x/L = (3/8)/(1/2) = 3/4$ and for $1/\text{Bi} = 5$

$$\frac{T\,(x,\tau) - T_\infty}{T\,(0,\tau) - T_\infty} = 0.95$$

$$T\,(x,\tau) = (0.95)\,(250 - 100) + 100$$

$$T\,(x,\tau) = 243°\text{F}$$

and we see only 7°F variation between the temperature at $x/L = 0$ (centerline) and at $x/L = 0.75$.

Method of Negligible Internal Resistance:

$$\frac{T - T_\infty}{T_o - T_\infty} = e^{-\left(\frac{hA}{\rho c V}\right)\tau}$$

$$T = T_\infty + (1000)\, e^{-\left[\frac{144}{(500)\,(1/24)\,(0.1)}\right](0.03)}$$

$$T = 100 + 1000 e^{-2.07}$$

$$T = 226\ ^\circ\text{F throughout the steel sheet}$$

Comparison: Even at Bi = 0.2, the negligible internal resistance approach agrees quite closely (within −24°F at the centerline) with the chart solutions. Consequently, it is suggested that for a *ball-park estimate*, this simplified approach can be used for Bi < 0.2. If Bi > 0.2, the charts should be used.

4-3.2 Plane Wall—Analytical Solution

The temperature distribution, shown in Figure 4-4, is obtained from the analytical solution to the governing differential equation. Referring to Figure 4-3, we start with the general three-dimensional heat conduction equation in Cartesian coordinates and proceed as follows:

$$\frac{\partial^2 T}{\partial x^2} + \frac{\partial^2 T}{\partial y^2} + \frac{\partial^2 T}{\partial z^2} + \frac{\dot{q}}{k} = \frac{1}{\alpha}\frac{\partial T}{\partial \tau}$$

since there are no heat sources $\dot{q}/k = 0$. Also, the temperature does not vary in the y or z direction, giving

$$\frac{\partial^2 T}{\partial y^2} = \frac{\partial^2 T}{\partial z^2} = 0$$

and the differential equation becomes

$$\frac{\partial^2 T}{\partial x^2} = \frac{1}{\alpha}\frac{\partial T}{\partial \tau} \tag{4-7}$$

This is called the heat or diffusion equation.

The basic problem to be discussed is one in which a slab experiencing no heat loss in the y and z directions and initially at a uniform temperature, T_o, is placed in a convective environment at temperature, T_∞. The convective heat transfer occurring at each face is characterized by a heat transfer coefficient equal to h on each face. Such a situation exists when a hot metal slab is removed from a heat-treating furnace and placed in a quenching oil. The Biot number is greater than 0.1 so that the negligible internal resistance technique cannot be used. The boundary conditions are that at $x = \pm L$, the heat conducted to the surface is convected away, or

$$\text{at } x = +L, \quad -kA\frac{\partial T}{\partial x} = hA\,(T - T_\infty) \tag{4-7a}$$

$$\text{at } x = -L, \quad +kA\frac{\partial T}{\partial x} = +hA\,(T - T_\infty) \tag{4-7b}$$

This problem is solved using the technique of separation of variables, which was discussed in Section 3-4.

The solution to this problem is, at best, long and tedious. It has been worked out by a number of people and put in the form of charts. The solution is

$$\frac{T(x, \tau) - T_\infty}{T_o - T_\infty} = 4\sum_{n=1}^{\infty}\left(\frac{\sin M_n}{2M_n + \sin 2M_n}\right)\exp\left[-M_n^2\left(\frac{\alpha\tau}{L^2}\right)\right]\cos\left(M_n\frac{x}{L}\right) \tag{4-8}$$

where the M_ns are obtained from the solution to the characteristic equation or eigenvalue problem. (See Reference 5 for details). The equation for the M_ns is

$$\cot\,(M_n) = M_n\left(\frac{k}{Lh}\right) \tag{4-8a}$$

4-3.3 Long Cylinder of Radius r_o

To determine the radial temperature distribution in a long (infinite) cylinder initially at temperature, T_o, and placed in a convective environment, analytical or chart solutions must be used if Bi > 0.1. The analytical solution is complex and contains Bessel functions. Chart solutions are available and will be utilized in this text. Figure 4-7 gives the axis (centerline) temperature for an infinite cylinder of radius r_o. Figure 4-8 gives the temperatures at other radial positions as a function of axis temperature, and Figure 4-9 gives the dimensionless heat loss. These charts are used in the same manner as those given for the plane wall of thickness, $2L$. As noted in Table 4-1, the characteristic length for a long cylinder is $(r_o/2)$, hence the Biot number for our purposes is defined as $(hr_o/2k)$.

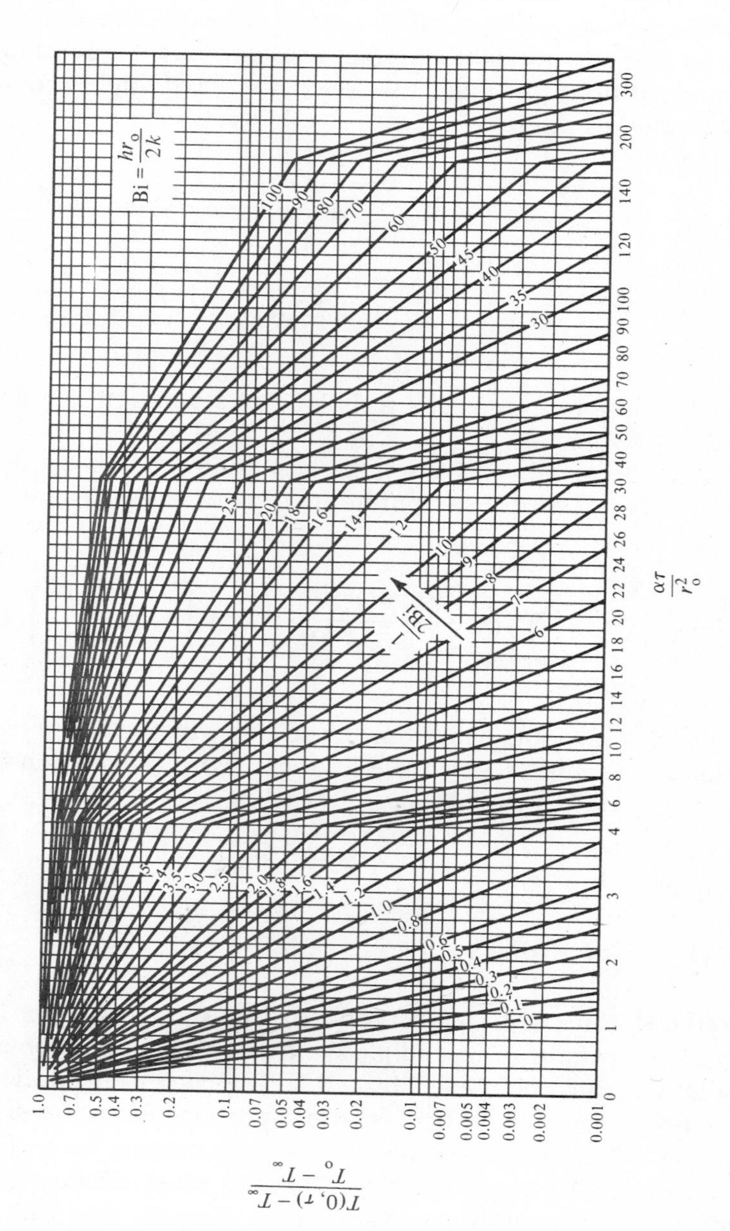

Figure 4-7 Axis temperature for a long cylinder of radius, r_o, from Heisler [3].

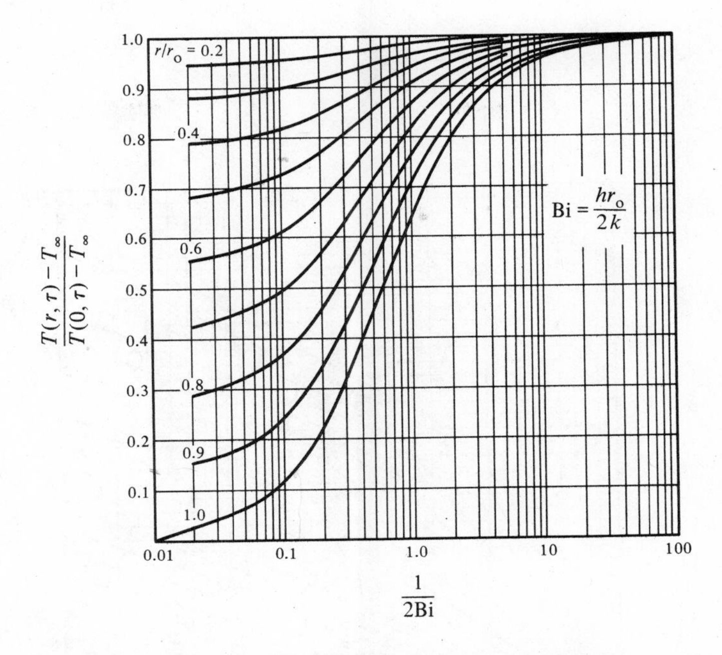

Figure 4-8 Temperature as a function of axis temperature in a long cylinder of radius, r_o, from Heisler [3].

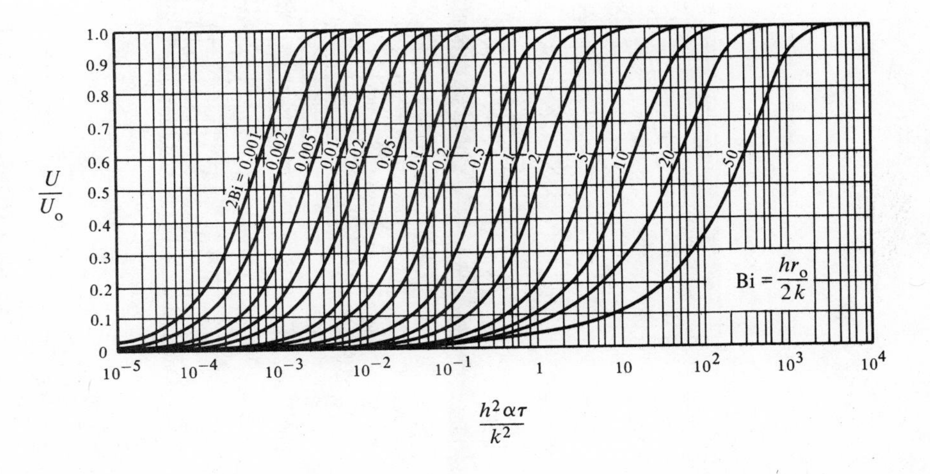

Figure 4-9 Dimensionless heat loss U/U_o of a long cylinder of radius, r_o, with time, from *Fundamentals of Heat Transfer* by H. Grober, S. Erk, and H. Grigull. Copyright 1961 by McGraw-Hill Book Company. Used with permission.

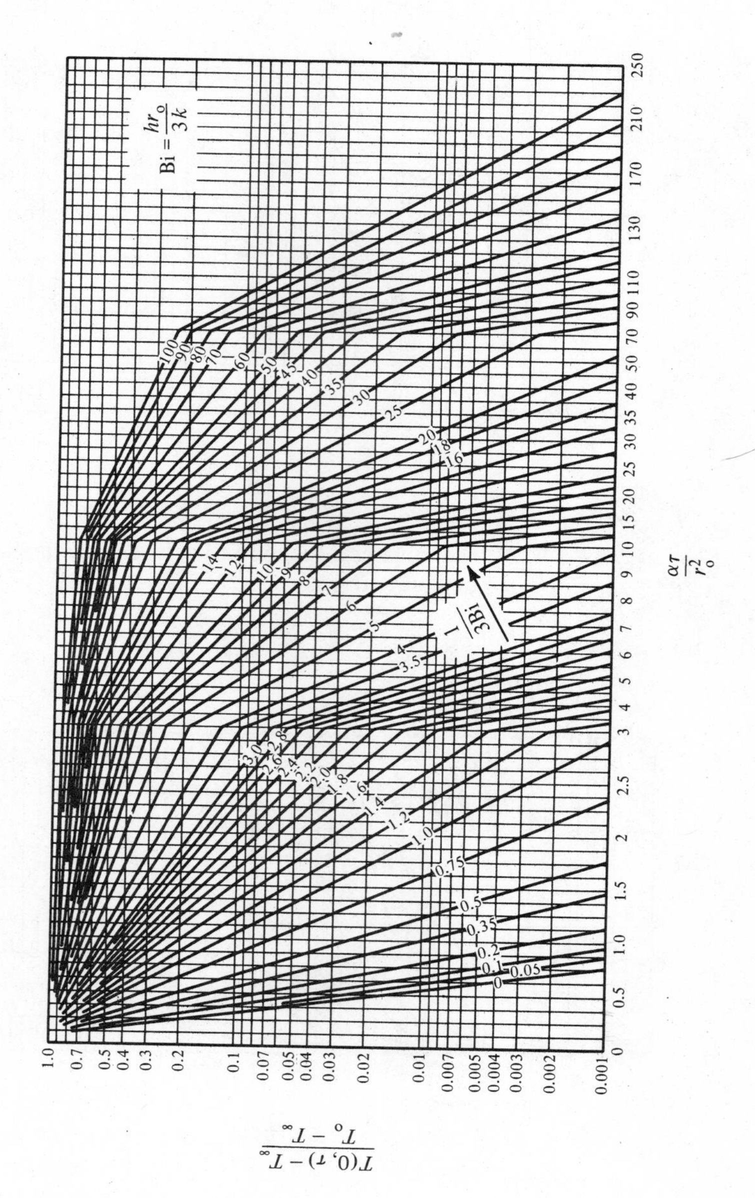

Figure 4-10 Center temperature for a sphere of radius, r_o, from Heisler [3].

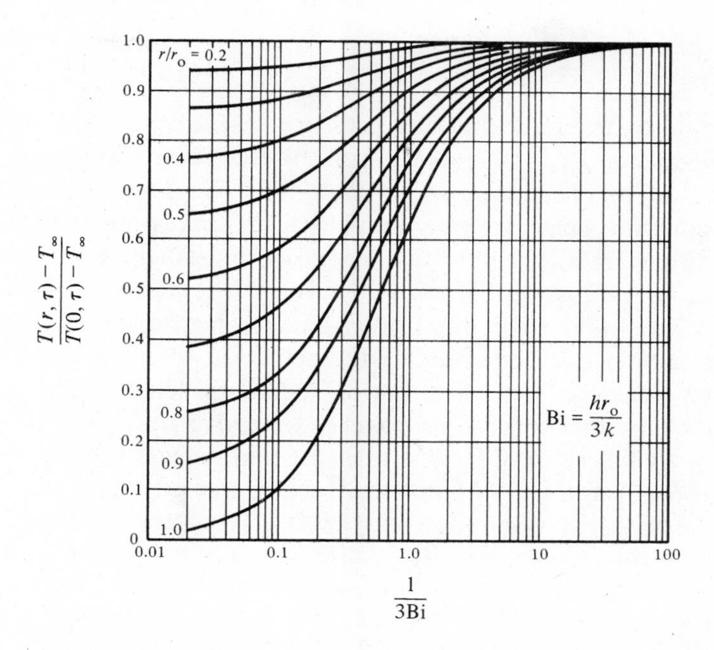

Figure 4-11 Temperature as a function of center temperature for a sphere of radius, r_o, from Heisler [3].

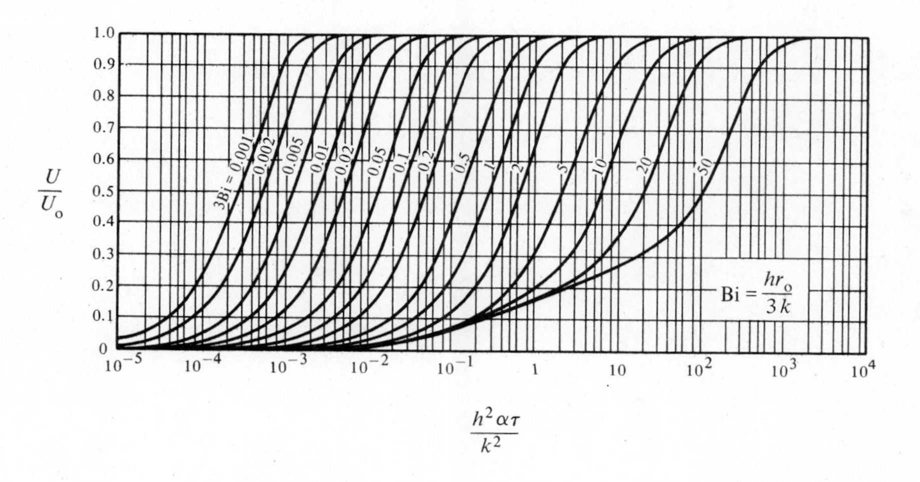

Figure 4-12 Dimensionless heat loss U/U_o of a sphere of radius, r_o, with time, from *Fundamentals of Heat Transfer* by H. Grober, S. Erk, and H. Grigull. Copyright 1961 by McGraw-Hill Book Company. Used with permission.

4-3.4 Solid Sphere of Radius r_o

To determine the radial temperature distribution in a solid sphere initially at temperature, T_o, and placed in a convective environment, it is possible to use analytical or chart solutions if Bi > 0.1. The analytical solution is also complex and contains Legendre polynomials in its expressions. Once again, we will therefore use chart solutions in this text. Figure 4-10 gives the center temperature for a sphere of radius r_o. Figure 4-11 gives the temperature at other radial positions, and Figure 4-12 gives the dimensionaless heat loss. These charts are used in the same manner as those given for the plane wall and the infinite cylinder.

As noted in Table 4-1, the characteristic length for a sphere is $(r_o/3)$, hence the Biot number for our purposes is defined as $(hr_o/3k)$.

In all the chart solutions presented, whenever the reciprocal of the Biot number (k/hL_c) for a problem equals zero, this situation corresponds to a value of $h \to \infty$ or to the case where the surface temperature of the body is immediately raised to the temperature, T_∞, of the surrounding fluid at time, τ, equal to zero. On the other hand, if the quantity (k/hL_c) is very large and approaches infinity, we have the case where $h \to 0$ or the situation where the surface is thermally insulated at time, τ, equal to zero.

Sample Problem 4-3 Consider a 25-mm O.D. solid steel (20% Cr) ball bearing at a temperature of 600°C and quenched in oil at 40°C. The convective heat transfer coeffficient between the bearing's surface and the oil is 1500 W/m²°C. Determine the center temperature and the temperature 1.25 mm from the surface after the bearing has been in the oil for one-half minute. Also, determine the heat lost by the spherical ball during the first half-minute.

Solution: The ball bearing properties are

$$\rho = 7689 \text{ kg/m}^3 \qquad c = 460 \text{ J/kg°C} \qquad k = 24.2 \text{ W/m°C}$$

$$\alpha = \frac{k}{\rho c} = 6.84 \times 10^{-6} \text{ m}^2/\text{sec} = 0.025 \text{ m}^2/\text{hr}$$

$$\text{Bi} = \frac{hr_o}{3k} = \frac{(1500)(0.0125)}{(3)(24.2)} = 0.26$$

$$\frac{1}{3\text{Bi}} = 1.29$$

$$\frac{\alpha\tau}{r_o^2} = \frac{(0.025)[(1/2)/60]}{(0.0125)^2} = 1.33$$

From Figure 4-10

$$\frac{T(0,\tau) - T_\infty}{T_o - T_\infty} = 0.09$$

Therefore

$$T(0,\tau) = (0.09)(600 - 40) + 40$$
$$T(0,\tau) = 50 + 40 = 90°C$$

To determine $T(r,\tau)$ we proceed as follows

$$r = 11.25 \text{ mm} \qquad r_o = 12.5 \text{ mm}$$

$$\frac{r}{r_o} = 0.9$$

$$\frac{1}{3\text{Bi}} = 1.29$$

From Figure 4-11

$$\frac{T(r,\tau) - T_\infty}{T(0,\tau) - T_\infty} = 0.73$$

Therefore

$$T(r,\tau) = 0.73(90 - 40) + 40$$
$$T(r,\tau) = 37 + 40 = 77°C$$

We note that the temperature at the center is 13°C higher than that at 1.25 mm from the surface. Since Bi > 0.1, the negligible internal resistance concept does not apply, resulting in this temperature differential.

To determine U, we proceed as follows

$$U_o = \rho c V (T_o - T_\infty)$$

$$U_o = (7689)(460)\frac{4}{3}\pi(0.0125)^3(600 - 40)$$

$$U_o = 1.62 \times 10^4 \text{ J}$$

Now

$$3\,\text{Bi} = 0.78$$

$$\frac{h^2 \alpha \tau}{k^2} = \frac{(1500)^2\,(0.025)\,(1/120)}{(24.2)^2} = 0.80$$

From Figure 4-12

$$\frac{U}{U_o} = 0.86 \qquad \text{and} \qquad U = 1.39 \times 10^4\ \text{J}$$

and 86 percent of the heat that is ultimately lost by the spherical bearing is lost in the first half-minute.

4-4 TRANSIENT HEAT FLOW IN SEMI-INFINITE BODIES

A semi-infinite body is one in which at a given time there is always a portion of the body where the temperature remains unchanged when a temperature change occurs on one of its boundaries. An example is the earth's crust. If the temperature on the earth's surface is changed, there is always some point below the surface that does not experience the effect of the change. Even at a distance of several feet below the surface, the surface temperature fluctuation may not be felt for a long time. This is noted in northern climates where the winter frost line does not reach 4 to 6 feet before April, at which time the surface temperature has warmed up. When it does reach this level, water and sewer lines oftentimes freeze up on what appears to be a pleasant spring day. Even the transient temperature distribution in a plane wall behaves like that of a semi-infinite solid until enough time has passed to allow any surface temperature changes to penetrate throughout the wall. Figure 4-13 represents a semi-infinite solid.

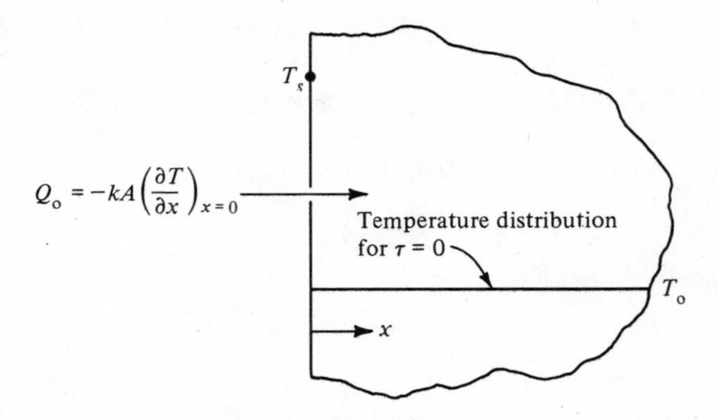

Figure 4-13 Semi-infinite solid.

As seen in Figure 4-13 the body extends to infinity in the $+x$ direction. This simply means that the body is of sufficient size that there is always some value of x at which any temperature change occurring at $x = 0$ is not felt.

Let us solve the semi-infinite body problem for the case where initially the entire body is at temperature, T_o, and at time, τ, equal to zero the face temperature at $x = 0$ is instantaneously raised to temperature, T_s. Starting with the diffusion equation, equation (4-7), we have

$$\frac{\partial^2 T}{\partial x^2} = \frac{1}{\alpha}\frac{\partial T}{\partial \tau}$$

The initial and boundary conditions are

$$T(x, 0) = T_o$$

$$T(0, \tau) = T_s \quad \text{for } \tau > 0$$

The problem is solved in Reference 5 and the solution is

$$\frac{T(x, \tau) - T_s}{T_o - T_s} = \text{erf}\left(\frac{x}{2\sqrt{\alpha\tau}}\right) \tag{4-9}$$

The right-hand side of the above equation is the Gauss error function, defined as

$$\text{erf}\left(\frac{x}{2\sqrt{\alpha\tau}}\right) = \frac{2}{\sqrt{\pi}}\int_0^{[x/(2\sqrt{\alpha\tau})]} e^{-\eta^2} d\eta$$

where η is a dummy variable. (See Table 4-2.)

Equation (4-9) can then be written as

$$\frac{T(x, \tau) - T_s}{T_o - T_s} = \frac{2}{\sqrt{\pi}}\int_0^{[x/(2\sqrt{\alpha\tau})]} e^{-\eta^2} d\eta \tag{4-10}$$

The surface heat flow Q_o at any instant of time can be calculated from Fourier's law.

$$Q_o(\tau) = -kA\left(\frac{\partial T}{\partial x}\right)_{x=0}$$

Differentiating equation (4-9), one has

$$\frac{\partial T}{\partial x} = (T_o - T_s)(2/\sqrt{\pi})e^{-(x^2/4\alpha\tau)}\frac{\partial}{\partial x}[x/(2\sqrt{\alpha\tau})]$$

giving

$$\left(\frac{\partial T}{\partial x}\right)_{x=0} = \frac{T_o - T_s}{\sqrt{\pi\alpha\tau}}$$

and

$$Q_o(\tau) = \frac{kA(T_s - T_o)}{\sqrt{\pi\alpha\tau}} \tag{4-11}$$

To determine the total heat added, U, to the semi-infinite solid during time, τ, one simply integrates Q_o over the time interval $(0 - \tau)$.
That is

$$U = \int_0^\tau Q_o\, d\tau = \int_0^\tau \frac{kA(T_s - T_o)}{\sqrt{\pi\alpha\tau}}\, d\tau$$

or

$$U = \frac{2kA(T_s - T_o)}{\sqrt{\pi\alpha/\tau}} \tag{4-12}$$

Values of the error function are tabulated in Table 4-2.

Sample Problem 4-4 A large steel slab, initially at a uniform temperature of 550°F, suddenly has its surface temperature lowered to 100°F. Calculate the time required for the temperature to reach 200°F at a depth of 1 inch. Also, determine the total heat removed from the slab per square foot during this time. The thermal diffusivity and the thermal conductivity of the steel are 0.45 ft²/hr and 25 Btu/hr-ft°F, respectively.

Solution: The pertinent properties and parameters are:

$$\alpha = 0.45 \text{ ft}^2/\text{hr};\ k = 25 \text{ Btu/hr-ft°F};\ x = 1/12 \text{ ft}$$

$$T_o = 550°\text{F};\ T_s = 100°\text{F};\ T(x,\tau) = 200°\text{F}$$

$$\frac{T(x,\tau) - T_s}{T_o - T_s} = \frac{200 - 100}{550 - 100} = 0.222 = \text{erf}\left(\frac{x}{2\sqrt{\alpha\tau}}\right)$$

TABLE 4-2 The Error Function

$\frac{x}{2\sqrt{\alpha\tau}}$	$\text{erf}\left(\frac{x}{2\sqrt{\alpha\tau}}\right)$	$\frac{x}{2\sqrt{\alpha\tau}}$	$\text{erf}\left(\frac{x}{2\sqrt{\alpha\tau}}\right)$	$\frac{x}{2\sqrt{\alpha\tau}}$	$\text{erf}\left(\frac{x}{2\sqrt{\alpha\tau}}\right)$
0.00	0.00000	0.76	0.71754	1.52	0.96841
0.02	0.02256	0.78	0.73001	1.54	0.97059
0.04	0.04511	0.80	0.74210	1.56	0.97263
0.06	0.06762	0.82	0.75381	1.58	0.97455
0.08	0.09008	0.84	0.76514	1.60	0.97635
0.10	0.11246	0.86	0.77610	1.62	0.97804
0.12	0.13476	0.88	0.78669	1.64	0.97962
0.14	0.15695	0.90	0.79691	1.66	0.98110
0.16	0.17901	0.92	0.80677	1.68	0.98249
0.18	0.20094	0.94	0.81627	1.70	0.98379
0.20	0.22270	0.96	0.82542	1.72	0.98500
0.22	0.24430	0.98	0.83423	1.74	0.98613
0.24	0.26570	1.00	0.84270	1.76	0.98719
0.26	0.28690	1.02	0.85084	1.78	0.98817
0.28	0.30788	1.04	0.85865	1.80	0.98909
0.30	0.32863	1.06	0.86614	1.82	0.98994
0.32	0.34913	1.08	0.87333	1.84	0.99074
0.34	0.36936	1.10	0.88020	1.86	0.99147
0.36	0.38933	1.12	0.88079	1.88	0.99216
0.38	0.40901	1.14	0.89308	1.90	0.99279
0.40	0.42839	1.16	0.89910	1.92	0.99338
0.42	0.44749	1.18	0.90484	1.94	0.99392
0.44	0.46622	1.20	0.91031	1.96	0.99443
0.46	0.48466	1.22	0.91553	1.98	0.99489
0.48	0.50275	1.24	0.92050	2.00	0.995322
0.50	0.52050	1.26	0.92524	2.10	0.997020
0.52	0.53790	1.28	0.92973	2.20	0.998137
0.54	0.55494	1.30	0.93401	2.30	0.998857
0.56	0.57162	1.32	0.93806	2.40	0.999311
0.58	0.58792	1.34	0.94191	2.50	0.999593
0.60	0.60386	1.36	0.94556	2.60	0.999764
0.62	0.61941	1.38	0.94902	2.70	0.999866
0.64	0.63459	1.40	0.95228	2.80	0.999925
0.66	0.64938	1.42	0.95538	2.90	0.999959
0.68	0.66278	1.44	0.95830	3.00	0.999978
0.70	0.67780	1.46	0.96105	3.20	0.999994
0.72	0.69143	1.48	0.96365	3.40	0.999998
0.74	0.70468	1.50	0.96610	3.60	1.000000

From Table 4-2

$$\frac{x}{2\sqrt{\alpha\tau}} = 0.20$$

and

$$\tau = 0.096 \text{ hrs} = 5.8 \text{ minutes}$$

From equation (4-12)

$$\frac{U}{A} = \frac{2k\,(T_s - T_o)}{\sqrt{\pi\alpha/\tau}} = \frac{(2)\,(25)\,(100 - 550)}{\sqrt{\pi\,(0.45)\,/\,(0.096)}}$$

or

$$\frac{U}{A} = -\ 5860 \text{ Btu/ft}^2$$

Consider next the case where the surface temperature of the semi-infinite solid is not instantaneously raised or lowered to temperature, T_s, but rather the surface is exposed to a convective environment at temperature, T_∞, at time, $\tau = 0$. For this situation, the solution becomes quite complicated and is given here in graphical form in Figure 4-14.

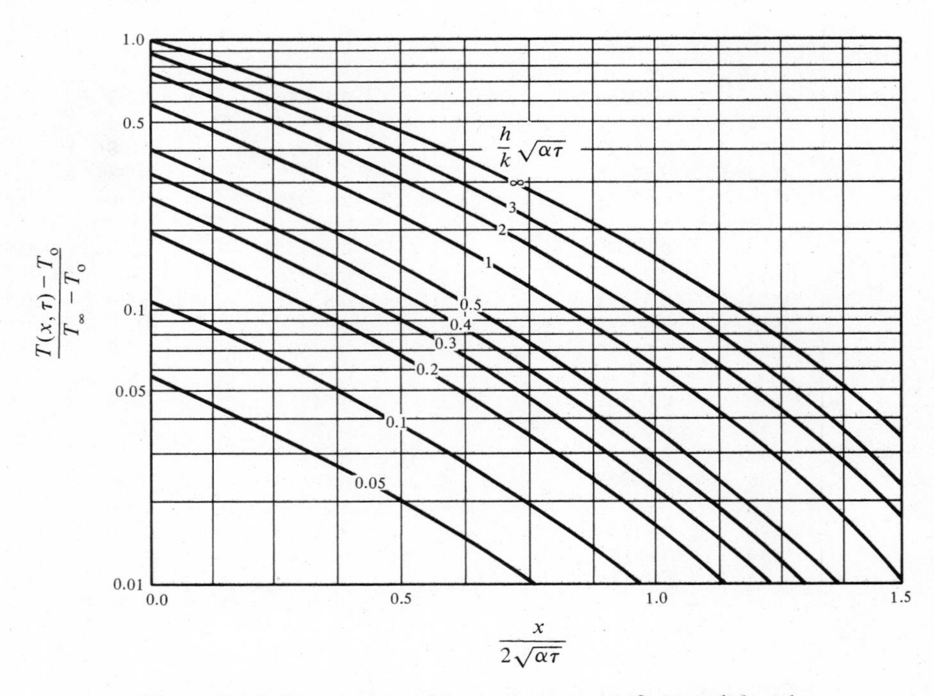

Figure 4-14 Temperature history in a semi-infinite solid with surface convection.

4-5 TWO- AND THREE-DIMENSIONAL TRANSIENT SYSTEMS

So far we have discussed only one-dimensional heat flow in walls, cylinders, and spheres. However, many practical problems involve two- and three-dimensional heat flow. Since the solution of such problems is often achieved from a product solution, it is possible to combine several of the one-dimensional solutions to obtain the solution to a two- or three-dimensional problem. Figure 4-15 shows the required product solution for the geometries indicated. In performing the analysis we will use the following notation:

$$C(r, \tau) = \frac{T(r, \tau) - T_\infty}{T_o - T_\infty}$$

$$P(x, \tau) = \frac{T(x, \tau) - T_\infty}{T_o - T_\infty}$$

$$S(x, \tau) = \frac{T(x, \tau) - T_\infty}{T_o - T_\infty}$$

where

$C(r, \tau)$ represents a transient solution for a cylindrical geometry

$P(x, \tau)$ represents a transient solution for a plane wall

$S(x, \tau)$ represents a transient solution for a semi-infinite body

In determining the solution to the different geometries [$C(r, \tau)$, $P(x, \tau)$, and $S(x, \tau)$], we must utilize the appropriate convective heat transfer coefficient, h, associated with the surface to be analyzed. It should be noted that the value of h may be different for the different geometries that comprise the overall solution.

Sample Problem 4-5 A short cylinder 75-mm O.D. and 10 cm long is at a uniform temperature of 250°C. At time equal to zero, it is placed in a convective environment where $h = 400$ W/m^2°C and T_∞ is 40°C. If the material properties are $\alpha = 0.046$ m^2/hr, and $k = 37$ W/m°C, determine the temperature at the center of the cylinder after 4 minutes.

Solution: From Figure 4-15, we see that the solution is

$$\frac{T(r, x, \tau) - T_\infty}{T_o - T_\infty} = C(r, \tau)P(x, \tau)$$

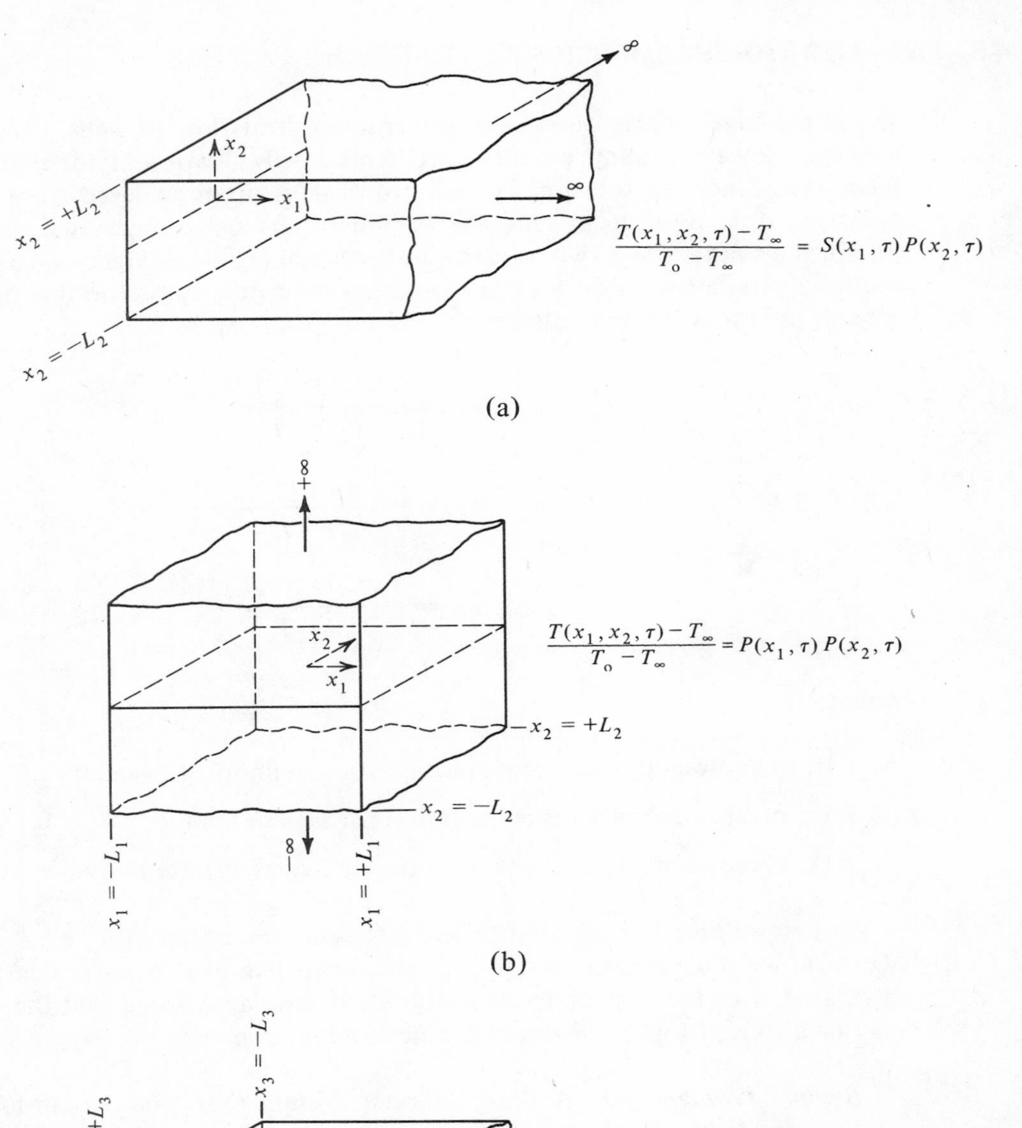
x_2
x_1
∞
∞
$x_2 = +L_2$
$x_2 = -L_2$
$\dfrac{T(x_1, x_2, \tau) - T_\infty}{T_o - T_\infty} = S(x_1, \tau)\,P(x_2, \tau)$
$+\infty$
$-\infty$
x_2
x_1
$\dfrac{T(x_1, x_2, \tau) - T_\infty}{T_o - T_\infty} = P(x_1, \tau)\,P(x_2, \tau)$
$x_2 = +L_2$
$x_2 = -L_2$
$x_1 = -L_1$
$x_1 = +L_1$

(a)

(b)

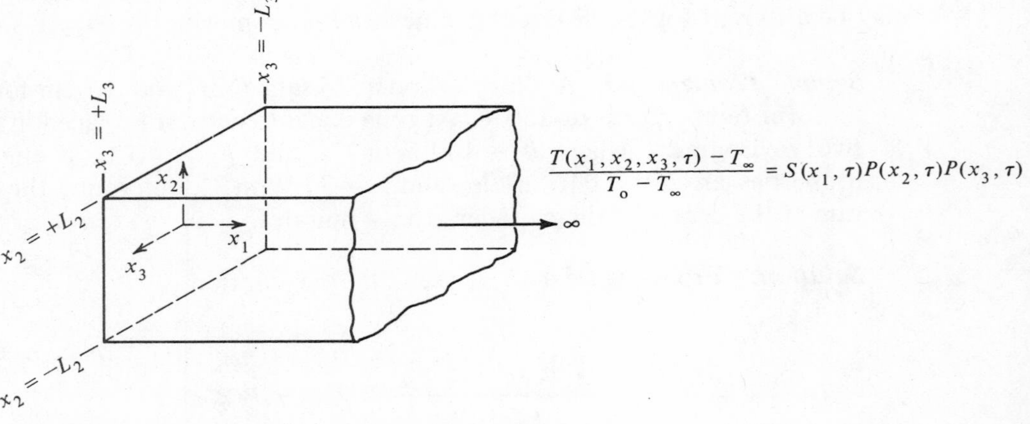
$x_3 = -L_3$
$x_3 = +L_3$
x_2
x_1
x_3
∞
$x_2 = +L_2$
$x_2 = -L_2$
$\dfrac{T(x_1, x_2, x_3, \tau) - T_\infty}{T_o - T_\infty} = S(x_1, \tau)P(x_2, \tau)P(x_3, \tau)$

(c)

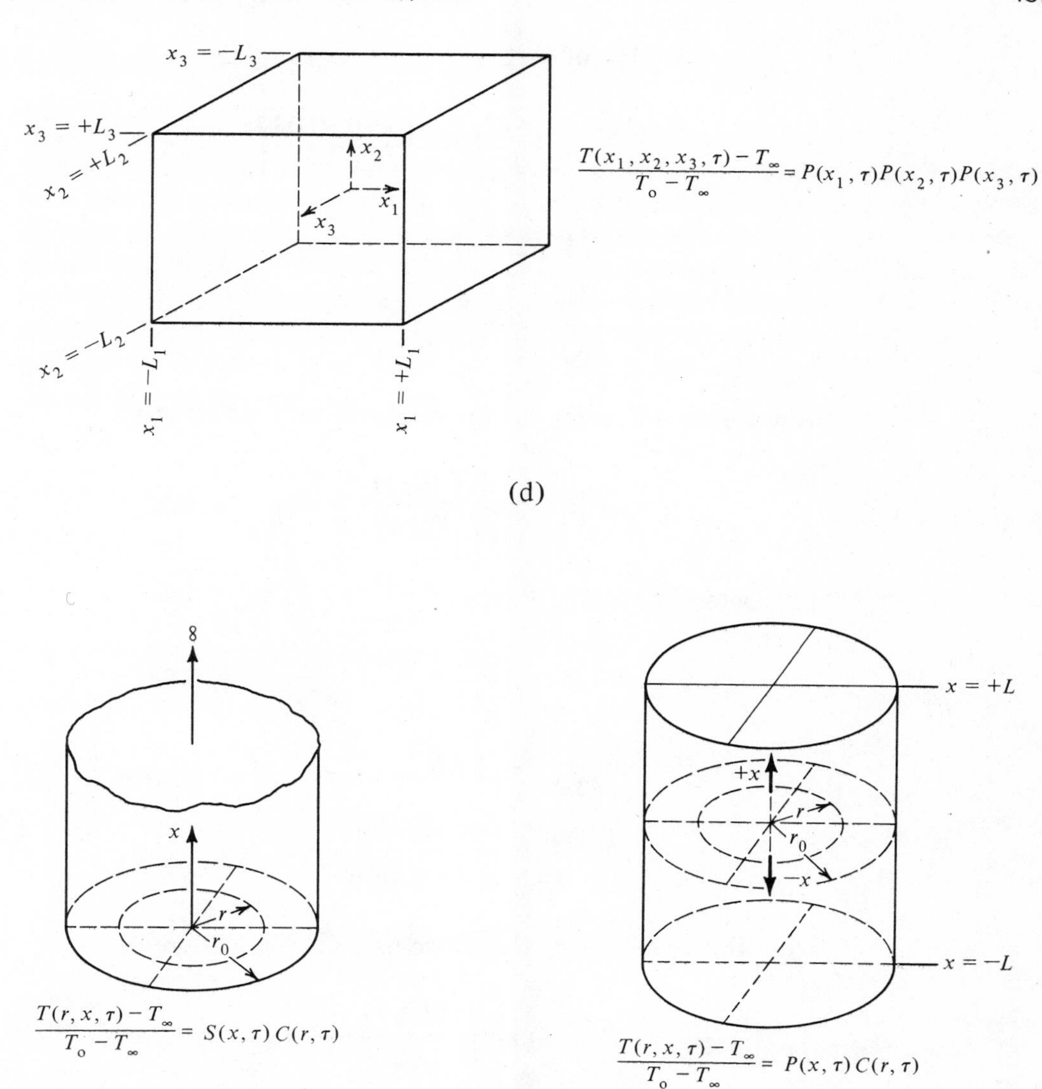

Figure 4-15 Product solutions for temperatures in multidimensional systems. In each case the body is initially at a uniform temperature equal to T_o, and is instantaneously placed in a convective environment at temperature T_∞. (a) Semi-infinite plate; (b) infinite rectangular bar; (c) semi-infinite rectangular bar; (d) rectangular parallelepiped; (e) semi-infinite cylinder; (f) short cylinder.

(1) Determination of $C(r, \tau)$

$$\mathrm{Bi} = \frac{hr_o}{2k} = \frac{(400)\,(0.0375)}{(2)\,(37)} = 0.2$$

$$\frac{1}{2\mathrm{Bi}} = \frac{1}{2\,(0.2)} = 2.5$$

$$\frac{\alpha\tau}{r_o^2} = \frac{(0.046)\,(4/60)}{(0.0375)^2} = 2.18$$

From Figure 4-7, using the above parameters, we obtain

$$C(r, \tau) = \left[\frac{T(0, \tau) - T_\infty}{T_o - T_\infty}\right]_{\text{cyl}} = 0.22$$

(2) Determination of $P(x, \tau)$

$$\mathrm{Bi} = \frac{hL}{k} = \frac{(400)\,(0.05)}{(37)} = 0.54$$

$$\frac{1}{\mathrm{Bi}} = 1.85$$

$$\frac{\alpha\tau}{L^2} = \frac{(0.046)(4/60)}{(0.05)^2} = 1.2$$

From Figure 4-4, using the above parameters, there results

$$P(x, \tau) = \left[\frac{T(0, \tau) - T_\infty}{T_o - T_\infty}\right]_{\text{slab}} = 0.65$$

Hence

$$\left[\frac{T(0, 0, \tau) - T_\infty}{T_o - T_\infty}\right]_{\substack{\text{short}\\\text{cylinder}}} = C(r, \tau)\,P(x, \tau) = (0.22)\,(0.65) = 0.14$$

and

$$T(0, 4 \text{ minutes}) = 0.16\,(T_o - T_\infty) + T_\infty = 0.16(250 - 40) + 40$$

or

$$T = 70°\text{C}$$

4-6 HEAT BALANCE INTEGRAL

During the discussion of convective heat transfer in Chapters 7 and 8, the Karman-Pohlhausen method employing momentum and energy integrals to solve boundary layer problems will be discussed. In spirit, this method is similar to the method of the *heat balance integral* for one-dimensional heat conduction problems. Consequently, the heat balance integral will be introduced here not only for its utility in solving one-dimensional conduction problems but also as an introduction to the techniques used in solving boundary layer equations. Application of this method to transient heat conduction in solids and in the melting of solids is discussed in References 6, 7, 8, and 9.

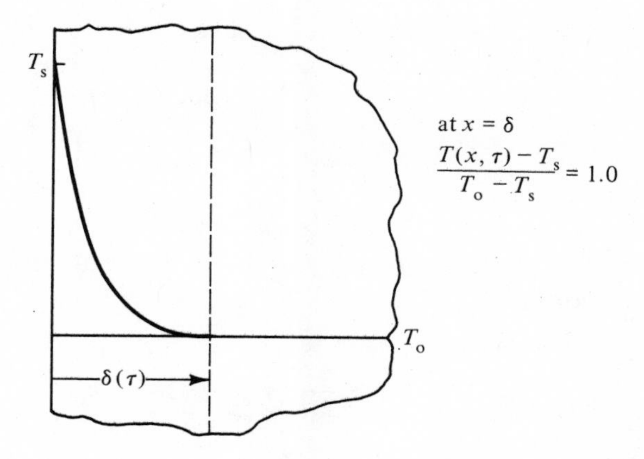

Figure 4-16 Application of heat balance integral to a semi-infinite solid.

Consider a semi-infinite solid as shown in Figure 4-16. The distance from the left-hand face of the solid where the effect of a temperature change on the surface is felt is designated by the quantity $\delta(\tau)$. It is a function of time, τ, and is defined as that position at which $T(x, \tau) = T_o$. That is, $\delta(\tau)$ is the value of x for which

$$\frac{T(x, \tau) - T_s}{T_o - T_s} = 1.0$$

The main difference between the analysis using the heat balance integral and the analytical solution presented in Section 4-4 is that the former requires that energy conservation be satisfied on the *average* across the entire layer, $x = 0$ to $x = \delta$, while the latter requires that the energy equation be satisfied at *every single point* in the body. The former is an integral approach while the latter is a differential approach.

We begin with an energy balance on the slab of width δ:

$$\left\{\begin{matrix}\text{Rate of heat conducted into}\\ \text{the layer at } x = 0\end{matrix}\right\} - \left\{\begin{matrix}\text{Rate of heat conducted out}\\ \text{of the layer at } x = \delta\end{matrix}\right\} = \left\{\begin{matrix}\text{Rate of change in inter-}\\ \text{nal energy of the layer}\\ \text{of thickness, } \delta\end{matrix}\right\}$$

The rate of change of internal energy of a thin slice of width, dx, and of area, A, measured normal to the x direction is $\rho c A dx(\partial T/\partial \tau)$. Hence

$$-kA\left(\frac{\partial T}{\partial x}\right)_{x=0} - (-kA)\left(\frac{\partial T}{\partial x}\right)_{x=\delta} = \int_0^\delta \underbrace{\rho c A\, dx}_{\text{volume}} \frac{\partial T}{\partial \tau}$$

Since k, ρ, c, and A are constant, we have

$$k\left(\frac{\partial T}{\partial x}\right)_{x=\delta} - k\left(\frac{\partial T}{\partial x}\right)_{x=0} = \rho c \int_0^\delta \frac{\partial T}{\partial \tau} dx$$

The initial temperature in the body is assumed to be a constant value, T_o. Since T_o is constant, $(\partial T/\partial \tau)$ can be written as

$$\frac{\partial T}{\partial \tau} = \frac{\partial}{\partial \tau}(T - T_o)$$

Also, since interchanging the order of differentiation and integration for a continuous function does not change the end result, we write

$$\int_0^\delta \frac{\partial T}{\partial \tau} dx = \int_0^\delta \frac{\partial}{\partial \tau}(T - T_o)\, dx = \frac{\partial}{\partial \tau}\int_0^\delta (T - T_o)dx$$

Therefore, we can write the energy balance equation as

$$k\left(\frac{\partial T}{\partial x}\right)_{x=\delta} - k\left(\frac{\partial T}{\partial x}\right)_{x=0} = \rho c \frac{\partial}{\partial \tau}\int_0^\delta (T - T_o)\, dx \tag{4-13}$$

This is the heat balance integral. It is a statement of the overall energy conservation for the finite element of width $x = 0$ to $x = \delta$.

Recalling that our objective in any heat transfer analysis is the calculation of the temperature distribution and the resulting heat fluxes, we proceed as follows: We postulate that $T(x, \tau)$ may be written in terms of a polynomial

$$T(x, \tau) = a(\tau) + b(\tau)x + c(\tau)x^2 + d(\tau)x^3 + \ldots$$

where the constants, a, b, c, d, etc. may be functions of time. It is necessary to eliminate the unknown constants. We do this by examining the temperature profile as shown in Figure 4-17.

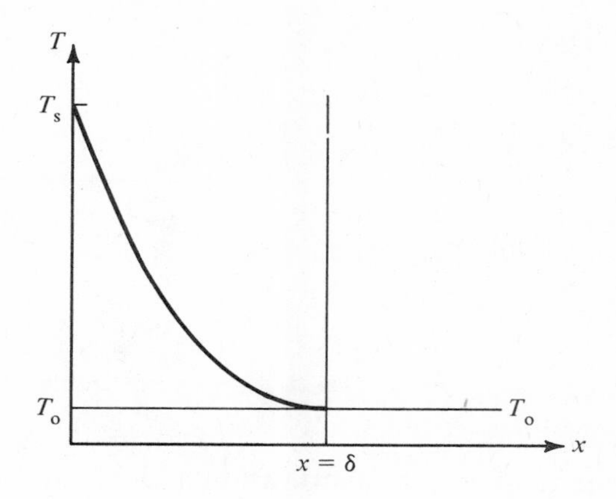

Figure 4-17 Temperature profile in a semi-infinite body with surface temperature changed to T_s at time, $\tau = 0$.

We see from Figure 4-17 the following conditions:

(1) At $x = 0$, $T = T_s$

(2) At $x = \delta$, $T = T_o$

(3) At $x = \delta$, $\dfrac{\partial T}{\partial x} = 0$ (to insure a smooth fit)

Since we have stated three conditions, we have at best a profile of the form

$$T(x, \tau) = a + bx + cx^2$$

Applying condition 1, at $x = 0$, $T = T_s$ we obtain

$$T_s = a + 0 + 0, \text{ or } a = T_s$$

Applying condition 2, at $x = \delta$, $T = T_o$, we get

$$T_o = T_s + b\delta + c\delta^2$$

Condition 3 results in

$$0 = b + 2c\delta$$

or $b = -2c\delta$. Combining with the results from condition 2, we have

$$T_o - T_s = c(-2\delta^2 + \delta^2)$$

or

$$c = \frac{-(T_o - T_s)}{\delta^2}$$

and

$$b = \frac{2(T_o - T_s)}{\delta}$$

The following temperature distribution is thus obtained.

$$\frac{T - T_s}{T_o - T_s} = \frac{2x}{\delta} - \frac{x^2}{\delta^2} \tag{4-14}$$

To determine δ, we now go back to the heat balance integral as given by equation (4-13).

$$k\left(\frac{\partial T}{\partial x}\right)_{x=\delta} - k\left(\frac{\partial T}{\partial x}\right)_{x=0} = \rho c \frac{\partial}{\partial \tau} \int_0^\delta (T - T_o)\,dx \tag{4-13}$$

From equation (4-14)

$$\frac{\partial T}{\partial x} = (T_o - T_s)\left(\frac{2}{\delta} - \frac{2x}{\delta^2}\right)$$

Substituting the above into the left hand side of equation (4-13), we have

$$k(0) - k(T_o - T_s)\left(\frac{2}{\delta}\right) = -k(T_o - T_s)(2/\delta)$$

Examining the right-hand side, we note from equation (4-14)

$$T - T_o = (T_o - T_s)\left(\frac{2x}{\delta} - \frac{x^2}{\delta^2}\right) + T_s - T_o$$

After integrating, the right-hand-side term becomes

$$\rho c \frac{\partial}{\partial \tau}\left[(T_o - T_s)\left(\frac{x^2}{\delta} - \frac{x^3}{3\delta^2}\right) - (T_o - T_s)x\right]_0^\delta$$

$$= \rho c \frac{\partial}{\partial \tau}\left[(T_o - T_s)\left(\delta - \frac{\delta}{3} - \delta - 0\right)\right]$$

$$= -\frac{1}{3}\rho c\,(T_o - T_s)\frac{\partial \delta}{\partial \tau}$$

Equating the right-hand and the left-hand sides, we obtain

$$k\,(T_o - T_s)\frac{2}{\delta} = \frac{1}{3}\rho c\,(T_o - T_s)\frac{d\delta}{d\tau}$$

Substituting, $\alpha = k/\rho c$, and cancelling $(T_o - T_s)$ we have

$$\frac{6\alpha}{\delta} = \frac{d\delta}{d\tau}$$

where $(d\delta/d\tau)$ is written as an ordinary derivative, since δ depends only on τ. Separation of the variables δ and τ and integration gives

$$\int_0^\delta \delta\, d\delta = 6\alpha \int_0^\tau d\tau$$

$$\frac{\delta^2}{2} = 6\alpha\tau$$

$$\delta = \sqrt{12\alpha\tau} = 3.46\sqrt{\alpha\tau} \qquad (4\text{-}15)$$

The temperature distribution is therefore given by

$$\frac{T - T_s}{T_o - T_s} = \frac{2x}{\delta} - \left(\frac{x}{\delta}\right)^2 \qquad (4\text{-}14)$$

where $\delta = 3.46\sqrt{\alpha\tau}$. The heat flux at $x = 0$ is

$$\frac{Q_o}{A} = -k\left(\frac{\partial T}{\partial x}\right)_{x=0}$$

$$\frac{Q_o}{A} = \frac{k\,(T_s - T_o)}{\sqrt{3\alpha\tau}} \tag{4-16}$$

The question may be asked: How well do the answers from the heat balance integral agree with the exact solution. The comparison is made below:

Exact Solution (Ref. 3)	*Heat Balance Integral*
$\delta = 3.66\sqrt{\alpha\tau}$	$\delta = 3.46\sqrt{\alpha\tau}$
$\dfrac{Q_o}{A} = \dfrac{k\,(T_s - T_o)}{\sqrt{\pi\alpha\tau}}$	$\dfrac{Q_o}{A} = \dfrac{k\,(T_s - T_o)}{\sqrt{3\alpha\tau}}$

As can be seen, they agree quite well. Utilizing other known facts, namely that the higher derivatives of temperature with respect to x are zero at $x = \delta$, the polynomial describing $T(x, \tau)$ may be expanded to give more accurate results.

The results obtained by the heat balance integral and by the exact solution may be applied to finite solids provided that the moving temperature front as defined by δ has not moved through the entire body. As an example, consider the wall shown in Figure 4-18.

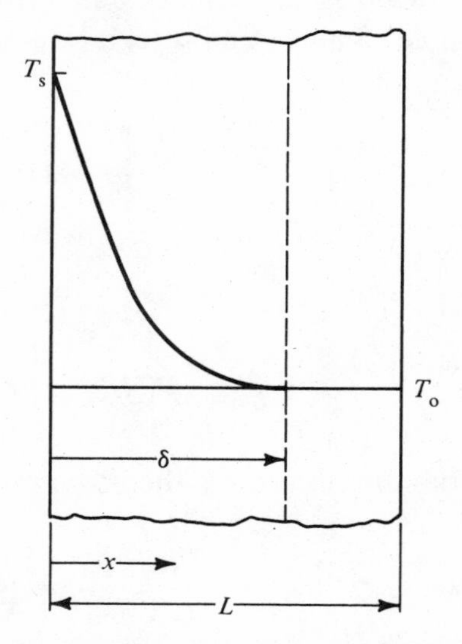

Figure 4-18 Temperature front moving through a plane wall of thickness, L.

For values of $\delta < L$, the temperature distribution

$$\frac{T - T_s}{T_o - T_s} = \frac{2x}{\delta} - \left(\frac{x}{\delta}\right)^2$$

will hold. Since $\delta = \sqrt{12\alpha\tau}$, the above temperature profile is valid for $0 < \tau \leqslant L^2/12\alpha$.

Sample Problem 4-6 A large steel slab at a uniform temperature of 550°F suddenly has its surface temperature lowered to 100°F. Calculate the time required for the temperature to reach 200°F at a depth of 1 inch. Also, determine the instantaneous heat removal rate from the slab per square foot at time, τ. The thermal diffusivity and the thermal conductivity of the steel are 0.45 ft²/hr and 25 Btu/hr-ft°F, respectively.

Solution: Equation (4-14) gives

$$\frac{T - T_s}{T_o - T_s} = \frac{2x}{\delta} - \left(\frac{x}{\delta}\right)^2$$

where

$$\delta = \sqrt{12\alpha\tau}$$

$$x = \frac{1}{12}\text{ ft};\ T = 200°\text{F};\ T_o = 550°\text{F};\ T_s = 100°\text{F}$$

$$\frac{T - T_s}{T_o - T_s} = \frac{200 - 100}{550 - 100} = \frac{(2)(1/12)}{\sqrt{(12)(0.45)\tau}} - \frac{(1/12)^2}{(12)(0.45)\tau}$$

Simplification gives

$$0.222 = \frac{0.072}{\sqrt{\tau}} - \frac{0.00128}{\tau}$$

and $\tau = 0.096$ hr, which is the same result as from Sample Problem 4-4.

Equation (4-16) gives

$$\frac{Q_o}{A} = \frac{k(T_s - T_o)}{\sqrt{3\alpha\tau}}$$

or

$$\frac{Q_o}{A} = \frac{(25)(100 - 550)}{\sqrt{(3)(0.45)(0.096)}} = -31,250 \text{ Btu/hr-ft}^2$$

PROBLEMS (English Engineering System of Units)

4-1. A metal cube initially at 1000°F is placed in a room maintained at 70°F. What will be the steady-state temperature of the cube?

4-2. Consider the plane wall problem with temperatures T_1 and T_2 maintained on the left- and right-hand faces, respectively. After steady state is reached, the entire wall is thermally insulated. What is the new steady-state temperature distribution in the wall?

4-3. A metal rod has an initial temperature distribution $T = f(x)$. At time equal to zero, the entire rod is thermally insulated. Write an expression for the steady-state temperature distribution in the rod.

4-4. A copper cube 2 inches on a side is originally at a temperature of 500°F and is suddenly immersed in a liquid at 100°F for which the convective heat transfer coefficient is 25 Btu/hr-ft^2°F. Determine the time required for the cube to reach a temperature of 200°F.

4-5. A long copper rod 1/2-inch O.D. originally at a temperature of 300°F is suddenly immersed in a liquid at 100°F for which the convective heat transfer coefficient is 20 Btu/hr-ft^2°F. Determine the time required for the rod to reach a temperature of 150°F.

4-6. A long steel (10 percent Cr) rod 1/2-inch O.D. originally at a temperature of 300°F is suddenly immersed in a liquid at 100°F for which the convective heat transfer coefficient is 20 Btu/hr-ft^2°F. Determine the temperature in the rod after 1 minute has elapsed.

4-7. The average heat transfer coefficients for flow of 100°F air over a flat plate are measured by observing the temperature time history of a 1-inch thick copper slab exposed to 100°F air. At any time, thermocouples on the surface and at the center of the slab record the same temperatures. In one test run, the initial temperature of the plate was 200°F, and in 5 minutes the temperature decreased by 30°F. Calculate the heat coefficient for this case.

4-8. Determine an expression giving the temperature as a function of time for a body of negligible internal resistance, originally at temperature, T_o, placed in a convective environment at temperature, T_∞, where the convective heat transfer coefficient is h. The body contains a uniformly distributed heat source $\dot{q}$ (energy/time vol). *Hint:* Your solution should follow closely the one presented in the text for equation (4-4).

4-9. A metal sphere initially at a uniform temperature, T_o, is immersed in a fluid that is heated by an electric heater in such a manner that, $T_\infty = T_o + 10\tau$, where temperature is measured in °F and time, τ, in hours. Derive an expression for the temperature of the sphere as a function of the convective heat transfer coefficient and time. Assume that the sphere possesses negligible internal resistance.

4-10. A copper cube 2 inches on a side originally at a temperature of 500°F is suddenly immersed in a liquid at 100°F for which the convective heat transfer coefficient is 100 Btu/hr-ft^2°F. Determine the temperature at the center of the cube after 1 minute has elapsed.

4-11. A slab of aluminum 1 inch thick initially at 500°F is suddenly immersed in a liquid at 100°F for which the convective heat transfer coefficient is 250 Btu/hr-ft^2°F. Determine the time at which the centerline temperature drops to 200°F.

4-12. A slab of aluminum 2 inches thick initially at 500°F is suddenly immersed in a liquid at 100°F for which the convective heat transfer coefficient is 250 Btu/hr-ft^2°F. Determine the time at which the centerline temperature drops to 200°F.

4-13. For the conditions of problem 4-12, determine the time at which the surface temperature drops to 200°F.

4-14. Estimate the time required to heat the center of a 6 lb roast in a 425°F oven to 275°F. Assume the roast has the properties of water and is approximately spherical in shape. The convective heat transfer coefficient between the roast and the oven air is 3 Btu/hr-ft^2°F.

4-15. A long cylindrical metal billet is prepared for a manufacturing process by heating it in an oven at 1200°F. If its center temperature at the time it is removed from the oven is to be 1000°F, determine the billet's residence time in the oven if h = 25 Btu/hr-ft^2°F, α = 0.15 ft^2/hr, and k = 12 Btu/hr-ft°F. The initial temperature of the billet is 100°F, and its radius is 3 inches.

4-16. A potato weighing 1/2 lb is dropped into boiling water at 212°F. If the potato is considered spherical in shape, possesses the thermophysical properties of water, and is initially at 72°F, determine the time required for its center temperature to reach 200°F. Assume h = 1000 Btu/hr-ft^2°F.

4-17. A long wooden 1/2 inch O.D. rod is exposed to air at 2000°F. If the ignition temperature of the wood is 800°F, determine the time of exposure necessary to cause combustion, given the initial temperature of the wood to be 50°F. Assume k = 0.1 Btu/hr-ft°F, h = 5 Btu/hr-ft^2°F, ρ = 50 lb/ft^3, and c = 0.6 Btu/°F lb$_m$.

4-18. A concrete cube 6 inches on a side originally at a temperature of 100°F is suddenly placed in air at 50°F for which the convective heat transfer coefficient is 50 Btu/hr-ft^2°F. Calculate the temperature at the center of the cube after 1 hour has elapsed.

4-19. A steel sphere 4 inches in diameter is suddenly immersed in a fluid at 500°F for which the convective heat transfer coefficient is 100 Btu/hr-ft^2°F. How long will it take for the center of the sphere to reach 250°F? T_o = 75°F.

4-20. A concrete cube 6 inches on a side originally at a temperature of 100°F is suddenly immersed in a fluid at 50°F for which the convective heat transfer coefficient is 5 Btu/hr-ft²°F. Calculate the temperature at the midpoint of one face after 1 hr has passed.

4-21. A thick layer of ice is at a temperature of 0°F throughout. Its surface temperature is suddenly raised to 30°F. How long will it take a point 1 inch below its surface to reach 10°F?

4-22. A thick layer of ice is at a temperature of 0°F throughout. It is suddenly exposed to air at 30°F with a heat transfer coefficient of 5 Btu/hr-ft²°F. How long will it take a point 1 inch below its surface to reach 10°F?

4-23. A sinusoidal expression for the temperature distribution in a semi-infinite solid is given below.

$$\frac{T - T_s}{T_o - T_s} = \sin\left(\frac{\pi x}{2\delta}\right)$$

Determine expressions for δ from the heat balance integral and the instantaneous heat flow rate at the free surface, which is given by

$$Q_o = -kA\left(\frac{\partial T}{\partial x}\right)_{x=0}$$

PROBLEMS (SI System of Units)

4-1. A metal cube initially at 800°C is placed in a room maintained at 20°C. What will be the steady-state temperature of the cube?

4-2. Consider the plane wall problem with temperatures T_1 and T_2 maintained on the left- and right-hand faces, respectively. After steady state is reached, the entire wall is thermally insulated. What is the new steady-state temperature distribution in the wall?

4-3. A metal rod has an initial temperature distribution, $T = f(x)$. At time equal to zero, the entire rod is thermally insulated. Write an expression for the steady-state temperature distribution in the rod.

4-4. An aluminum cube 5 cm on a side is originally at a temperature of 500°C. It is suddenly immersed in a liquid at 100°C for which the convective heat transfer coefficient is 120 W/m²°C. Determine the time required for the cube to reach a temperature of 200°C.

4-5. A stainless steel rod 8-mm O.D. originally at a temperature of 300°C is suddenly immersed in a liquid at 100°C for which the convective heat transfer coefficient is 100 W/m^2°C. Determine the time required for the rod to reach a temperature of 150°C.

4-6. A stainless steel rod 8-mm O.D. originally at a temperature of 300°C is suddenly immersed in a liquid at 100°C for which the convective heat transfer coefficient is 100 W/m^2°C. Determine the temperature in the rod after 1 minute has elapsed.

4-7. The average heat transfer coefficients for flow of 100°C air over a flat plate are measured by observing the temperature time history of a 1-inch thick copper slab exposed to 100°C air. At any time, thermocouples on the surface and at the center of the slab record the same temperatures. In one test turn, the initial temperature of the plate was 200°C, and in 5 minutes the temperature decreased by 30°C. Calculate the heat transfer coefficient for this case.

4-8. Determine an expression giving the temperature as a function of time for a body of negligible internal resistance; originally the temperature, T_o, placed in a convective environment at temperature, T_∞, where the convective heat transfer coefficient is h. The body contains a uniformily distributed heat source $\dot{q}$ (energy/time vol). *Hint:* Your solution should follow closely the one presented in the text for equation (4-4).

4-9. A metal sphere initially at a uniform temperature, T_o, is immersed in a fluid that is heated by an electric heater in such a manner that, $T_\infty = T_o + 10\tau$. Derive an expression for the temperature of the sphere as a function of the convective heat transfer coefficient and time. Assume that the sphere possesses negligible internal resistance.

4-10. A stainless steel cube 5 cm on a side originally at a temperature of 500°C is suddenly immersed in a liquid at 100°C for which the convective heat transfer coefficient is 120 W/m^2°C. Determine the temperature at the center of the cube after 1 minute has elapsed.

4-11. A slab of aluminum 9 cm thick initially at 500°C is suddenly immersed in a liquid at 100°C for which the convective heat transfer coefficient is 1200 W/m^2°C. Determine the time at which the centerline temperature drops to 200°C.

4-12. A slab of aluminum 2.5 cm thick initially at 500°C is suddenly immersed in a liquid at 100°C for which the convective heat transfer coefficient is 1200 W/m^2°C. Determine the time at which the centerline temperature drops to 200°C.

4-13. For the conditions of problems 4-11 determine the time at which the surface temperature drops to 200°C.

4-14. Estimate the time required to heat the center of a 3 kg roast in a 225°C oven to 130°C. Assume the roast has the properties of water and is approximately spherical in shape. The convective heat transfer coefficient between the roast and the oven air is 15 W/m^2°C.

4-15. A long cylindrical metal billet is prepared for a manufacturing process by heating it in an oven at 800°C. If its center temperature at the time it is removed from the oven is to be 675°C, determine the billet's residence time in the oven if: h = 100 W/m²°C, α = 0.014 m²/hr, k = 20 W/m°C. The initial temperature of the billet is 100°C, and its radius is 75 mm.

4-16. A potato weighing 0.25 kg is dropped into boiling water at 100°C. If the potato is considered spherical in shape, possesses the thermophysical properties of water, and is initially at 20°C, determine the time required for its center temperature to reach 95°C. Assume h = 5000 W/m²°C.

4-17 A long wooden 12-mm O.D. rod is exposed to air at 1400°C. If the ignition temperature of the wood is 425°C, determine the time of exposure necessary to cause combustion, given the initial temperature of the wood to be 10°C. Assume k = 0.15 W/m°C, h = 16 W/m²°C, ρ = 730 kg/m³, and c = 2.5 J/gm °C.

4-18. A cinderblock cube 15 cm on a side originally at a temperature of 100°C is suddenly immersed in air at 50°C for which the convective heat transfer coefficient is 25 W/m²°C. Calculate the temperature at the center of the cube after 1 hr has elapsed.

4-19. A stainless steel sphere 10 cm in diameter is suddenly immersed in a fluid at 500°C for which the convective heat transfer coefficient is 500 W/m²°C. How long will it take for the center of the sphere to reach 250°C? T_o = 20°C.

4-20. A cinder-block cube 15 cm on a side originally at a temperature of 100°C is suddenly immersed in a fluid at 50°C for which the convective heat transfer coefficient is 25 W/m²°C. Calculate the temperature at the midpoint of one face after 1 hr has passed.

4-21. A thick layer of ice is at a temperature of − 20°C throughout. Its surface temperature is suddenly raised to − 2°C. How long will it take a point 1 cm below its surface to reach − 10°C?

4-22. A thick layer of ice is at a temperature of − 20°C throughout. It is suddenly exposed to air at − 2°C with a heat transfer coefficient of 25 W/m²°C. How long will it take a point 1 cm below its surface to reach − 10°C?

4-23. Using the sinusoidal expression for the temperature distribution in a semi-infinite solid as given below, determine expressions for δ from the heat balance integral and the instantaneous heat flow rate at the free surface, $Q_o = -kA(\partial T/\partial x)_{x=0}$.

$$\frac{T - T_s}{T_o - T_s} = \sin\left(\frac{\pi x}{2\delta}\right)$$

REFERENCES

[1] Kreith, F., *Principles of Heat Transfer*, 3rd ed., Intext Educational Publishers, New York, (1973).

[2] Holman, J. P., *Heat Transfer*, 3rd ed., McGraw-Hill Book Company, New York, (1972).

[3] Heisler, M. P., "Temperature Charts for Induction and Constant Temperature Heating," *Trans. ASME,* **69**, (1947), p. 227.

[4] Grober, H., S. Erk, and U. Grigull, *Fundamentals of Heat Transfer*, McGraw-Hill Book Company, New York, (1961).

[5] Schneider, P. J., *Conduction Heat Transfer*, Addison-Wesley Publishing Company, Inc., Reading, Mass., (1955).

[6] Goodman, T. R., "The Heat Balance Integral and its Applications to Problems Involving Change of Phase," *Trans. ASME*, **80**, (1958), p. 335.

[7] Reynolds, N. C. and T. A. Dolton, "Use of Integral Methods in Transient Heat Transfer Analysis," *ASME* paper *58-A-248,* (1958).

[8] Yang, K. T. and A. Szewczyk, 'An Approximate Treatment of Unsteady Heat Transfer in Semi-Infinite Solids with Variable Thermal Properties', *Trans. ASME*, J. of Heat Transfer, Vol. **81C**, 1959, p 251.

[9] Goodman, T. R., and J. J. Shea, "The Melting of Finite Solids," *Trans. ASME*, Journal Applied Mechanics, **27**, (1960), p. 16.

5

Numerical Methods in Heat Conduction

5-1 INTRODUCTION

The use of numerical methods for solving heat transfer problems is a result of the complexity of the analytical solutions associated with practical engineering problems. Oftentimes, analytical solutions are impossible. Factors that bring about the use of numerical methods are complex geometry, nonuniform boundary conditions, time-dependent boundary conditions, and temperature-dependent properties. Examples of complex geometry are a turbine blade, the cylinder head of an internal combustion engine, and the supporting structure for a pipeline carrying hot fluids. Convective heat transfer coefficients, which are involved in the boundary conditions for conduction problems, generally vary with position and in natural convection problems may even depend on the surface-to-fluid temperature difference. When large temperature changes are present within a body, the thermal conductivity is usually not constant but is expected to vary significantly within the body.

In some cases, analytical solutions are possible, in principle, but the mechanics of obtaining the solution may be much more difficult than the task of solving the problem numerically. For example, in the case of a composite body of several layers of materials undergoing a transient heat transfer process, it is relatively easy to set up the differential equations. The solution, however, is extremely complex, because it is necessary to deal with simultaneous partial differential equations.

In all such cases and many others, if one is equipped with the fundamental principles of heat transfer discussed earlier in this text and a knowledge of numerical methods and of computer programming (usually Fortran IV), the required solution can be successfully obtained. Numerical methods to be presented in this chapter are based on the finite difference technique. The finite difference equations will be derived by applying the principle of conservation of energy. This approach, which keeps in touch with the physics of the problem, is believed preferable to the alternate approach of starting with the existing differential equations of energy conservation and then breaking them down into finite difference form. For what follows, it will be assumed that the student has had a first course in Fortran IV programming.

In exact methods of analysis, one seeks a mathematical function of the spatial variables (x, y, z) and time (τ), which will give a value of temperature (or heat flux) at any given point in the body at any given time. In numerical methods, attention is focused on a finite number of discrete points within the body and on its surface. Typically, steps involved in a numerical solution are as follows:

(1) Assemble all relevant information about the given problem including geometry, boundary conditions (prescribed temperatures, prescribed heat fluxes, insulated boundaries, radiation at the boundaries, natural or forced convection at the boundaries), and physical properties.

(2) Divide the body into a grid pattern, which subdivides it into a finite number of elements and points called nodes. As the grid is made finer, the accuracy increases; however, the time required for solution increases.

(3) Assume that (a) the temperature of an element is represented by that at the node, (b) the temperature distribution between two adjacent nodes is linear, (c) the thermal conductivity to be used for the heat flow between two adjacent nodes is evaluated at the temperature of the interface of the two adjacent elements, and (d) the area available for the heat conduction between two adjacent nodes is the area of the interface of the two elements.

(4) Perform an energy balance on each element leading to an algebraic equation for the node representing the element.

(5) Arrange all the equations in a suitable form so that they can be solved by an iterative procedure or by some other method, and obtain a solution using a computer.

As the reader has already seen from the chapters on heat conduction, the degree of complexity increases as one goes from one-dimensional heat transfer in bodies with constant conductivity to unsteady-state heat transfer problems in bodies with varying properties. In this chapter, problems of varying complexities are successively discussed.

In discussing numerical solutions for steady-state problems, the problem of a rectangular fin is presented first. A rectangular fin (Figure 5-1) is simply a small thin plate attached to a wall whose temperature is constant. Fins are used to increase the rate of heat transfer from a surface. Heat energy is conducted into the fin at its base or root from the wall. It is then carried away by convection from the top and bottom surfaces of the fin. If the thickness of the fin is uniform, it is called a rectangular fin. The rectangular fin problem is essentially a case of one-dimensional heat transfer.*

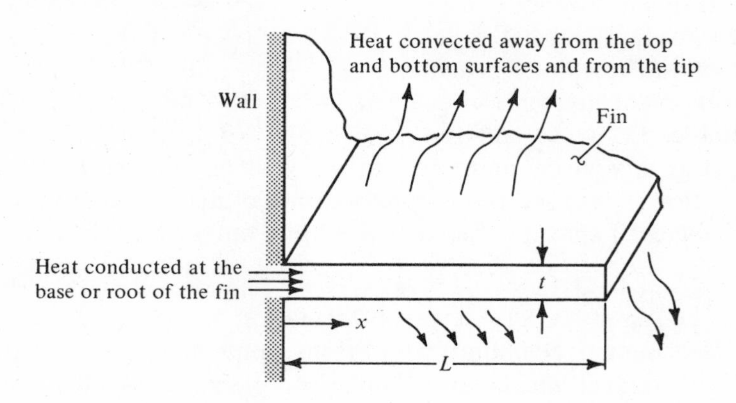

Figure 5-1 A rectangular fin.

The treatment of other geometries and of cases that account for variable physical properties in the one-dimensional case is taken up next. Section 5-4 deals with the numerical solution to two-dimensional problems with a variety of boundary conditions. Finally, unsteady-state situations with explicit and implicit formulations are discussed. Simplified numerical stability criteria for the unsteady-state cases are also presented.

5-2 ONE-DIMENSIONAL FORMULATION

In a typical fin problem, the amount of heat conducted across a section progressively decreases as one goes from the base of the fin to the tip of the fin. This is due to the convective heat flowing from the fin surface to the ambient fluid. When specified values of the base temperature, the ambient fluid temperature, and the convective heat transfer coefficients on the top and the bottom surface of the fin are given, one is usually interested in determining the temperature distribution in the fin and the rate of heat flow at its base. This heat flow is equal to the energy removed from the surface to which the fin is attached. In this section, we will apply numerical methods to a straight fin with rectangular cross-section and

*See Chapter 10, Section 10-2, for a thorough discussion of a rectangular fin.

obtain a solution that will be compared with an existing analytical solution. At this point, we do not need to know the method of obtaining the analytical solution; we will simply use the expressions for the temperature distribution and the rate of heat flow as given in Chapter 10.

We consider a rectangular fin of length, L, and of uniform thickness, t. The temperature at its base ($x = 0$) is T_o and the tip of the fin ($x = L$) loses heat by convection. Let the ambient temperature be T_∞. The convective heat transfer coefficients for the top surface, the bottom surface, and the tip are assumed to be identical and are designated as h. We also assume that the fin material has constant conductivity, k.

Figure 5-2 shows the length, L, of the fin divided into $(M - 1)$ equal parts of length Δx, resulting in M nodes. The value of M will depend on the accuracy desired.

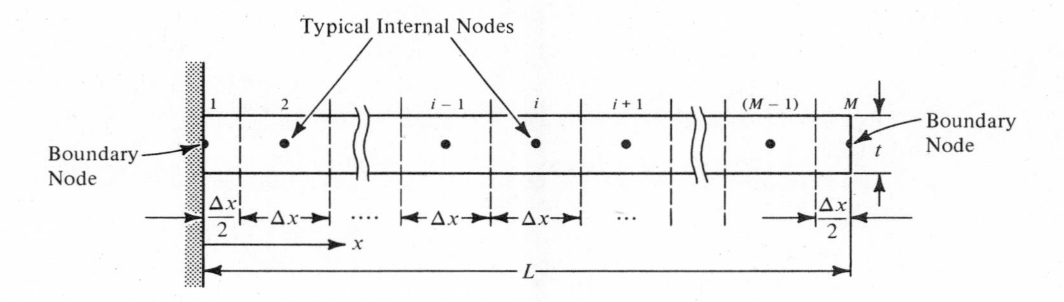

Figure 5-2 Grid for a straight fin.

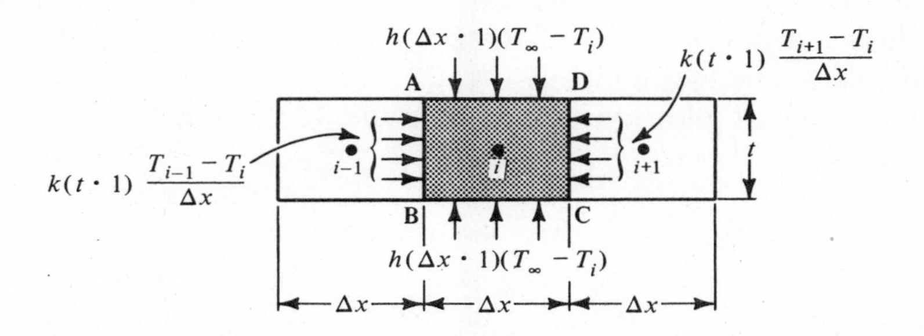

Figure 5-3 Energy balance on an element associated with ith internal node.

The domain of each *internal node* is considered to range a distance of $(\pm\Delta x/2)$ on each side of its location. Thus, an internal node is representative of a block or element of dimensions $(\Delta x \cdot t \cdot 1)$, where a unit depth into the plane of the paper is considered. For the purpose of the numerical method, it is assumed that the mean temperature of this element is T_i and that the mean temperature changes to T_{i+1} and T_{i-1} for the two adjacent elements as shown in Figure 5-3. The x

coordinates of these elements, may be designated as x_i, x_{i+1}, x_{i-1}, etc., and the heavy center points of these elements, that is, the nodal points, may likewise be designated as i, $(i + 1)$, $(i - 1)$, etc. The nodes at the boundary, the first and the Mth nodes in Figure 5-2, are the *boundary nodes*, and they represent elements of dimensions $(\Delta x/2)(t \cdot 1)$.

5-2.1 Temperature Distribution

Thus far, we have taken care of steps 1 and 2 outlined in Section 5-1. Next, in order to determine the temperature distribution in the fin, we perform an energy balance on the ith node, which is typical of all internal nodes from $i = 2$ to $i = (M - 1)$. Nodes 1 and M will be treated separately.

For convenience in the derivation, it will be assumed that heat energy flows *into* the ith element by conduction from the left and from the right, and by convection from the top and from the bottom. The algebraic sum of all four of these contributions must be zero under steady-state conditions or be equal to the change of internal energy of the ith element under unsteady conditions. We now proceed by writing expressions for the heat flow into the element from the various directions.

The rate of heat conducted from the $(i - 1)$th node to the ith node is

$$Q_{(i-1)\to i} = k\,(t \cdot 1)\,\frac{T_{i-1} - T_i}{\Delta x} \tag{5-1}$$

The above equation is based on the assumptions that the temperature varies linearly from $(i - 1)$ to i and that the area available for the flow of heat is $(t \cdot 1)$, where t is the thickness of the fin. We also note that Δx is the distance between the nodes $(i - 1)$ and i.

The rate of heat conducted from the $(i + 1)$th node to the ith node is

$$Q_{(i+1)\to i} = k\,(t \cdot 1)\,\frac{T_{i+1} - T_i}{\Delta x} \tag{5-1a}$$

From the physics of the problem, we know that heat should be conducted from i to $(i + 1)$ if node i is at a higher temperature. However, it is convenient to employ the form of equation (5-1a) to keep the bookkeeping simple. If, in actuality, T_{i+1} is less than T_i, equation (5-1a) will automatically yield a negative quantity, forcing us to reverse the direction of the arrow in $Q_{(i+1)\to i}$.

For the convective heat inflows at the top and the bottom faces, the expressions are

$$Q_{\text{top}\to i} = h\,(\Delta x \cdot 1)\,(T_\infty - T_i) \tag{5-1b}$$

and

$$Q_{\text{bot}\to i} = h\,(\Delta x \cdot 1)\,(T_\infty - T_i) \tag{5-1c}$$

respectively.

Under steady-state conditions, the net energy entering the element i must be zero.

$$Q_{(i-1)\to i} + Q_{(i+1)\to i} + Q_{\text{top}\to i} + Q_{\text{bot}\to i} = 0$$

Substitution of equations (5-1, a, b, and c) yields

$$kt\,\frac{(T_{i+1} - T_i)}{\Delta x} + kt\,\frac{(T_{i-1} - T_i)}{\Delta x} + h\,\Delta x\,(T_\infty - T_i)$$
$$+ h\,\Delta x\,(T_\infty - T_i) = 0 \tag{5-2}$$

Multiplying the above equation by $(\Delta x/kt)$ and rearranging, one obtains

$$T_{i-1} + \left\{-2\left[\frac{h\,(\Delta x)^2}{kt} + 1\right]\right\}T_i + T_{i+1} = \left[-\frac{2h\,(\Delta x)^2}{kt}\right]T_\infty$$
$$\text{for } i = 2, 3, \ldots, (M-1) \tag{5-2a}$$

Equation (5-2a) contains three unknown temperatures, namely T_{i-1}, T_i, and T_{i+1}, and the known ambient temperature, T_∞. The coefficients of these unknowns and the quantity on the right side of the equation are known from the dimensions of the grid and the physical properties of the fin material. The heat transfer coefficient, h, is usually prescribed or its value can be determined from relationships for forced and natural convection.

We observe that equation (5-2a) is valid for all internal nodes of our configuration. It is referred to as the equation for the ith node, because it is the result of performing an energy balance on the ith element. For example, when the equation is written for the second and the $(M-1)$th nodes, one has

$$T_1 + \left\{-2\left[\frac{h\,(\Delta x)^2}{kt} + 1\right]\right\}T_2 + T_3 = -\frac{2h\,(\Delta x)^2}{kt}\,T_\infty$$

and

$$T_{(M-2)} + \left\{-2\left[\frac{h\,(\Delta x)^2}{kt} + 1\right]\right\}T_{(M-1)} + T_M = -\frac{2h\,(\Delta x)^2}{kt}\,T_\infty$$

Clearly, we have a situation here of $(M - 2)$ equations, as given by equation (5-2a), and M unknowns (T_1 through T_M), indicating that two more equations are needed to determine a solution. The two missing equations are provided by the boundary conditions of the given problem, which are

$$\text{at } x = 0, T = T_o \tag{5-3}$$

where T_o is the temperature at the base of the fin, and

$$\text{at } x = L, -k\frac{dT}{dx} = h(T - T_\infty) \tag{5-3a}$$

since all of the heat conducted to the end of the fin is convected from the end of the fin to the surrounding fluid. The first condition immediately results in the equation

$$T_1 = T_o \tag{5-4}$$

It should be observed that nothing is to be gained at this point by performing an energy balance on node 1, since the quantity of heat, Q_{fin}, flowing into the fin at its root is unknown. In fact, one of the objectives of the solution is the determination of Q_{fin} since this quantity is the heat dissipated by the fin when it is attached to the surface, which is at temperature, T_o. The second boundary condition, equation (5-3a), states that the heat which arrives at the tip of the fin by conduction is carried away by convection from the tip to the surrounding fluid. To implement this condition in the approximate numerical method, we have to perform an energy balance on the Mth element (Figure 5-4). It yields

$$Q_{cond} = k(t \cdot 1)\frac{T_{M-1} - T_M}{\Delta x}$$

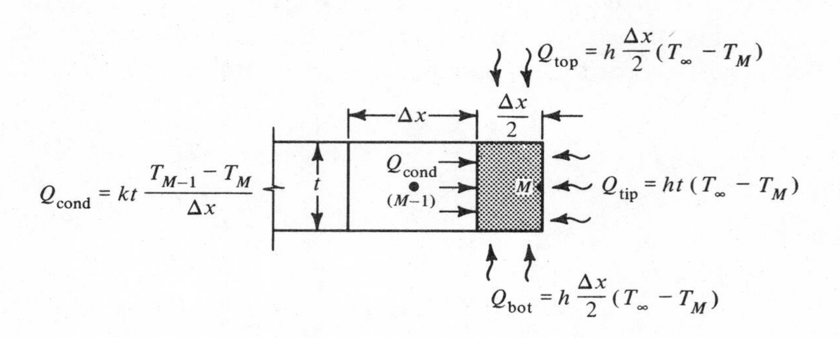

Figure 5-4 Energy balance on the boundary node.

Assuming the entire element of length $(\Delta x/2)$ is at T_M, we obtain

$$Q_{\text{top}} = Q_{\text{bot}} = h\left(\frac{\Delta x}{2} \cdot 1\right)(T_\infty - T_M)$$

and

$$Q_{\text{tip}} = h\,(t \cdot 1)\,(T_\infty - T_M)$$

Under steady-state conditions

$$Q_{\text{cond}} + Q_{\text{top}} + Q_{\text{bot}} + Q_{\text{tip}} = 0$$

or

$$k\,\frac{T_{M-1} - T_M}{\Delta x}\,t + h\,\frac{\Delta x}{2}\,(T_\infty - T_M) + h\,\frac{\Delta x}{2}\,(T_\infty - T_M)$$
$$+\, ht\,(T_\infty - T_M) = 0 \qquad (5\text{-}5)$$

so that

$$T_{M-1} + \left[-\left\{1 + \frac{h(\Delta x)^2}{kt} + \frac{h\Delta x}{k}\right\}\right]T_M = -\left[\frac{h(\Delta x)^2}{kt} + \frac{h\Delta x}{k}\right]T_\infty \qquad (5\text{-}5a)$$

Equations (5-4) and (5-5a) are referred to as the equations for the boundary nodes. These, together with equation (5-2a) represent a complete set of M simultaneous equations in the M unknown temperatures.

Most of today's computers have a library of programs, called subroutines, to do specific tasks such as finding the roots of a polynomial, inverting matrices, solving simultaneous linear algebraic equations, etc. The IBM computers have, for instance, a subroutine called SIMQ, which solves a set of simultaneous linear algebraic equations. Use of the subroutine SIMQ requires that the given set of equations be arranged in a matrix form as follows:

$$[A]\,\{T\} = \{B\} \qquad (5\text{-}6)$$

where

[A] is the $M \times M$ coefficient matrix

{T} is the column vector of the unknown temperatures, and

{B} is the column vector of the constants

For the present problem

$$[A] = \begin{bmatrix} 1 & 0 & 0. \ldots\ldots\ldots .0 & 0 & 0 \\ 1 & -\beta & 1. \ldots\ldots\ldots .0 & 0 & 0 \\ \cdot & \cdot & \cdots\cdots\cdots\cdots & \cdot & \cdot \\ \cdot & \cdot & \cdots\cdots\cdots\cdots & \cdot & \cdot \\ \cdot & \cdot & \cdots\cdots\cdots\cdots & \cdot & \cdot \\ 0 & 0 & 0. \ldots\ldots\ldots .1 & -\beta & 1 \\ 0 & 0 & 0. \ldots\ldots\ldots .0 & 1 & -(1+\gamma) \end{bmatrix}$$

where

$$\beta = 2\left[\frac{h\,(\Delta x)^2}{kt} + 1\right] \text{ and } \gamma = \left[\frac{h\,(\Delta x)^2}{kt} + \frac{h\Delta x}{k}\right]$$

$$\{T\} = \begin{Bmatrix} T_1 \\ T_2 \\ \cdot \\ \cdot \\ \cdot \\ \cdot \\ T_{M-1} \\ T_M \end{Bmatrix} \quad \text{and} \quad \{B\} = \begin{Bmatrix} T_o \\ (2-\beta)T_\infty \\ (2-\beta)T_\infty \\ \cdot \\ \cdot \\ \cdot \\ \cdot \\ -\gamma T_\infty \end{Bmatrix}$$

To solve equation (5-6) using a computer, we write a program designed to calculate the elements of the matrix [A] and the vector {B} . The program then feeds the numerical values of the matrix and the vector to a subroutine such as SIMQ. The subroutine subsequently returns the values of the unknown T_is to the main program, which can then be printed out.

Another way to solve equation (5-6) would be to multiply both sides of equation (5-6) by the matrix, $[A]^{-1}$, which is the inverse of the matrix [A] to obtain

$$[A]^{-1}\,[A]\,\{T\} = [A]^{-1}\,\{B\}$$

or

$$\{T\} = [A]^{-1} \{B\} \tag{5-6a}$$

Equation (5-6a) states that the vector of the unknown temperatures, T_is, equals the product of the matrix $[A]^{-1}$ and the vector $\{B\}$. A subroutine, such as MINV on the IBM computer systems, can be used to obtain the elements of the inverse of the matrix [A]. Finally, a subroutine, such as GMPRD on the IBM computer systems, can be used to multiply $[A]^{-1}$ and $\{B\}$ giving the values of the unknown T_is.

As the number of equations becomes very large, as in the case of a three-dimensional unsteady-state problem, the subroutines require a large storage space in the computer memory. When the number of equations exceeds a few hundred, it becomes impractical to use subroutines. The iterative method, shortly to be discussed, does not require any subroutines or a large storage space. It can quite comfortably handle thousands of equations. This, however, does not preclude the use of the iterative method for a few equations.

In the iterative method of solution, all the equations are arranged in the following form:

$$\left.\begin{aligned} T_1 &= f_1(T_2, T_3, T_4, \quad \ldots \quad, T_M) \\ T_2 &= f_2(T_1, T_3, T_4, \quad \ldots \quad, T_M) \\ &\cdot \\ &\cdot \\ &\cdot \\ T_i &= f_i(T_1, T_2, \ldots, T_{i-1}, T_{i+1}, \quad \ldots \quad, T_M) \\ &\cdot \\ &\cdot \\ &\cdot \\ T_M &= f_M(T_1, T_2, T_3, \quad \ldots \quad, T_{M-1}) \end{aligned}\right\} \tag{5-7}$$

Observe that in the first equation above, T_1 appears only on the left side of the equation and does not appear in the arguments of the function f_1.* In general, the ith equation expresses T_i as a function of all temperatures except T_i.

*In dealing with nonlinear simultaneous equations, every attempt is made to retain the nonlinear terms on the right side of an equation even though it means that T_i appears on both sides of the ith equation.

A trial solution is assumed to initiate the iterations. From the physics of a given problem, it is usually not too difficult to assume a trial solution. For example, for the fin problem, where the extreme temperatures in the system are T_o and T_∞, one may assume a linear variation of temperature along the fin with $T = T_o$ at $x = 0$ and $T = T_\infty$ at $x = L$. The assumed values of temperatures are fed into the right side of all the equations, i.e., in equations (5-7), and a new set of the T_is is obtained. Next, the values of the T_is are compared with the originally assumed values. Typically, after the first iteration, there will be large differences between the assumed values of the T_is and the values of the T_is obtained after the first iteration.

To start the second iteration, the originally assumed values of the T_is are replaced by the iterated values, and a new set of values of the T_is is obtained by once again employing equations (5-7). A comparison is then made between the set of values obtained before and after the second iteration. The iterative process is repeated until the results of two successive iterations are within a prescribed tolerence at which point the iterative process is stopped and the final values of the temperatures are printed out.

A variation of the above method employs the most recently calculated values of temperatures in the right side of equation (5-7). This method is called the Gauss-Seidel method.

Before the advent of computers, Southwell's relaxation procedure (Reference 1) was widely used for solving simultaneous equations. It is basically a trial-and-error procedure. Interested readers may see References 2 and 3.

5-2.2 Rate of Heat Flow

Once the temperature distribution is obtained, the amount of heat that must flow into the fin in order to maintain steady state is given by

$$Q_{\text{fin}} = -k\,(t \cdot 1)\left(\frac{dT}{dx}\right)_{x=0} \tag{5-8}$$

In the approximate numerical method, the rate of heat conducted into the element associated with the second node, is given by

$$Q_{1\to 2} = k\,(t \cdot 1)\,\frac{T_1 - T_2}{\Delta x}$$

In accordance with the approximate finite difference method used in developing equations (5-1, a), the quantity $Q_{1\to 2}$ is really an approximation for heat conducted at $x = \Delta x/2$. Therefore, the heat energy convected away from the top and the bottom surfaces of the fin element of length $(\Delta x/2)$, located between

$x = 0$ and $x = (1/2)\Delta x$, must be accounted for when we calculate Q_{fin}. The mean temperature of this element is

$$\{(1/2)\,[(1/2)\,(T_1 + T_2) + T_1\,]\} \text{ or } [(3/4)\,T_1 + (1/4)\,T_2]$$

Thus

$$Q_{fin} = k\,(t \cdot 1)\,\frac{T_1 - T_2}{\Delta x} + 2\left(h \cdot \frac{1}{2}\Delta x\right)\left[\left(\frac{3}{4}T_1 + \frac{1}{4}T_2\right) - T_\infty\right]$$

or

$$Q_{fin} = kt\,\frac{T_1 - T_2}{\Delta x} + h\Delta x\left(\frac{3}{4}T_1 + \frac{1}{4}T_2 - T_\infty\right) \tag{5-9}$$

Another way to determine the rate of heat flow at the base of the fin would be to fit a polynomial of degree n to the temperature values already obtained. This can be accomplished by using a subroutine from the computer library, resulting in an expression like

$$T = a_o + a_1 x + a_2 x^2 + a_3 x^3 + \ldots + a_n x^n$$

where the a_is are the coefficients of the nth-degree polynomial. When the degree, n, of the polynomial is specified to the subroutine, it will return the values of the a_is. Substituting the polynomial in equation (5-8), we obtain

$$Q_{fin} = -kta_1 \tag{5-10}$$

Sample Problem 5-1 A steel fin of 6-mm thickness and 7-cm length has a root temperature of 170°C. The ambient air temperature is 20°C, and the convective heat transfer coefficient is 42.5 W/m^2°K for the flat surface as well as for the tip of the fin. Determine the heat dissipated by the fin using the numerical method developed previously in this chapter. Compare the temperature profile obtained by using the numerical method with that of the exact solution. The thermal conductivity of steel may be taken taken as 50 W/m°K.

Solution: The solution can be broken down into four steps:

(1) Set up equations.
(2) Write various steps involved in the computational sequence.
(3) Write a program and execute it.
(4) Plot the results for the temperature profile.

We have already carried out step 1 in Section 5-2. The equations are reproduced here for convenience.

$$T_1 = T_o = 170°C$$

$$T_{i-1} + \left[-2\left\{\frac{h(\Delta x)^2}{kt} + 1\right\}\right]T_i + T_{i+1} = \left[-\frac{2h(\Delta x)^2}{kt}\right] T_\infty$$

for $i = 2, 3, 4, \ldots, (M - 1)$

$$T_{M-1} + \left[-\left\{1 + \frac{h(\Delta x)^2}{kt} + \frac{h\Delta x}{k}\right\}\right] T_M = -\left[\frac{h(\Delta x)^2}{kt} + \frac{h\Delta x}{k}\right] T_\infty$$

for $i = M$

and

$$Q_{\text{fin}} = k(t \cdot 1)\frac{T_1 - T_2}{\Delta x} + h\Delta x\left(\frac{3}{4}T_1 + \frac{1}{4}T_2 - T_\infty\right)$$

The analytical solution to this problem is given by equation (10-9) of Chapter 10 as

$$\frac{T - T_\infty}{T_o - T_\infty} = \frac{\cosh[m(L - x)] + (h/mk)\sinh[m(L - x)]}{\cosh(mL) + (h/mk)\sinh(mL)}$$

$$Q_{\text{fin}} = \sqrt{2hkt}\,(T_o - T_\infty)\frac{\sinh(mL) + (h/mk)\cosh(mL)}{\cosh(mL) + (h/mk)\sinh(mL)}$$

where $m^2 = 2h/kt$.

We will write two computer programs, *Program A* and *Program B*. *Program A* will employ the subroutine SIMQ, and *Program B* will be based on the iterative method of solution. The program for computing the temperatures from the analytical solution will be incorporated in *Program A*. *Program A* will compute the rate of heat dissipation from the fin, Q_{fin}, from equation (5-9). *Program B* will call in a subroutine to fit a polynomial to the temperatures obtained and will compute Q_{fin} from equations (5-9) and (5-10). Both the programs will be written in such a way that different values of M can be used. The following is an outline of the principle operations in the two programs.

Program A

(1) Read in the physical properties, the geometric dimensions, and the value of M to be used, and print them.
(2) Compute parameters such as Δx, $(h\Delta x/k)$, etc., and multiply variables by appropriate conversion factors.
(3) Put the appropriate zeros in the coefficient matrix [A] (this operation is not necessary with all computers).

(4) Compute the elements of [A] and {B} and call the subroutine SIMQ.
(5) Compute Q_{fin} from equation (5-9).
(6) Print values of T_is and Q_{fin}.
(7) Read a new value of M. If it is nonzero, go to step 3. If it is zero, go to step 8.
(8) Compute the temperatures and Q_{fin} by employing analytical equations.

Program B

(1) and (2) These steps are the same as those in *Program A*.
(3) Generate a first set of trial values of temperatures.
(4) Generate a new set of iterated temperatures.
(5) Compare the values of the new set with those of the previous set. If the difference is less than the prescribed tolerance, go to step 7; otherwise, go to step 6.
(6) Replace the previous set with the new set, and go to step 4.
(7) Call in a subroutine POLYFT (Xerox Data Systems computer library) to fit a polynomial to the latest set of temperature values.
(8) Compute Q_{fin} employing equations (5-9) and (5-10).
(9) Read a new value of M. If $M = 0$, stop; otherwise, go to step 3.

The principle variables used in the programs are defined below:

T(I): Temperatures T_i, $i = 1$ to M, °C

TT(I): New set of iterated values of temperatures, °C

TZERO: Temperature, T_0, at the base of the fin, °C

TAIR: Temperature of ambient air, T_∞, °C

H: Convective heat transfer coefficient, h, W/m^2 °K

X: Coordinate x, meter

DELX: Width of an element, Δx, meter

L: Length of the fin, L, meter (Input in cm)

K: Thermal conductivity, k, W/m°K

THICK: Thickness of the fin, t, meter

QFIN: Heat conducted at the root of the fin, Q_{fin}, watt

A2: The dimensionless parameter, $m = \sqrt{2h/kt}$

TOL: Tolerance value used in *Program B* to determine if iterative process is to be continued. When the absolute value of the fractional change in temperatures, T_is, from one iteration to the next (for all values of i) becomes less than TOL, the iterative process is stopped.

Listed below are *Program A* and *Program B.*

PROGRAM A

```
 1.    C  THIS PROGRAM USES SUBROUTINE SIMQ FOR FIN ANALYSIS
 2.          DIMENSION T(60),A(60,60),B(60),AA(3600)
 3.          REAL K,L
 4.          READ(105,100) TZERO,H,K,THICK,TAIR,L
 5.      100 FORMAT(F5.1)
 6.          WRITE(108,102) TZERO,H,K,THICK,TAIR,L
 7.      102 FORMAT(1H1,/,
 8.         1'TEMPERATURE AT THE BASE OF THE FIN   =',F7.2,'  DEG.C',//,
 9.         2'CONVECTIVE HEAT TRANSFER COEFF,      =',F7.2,'  WATTS/SQ.M.,K',//,
10.         3'THERMAL CONDUCTIVITY                 =',F7.2,'  WATTS/M.K.',//,
11.         4'THICKNESS OF THE FIN                 =',F5.2,'    MM',//,
12.         5'TEMPERATURE OF THE AMBIENT AIR       =',F6.2,'   DEG.C',//,
13.         6'LENGTH OF THE FIN                    =',F6.2,'   CMS',//)
14.          CONTINUE
15.      103 FORMAT(I2)
16.          L=L/100.                                    } Conversion of units.
17.          THICK=THICK/1000.
18.      501 READ (105,103) M
19.          IF(M.EQ.0) GO TO 500                        } M is the number of equations.
20.          DELX=L/(M-1.)                                 If M is zero, obtain exact solution.
21.          A1=H*DELX*DELX/(K*THICK)
22.          DO 200 I=1,M                                } Elements of coefficient matrix
23.          DO 200 J=1,M                                  are assigned zero value.
24.      200 A(I,J)=0.
25.          A(1,1)=1.
26.          B(1)=TZERO
27.          MM=M-1
28.          DO 201 I=2,MM
29.          A(I,I)=-2.*(A1+1.)                          } Nonzero coefficients and the right-hand
30.          II=I-1                                        constant in equation numbers 2 to
31.          A(I,II)=1.                                    (M - 1) are calculated.
32.          II=I+1
33.          A(I,II)=1.
34.      201 B(I)=-2.*A1*TAIR
35.          A(M,MM)=1.
36.          A(M,M)=-(1.+A1+H*DELX/K)
37.          B(M)=-(A1+H*DELX/K)*TAIR
38.          DO 90 J=1,M
39.          DO 90 I=1,M                                 } Transforms the coefficient matrix
40.          N=(I+(J-1)*M)                                 into a column vector (one-dimensional
41.       90 AA(N)=A(I,J)                                  array).
42.          CALL SIMQ(AA,B,M,IR)
43.          QFIN=K*THICK*(B(1)-B(2))/DELX+(.75*B(1)+.25*B(2)-TAIR)*H*DELX
44.          WRITE(108,104)
45.      104 FORMAT(1H1,11X,'I',9X,'T(I)')
46.          WRITE(108,105)((I,B(I)),I=1,M)
47.      105 FORMAT(/,10X,I2,4X,F10.3)
48.          WRITE(108,106) M,QFIN
49.      106 FORMAT(////,10X,'NO. OF NODES- ',I2,10X,'RATE OF HEAT TRANSFER='
50.         1,F7.2,'  WATTS')
51.          GO TO 501
52.      500 M=21 ------------------------------------>  { Values of temperatures are obtained for
53.          DELX=L/(M-1.0)                                21 values of x from analytical solution.
54.          A2=SQRT(2.*H/(K*THICK))
55.          A3=H/(A2*K)
56.          DEN=COSH(A2*L)+ SINH(A2*L)*A3
57.
58.          A4=SQRT(2.*H*K*THICK)*(TZERO-TAIR)
59.          QFIN=(A4/DEN)*(SINH(A2*L)+A3*COSH(A2*L))
60.          T(1)=TZERO
61.          DO 202 I=2,M
62.          X=(I-1)*DELX
63.          T(I)=TAIR+((TZERO-TAIR)*(COSH(A2*(L-X))+A3*SINH(A2*(L-X)))/DEN
64.         1)
65.      202 CONTINUE
66.          WRITE(108,108)
```

```
67.      108 FORMAT(1H1,///,'         ANALYTICAL  SOLUTION',///)
68.          WRITE(108,104)
69.          WRITE(108,105)((I,T(I)),I=1,M)
70.          WRITE (108,106) M,QFIN
71.          STOP
72.          END
```

PROGRAM B

```
 1.    C  THIS IS A PROGRAM FOR FIN ANALYSIS USING ITERATIVE METHOD
 2.          DIMENSION T(50), TT(50),X(50),Z(50),P(50),Q(50)
 3.          DIMENSION C(50)
 4.          REAL K,L
 5.          READ(105,100) TZERO,H,K,THICK,TAIR,L
 6.      100 FORMAT(F5.1)
 7.          WRITE(108,102) TZERO,H,K,THICK,TAIR,L
 8.      102 FORMAT(1H1,/,
 9.         1'TEMPERATURE AT THE BASE OF THE FIN  =',F7.2,'  DEG.C',//,
10.         2'CONVECTIVE HEAT TRANSFER COEFF.     =',F7.2,'  WATTS/SQ.M.,K',//,
11.         3'THERMAL CONDUCTIVITY                =',F7.2,'  WATTS/M.K.',//,
12.         4'THICKNESS OF THE FIN                =',F5.2,'    MM',//,
13.         5'TEMPERATURE OF THE AMBIENT AIR      =',F6.2,'   DEG.C',//,
14.         6'LENGTH OF THE FIN                   =',F6.2,'   CMS',//)
15.      103 FORMAT (I2)
16.          CONTINUE
17.          L=L/100.                                        } Conversion of units.
18.          THICK=THICK/1000.
19.      501 READ (105,103)M                                 } M is the number of equations.
20.          IF(M.EQ.0) GO TO 500                            } If M is zero, terminate the program.
21.          DELX=L/(M-1.)
22.          A1=H*DELX*DELX/(K*THICK)
23.          READ(105,506) TOL  ------------------------->   Suggested value: TOL≃10^-5.
24.      506 FORMAT(F10.6)
25.          MM=M-1
26.          DO 203 I=1,M                                    } Initial trial values of T_is
27.          X(I)=DELX*(I-1)                                 } are generated by assuming a
28.      203 T(I)=TZERO +(TAIR-TZERO)*((I-1.)*DELX/L)        } linear distribution.
29.          KOUNT=0
30.      300 TT(1)=TZERO
31.          DO 204 I=2,MM                                   } New T_is are
32.      204 TT(I)=(2.*A1*TAIR + T(I-1) + T(I+1) )/(2.*(A1+1.)) } generated.
33.          TT(M)=( (A1+H*DELX/K)*TAIR + T(MM) )/(1.+A1+H*DELX/K)
34.          KOUNT= KOUNT+1
35.          DO 205 I=2,M
36.      205 IF(ABS((TT(I)-T(I))/T(I)) . GT . TOL) GO TO 12  } Tolerence test is
37.          GO TO 302                                       } applied to each iterated
38.       12 DO 333 I=1,M                                    } value of T_i.
39.      333 T(I)=TT(I)
40.          GO TO 300
41.      302 CONTINUE                                        } A subroutine that fits a polynomial
42.          CALL POLYFT(X,TT,Z,P,Q,C,M,MM,MN,STD,1.)        } to a set of points.
43.          QFIN1=-C(2)*K*THICK*100.
44.          QFIN2=K*THICK*(T(1)-T(2))/DELX+(.75*T(1)+.25*T(2)-TAIR)*H*DELX
45.          WRITE(108,604)
46.      604 FORMAT(1H1,'   I       X(I)       T(I)    POLYFIT T(I)  COFFS.',/)
47.          WRITE(108,605)((I,X(I),TT(I),Z(I),C(I)),I=1,M)
48.      605 FORMAT(I4,3X,F8.2,3X,F8.2,3X,F8.2,3X,F8.2)
49.          WRITE(108,609) STD
50.      609 FORMAT(//,'STANDARD DEVIATION OF THE POLYNOMIAL FIT IS ',F6.3,//)
51.          TOL=TOL*100.
52.          WRITE(108,555) TOL
53.      555 FORMAT(/,'THE LAST TWO ITERATED VALUES ARE WITHIN ',F7.2,'%',//)
54.          WRITE(108,607) KOUNT
55.      607 FORMAT(//,'SOLUTION WAS OBTAINED AFTER',I4,' ITERATIONS',//)
56.          WRITE(108,606) QFIN1,QFIN2
57.      606 FORMAT('RATE OF HT.TR.(POLY. FIT TO TEMPS.) ',F8.2,' WATTS',
58.         1//,'RATE OF HT.TR.(FINITE DIFF.) ',F8.2,' WATTS')
59.          CONTINUE
60.          GO TO 501
61.      500 STOP
62.          END
```

A value of 10^{-5} assigned to TOL in *Program B* resulted in values of temperatures that were within 0.1% of those obtained from *Program A*. The number of iterations required to achieve the desired accuracy was 116 for $M = 6$ and increased to 2387 for $M = 30$. Temperature distributions obtained for $M = 6$ and for $M = 30$ and the results from the analytical solution are plotted in Figure 5-5. It is clear from the plot, that with as few as six nodes, the resulting temperature distribution is in close agreement with the one obtained from the analytical solution.

The values of Q_{fin} obtained from the two programs for different values of M are summarized below.

Q_{fin} in Watts

M	*Program A* *Eq. (5-9)*	*Program B* *Eq. (5-9)*	*Eq. (5-10)*
6	637.85	637.89	635.65
10	637.84	637.97	637.37
14	637.88	638.18	637.96
18	637.76	638.26	638.18
22	637.73	638.46	638.38
26	637.61	638.72	638.63
30	637.56	639.05	639.06

Analytical solution: $Q_{fin} = 637.96$ watts

We can draw some conclusions based on our experience with this problem. As the total number of nodes is increased, the number of iterations needed to achieve the desired accuracy increases almost in geometric proportion, and therefore the computer time required for the solution also increases accordingly. The rate of heat dissipation from the fin, as calculated from equations (5-9) and (5-10), is in excellent agreement with the analytical solution.

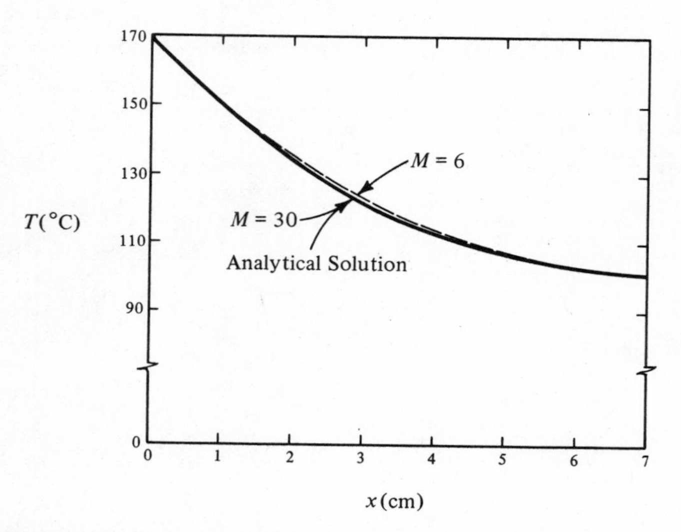

Figure 5-5 Temperature distribution for Sample Problem 5-1.

5-3 EXTENDED ONE-DIMENSIONAL FORMULATION

In the formulation of the equations in the preceding section, only the temperatures changed from one element to the next. In this section, we will consider variations in the thickness, t, the heat transfer coefficient, h, and the thermal conductivity, k.

5-3.1 Variable Thickness

We begin by analyzing a trapezoidal fin for which the exact solution leads to Bessel functions. To do so, we will formulate the equation for the ith node. Figure 5-6 shows a trapezoidal fin and an expanded view of the ith element. Once again, we need expressions for $Q_{(i-1)\to i}$, Q_{top}, etc. Consider $Q_{(i-1)\to i}$. With one-dimensional heat flow from node $(i-1)$ to node i over a distance Δx and through a mean cross-sectional area of $[(t_i + t_{i-1})/2] \cdot 1$, the heat flow is

$$Q_{(i-1)\to i} = k\frac{t_{i-1} + t_i}{2} \cdot \frac{T_{i-1} - T_i}{\Delta x} \tag{5-11}$$

The thickness t_i is to be determined as follows: At any x coordinate, the thickness is given by

$$t = t_1 - \left(\frac{t_1 - t_L}{L}\right) x \qquad \text{or} \qquad t_i = t_1 - \left(\frac{t_1 - t_L}{L}\right) x_i$$

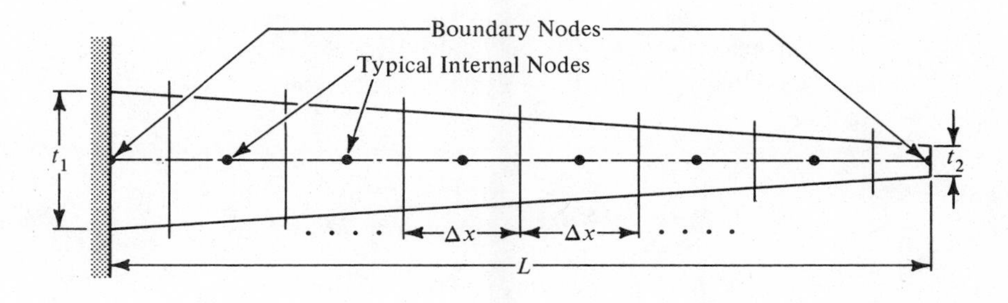

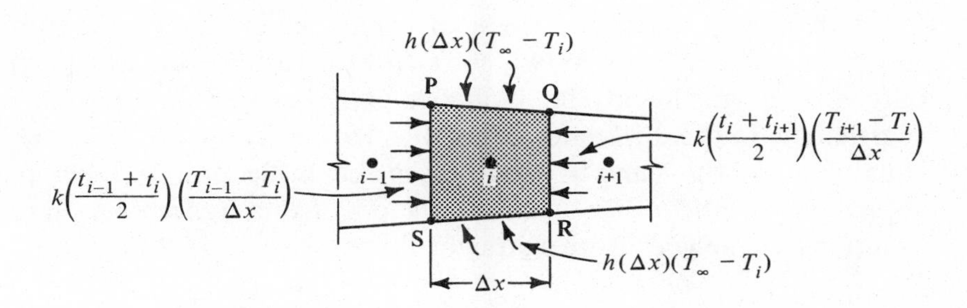

Figure 5-6 Grid for trapezoidal fin.

Since $x_i = (i-1)\Delta x$, and $L = (M-1)\Delta x$, we have

$$t_i = t_1 - \left[\frac{t_1 - t_L}{(M-1)\Delta x}\right](i-1)\,\Delta x = t_1 - (t_1 - t_L)\left(\frac{i-1}{M-1}\right) \quad (5\text{-}12)$$

Observe that we have expressed the thickness, t_i, in terms of t_1, t_L, and M, which are fixed quantities, and i, which is a variable. Such a form greatly facilitates programming the equation for a computer.

In a similar fashion, we can write

$$Q_{(i+1)\to i} = k\frac{t_{i+1} + t_i}{2} \cdot \frac{T_{i+1} - T_i}{\Delta x} \quad (5\text{-}11a)$$

The expression for Q_{top} and Q_{bot} for convective gains will be the same as before [equations (5-1b, c)], since the face-width **PQ** (or **SR**) is approximately equal to Δx for a slightly tapered fin such as the one analyzed in this problem. Thus, the equation for the ith node in a trapezoidal fin is

$$\frac{1}{2}k\,(t_{i-1} + t_i)\frac{T_{i-1} - T_i}{\Delta x} + \frac{1}{2}k\,(t_{i+1} + t_i)\frac{T_{i+1} - T_i}{\Delta x}$$
$$+ 2h\Delta x\,(T_\infty - T_i) = 0 \quad (5\text{-}13)$$

The equations for the boundary nodes are

$$T_1 = T_o \quad (5\text{-}14)$$

and

$$\frac{1}{2}k\,(t_{M-1} + t_M)\left(\frac{T_{M-1} - T_M}{\Delta x}\right) + 2\left[h\frac{\Delta x}{2}(T_\infty - T_M)\right]$$
$$+ ht_M\,(T_\infty - T_M) = 0 \quad (5\text{-}15)$$

With known inputs for the coefficients, equations (5-13) through (5-15) can easily be programmed. As we shift from a straight fin to a trapezoidal fin, the only change arising in the numerical approach is a somewhat complex set of coefficients in the difference equations. However, the procedures for solving these equations are the same as those discussed in Section 5-2. The purely analytical approach, on the other hand, takes one from hyperbolic functions to Bessel functions as one goes from a straight fin to a trapezoidal fin.

5-3.2 Variable Convective Heat Transfer Coefficient

We next consider the variations in the values of the heat transfer coefficient, h. If h is known to vary according to a definite functional relationship with x, then all that has to be done is to evaluate h for the different x_is and thus compile a set of values of h_i, $i = 1,2, \ldots, M$. From hereon, we simply replace h by h_i whenever h is encountered in the nodal equations. Thus equations (5-13) and (5-15) would become

$$\frac{1}{2}k\,(t_{i-1} + t_i)\,\frac{T_{i-1} - T_i}{\Delta x} + \frac{1}{2}k\,(t_{i+1} + t_i)\,\frac{T_{i+1} - T_i}{\Delta x} + 2h_i\,\Delta x\,(T_\infty - T_i) = 0 \tag{5-16}$$

and

$$\frac{1}{2}k\,(t_{M-1} + t_M)\frac{T_{M-1} - T_M}{\Delta x} + 2\left[h_M\,\frac{\Delta x}{2}\,(T_\infty - T_M)\right] + h_L t_M\,(T_\infty - T_M) = 0 \tag{5-17}$$

where h_L is the heat transfer coefficient at the tip of the fin ($x = L$) which could be different from h_i for $i = M$, the heat transfer coefficient at the surface of the Mth element.

In practical situations involving natural convection, the heat transfer coefficient, h, is prescribed by the following relationship

$$h = C\,(T - T_\infty)^n \tag{5-18}$$

with C and n being specified. Inasmuch as the temperatures are not known in advance, it is impossible to calculate the values of h as input. An iterative procedure is therefore required. The steps involved in such a procedure are listed below.

(1) Calculate h for $T = T_o$ using equation (5-18).
(2) Set all h_is equal to the value calculated in step 1.
(3) Solve equations (5-14), (5-16), and (5-17) for the T_is.
(4) Recalculate all h_i values using equation (5-18), employing the temperatures found in step 3.
(5) Use the latest h_i values, and repeat step 3.
(6) Compare the temperatures obtained now with those obtained in step 3. If the differences are less than a prescribed tolerance, stop; otherwise, repeat steps 4, 5, and 6.

The convergence is usually fairly rapid. It should be noted that an analytical solution is virtually impossible.

5-3.3 Variable Thermal Conductivity

The variations in thermal conductivity, k, due to temperature changes can be accounted for in the following manner. Recall that equation (5-1) gives the heat flux at the left face of the element, associated with the ith node. The conductivity, k, appearing in that equation must therefore be evaluated at the mean temperature of the left face of the element. If the nodal spacing is uniform, we evaluate the conductivity at a temperature that is the mean of the temperatures at the $(i-1)$th node and at the ith node. Referring to this value of conductivity as $k_{i-1/2}$, the expression for $Q_{(i-1)\to i}$ becomes

$$Q_{(i-1)\to i} = k_{i-1/2}\, t \left(\frac{T_{i-1} - T_i}{\Delta x}\right)$$

If we proceed in the above manner, the equation for the ith internal node in a trapezoidal fin becomes

$$\frac{1}{2} k_{i-1/2} (t_{i-1} + t_i)\left(\frac{T_{i-1} - T_i}{\Delta x}\right) + \frac{1}{2} k_{i+1/2} (t_{i+1} + t_i)\frac{T_{i+1} - T_i}{\Delta x} + 2h\Delta x (T_\infty - T_i) = 0 \tag{5-19}$$

The equations for the boundary nodes are

$$T_1 = T_o \tag{5-20}$$

and

$$\frac{1}{2} k_{M-1/2} (t_{M-1} + t_M)\left(\frac{T_{M-1} - T_M}{\Delta x}\right) + 2\left[h\frac{\Delta x}{2} (T_\infty - T_M)\right] + ht_M (T_\infty - T_M) = 0 \tag{5-21}$$

The variation of the conductivity with temperature may be specified as a functional relationship or as a set of tabulated values. In either case, one can use an iterative procedure similar to the one given for the case of a variable heat transfer coefficient. Care should be exercised in compiling values of the conductivity since it is $k_{i-1/2}$ and $k_{i+1/2}$ that is required in the equations and not k_i.

Sample Problem 5-2 A radial fin of rectangular section is shown below. It is 5/16 inch in thickness and has inner and outer radii of 2 inches and 4 inches, respectively. The temperature at the base of the fin is 200°F, and the surrounding fluid is at 100°F. The heat transfer coefficient is given by

$$h = 0.29\,(T - T_\infty)^{0.25}$$

where h is in Btu/hr-ft^2 °F, and T and T_∞ are in °F. Using a 19-node model for finite differences, determine the temperature distribution in the fin. The thermal conductivity of the fin material is 1.0 Btu/hr-ft°F.

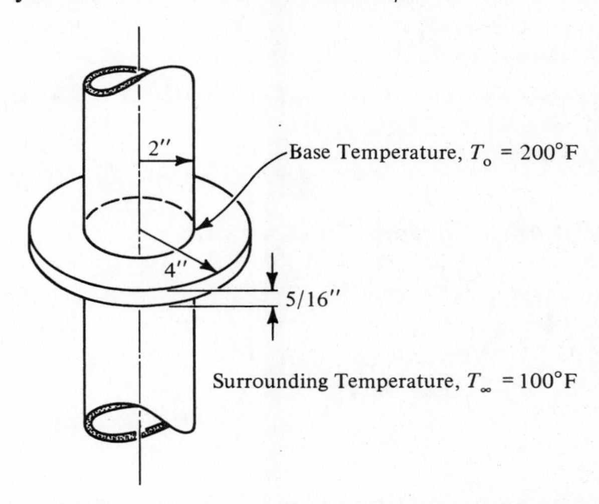

Solution: We observe that the fin is circular and the heat transfer coefficient is temperature-dependent. The grid layout for the fin is as shown below.

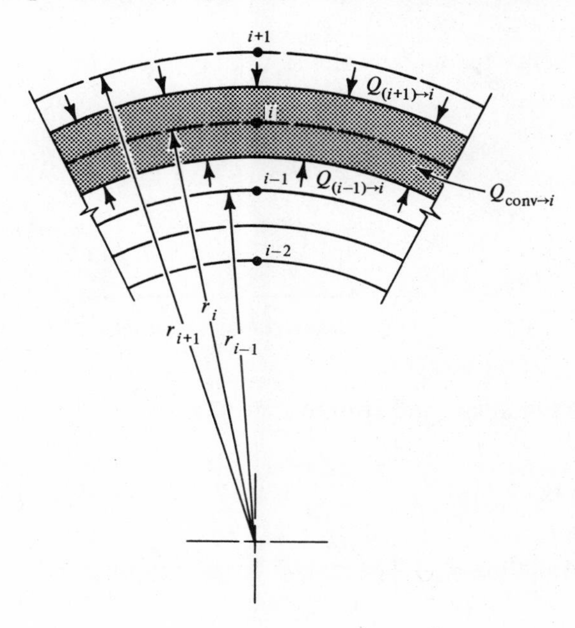

Performing an energy balance on the circular ring of width, Δr, and thickness, t, associated with the ith node, we have

$$Q_{(i-1)\to i} = kt2\pi\left(\frac{r_{i-1} + r_i}{2}\right)\left(\frac{T_{i-1} - T_i}{\Delta r}\right)$$

$$Q_{(i+1)\to i} = kt2\pi\left(\frac{r_i + r_{i+1}}{2}\right)\left(\frac{T_{i+1} - T_i}{\Delta r}\right)$$

$$Q_{\text{conv}\to i} = 2h_i 2\pi r_i \Delta r\,(T_\infty - T_i)$$

Conservation of energy requires, for steady state that

$$Q_{(i-1)\to i} + Q_{(i+1)\to i} + Q_{\text{conv}\to i} = 0$$

Substituting and simplifying, we obtain

$$kt\left(\frac{r_{i-1} + r_i}{\Delta r}\right)T_{i-1} - \left[\frac{kt}{\Delta r}(r_{i-1} + 2r_i + r_{i+1}) + 4h_i r_i \Delta r\right]T_i$$

$$+ kt\left(\frac{r_{i+1} + r_i}{\Delta r}\right)T_{i+1} = -4h_i r_i \Delta r T_\infty, \text{ for } i = 2, 3, \ldots, 18$$

The boundary condition at $r_1 = 2$ inches gives

$$T_1 = 200°\text{F}$$

An energy balance on the boundary node, at $r_{19} = 4$ inches, results in

$$kt2\pi\left(\frac{r_{18} + r_{19}}{2}\right)\left(\frac{T_{18} - T_{19}}{\Delta r}\right) + h_{19}2\pi r_{19}t\,(T_\infty - T_{19})$$

$$+ 2h_{19}2\pi \underbrace{\left[\frac{1}{2}\left(\frac{r_{18} + r_{19}}{2} + r_{19}\right)\right]}_{\text{Mean radius of the boundary element}} \frac{\Delta r}{2}(T_\infty - T_{19}) = 0$$

Rearrangement and simplification gives

$$\frac{kt}{\Delta r}(r_{19} + r_{18})T_{18} - \left[\frac{kt}{\Delta r}(r_{19} + r_{18}) + 2h_{19}r_{19}t + 2h_{19}\left(\frac{1}{4}r_{18} + \frac{3}{4}r_{19}\right)\cdot\right.$$

$$\left.\Delta r\right]T_{19} = -\left[2h_{19}r_{19}t + 2h_{19}\left(\frac{1}{4}r_{18} + \frac{3}{4}r_{19}\right)\Delta r\right]T_\infty$$

The above equations, when programmed and solved on a computer, will give the temperature distribution in the radial fin. It should be observed, however, that the convective heat transfer coefficient is not a constant quantity but varies with the surface temperature in a nonlinear fashion. The solution has to be iterative as already described. In an iterative method of solution for this problem, we need tolerences on the differences in the T_i values as well as on the differences in the h_i values. At the start, noting that T_o and T_∞ are known, assume $h = 0.29\,(T_o - T_\infty)^{0.25} = h_o$, and calculate all the temperatures in the fin. At this point, it would not be expected that the given tolerence on the difference in the T_is of two successive iterations would be met, since we have used a rather crude approximation for the values of the h_is. Next, recalculate h_1 through h_{19} on the basis of the newly calculated temperatures. A comparison should be made with the two sets of values of the h_is to determine if the differences between the two sets are within the prescribed tolerance. At this stage, the differences are not expected to be small enough. Therefore, once again, the temperatures are calculated using the new set of h_is. When the entire procedure is repeated a number of times, we are able to obtain a solution that meets the prescribed tolerances on the values of the h_is and T_is.

Given below is a list of variables to be used in the computer program.

T(I): Temperatures T_i, i = 1 to 19, °F

TN(I): New set of iterated values of temperatures, °F

TZERO: Temperature T_o, at the base of the fin, °F

TINF: Temperature of ambient air, T_∞, °F

H(I): Convective heat transfer coefficients, h_i, for various nodes, Btu/hr-ft^2 °F

HH(I): New set of iterated values of heat transfer coefficients, Btu/hr-ft^2 °F

R(I): Radius of the ith node, r_i, ft (Input in inches)

DELR: Radial distance between two successive nodes, Δr, ft

TH: Thickness of the radial fin, t, ft (Input in inches)

TOL: Tolerence on the values of T_is

HTOL: Tolerence on the values of h_is

QFIN: Rate of heat dissipation from the radial fin, Btu/hr

A computer program to solve the problem using the iterative method is given below.

```
 1.    C  RADIAL FIN ANALYSIS - VARIABLE HEAT TRANSFER COEFFICIENT
 2.          DIMENSION R(19),T(19),TN(19),H(19),HH(19)
 3.          REAL K
 4.          PI=3.141592654
 5.          IR=105
 6.          IW=108
 7.          READ (IR,100) TH,R(1),R(19),TZERO,TINF,K,TOL,HTOL
 8.      100 FORMAT(F10.4)
 9.          WRITE (IW,101) TH,R(1),R(19),TZERO,TINF,K
10.      101 FORMAT (1H1, 'THICKNESS OF THE FIN      =',F6.4,'   INCH',//
11.         1,'INNER RADIUS OF THE FIN       =',F6.2,' INCHES',//
12.         2,'OUTER RADIUS OF THE FIN       =',F6.2,' INCHES',//
13.         3,'FIN-BASE TEMPERATURE          =',F6.2,' DEG. F',//
14.         4,'AMBIENT TEMPERATURE           =',F6.2,' DEG. F',//
15.         5,'THERMAL CONDUCTIVITY          =',F7.2,' BTU/DEG.F,HR,FT',//)
16.          ERR=0.01
17.          ITEST=1
18.          ICOUNT=0
19.          T(1)=TZERO
20.          TN(1)=TZERO
21.          R(1)=R(1)/12.
22.          R(19)=R(19)/12.
23.          TH=TH/12.
24.          DELR=(R(19)-R(1))/18.
25.          DO 200 I=1,19
26.          T(I)=T(1)+(TINF-TZERO)*(I-1)/18.
27.          R(I)=R(1)+(I-1)*DELR
28.      200 H(I)=0.29*((T(I)-TINF)**0.25)
29.      600 DO 201 I=2,18
30.          C1=K*TH*(R(I-1)+R(I))/DELR
31.          C2=K*TH*(R(I+1)+R(I))/DELR
32.          C3=4.*H(I)*R(I)*DELR
33.          C4=C1+C2+C3
34.      201 TN(I)=(C1*T(I-1)+C2*T(I+1)+C3*TINF)/C4
35.          C1=K*TH*(R(19)+R(18))/DELR
36.          C2=2.*H(19)*R(19)*TH+H(19)*(.25*R(18)+.75*R(19))*DELR*2.
37.          C3=C1+C2
38.          TN(19)=(C1*T(18)+C2*TINF)/C3
39.          DO 202 J=2,19
40.      202 IF (ABS((TN(J)-T(J))/TN(J)) .GT. ERR) GO TO 310
41.          IF (ITEST.EQ.0) GO TO 700
42.          DO 203 J=1,19
43.      203 HH(J)=.29*((TN(J)-TINF)**.25)
44.          DO 204 KK=1,19
45.      204 IF (ABS((HH(KK)-H(KK))/HH(KK)) .GT. HTOL) GO TO 300
46.          ITEST=0
47.          ERR=TOL
48.      300 DO 250 J=1,19
49.      250 H(J)=HH(J)
50.      310 DO 205 I=2,19
51.      205 T(I)=TN(I)
52.          ICOUNT=ICOUNT+1
53.          GO TO 600
54.      700 WRITE(IW,102)
55.      102 FORMAT(///,'   I        T(I)',//)
56.          WRITE (IW,103)((I,TN(I)),I=1,19)
57.      103 FORMAT(I4,4X,F8.2,/)
58.          TZ=.75*T(1)+.25*T(2)
59.          QFIN=((T(1)-T(2))/DELR)*K*TH*2*PI*(R(1)+R(2))/2.
60.         1+.29*((TZ-TINF)**.25)*(TZ-TINF)*4.*PI*(.75*R(1)
61.         2+.25*R(2))*DELR
62.          TOL=TOL*100.
63.          WRITE(IW,104) ICOUNT,TOL
64.      104 FORMAT(//,'SOLUTION IS OBTAINED AFTER  ',I6,'  ITERATIONS
65.         1',//,'SUCCESSIVE TEMPERATURE DISTRIBUTIONS ARE WITHIN  ',
66.         2F6.3,'  PERCENT',//)
67.          WRITE(IW,105) QFIN
68.      105 FORMAT(//,'RATE OF HEAT TRANSFER IS  ',F8.2,'  BTUS/HR')
69.          STOP
70.          END
```

Lines 21–23: Conversion of units.

Line 24: Defines Δr.

Line 26: Calculates initial guess for T_is.

Line 27: Calculates r_is.

Line 28: Calculates initial guess for h_is.

Lines 30–33: Coefficients in the equations for internal nodes are calculated.

Lines 42–43: Calculate new h_is.

Lines 48–53: Sets h_{old} equal to h_{new} and T_{old} equal to T_{new}.

Notes on the above program:

Line No.

16. ERR, which is used for the convergence criterion of T_is for given h_is, is selected as 0.01. It is set equal to TOL in line 46 when sufficient convergence on h_is is obtained.
17. ITEST is assigned a value of 1 until convergence on h_is is obtained, at which point it is assigned a value of zero and ERR is set equal to TOL.
18. ICOUNT is a counter for number of iterations.

29. to 38. T_2 to T_{19} are calculated.

39. and 40. Convergence test is applied to all iterated values of T_is.

44. and 45. Convergence test is applied to all iterated values of h_is.

Results obtained from the above program are presented below.

```
THICKNESS OF THE FIN          = .3125 INCH

INNER RADIUS OF THE FIN       = 2.00 INCHES

OUTER RADIUS OF THE FIN       = 4.00 INCHES

FIN-BASE TEMPERATURE          = 200.00 DEG. F

AMBIENT TEMPERATURE           = 100.00 DEG. F

THERMAL CONDUCTIVITY          = 1.00 BTU/DEG.F,HR,FT
```

I	T(I)	I	T(I)
1	200.00	10	154.03
2	192.00	11	151.51
3	184.93	12	149.34
4	178.69	13	147.48
5	173.18	14	145.92
6	168.32	15	144.64
7	164.02	16	143.61
8	160.24	17	142.82
9	156.03	18	142.26
		19	141.92

```
SOLUTION IS OBTAINED AFTER   1441 ITERATIONS

SUCCESSIVE TEMPERATURE DISTRIBUTIONS ARE WITHIN   .0001 PERCENT

RATE OF HEAT TRANSFER IS     25.99  BTUS/HR
```

5-4 TWO-DIMENSIONAL STEADY-STATE SYSTEMS

The approach for solving a two-dimensional steady-state problem using numerical methods is essentially the same as that used in the one-dimensional case. The steps to be followed are identical to those outlined in the previous section.

Consider a rectangular plate in which the temperature distribution is desired. Let the physical properties be constant and uniform and the boundary conditions be as shown in Figure 5-7.

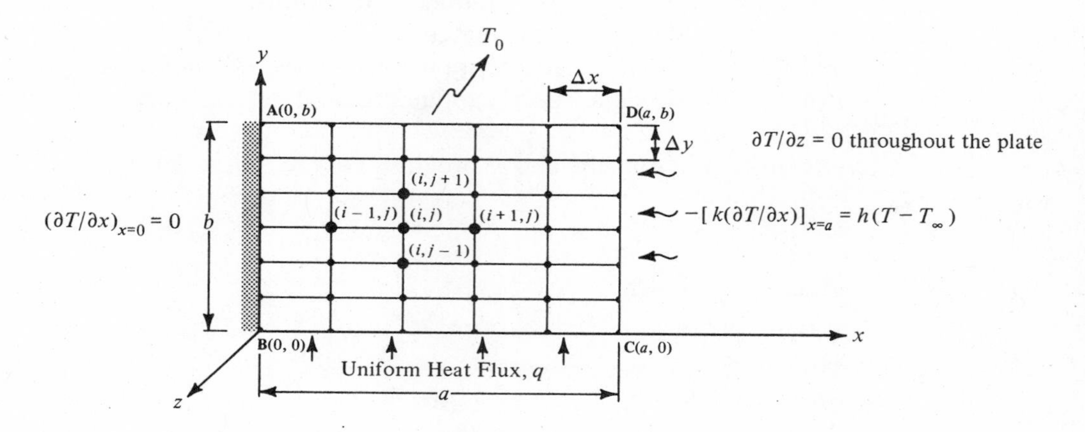

Figure 5-7 Grid layout for rectangular plate.

At

$$x = 0, \frac{\partial T}{\partial x} = 0 \tag{5-22}$$

$$x = a, -k\frac{\partial T}{\partial x} = h\,(T - T_\infty) \tag{5-22a}$$

$$y = 0, -k\frac{\partial T}{\partial y} = q \tag{5-22b}$$

$$y = b, T = T_o \tag{5-22c}$$

The above equations imply that face **AB** is insulated, face **BC** receives a uniform heat flux, face **CD** loses heat by convection, and face **DA** is maintained at a constant temperature, T_o. There is no flow of heat in the direction normal to the plane of the figure. Although the problem is analytically tractable, its solution would involve superposition of several solutions, since the boundary conditions are nonhomogeneous. To handle the problem numerically, we divide the width a of the plate into $(M - 1)$ equal parts of length, Δx, and the height b into $(N - 1)$ equal parts of length, Δy. The resulting grid will have M times N nodal points. All the points lying on the boundary **A-B-C-D-A** are called boundary nodes, and the rest of the points of the grid are called internal nodes. It is convenient to refer to a nodal point by its nodal coordinates (i, j). The relationship between the nodal coordinates (i, j) and the physical coordinates (x, y) is

$$(i - 1)\Delta x = x_i \quad \text{and} \quad (j - 1)\Delta y = y_j$$

We observe that i and j will take on integer values starting from 1 and terminating in M and N, respectively.

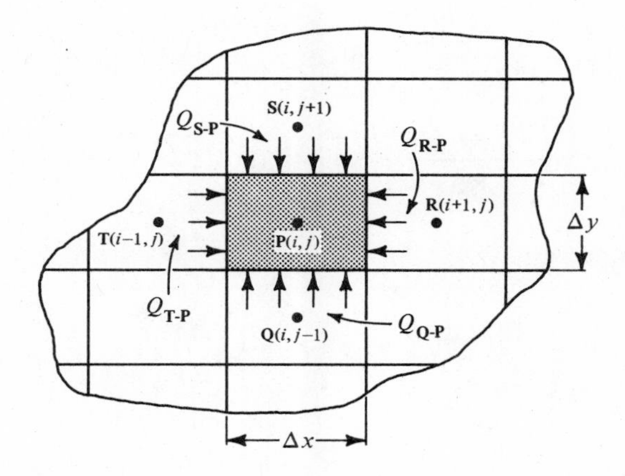

Figure 5-8 Energy balance on two-dimensional element.

We now perform an energy balance (Figure 5-8) on an internal node (i, j). Each internal node has associated with it a volume of thickness unity, the node itself being at the center of the volume of the element. Temperatures between two adjacent nodes, vertical or horizontal, are assumed to vary linearly. The flow of heat from adjacent elements to the element being examined is assumed to be of the form shown in Figure 5-8. Conservation of energy requires that the algebraic sum of the heat flowing into the element with **P** as its center be zero for steady-state conditions. Heat conducted into the element from various directions is given by

$$Q_{\mathrm{T \to P}} = k(\Delta y \cdot 1)\frac{T_{i-1,j} - T_{i,j}}{\Delta x}$$

$$Q_{\mathrm{R \to P}} = k(\Delta y \cdot 1)\frac{T_{i+1,j} - T_{i,j}}{\Delta x}$$

$$Q_{\mathrm{Q \to P}} = k(\Delta x \cdot 1)\frac{T_{i,j-1} - T_{i,j}}{\Delta y}$$

$$Q_{\mathbf{S}\to\mathbf{P}} = k(\Delta x \cdot 1)\frac{T_{i,j+1} - T_{i,j}}{\Delta y}$$

Since, for conservation of energy

$$Q_{\mathbf{T}\to\mathbf{P}} + Q_{\mathbf{R}\to\mathbf{P}} + Q_{\mathbf{Q}\to\mathbf{P}} + Q_{\mathbf{S}\to\mathbf{P}} = 0$$

we have

$$k\left[\Delta y\left(\frac{T_{i-1,j} - T_{i,j}}{\Delta x} + \Delta y\,\frac{T_{i+1,j} - T_{i,j}}{\Delta x}\right) + \Delta x\left(\frac{T_{i,j-1} - T_{i,j}}{\Delta y}\right)\right.$$
$$\left. + \Delta x\left(\frac{T_{i,j+1} - T_{i,j}}{\Delta y}\right)\right] = 0 \qquad (5\text{-}23)$$

Rearrangement gives

$$\frac{\Delta y}{\Delta x}(T_{i-1,j} - 2T_{i,j} + T_{i+1,j}) + \frac{\Delta x}{\Delta y}(T_{i,j-1} - 2T_{i,j} + T_{i,j+1}) = 0$$

or

$$\lambda T_{i-1,j} + \lambda T_{i+1,j} + \frac{1}{\lambda}T_{i,j-1} + \frac{1}{\lambda}T_{i,j+1} - 2\left(\lambda + \frac{1}{\lambda}\right)T_{i,j} = 0$$
$$\text{for } i = 2, 3, \ldots, M-1$$
$$j = 2, 3, \ldots, N-1 \qquad (5\text{-}24)$$

where $\lambda = \Delta y/\Delta x$. If $\Delta y = \Delta x$ or $\lambda = 1$, equation (5-24) simplifies to

$$T_{i-1,j} + T_{i+1,j} + T_{i,j-1} + T_{i,j+1} - 4T_{i,j} = 0$$
$$\text{for } i = 2, 3, \ldots, M-1$$
$$j = 2, 3, \ldots, N-1 \qquad (5\text{-}24\text{a})$$

Equation (5-24 or 5-24a) is referred to as the equation for node (i, j), since it is obtained by performing an energy balance on that element. Deriving equations for temperatures at the internal nodes alone is not enough since equation (5-24 or 5-24a) represents $(M-2)(N-2)$ equations with $M \times N$ unknowns. A complete set of equations will be obtained by writing an energy balance for the boundary nodes. Special care must be exercised when handling corner nodes at **A**, **B**, **C**, and **D**.

Face **AB**. Let us examine an arbitrary node, **P**, on face **AB**, surrounded by nodes **R**, **S**, and **T** [Figure 5-9(a)]. The expression for heat conducted from **S** to **P** will be similar to the one used for heat conduction in internal nodes. When heat conduction from the boundary nodes **R** or **T** to **P** is examined, we observe that the area available for heat flow is only $[(\Delta x/2) \cdot 1]$, although the distance across which heat is conducted is still Δy. Thus we have

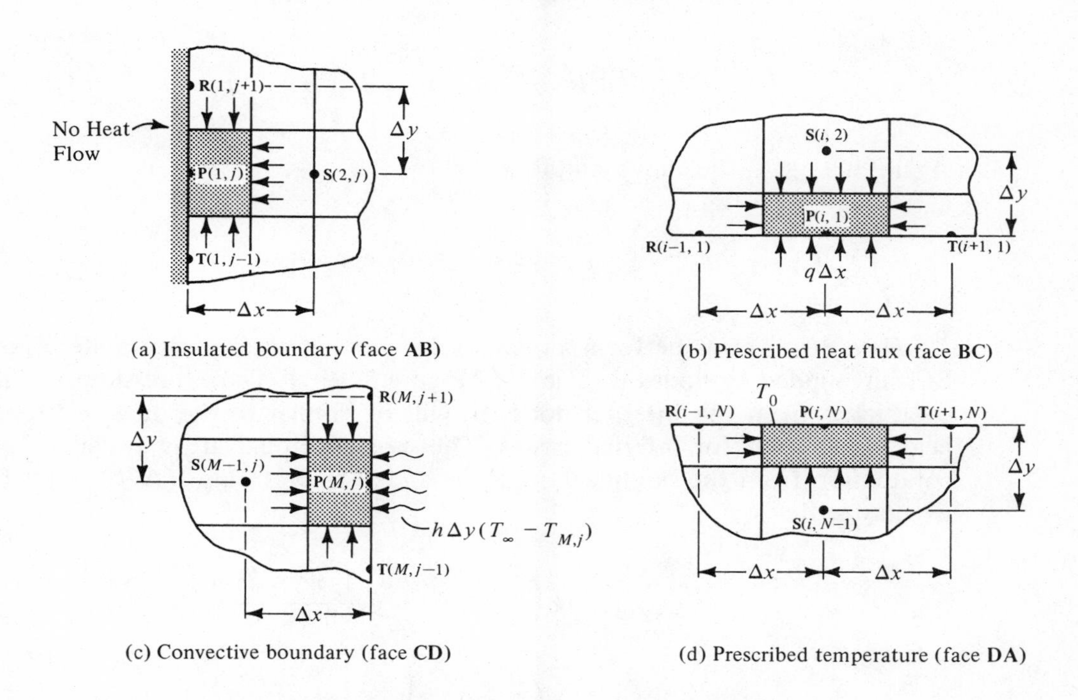

Figure 5-9 Energy balance on boundary nodes of Figure 5-7

$$Q_{\mathbf{S}\to\mathbf{P}} = k\,(\Delta y \cdot 1)\,\frac{T_{2,j} - T_{1,j}}{\Delta x}$$

$$Q_{\mathbf{R}\to\mathbf{P}} = k\left(\frac{\Delta x}{2} \cdot 1\right)\frac{T_{1,j+1} - T_{1,j}}{\Delta y}$$

$$Q_{\mathbf{T}\to\mathbf{P}} = k\left(\frac{\Delta x}{2} \cdot 1\right)\frac{T_{1,j-1} - T_{1,j}}{\Delta y}$$

Since the face **AB** is insulated, there is no flow of heat across it. Hence, for steady state

$$Q_{\mathbf{S}\to\mathbf{P}} + Q_{\mathbf{R}\to\mathbf{P}} + Q_{\mathbf{T}\to\mathbf{P}} = 0$$

Substitution gives

$$\Delta y\left(\frac{T_{2,j} - T_{1,j}}{\Delta x}\right) + \frac{\Delta x}{2}\left(\frac{T_{1,j+1} - T_{1,j}}{\Delta y}\right) + \frac{\Delta x}{2}\left(\frac{T_{1,j-1} - T_{1,j}}{\Delta y}\right) = 0$$

After some rearrangement and after substituting $\lambda = \Delta y/\Delta x$, we have

$$\lambda T_{2,j} + \frac{1}{2\lambda}T_{1,j+1} + \frac{1}{2\lambda}T_{1,j-1} - \left(\lambda + \frac{1}{\lambda}\right)T_{1,j} = 0 \tag{5-25}$$

Again, if $\lambda = 1$, the above simplifies to

$$T_{2,j} + \frac{1}{2}T_{1,j+1} + \frac{1}{2}T_{1,j-1} - 2T_{1,j} = 0 \quad \text{for } j = 2, \ldots, N-1 \tag{5-25a}$$

Face **BC**. Let us perform an energy balance on an arbitrary node, **P**, on face **BC**, surrounded by nodes **R**, **S**, and **T** [Figure 5-9(b)]. The expression for the heat conducted from the internal node, **S**, will be similar to the ones developed for heat conduction for internal nodes. The cross-sectional areas available for heat conduction from the boundary nodes **R** and **T** to **P** is only $[(\Delta y/2) \cdot 1]$. Hence, we may write

$$Q_{\mathbf{R}\to\mathbf{P}} = k\left(\frac{1}{2}\Delta y \cdot 1\right)\frac{T_{i-1,1} - T_{i,1}}{\Delta x}$$

$$Q_{\mathbf{T}\to\mathbf{P}} = k\left(\frac{1}{2}\Delta y \cdot 1\right)\frac{T_{i+1,1} - T_{i,1}}{\Delta x}$$

$$Q_{\mathbf{S}\to\mathbf{P}} = k(\Delta x \cdot 1)\frac{T_{i,2} - T_{i,1}}{\Delta y}$$

$$Q_{\text{ext}\to\mathbf{P}} = q(\Delta x \cdot 1)$$

In the last equation, above, $Q_{\text{ext}\to\mathbf{P}}$ represents heat flow due to the external heat flux into the body.

For steady state we have

$$Q_{\mathbf{R}\to\mathbf{P}} + Q_{\mathbf{T}\to\mathbf{P}} + Q_{\mathbf{S}\to\mathbf{P}} + Q_{\text{ext}\to\mathbf{P}} = 0$$

Substitution gives

$$\frac{1}{2}\Delta y\left(\frac{T_{i-1,1} - T_{i,1}}{\Delta x}\right) + \frac{1}{2}\Delta y\left(\frac{T_{i+1,1} - T_{i,1}}{\Delta x}\right) + \Delta x\left(\frac{T_{i,2} - T_{i,1}}{\Delta y}\right) + \frac{q\Delta x}{k} = 0 \tag{5-26}$$

or recalling that $\lambda = \Delta y/\Delta x$, equation (5-26) becomes

$$\frac{1}{2}\lambda T_{i-1,1} + \frac{1}{2}\lambda T_{i+1,1} + \frac{1}{\lambda}T_{i,2} - \left(\lambda + \frac{1}{\lambda}\right)T_{i,1} = -\frac{q\Delta x}{k} \quad \text{for } i = 2, \ldots, M-1 \tag{5-26a}$$

If $\lambda = 1$, one has

$$\frac{1}{2}T_{i-1,1} + \frac{1}{2}T_{i+1,1} + T_{i,2} - 2T_{i,1} = -\frac{q\Delta x}{k} \quad \text{for } i = 2, \ldots, M-1 \tag{5-26b}$$

Face **CD**. Observing from Figure 5-9(c) that this face receives energy from the ambient fluid by convection and that the area $[(\Delta x/2) \cdot 1]$ is available for the heat flow from the boundary nodes, we can write

$$k\left(\frac{\Delta x}{2} \cdot 1\right)\frac{T_{M,j+1} - T_{M,j}}{\Delta y} + k\left(\frac{\Delta x}{2} \cdot 1\right)\frac{T_{M,j-1} - T_{M,j}}{\Delta y} + k(\Delta y \cdot 1)\frac{T_{M-1,j} - T_{M,j}}{\Delta x} + h(\Delta y \cdot 1)(T_\infty - T_{M,j}) = 0 \tag{5-27}$$

Substituting $\Delta y/\Delta x = \lambda$ and rearranging, we obtain

$$\frac{1}{2\lambda}T_{M,j+1} + \frac{1}{2\lambda}T_{M,j-1} + \lambda T_{M-1,j} - \left(\frac{1}{\lambda} + \lambda + \frac{h\Delta y}{k}\right)T_{M,j} = -\frac{h}{k}\Delta y T_\infty \quad \text{for } j = 2, 3, \ldots, N-1 \tag{5-27a}$$

Again, if $\lambda = 1$, one has

$$\frac{1}{2}T_{M,j+1} + \frac{1}{2}T_{M,j-1} + T_{M-1,j} - \left(2 + \frac{h}{k}\Delta x\right)T_{M,j} = -\frac{h\Delta x}{k}T_\infty \quad \text{for } j = 2, 3, \ldots, N-1 \tag{5-27b}$$

Face **DA**. The boundary condition imposed on this face is the simplest one to handle, since the temperature itself is prescribed to be T_o. Thus, there is no need to perform any energy balance, and we have

$$T_{i,N} = T_o \quad \text{for } i = 2, 3, \ldots, M - 1 \tag{5-28}$$

Corner Nodes. At each corner, two edges meet, and it would, therefore, seem that we have to satisfy two types of boundary conditions at the same point. As long as a corner node is not required to have two different temperatures at the same time by virtue of its being the common point of two isothermal boundaries, there is no serious difficulty in handling a corner node. A thorough discussion of this point is beyond the scope of this text. Consequently, we will use that boundary condition for a corner that leads to a relatively simple equation. Thus, at corners **A** and **D**, which are parts of the face **AD**, since the temperature on the entire face **AD** is prescribed to be T_o, we require

$$T_{1,N} = T_o \tag{5-28a}$$

and

$$T_{M,N} = T_o \tag{5-28b}$$

We now have to obtain equations for corner nodes **B** and **C**, (Figure 5-7). The grid layout for these corners is shown in Figure 5-10. An energy balance for corner **B** gives

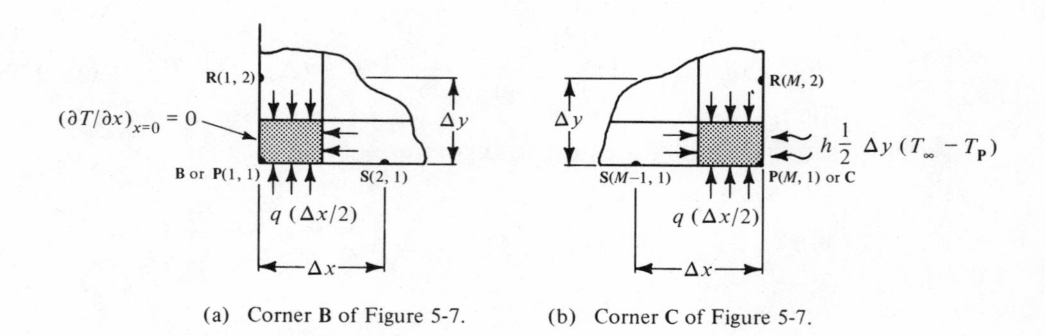

(a) Corner **B** of Figure 5-7. (b) Corner **C** of Figure 5-7.

Figure 5-10 Energy balance on corner nodes.

$$Q_{\mathbf{R}\to\mathbf{B}} = k\left(\frac{1}{2}\Delta x \cdot 1\right)\frac{T_{1,2} - T_{1,1}}{\Delta y}$$

$$Q_{\mathbf{S}\to\mathbf{B}} = k\left(\frac{1}{2}\Delta y \cdot 1\right)\frac{T_{2,1} - T_{1,1}}{\Delta x}$$

$$Q_{\text{left}\to\mathbf{B}} = 0 \quad \text{(boundary } \mathbf{BR} \text{ is insulated)}$$

$$Q_{\text{bot}\to\mathbf{B}} = q\left(\frac{1}{2}\Delta x \cdot 1\right)$$

For conservation of energy and steady state, we have

$$k\frac{1}{2}\Delta x\left(\frac{T_{1,2} - T_{1,1}}{\Delta y}\right) + k\frac{1}{2}\Delta y\left(\frac{T_{2,1} - T_{1,1}}{\Delta x}\right) + \frac{1}{2}\Delta x q = 0 \quad \text{(5-29)}$$

or

$$\frac{1}{\lambda}T_{1,2} + \lambda T_{2,1} - \left(\lambda + \frac{1}{\lambda}\right) T_{1,1} = -\frac{q\Delta x}{k} \quad \text{(5-29a)}$$

If $\lambda = 1$, the above equation takes the form

$$T_{1,2} + T_{2,1} - 2T_{1,1} = -\frac{q\Delta x}{k} \quad \text{(5-29b)}$$

Examination of corner **C** indicates [Figure 5-10(b)]

$$Q_{\mathbf{S}\to\mathbf{C}} = k\left(\frac{1}{2}\Delta y \cdot 1\right)\frac{T_{M-1,1} - T_{M,1}}{\Delta x}$$

$$Q_{\mathbf{R}\to\mathbf{C}} = k\left(\frac{1}{2}\Delta x \cdot 1\right)\frac{T_{M,2} - T_{M,1}}{\Delta y}$$

$$Q_{\text{right}\to\mathbf{C}} = h\left(\frac{1}{2}\Delta y \cdot 1\right)(T_\infty - T_{M,1})$$

$$Q_{\text{bot}\to\mathbf{C}} = q\left(\frac{1}{2}\Delta x \cdot 1\right)$$

For conservation of energy and steady state

$$Q_{\mathbf{S}\to\mathbf{C}} + Q_{\mathbf{R}\to\mathbf{C}} + Q_{\text{right}\to\mathbf{C}} + Q_{\text{bot}\to\mathbf{C}} = 0$$

Therefore

$$k\frac{1}{2}\Delta y\left(\frac{T_{M-1,1} - T_{M,1}}{\Delta x}\right) + k\frac{1}{2}\Delta x\left(\frac{T_{M,2} - T_{M,1}}{\Delta y}\right) + h\frac{1}{2}\Delta y\,(T_\infty - T_{M,1}) + q\frac{1}{2}\Delta x = 0 \quad \text{(5-30)}$$

Rearranging and substituting $\lambda = \Delta y/\Delta x$, there results

$$\lambda T_{M-1,1} + \frac{1}{\lambda} T_{M,2} - \left(\lambda + \frac{1}{\lambda} + \frac{h}{k}\Delta y\right) T_{M,1} = -\frac{q\Delta x}{k} - \frac{h}{k}\Delta y\, T_\infty \quad \text{(5-30a)}$$

If $\lambda = 1$, the above becomes

$$T_{M-1,1} + T_{M,2} - \left(2 + \frac{h}{k}\Delta x\right) T_{M,1} = -\frac{q\Delta x}{k} - \frac{h}{k}\Delta x T_\infty \quad \text{(5-30b)}$$

Equation (5-24) for the internal nodes and equations (5-25) to (5-30) for the boundary nodes represent a complete set of $(M \times N)$ equations in $(M \times N)$ unknown temperatures. These equations can be solved by the methods discussed in Section 5-2. In a two-dimensional problem, one often finds it necessary to use the iterative method of solution as the number of equations increases.

Sample Problem 5-3 Consider a rectangular plate of dimensions 24 cm by 40 cm and 1 cm thick. Thermal boundary conditions are as shown in the following sketch. There is no flow of heat in the direction normal to the plane of the figure. Determine the steady-state temperature distribution in the plate. The thermal conductivity of the plate is 25 W/m°K. The ambient temperature is 25°C.

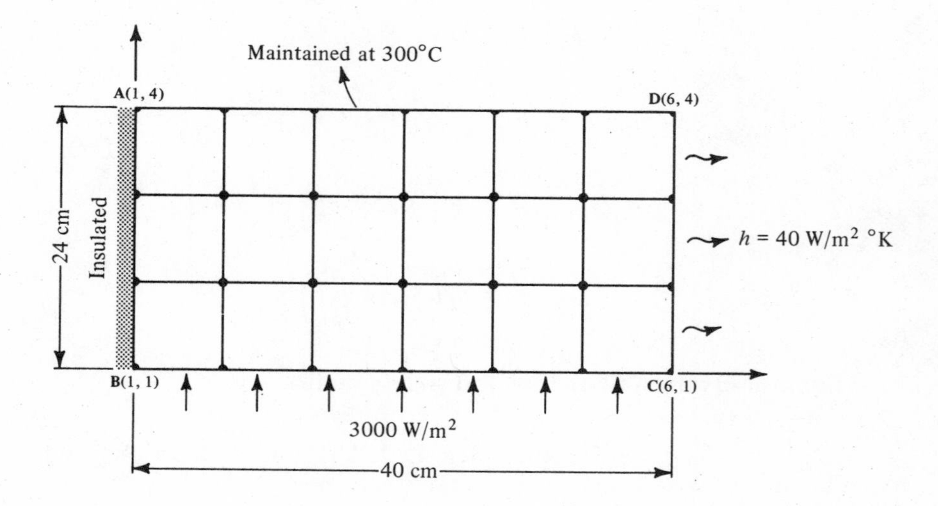

Solution: Since this problem is intended to demonstrate the application of the numerical method to a two-dimensional problem, we will divide the plate into only 15 elements of size 8 cm $\times$ 8 cm, resulting in 20 nodal points. Of these, there are only 8 internal nodes and 12 boundary nodes. It is to be expected, therefore, that the results based on such a grid pattern will have large errors. If a similar problem were to arise in practice, the nodal spacings should be reduced by a factor of at least 2 (resulting in 80 equations) so that reliable results could be obtained.

Equations derived in the preceding section are directly applicable to this problem. Inputs, pertinent to this problem, are:

$$q = 3000 \text{ W/m}^2 \qquad a = 40 \text{ cm}, b = 24 \text{ cm}$$

$$h = 40 \text{ W/m}^2\,°\text{K} \qquad \Delta x = \Delta y = 8 \text{ cm}, M = 6, N = 4$$

$$k = 25 \text{ W/m°K} \qquad T_\infty = 25°\text{C}, T_o = 300°\text{C}$$

We give below the governing nodal equations wherein various coefficients have been calculated. In general, they will consist of algebraic expressions.

Internal nodes, equation (5-24a)

$$i = 2, 3, 4, 5 \qquad j = 2, 3$$

$$T_{i-1,j} + T_{i+1,j} + T_{i,j-1} + T_{i,j+1} - 4T_{i,j} = 0$$

Nodes on face **AB**, equation (5-25a)

$$i = 1 \qquad j = 2, 3$$

$$T_{2,j} + \frac{1}{2}T_{1,j+1} + \frac{1}{2}T_{1,j-1} - 2T_{1,j} = 0$$

Nodes on face **BC**, equation (5-26b)

$$i = 6 \qquad j = 2, 3$$

$$\frac{1}{2}T_{i-1,1} + \frac{1}{2}T_{i+1,1} + T_{i,2} - 2T_{i,1} = -9.6$$

Nodes on face **CD**, equation (5-27b)

$$i = 2, 3, 4, 5 \qquad j = 1$$

$$\frac{1}{2}T_{6,j+1} + \frac{1}{2}T_{6,j-1} + T_{5,j} - 2.128\,T_{6,j} = -3.2$$

Nodes on face **DA**, equation (5-28)

$$i = 1, 2, 3, 4, 5, 6 \qquad j = 4$$

$$T_{i,j} = 300°\text{C}$$

Corner **B**, equation (5-29b)

$$T_{1,2} + T_{2,1} - 2T_{1,1} = -9.6$$

Corner **C**, equation (5-30b)

$$T_{5,1} + T_{6,2} - 2.128\, T_{6,1} = -12.8$$

The above equations for the internal and the boundary nodes are arranged on page 184 in matrix form. Although an iterative method that does not require a matrix form of solution may be used, it is believed that the arrangement of equations in a matrix form will help the student appreciate the scope of the problem.

The following computer program is designed to solve the 24 equations for the rectangular plate problem. It employs the subroutine SIMQ.

```
 1.          DIMENSION A(24,24),B(24),AA(576)
 2.          REAL K
 3.          Q = 3000.
 4.          H=40.
 5.          K=25.
 6.          XA=40./100.
 7.          YB=24./100.
 8.          TAIR=25.
 9.          TZERO=300.
10.          DELX=8./100.
11.          DO 200 I=1,24                          }  Elements of A(I, J) and B(I)
12.          B(I)=0.                                }  are set equal to zero.
13.          DO 200 J=1,24
14.      200 A(I,J)=0.
15.          DO 201 I=1,18
16.          J=I+6
17.          A(I,J)=1.
18.          J=I+1
19.          A(I,J)=1.                              Those coefficients that are
20.          IF(I.LT.2)  GO TO 201                  unity are generated using a
21.          J=I-1                                  DO loop.
22.          A(I,J)=1.
23.          IF(I.LT.8) GO TO 201
24.          J=I-6
25.          A(I,J)=1.
26.      201 CONTINUE
27.          DO 202 I=19,24
28.      202 A(I,I)=1.
29.          A(1,1)=-2.
30.          A(2,1)=.5
31.          A(2,2)=-2.
32.          A(2,3)=.5
33.          A(3,2)=.5
34.          A(3,3)=-2.
35.          A(3,4)=.5
36.          A(4,3)=.5
37.          A(4,4)=-2.
38.          A(4,5)=.5
39.          A(5,4)=.5
40.          A(5,5)=-2.
41.          A(5,6)=.5
42.          A(6,6)=-2.128
43.          A(6,7)=0.
44.          A(7,1)=.5
45.          A(7,6)=0.
46.          A(7,7)=-2.
47.          A(7,13)=.5
48.          DO 203 I=8,17
49.      203 A(I,I)=-4.
```

```
50.           A(12,12)=-2.128
51.           A(13,13)=-2.
52.           A(18,18)=-2.128
53.           A(12,6)=.5
54.           A(13,7)=.5
55.           A(18,12)=.5
56.           A(13,12)=0.
57.           A(12,13)=0.
58.           A(18,19)=0.
59.           A(12,18)=.5
60.           A(13,19)=.5
61.           A(18,24)=.5
62.           DO 210 KK=1,5
63.       210 B(KK)=-9.6
64.           B(6)=-12.8
65.           B(12)=-3.2
66.           B(18)=-3.2
67.           DO 204 I=19,24
68.       204 B(I)=300.
69.           WRITE(108,100)  (I,I=1,24)
70.           DO 205 I=1,24
71.       205 WRITE (108,101) I,(A(I,J),J=1,24)
72.       100 FORMAT(//,'  I   ',24I5,//)
73.       101 FORMAT(I3,2X,24F5·2,/)
74.           DO 70 I=1,24
75.           DO 70 J=1,24
76.           N=(J+I-1)*24)
77.        70 AA(N)=A(J,I)
78.           CALL SIMQ(A,B,24,IR)
79.           WRITE(108,206)
80.       206 FORMAT(/////,'TEMPERATURES AT VARIOUS NODES:',///)
81.           WRITE(108,207)(I,I=1,6)
82.       207 FORMAT(10X,6(I5,5X))
83.           WRITE(108,208) (B(J),J=1,24)
84.       208 FORMAT(//,(10X,6(F8·2,2X)))
85.           END
```

The program computes steady-state temperatures for the specified inputs. If the program is to be used for different sets of properties, then READ statements should be used to read each set of inputs as in Sample Problem 5-1. Wherever possible, DO loops are used to generate various repetitive elements of the coefficient matrix. In Sample Problem 5-1, the grid could be made finer, because the one-dimensional nature of the problem results in only two equations for the boundary nodes independent of the total number of nodal points. On the other hand, in a two-dimensional problem there are 2 $(M + N)$ boundary nodes most of which require individual equations. Also, each time the nodal spacing is halved, the number of boundary nodes increases by a factor of two. It would, therefore, require many more Fortran statements to make the program flexible enough to accommodate varying grid size.

The results obtained from the computer for the present problem are given below.

Temperatures at Various Nodes

	I = 1	2	3	4	5	6
J = 1	317.84	316.39	311.67	302.53	286.99	262.31
2	309.70	308.43	304.27	296.14	281.95	258.42
3	304.10	303.34	300.86	295.81	286.25	267.23
4	300.00	300.00	300.00	300.00	300.00	300.00

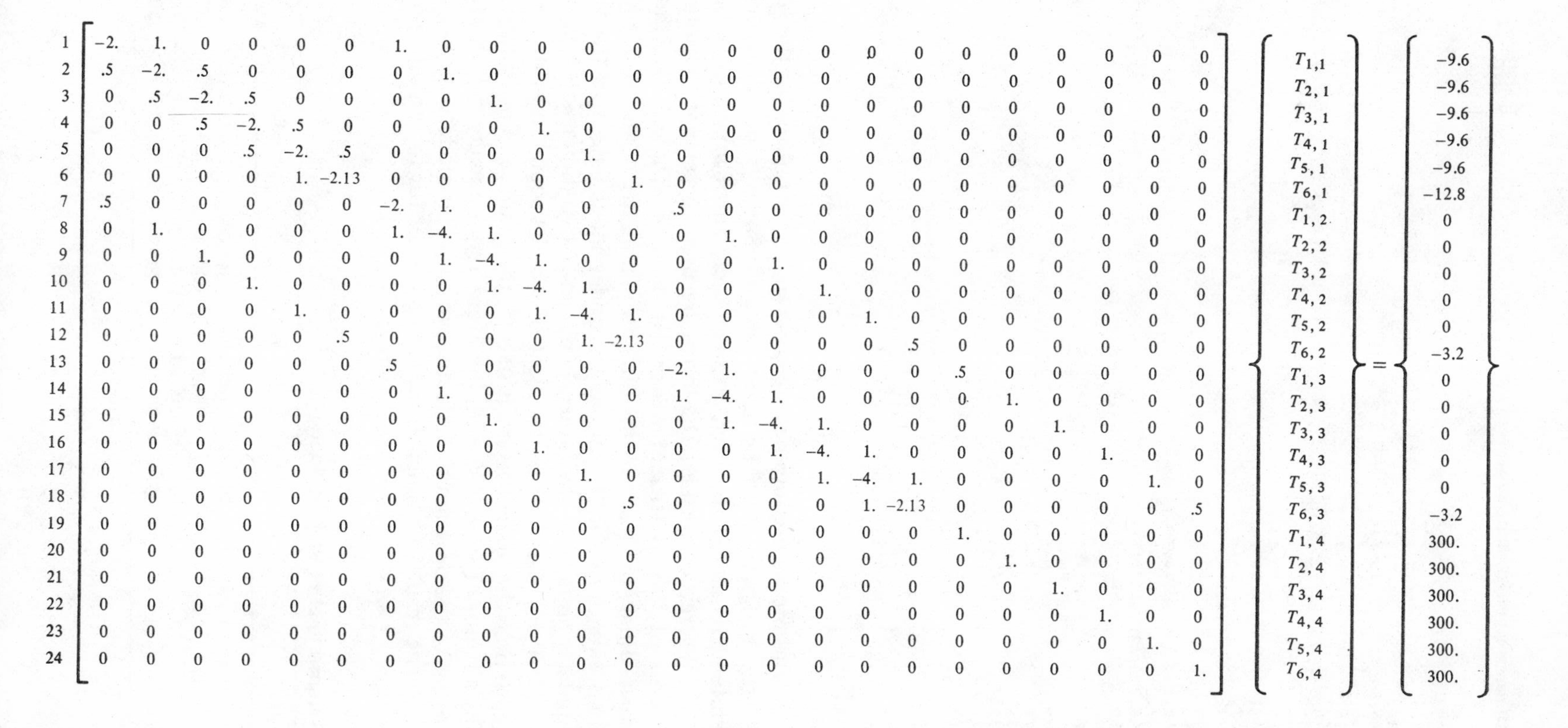

Row	1	2	3	4	5	6	7	8	9	10	11	12	13	14	15	16	17	18	19	20	21	22	23	24	{T}	{B}
1	−2.	1.	0	0	0	0	1.	0	0	0	0	0	0	0	0	0	0	0	0	0	0	0	0	0	$T_{1,1}$	−9.6
2	.5	−2.	.5	0	0	0	0	1.	0	0	0	0	0	0	0	0	0	0	0	0	0	0	0	0	$T_{2,1}$	−9.6
3	0	.5	−2.	.5	0	0	0	0	1.	0	0	0	0	0	0	0	0	0	0	0	0	0	0	0	$T_{3,1}$	−9.6
4	0	0	.5	−2.	.5	0	0	0	0	1.	0	0	0	0	0	0	0	0	0	0	0	0	0	0	$T_{4,1}$	−9.6
5	0	0	0	.5	−2.	.5	0	0	0	0	1.	0	0	0	0	0	0	0	0	0	0	0	0	0	$T_{5,1}$	−9.6
6	0	0	0	0	1.	−2.13	0	0	0	0	0	1.	0	0	0	0	0	0	0	0	0	0	0	0	$T_{6,1}$	−12.8
7	.5	0	0	0	0	0	−2.	1.	0	0	0	0	.5	0	0	0	0	0	0	0	0	0	0	0	$T_{1,2}$	0
8	0	1.	0	0	0	0	1.	−4.	1.	0	0	0	0	1.	0	0	0	0	0	0	0	0	0	0	$T_{2,2}$	0
9	0	0	1.	0	0	0	0	1.	−4.	1.	0	0	0	0	1.	0	0	0	0	0	0	0	0	0	$T_{3,2}$	0
10	0	0	0	1.	0	0	0	0	1.	−4.	1.	0	0	0	0	1.	0	0	0	0	0	0	0	0	$T_{4,2}$	0
11	0	0	0	0	1.	0	0	0	0	1.	−4.	1.	0	0	0	0	1.	0	0	0	0	0	0	0	$T_{5,2}$	0
12	0	0	0	0	0	.5	0	0	0	0	1.	−2.13	0	0	0	0	0	.5	0	0	0	0	0	0	$T_{6,2}$	−3.2
13	0	0	0	0	0	0	.5	0	0	0	0	0	−2.	1.	0	0	0	0	.5	0	0	0	0	0	$T_{1,3}$	0
14	0	0	0	0	0	0	0	1.	0	0	0	0	1.	−4.	1.	0	0	0	0	1.	0	0	0	0	$T_{2,3}$	0
15	0	0	0	0	0	0	0	0	1.	0	0	0	0	1.	−4.	1.	0	0	0	0	1.	0	0	0	$T_{3,3}$	0
16	0	0	0	0	0	0	0	0	0	1.	0	0	0	0	1.	−4.	1.	0	0	0	0	1.	0	0	$T_{4,3}$	0
17	0	0	0	0	0	0	0	0	0	0	1.	0	0	0	0	1.	−4.	1.	0	0	0	0	1.	0	$T_{5,3}$	0
18	0	0	0	0	0	0	0	0	0	0	0	.5	0	0	0	0	1.	−2.13	0	0	0	0	0	.5	$T_{6,3}$	−3.2
19	0	0	0	0	0	0	0	0	0	0	0	0	0	0	0	0	0	0	1.	0	0	0	0	0	$T_{1,4}$	300.
20	0	0	0	0	0	0	0	0	0	0	0	0	0	0	0	0	0	0	0	1.	0	0	0	0	$T_{2,4}$	300.
21	0	0	0	0	0	0	0	0	0	0	0	0	0	0	0	0	0	0	0	0	1.	0	0	0	$T_{3,4}$	300.
22	0	0	0	0	0	0	0	0	0	0	0	0	0	0	0	0	0	0	0	0	0	1.	0	0	$T_{4,4}$	300.
23	0	0	0	0	0	0	0	0	0	0	0	0	0	0	0	0	0	0	0	0	0	0	1.	0	$T_{5,4}$	300.
24	0	0	0	0	0	0	0	0	0	0	0	0	0	0	0	0	0	0	0	0	0	0	0	1.	$T_{6,4}$	300.

$$[A]\{T\} = \{B\}$$

5-4.1 Representation of Boundary Conditions at a Reentrant Corner

If we have a two-dimensional configuration that looks like the cross section of two walls meeting at a corner, the nodal point at the corner so formed is called a *reentrant corner.* The node **P** in Figure 5-11 is located at a reentrant corner. The nodes at corners **A, B, C,** and **D** in Figure 5-7 are called *exterior corner nodes.* A boundary that looks like the cross section of a staircase would have a number of reentrant and exterior corner nodes. In the previous section, we examined the boundary conditions for an edge and an exterior corner in a two-dimensional problem. An isothermal boundary, a convective boundary, an insulated boundary, and a boundary with uniform heat flux were discussed. We present below the treatment of these boundary conditions for a reentrant corner.

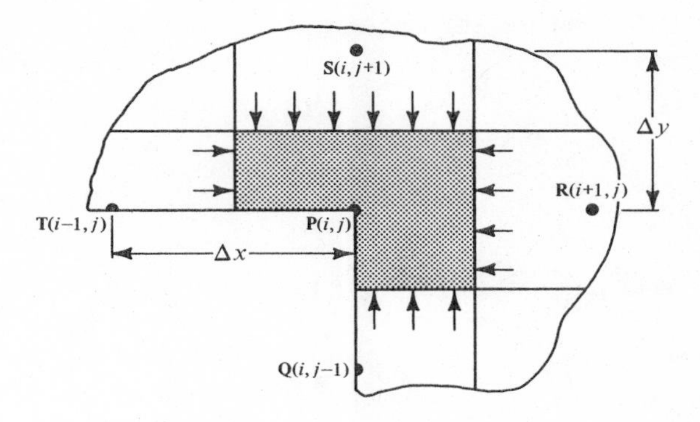

Figure 5-11 Node at reentrant corner.

Observing that the areas available for the flow of heat from **T** to **P** and from **Q** to **P** in Figure 5-11 are $[(1/2)\,(\Delta y \cdot 1)]$ and $[(1/2)\,(\Delta x \cdot 1)]$, respectively, we have

$$Q_{\mathbf{T}\rightarrow\mathbf{P}} = k\left(\frac{1}{2}\,\Delta y \cdot 1\right)\frac{T_{i-1,j} - T_{i,j}}{\Delta x} \tag{5-31}$$

$$Q_{\mathbf{Q}\rightarrow\mathbf{P}} = k\left(\frac{1}{2}\,\Delta x \cdot 1\right)\frac{T_{i,j-1} - T_{i,j}}{\Delta y} \tag{5-31a}$$

$$Q_{\mathbf{R}\rightarrow\mathbf{P}} = k\,(\Delta y \cdot 1)\,\frac{T_{i+1,j} - T_{i,j}}{\Delta x} \tag{5-31b}$$

$$Q_{\mathbf{S}\rightarrow\mathbf{P}} = k\,(\Delta x \cdot 1)\,\frac{T_{i,j+1} - T_{i,j}}{\Delta y} \tag{5-31c}$$

We now define

$$Q_{\mathrm{INT}\to\mathrm{P}} = Q_{\mathrm{T}\to\mathrm{P}} + Q_{\mathrm{Q}\to\mathrm{P}} + Q_{\mathrm{R}\to\mathrm{P}} + Q_{\mathrm{S}\to\mathrm{P}} \tag{5-32}$$

We will now examine various boundary conditions for node **P**.

Prescribed Temperature at **P**. This is the simplest condition requiring

$$T_{i,j} = T_o$$

Equations (5-31) are not needed in this case.

Insulated Boundary. There cannot be any flow of heat from the reentrant surface to **P**. Therefore, for conservation of energy, we simply have

$$Q_{\mathrm{INT}\to\mathrm{P}} = 0$$

or

$$\frac{1}{2}\lambda\,(T_{i-1,j} - T_{i,j}) + \frac{1}{2\lambda}(T_{i,j-1} - T_{i,j}) + \lambda\,(T_{i+1,j} - T_{i,j}) + \frac{1}{\lambda}(T_{i,j+1} - T_{i,j}) = 0$$

or

$$\frac{1}{2}\lambda T_{i-1,j} + \frac{1}{2\lambda}T_{i,j-1} + \lambda T_{i+1,j} + \frac{1}{\lambda}T_{i,j+1} - \frac{3}{2}\left(\lambda + \frac{1}{\lambda}\right)T_{i,j} = 0 \tag{5-33}$$

If $\lambda = 1$, we obtain

$$\frac{1}{2}T_{i-1,j} + \frac{1}{2}T_{i,j-1} + T_{i+1,j} + T_{i,j+1} - 3T_{i,j} = 0 \tag{5-33a}$$

Convective Boundary. Here, we have one more contribution to the energy balance due to convection from the ambient fluid. Note that the area across which such gain takes place is $[(1/2)(\Delta x \cdot 1) + (1/2)(\Delta y \cdot 1)]$. Thus

$$Q_{\mathrm{conv}\to\mathrm{P}} = h\frac{1}{2}(\Delta x + \Delta y)(T_\infty - T_{i,j})$$

Conservation of energy under steady state would require

$$Q_{\text{INT}\to\text{P}} + Q_{\text{conv}\to\text{P}} = 0$$

Upon substitution and simplification, one obtains

$$\frac{1}{2}\lambda T_{i-1,j} + \frac{1}{2\lambda}T_{i,j-1} + \lambda T_{i+1,j} + \frac{1}{\lambda}T_{i,j+1} - \left[\frac{3}{2}\left(\lambda + \frac{1}{\lambda}\right) + \frac{h\Delta x}{2k}(1 + \lambda)\right]T_{i,j} = -\frac{h\Delta x}{2k}(1 + \lambda)T_{\infty} \tag{5-34}$$

If $\Delta x = \Delta y$, the above simplifies to

$$\frac{1}{2}T_{i-1,j} + \frac{1}{2}T_{i,j-1} + T_{i+1,j} + T_{i,j+1} - \left[3 + \frac{h\Delta x}{k}\right]T_{i,j} = -\frac{h\Delta x}{k}T_{\infty} \tag{5-34a}$$

Prescribed Heat Flux. A heat flux of q(Btu/hr-ft^2 or W/m^2) at the boundary would contribute the following to the energy balance

$$Q_{\text{ext}\to\text{P}} = \frac{1}{2}(\Delta x + \Delta y)q$$

and the energy balance would lead to the following equation

$$\frac{1}{2}\lambda T_{i-1,j} + \frac{1}{2\lambda}T_{i,j-1} + \lambda T_{i+1,j} + \frac{1}{\lambda}T_{i,j+1} - \left[\frac{3}{2}\left(\lambda + \frac{1}{\lambda}\right)\right]T_{i,j} = -\frac{1}{2}(\Delta x + \Delta y)q \tag{5-35}$$

and for $\lambda = 1$

$$\frac{1}{2}T_{i-1,j} + \frac{1}{2}T_{i,j-1} + T_{i+1,j} + T_{i,j+1} - 3T_{i,j} = -\Delta xq \tag{5-35a}$$

5-4.2 Curved Boundary

Many practical problems involve boundaries that are not necessarily straight lines. One of the simplest ways to accommodate such boundaries is to replace the actual boundary with a jagged boundary as shown in Figure 5-12, where, for convenience, Δx and Δy are assumed to be equal. By introducing such a jagged boundary, we have added areas in some regions and removed areas in others. As far as possible, every attempt is made to have a trade-off between the surplus and the deficit areas. The modified boundary is of the type that we have discussed before inasmuch as it consists of exterior and reentrant corner nodes only.

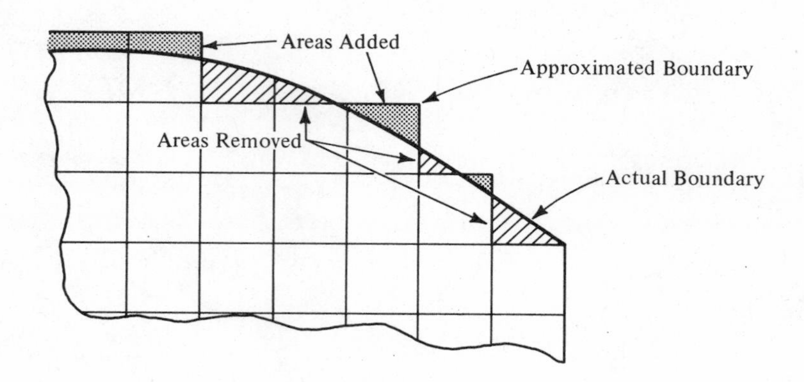

Figure 5-12 Approximation for a curved boundary.

Needless to say, the computed temperatures at points on the fictitious boundary will be higher or lower than the true temperatures because of exclusion or inclusion of material in the approximation. A more accurate method of accounting for the curved boundary is given in Figure 5-13.

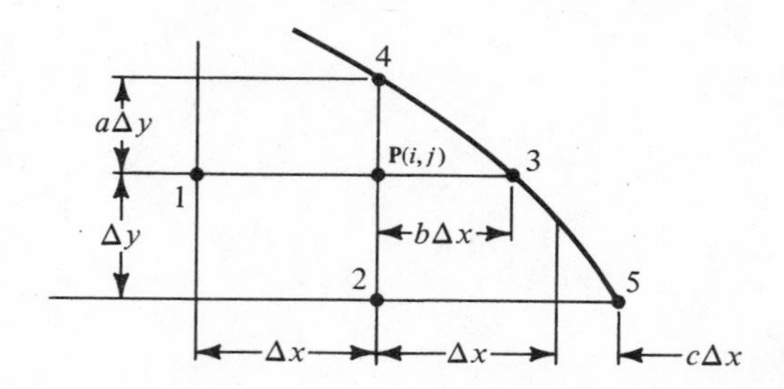

Figure 5-13 Grid for an internal node near a curved boundary.

For this method, the nodal equation for **P**(i, j) is best arrived at by considering a Taylor series expansion of T_1, T_2, T_3, and T_4 about $T_\mathbf{P}$. The Taylor series expansions are

$$T_1 = T_\mathbf{P} - \Delta x \left(\frac{\partial T}{\partial x}\right)_\mathbf{P} + \frac{(-\Delta x)^2}{2!}\left(\frac{\partial^2 T}{\partial x^2}\right)_\mathbf{P} + O(\Delta x)^3 \qquad (5\text{-}36)$$

$$T_2 = T_P - \Delta y \left(\frac{\partial T}{\partial y}\right)_P + \frac{(-\Delta y)^2}{2!}\left(\frac{\partial^2 T}{\partial y^2}\right)_P + O\,(\Delta y)^3 \qquad (5\text{-}36a)$$

$$T_3 = T_P + b\Delta x \left(\frac{\partial T}{\partial x}\right)_P + \frac{(b\Delta x)^2}{2!}\left(\frac{\partial^2 T}{\partial x^2}\right)_P + O(\Delta x)^3 \qquad (5\text{-}36b)$$

$$T_4 = T_P + a\Delta y \left(\frac{\partial T}{\partial y}\right)_P + \frac{(a\Delta y)^2}{2!}\left(\frac{\partial^2 T}{\partial y^2}\right) + O(\Delta y)^3 \qquad (5\text{-}36c)$$

In the above equations, the last terms on the right are meant to indicate terms of the order (O) of $(\Delta x)^3$ or $(\Delta y)^3$ or higher.

To proceed, we multiply equation (5-36) by b and add the result to equation (5-36b) to eliminate $(\partial T/\partial x)$; likewise we multiply equation (5-36a) by a and add the result to equation (5-36c) to eliminate $(\partial T/\partial y)$. This results in equations for

$$\left(\frac{\partial^2 T}{\partial x^2}\right)_P \quad \text{and} \quad \left(\frac{\partial^2 T}{\partial y^2}\right)_P$$

Adding these two resulting equations gives

$$\left(\frac{\partial^2 T}{\partial x^2}\right)_P + \left(\frac{\partial^2 T}{\partial y^2}\right)_P = \frac{2}{(\Delta x)^2}\left[\frac{T_4}{a\,(a+1)} + \frac{T_3}{b\,(b+1)} + \frac{T_2}{a+1} + \frac{T_1}{b+1} - \frac{a+b}{ab}T_P\right] + O(\Delta x)$$

Observing that $\nabla^2 T \equiv \partial^2 T/\partial x^2 + \partial^2 T/\partial y^2 = 0$ at point **P**, we set the above expression equal to zero. After neglecting higher-order terms, the nodal equation for **P** becomes

$$\frac{T_1}{b+1} + \frac{T_2}{a+1} + \frac{T_3}{b(b+1)} + \frac{T_4}{a(a+1)} - \left(\frac{a+b}{ab}\right) T_P = 0 \qquad (5\text{-}37)$$

If a convective boundary exists at node 3, the nodal equation for node 3 is given by

$$\left(\frac{b}{\delta_1}\right) T_4 + \left(\frac{b}{\delta_2}\right) T_5 + \left(\frac{a+1}{b}\right) T_P + \frac{h\Delta x}{k}(\delta_1 + \delta_2)T_\infty - \left[\frac{b}{\delta_1} + \frac{b}{\delta_2} + \frac{a+1}{b} + (\delta_1 + \delta_2)h\frac{\Delta x}{k}\right] T_3 = 0 \qquad (5\text{-}38)$$

where $\delta_1 = \sqrt{a^2 + b^2}$ and $\delta_2 = \sqrt{c^2 + 1}$.

The various types of boundary conditions arising in a two-dimensional problem are summarized in Table 5-1.

TABLE 5-1 Nodal Equations for Different Boundary Conditions

Physical Situation	*Nodal Equation for $\Delta x = \Delta y$*

(a) Internal node

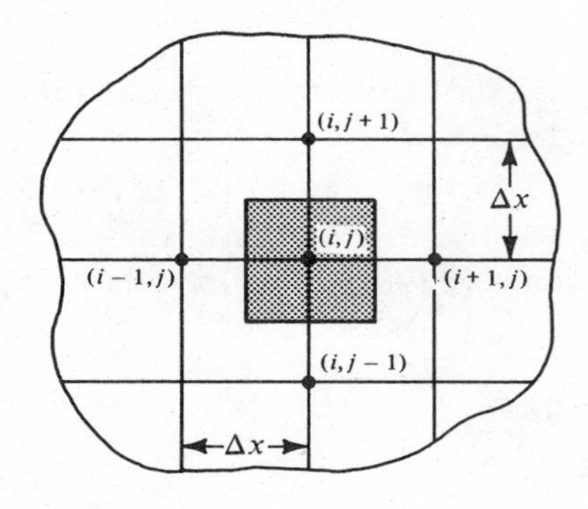

$$T_{i+1,j} + T_{i-1,j} + T_{i,j+1} + T_{i,j-1} - 4T_{i,j} = 0$$

(b) Convective boundary node

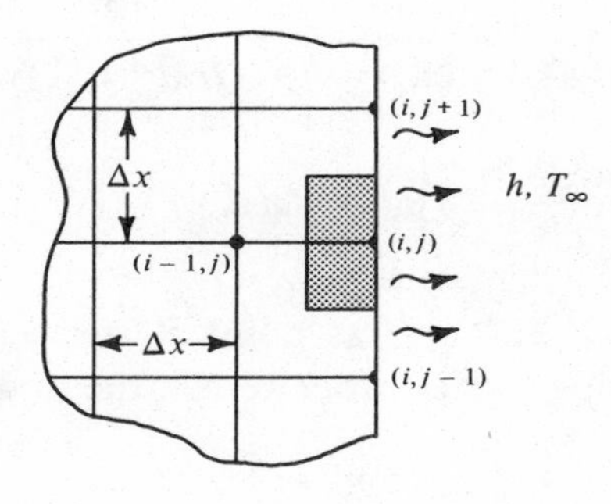

$$\left(\frac{h\Delta x}{k}\right) T_\infty + \frac{1}{2}(2T_{i-1,j} + T_{i,j+1} + T_{i,j-1}) - \left(\frac{h\Delta x}{k} + 2\right) T_{i,j} = 0$$

(c) Exterior corner with convective boundary

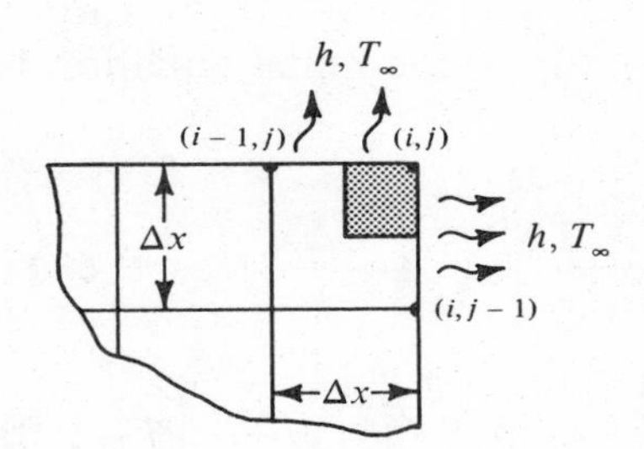

$$2\left(\frac{h\Delta x}{k}\right) T_\infty + (T_{i-1,j} + T_{i,j-1}) - 2\left[\left(\frac{h\Delta x}{k}\right) + 1\right] T_{i,j} = 0$$

(d) Reentrant corner with convective boundary

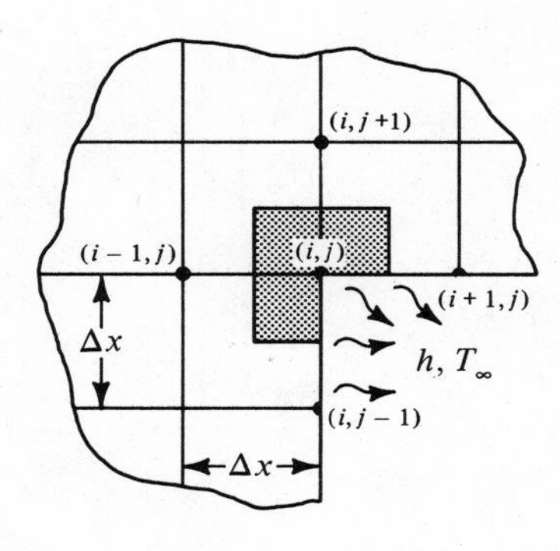

$$2\left(\frac{h\Delta x}{k}\right) T_\infty + 2T_{i-1,j} + 2T_{i,j+1} + T_{i+1,j} + T_{i,j-1} - 2\left(3 + \frac{h\Delta x}{k}\right) T_{i,j} = 0$$

(e) Insulated boundary

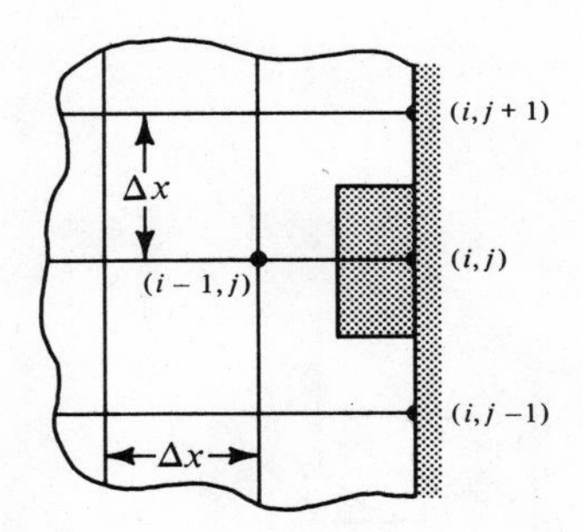

$$T_{i,j+1} + T_{i,j-1} + 2T_{i-1,j} - 4T_{i,j} = 0$$

(f) Internal node near curved boundary

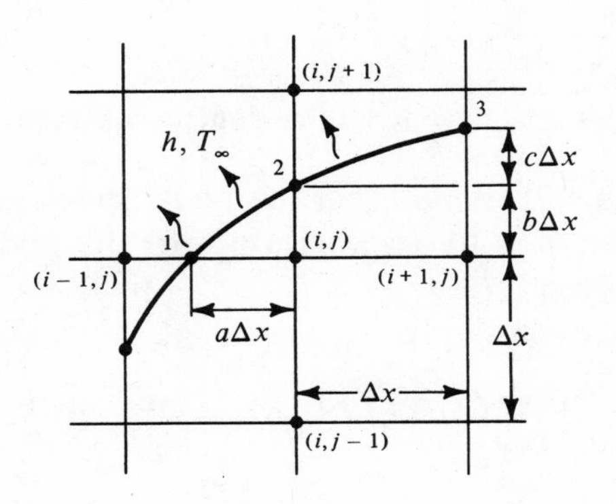

$$\frac{2}{b(b+1)} T_2 + \frac{2}{a+1} T_{i+1,j} + \frac{2}{b+1} T_{i,j-1} + \frac{2}{a(a+1)} T_1 - 2\left(\frac{1}{a} + \frac{1}{b}\right) T_{i,j} = 0$$

TABLE 5-1 Nodal Equations for Different Boundary Conditions (contd.)

(g) Boundary node with convection along curved boundary-node 2. for (f) on previous page

$$\frac{b}{\delta_1}T_1 + \frac{b}{\delta_2}T_3 + \frac{a+1}{b}T_{i,j} + \frac{h\Delta x}{k}(\delta_1+\delta_2)T_\infty$$

$$-\left[\frac{b}{\delta_1} + \frac{b}{\delta_2} + \frac{a+1}{b} + (\delta_1+\delta_2)\frac{h\Delta x}{k}\right]T_2 = 0$$

where

$$\delta_1 = \sqrt{a^2+b^2} \text{ and } \delta_2 = \sqrt{c^2+1}$$

5-5 THREE-DIMENSIONAL STEADY-STATE SYSTEMS

The procedure for handling three-dimensional heat conduction problems numerically is exactly the same as that for two-dimensional cases; the only additional complexity is that a three-dimensional grid must be made. For a three-dimensional problem, there will be six neighboring nodes for any internal node, as shown in Figure 5-14.

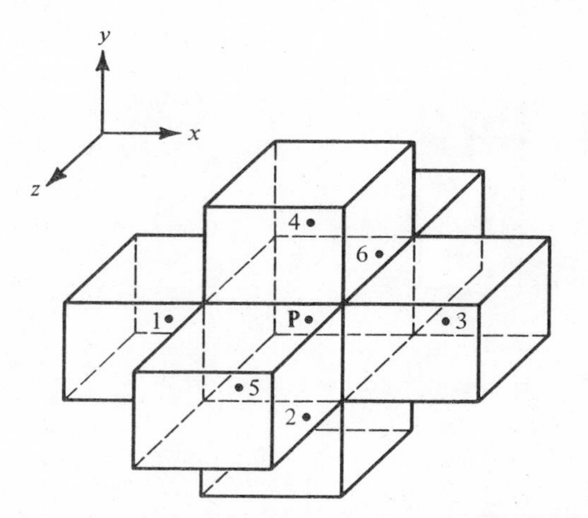

Figure 5-14 Grid for an internal node in a three-dimensional case.

To keep the algebra simple, we will assume that the nodal spacings in the x, y, and z directions are identical. Heat flow by conduction into the node **P** from all the neighboring six nodes will be given by

$$Q_{1\to P} = k(\Delta y \Delta z)\frac{T_1 - T_P}{\Delta x} = k\Delta x\,(T_1 - T_P)$$

$$Q_{2\to P} = k(\Delta z \Delta x)\frac{T_2 - T_P}{\Delta y} = k\Delta x\,(T_2 - T_P)$$

$$Q_{3\to P} = k(\Delta y \Delta z)\frac{T_3 - T_P}{\Delta x} = k\Delta x\,(T_3 - T_P)$$

$$Q_{4\to P} = k(\Delta z \Delta x)\frac{T_4 - T_P}{\Delta y} = k\Delta x\,(T_4 - T_P)$$

$$Q_{5\to P} = k(\Delta x \Delta y)\frac{T_5 - T_P}{\Delta z} = k\Delta x\,(T_5 - T_P)$$

$$Q_{6\to P} = k(\Delta x \Delta y)\frac{T_6 - T_P}{\Delta z} = k\Delta x\,(T_6 - T_P)$$

Conservation of energy requires that

$$Q_{1\to P} + Q_{2\to P} + Q_{3\to P} + Q_{4\to P} + Q_{5\to P} + Q_{6\to P} = 0$$

or

$$T_1 + T_2 + T_3 + T_4 + T_5 + T_6 - 6T_P = 0 \tag{5-39}$$

Equations for the boundary nodes can be written by performing appropriate energy balances.

5-6 DIFFERENCE EQUATIONS FROM DIFFERENTIAL ENERGY EQUATION

It is interesting to derive the difference equations in a more formal but less physical manner. The governing equation for two-dimensional steady-state heat conduction with constant properties and without any heat sources is

$$\nabla^2 T = \frac{\partial^2 T}{\partial x^2} + \frac{\partial^2 T}{\partial y^2} = 0$$

or

$$\frac{\partial}{\partial x}\left(\frac{\partial T}{\partial x}\right) + \frac{\partial}{\partial y}\left(\frac{\partial T}{\partial y}\right) = 0$$

When central difference formulas are used (Figure 5-15)

$$\frac{\partial T}{\partial x} \simeq \frac{T_{i+1,j} - T_{i-1,j}}{2\Delta x} \quad \text{at } x = x_i,\ y = y_j$$

$$\frac{\partial^2 T}{\partial x^2} \simeq \frac{\partial}{\partial x}\left(\frac{T_{i+1,j} - T_{i-1,j}}{2\Delta x}\right) = \frac{\left(\dfrac{T_{i+1,j} - T_{i,j}}{\Delta x} - \dfrac{T_{i,j} - T_{i-1,j}}{\Delta x}\right)}{\Delta x}$$

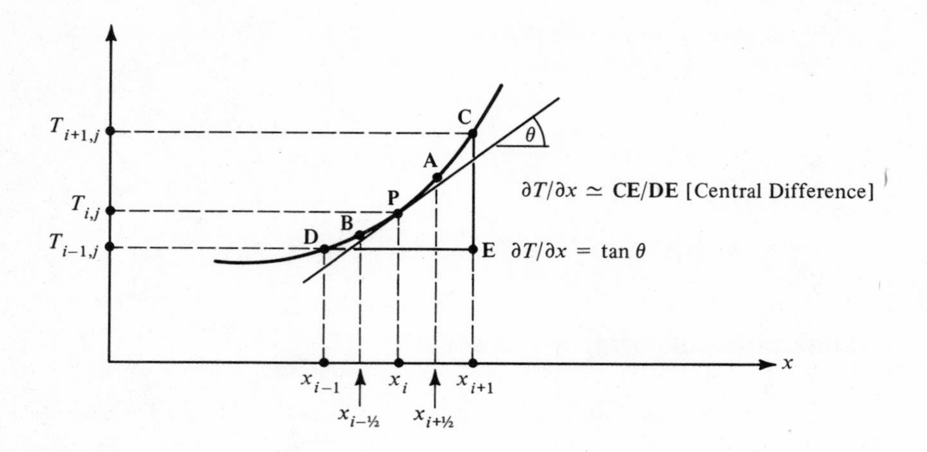

Figure 5-15 Finite difference approximation of a derivative.

Observe that the first quantity within the parentheses is the central difference approximation for $(\partial T/\partial x)$ at **A**, while the second term is the central difference approximation for $(\partial T/\partial x)$ at **B** as shown in Figure 5-15. Hence

$$\frac{\partial^2 T}{\partial x^2} \simeq \frac{(T_{i+1,j} - 2T_{i,j} + T_{i-1,j})}{(\Delta x)^2}$$

Likewise, it can be shown that

$$\frac{\partial^2 T}{\partial y^2} \simeq \frac{T_{i,j+1} - 2T_{i,j} + T_{i,j-1}}{(\Delta y)^2}$$

The governing differential equation, $\nabla^2 T = 0$, then takes the following form (the finite difference form)

$$\frac{T_{i+1,j} - 2T_{i,j} + T_{i-1,j}}{(\Delta x)^2} + \frac{T_{i,j+1} - 2T_{i,j} + T_{i,j-1}}{(\Delta y)^2} = 0$$

If $\Delta x = \Delta y$, the above simplifies to

$$T_{i+1,j} + T_{i-1,j} + T_{i,j+1} + T_{i,j-1} - 4T_{i,j} = 0$$

It is not surprising that this is the same as equation (5-24a). It is recommended that a physical energy balance be performed on the boundary nodes to obtain appropriate difference equations. Most of the boundary conditions commonly encountered were handled in Sections 5-4 and 5-4.1 using an energy balance technique.

Now we will examine the order of magnitude of the error introduced by the finite difference approximation. To this end, we expand $T(x_{i+1}, y_j)$ and $T(x_{i-1}, y_j)$ about (x_i, y_j) using a Taylors series.

$$T_{i+1,j} = T(x_{i+1}, y_j) = T(x_i, y_j) + \left(\frac{\partial T}{\partial x}\right)_P \Delta x + \left(\frac{\partial^2 T}{\partial x^2}\right)_P \frac{(\Delta x)^2}{2!} + \left(\frac{\partial^3 T}{\partial x^3}\right)_P \frac{(\Delta x)^3}{3!} + \left(\frac{\partial^4 T}{\partial x^4}\right)_P \frac{(\Delta x)^4}{4!} + \cdots$$

$$T_{i-1,j} = T(x_{i-1}, y_j) = T(x_i, y_j) + \left(\frac{\partial T}{\partial x}\right)_P (-\Delta x) + \left(\frac{\partial^2 T}{\partial x^2}\right)_P \frac{(-\Delta x)^2}{2!} + \left(\frac{\partial^3 T}{\partial x^3}\right)_P \frac{(-\Delta x)^3}{3!} + \left(\frac{\partial^4 T}{\partial x^4}\right)_P \frac{(-\Delta x)^4}{4!} + \cdots$$

Adding and rearranging these two equations, we have

$$\left(\frac{\partial^2 T}{\partial x^2}\right)_P \simeq \frac{T_{i-1,j} - 2T_{i,j} + T_{i+1,j}}{(\Delta x)^2} - \left(\frac{\partial^4 T}{\partial x^4}\right)_P \frac{(\Delta x)^2}{12}$$

where $T_{i,j}$ has been substituted for $T(x_i, y_j)$.

The above equation shows that the use of the finite difference method, based on the central difference formulation, involves an error that is of the order of $[(\Delta x)^2/12$. The Taylor series expansion of $T_{i+1,j}$ also shows that the finite difference approximation of the first derivative, $(\partial T/\partial x)_P$, involves an error of the order of $(\Delta x/2)$. Thus, when the rate of heat flow, $-kA(dT/dx)$, is evaluated using the finite difference techique, it involves an error of the order of $(\Delta x/2)$. On the other hand, when a temperature distribution is obtained by using the same technique, an error of the order of $[(\Delta x)^2/12]$ is involved.

5-7 UNSTEADY-STATE SYSTEMS

When dealing with unsteady-state problems using numerical methods, one has to use a finite difference grid for the time as well as for the spatial variables. In fact, the solution is advanced from "present" to "future" by using small increments of time. The smaller the time increment used, the greater the resulting accuracy. If an excessively large time step is used, numerical instability may result; or, in other words, no meaningful solution can be obtained. Reasons for such possible behavior will be discussed later.

In dealing with steady-state systems, a solution of simultaneous equations is required only once. In the case of unsteady-state problems, new solutions to either algebraic equations or simultaneous equations must be obtained for each successive time increment until the desired time is reached. If a steady-state solution is possible for a problem, it is obtained when the process is carried out for a sufficiently large number of time steps such that the nodal temperatures do not change for additional time increments.

In setting up the equations for all the nodal points, the principle of conservation of energy is once again used. The net flow of heat into an element during a small interval of time, $\Delta\tau$, is set equal to the change of internal energy of that element, thus incorporating both present and future temperatures at the node. We will now discuss a one-dimensional unsteady-state case.

5-7.1 One-Dimensional System

Consider a slab of width, L, in the x direction and having no temperature gradients in the y and the z directions. Let the slab be initially at a uniform temperature, T_o. From time $\tau = 0$ onwards, the left-face is maintained at a steady temperature T^* and the right face is exposed to a convective heat loss. We are interested in obtaining a numerical solution to this problem. The slab width, L, is divided into $(M - 1)$ equal parts resulting in M nodal planes as shown in Figure 5-16. The spatial grid layout is exactly the same as that for a steady-state situation. All the internal nodes have material of width $(\Delta x/2)$ associated with them on either side of their center plane, while the boundary nodes have material of width $(\Delta x/2)$ on one side only.

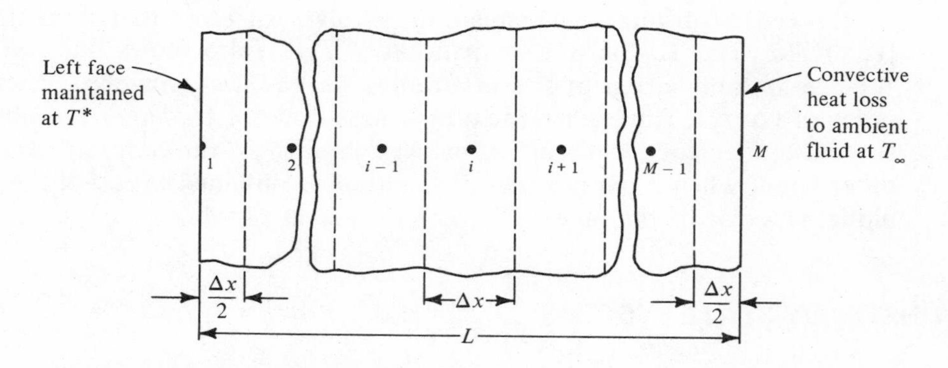

Figure 5-16 Grid for a slab of width, L, initially at a temperature T_o.

The energy conducted into node i across a cross-sectional area, A, during a unit time is given by

$$Q_{(i-1)\to i} = kA \frac{T_{i-1}^n - T_i^n}{\Delta x}$$

$$Q_{(i+1)\to i} = kA\frac{T_{i+1}^n - T_i^n}{\Delta x}$$

where the superscript n denotes temperatures at time $\tau = n\Delta\tau$. We use a superscript to represent time to emphasize the unsteady nature of the problem and to maintain a distinction between the space coordinates and the time coordinate.

In view of the unsteady nature of the problem, there is a change of internal energy, ΔU_i in time $\Delta\tau$ which is given by

$$\Delta U_i = \rho(A \cdot \Delta x)c\left(T_i^{n+1} - T_i^n\right)$$

where ρ is the density of the slab, c is the specific heat of the slab, and T_i^{n+1} denotes the temperature of node i at time $(\tau + \Delta\tau)$. To satisfy conservation of energy

$$\Delta U_i = (Q_{(i-1)\to i} + Q_{(i+1)\to i})\,\Delta\tau$$

or

$$\rho A\Delta x c\left(T_i^{n+1} - T_i^n\right) = kA\frac{T_{i-1}^n - T_i^n}{\Delta x}\Delta\tau + kA\frac{T_{i+1}^n - T_i^n}{\Delta x}\Delta\tau \quad (5\text{-}40)$$

where the superscript $(n + 1)$ denotes the temperature at time $(\tau + \Delta\tau)$. Solving for T_i^{n+1}, we obtain

$$T_i^{n+1} = \frac{k\Delta\tau}{\rho c(\Delta x)^2}\left(T_{i-1}^n - 2T_i^n + T_{i+1}^n\right) + T_i^n \quad (5\text{-}40a)$$

or

$$T_i^{n+1} = \frac{1}{\beta}\left[\left(T_{i-1}^n + T_{i+1}^n\right) + (\beta - 2)\,T_i^n\right]$$

where

$$\frac{1}{\beta} = \frac{k}{\rho c}\frac{\Delta\tau}{(\Delta x)^2} \quad \text{or} \quad \beta = \frac{(\Delta x)^2}{\alpha\Delta\tau} \quad (5\text{-}40b)$$

Equation (5-40a) expresses the temperature at node i at a time $(\tau + \Delta\tau)$ in terms of the temperatures at time τ. In our case, since the temperatures for $\tau = 0$ are all equal to T_o and are known, future temperatures at time $\Delta\tau$ at all the internal nodes can be computed with a preselected value of β. Once the temperatures at all the nodes for time $\Delta\tau$ are calculated, their values are used as input when computing the temperatures for time $2\Delta\tau$, etc.

The equation for the boundary node 1 is

$$T_1^n = T^* \quad \text{for all } \tau > 0 \text{ or for } n = 1, 2, 3 \ldots$$

The equation for the boundary node M can be obtained by performing an energy balance on the Mth node. It yields

$$\Delta\tau \left(Q_{(M-1)\to M} + Q_{\text{conv}\to M}\right) = \Delta U_M$$

When appropriate expressions are substituted for each term in the above equation, we obtain

$$kA \frac{T_{M-1}^n - T_M^n}{\Delta x} \Delta\tau + hA\left(T_\infty - T_M^n\right)\Delta\tau = \rho\left(A \frac{1}{2}\Delta x\right)c\left(T_M^{n+1} - T_M^n\right)$$

Rearrangement gives

$$T_M^{n+1} = \frac{2}{\beta}\left(T_{M-1}^n - T_M^n\right) + \frac{2h\Delta\tau}{\rho c \Delta x}\left(T_\infty - T_M^n\right) + T_M^n$$

or

$$T_M^{n+1} = \frac{2}{\beta} T_{M-1}^n + \frac{2h\Delta\tau}{\rho c\Delta x} T_\infty + \left(1 - \frac{2}{\beta} - \frac{2h\Delta\tau}{\rho c\Delta x}\right) T_M^n \tag{5-41}$$

The temperature at node 1 will be fixed, whereas that at node M will continually change until steady state is reached. This form of the difference equations is known as the *explicit form*, since the temperatures, T_i^{n+1}, corresponding to time $(\tau + \Delta\tau)$ can be solved for explicitly; only the temperatures, T_i^n, corresponding to time, τ, appear in the right-hand sides of equations (5-40a) and (5-41). This is due to the use of T_i^n in the energy balance for all the nodes.

5-7.2 Stability Criterion

A close examination of equation (5-40a) reveals that if β is less than 2, the coefficient of T_i^n will be negative. A negative coefficient of T_i^n can lead to fluctuating values of temperatures as the computation is repeated for successive $\Delta\tau$s.

The fluctuations can be so wild that the laws of thermodynamics are violated, and the number tolerence of the computer is exceeded.

As an example, consider a slab originally at a temperature of 100°F. From $\tau = 0$ onward, let the left face of the slab be maintained at 400°F and the right face be insulated. If we choose $\beta = 1$, equation (5-40a) becomes

$$T_i^{n+1} = T_{i-1}^n + T_{i+1}^n - T_i^n$$

Successive computations lead to the following values of temperatures:

	Node				
Time	1	2	3	4	5 . . .
0	400	100	100	100	100
$\Delta\tau$	400	400	100	100	100
$2\Delta\tau$	400	100	400	100	
$3\Delta\tau$	400	700	−200	400	
$4\Delta\tau$	400	−500	1300		

From the physics of the problem, we know that the temperature at any point in the slab must be in the range of 100°F to 400°F. Yet, the above table shows that from $\tau = 3\Delta\tau$ onward, we obtain values of temperatures that are in direct violation of the laws of thermodynamics.

Consequently, a negative coefficient of T_i^n would not be permissible if a physically meaningful solution is to be obtained. Such a requirement automatically implies

$$\beta \geqslant 2$$

or

$$\Delta\tau \leqslant \frac{(\Delta x)^2}{2\alpha} \tag{5-42}$$

More sophisticated criteria for stability of numerical solutions of partial differential equations are discussed in References 4 and 5. The simple criterion given above immediately restricts the value of time step, $\Delta\tau$, that can be used once the grid (and, hence, Δx) is fixed. A similar examination of equation (5-41) indicates that the coefficient of T_m^n should be positive. The stability criterion then becomes

$$\left(1 - \frac{2}{\beta} - \frac{2h\Delta\tau}{\rho c \Delta x}\right) \geqslant 1$$

After substituting for β, one obtains

$$\Delta\tau \leqslant \frac{1}{\dfrac{2\alpha}{(\Delta x)^2} + \dfrac{2h}{\rho c \Delta x}} \tag{5-43}$$

When we compare conditions (5-42) and (5-43), we find that the latter would result in a smaller time step. We therefore use condition (5-43) to select the time step $\Delta\tau$.

If $\beta = 2$, i.e., $\alpha\Delta\tau = (\Delta x)^2$, then one obtains from equation (5-40a)

$$T_i^{n+1} = \frac{1}{2}\left(T_{i-1}^n + T_{i+1}^n\right) \tag{5-44}$$

Or, the $(\tau + \Delta\tau)$ temperature at a node is the arithmetic mean of the τ temperatures at the neighboring nodes. Equation (5-44) readily leads to a graphical solution known as a Schmidt plot. With the use of computers, one can generate solutions that are much more accurate than those obtained by Schmidt plots.

Sample Problem 5-4 Set up the nodal equations for the straight rectangular fin assuming a uniform initial temperature, T_o, throughout the fin with the fin-root temperature suddenly raised to T^* at time, $\tau = 0$. The fin surface is losing heat by convection to ambient air at temperature, T_∞.

Solution: The grid layout for a straight fin of thickness, t, is as shown in Figure 5-2. The left side of equation (5-2) represents the net heat transported per unit time into the element, associated with the ith node. It is repeated here for convenience.

$$kt\frac{T_{i+1}^n - T_i^n}{\Delta x} + kt\frac{T_{i-1}^n - T_i^n}{\Delta x} + h\Delta x\left(T_\infty - T_i^n\right) + h\Delta x\left(T_\infty - T_i^n\right) \tag{5-2}$$

This quantity must be equal to the rate of change of internal energy of the element, which is

$$\rho c\,(\Delta x \cdot 1 \cdot t)\frac{T_i^{n+1} - T_i^n}{\Delta\tau}$$

Equating the above two expressions and simplifying, we have

$$\left(T_{i+1}^{n} - 2T_{i}^{n} + T_{i-1}^{n}\right) + \frac{2h(\Delta x)^2}{kt}\left(T_{\infty} - T_{i}^{n}\right) = \frac{\rho c(\Delta x)^2}{k\Delta\tau}\left(T_{i}^{n+1} - T_{i}^{n}\right)$$

or

$$T_{i}^{n+1} = \frac{\alpha\Delta\tau}{(\Delta x)^2}\left(T_{i+1}^{n} + T_{i-1}^{n}\right) + \frac{2h\alpha\Delta\tau}{kt}T_{\infty} + \left(1 - \frac{2\alpha\Delta\tau}{(\Delta x)^2} - \frac{2h\alpha\Delta\tau}{kt}\right)T_{i}^{n} \tag{5-45}$$

The temperatures at node 1 is given by

$$T_{1}^{n+1} = T^{*} \quad \text{for all } \tau > 0 \text{ or for } n = 1, 2, 3 \ldots \tag{5-46}$$

The temperature at node M at time $(\tau + \Delta\tau)$ can be found by equating the net energy transported per unit time into the element associated with the Mth node [the left side of equation (5-5)] with the rate of change of energy of the element.

$$kt\left(\frac{T_{M-1}^{n} - T_{M}^{n}}{\Delta x}\right) + h\frac{\Delta x}{2}\left(T_{\infty} - T_{M}^{n}\right) + h\frac{\Delta x}{2}\left(T_{\infty} - T_{M}^{n}\right) + ht\left(T_{\infty} - T_{M}^{n}\right) = \rho c\left(\frac{1}{2}\Delta x \cdot t \cdot 1\right)\frac{T_{M}^{n+1} - T_{M}^{n}}{\Delta\tau}$$

Rearrangement and simplification yields

$$T_{M}^{n+1} = \left[\frac{2\alpha\Delta\tau}{(\Delta x)^2}\right]T_{M-1}^{n} + \frac{2h\alpha\Delta\tau}{k}\left(\frac{1}{t} + \frac{1}{\Delta x}\right)T_{\infty} + \left[1 - 2\alpha\Delta\tau\left(\frac{1}{(\Delta x)^2} + \frac{h}{kt} + \frac{h}{k\Delta x}\right)\right]T_{M}^{n} \tag{5-47}$$

Equations (5-45) through (5-47), when solved, will yield temperature values at $(\tau + \Delta\tau)$ at the various nodes. Observe that they are algebraic equations but are not simultaneous equations.

To ensure stability of the numerical solution, we once again require that the coefficients of T_i in equation (5-45) and of T_M in equation (5-47) should not be negative. This leads to the following conditions

$$\Delta\tau \left[\frac{2\alpha}{(\Delta x)^2} + \frac{2h\alpha}{kt}\right] \leqslant 1 \tag{5-48}$$

and

$$\Delta\tau \left[\frac{2\alpha}{(\Delta x)^2} + \frac{2h\alpha}{kt} + \frac{2h\alpha}{k\Delta x}\right] \leqslant 1 \tag{5-48a}$$

Of these two requirements, condition (5-48a) is more restrictive than condition (5-48), and therefore we should select a value of $\Delta\tau$ to satisfy condition (5-48a).

5-7.3 Implicit Formulation

The requirement that $\Delta\tau$ should be restricted in size to ensure stability sometimes results in an extremely small time step, of the order of a fraction of a second, especially when transient conduction in multiple layers is involved. The implicit formulation eliminates this restriction, but it involves solving simultaneous equations at each time step. Consider equation (5-40), which is reproduced below.

$$\rho A \Delta x c \left(T_i^{n+1} - T_i^n\right) = kA \frac{T_{i-1}^n - T_i^n}{\Delta x} \Delta\tau + kA \frac{T_{i+1}^n - T_i^n}{\Delta x} \Delta\tau \tag{5-40}$$

The left-hand side represents the change of the internal energy due to the flow of heat associated with the "present" temperature gradients at time, τ, as written on the right-hand side. It is equally plausible that the change in internal energy is due to the flow of heat associated with temperature gradients at time $(\tau + \Delta\tau)$ leading to the following equation.

$$\rho \Delta x c \left(T_i^{n+1} - T_i^n\right) = k \frac{T_{i-1}^{n+1} - T_i^{n+1}}{\Delta x} \Delta\tau + k \frac{T_{i+1}^{n+1} - T_i^{n+1}}{\Delta x} \Delta\tau$$

The above equation contains only one temperature at time τ, that is, T_i^n, whereas all the rest of the terms contain temperatures at time $(\tau + \Delta\tau)$. Therefore, temperatures at $(\tau + \Delta\tau)$ cannot be solved for explicity; instead, one has to solve the equations for all the nodes simultaneously after setting up equations for the internal and the boundary nodes. Such an implicit formulation is stable regardless of the value of the time increment, $\Delta\tau$ that is chosen. An excessively large value of $\Delta\tau$ will cause large errors inherently associated with the finite difference method. For a discussion of stability, convergence, and errors in the implicit and explicit formultions, the reader is referred to References 4, 5, 6, and 7.

5-7.4 Two- and Three-Dimensional Systems

The rectangular fin and the slab of finite width, L, were both one-dimensional transient problems. Transient temperature distributions in two-dimensional systems such as axisymmetric bodies or rectangular plates are of importance for two major reasons. First, they determine the warm-up period—for example, the time of residence of a metal billet in a furnace before it is put through a rolling mill. Second, during the unsteady state, large thermal stresses are frequently created in the body, and the transient stress distribution is of particular importance from a structural point of view. Sometimes the transients arise due to the time-dependent nature of the boundary conditions as in the case of the reentry of a space vehicle.

The basis for formulating the nodal equations is once again the principle of energy conservation. In Section 5-4, the net rate of energy arriving at an internal or a boundary node was set equal to zero. Now, we set it equal to the rate of change of internal energy of the element, that is (see Figure 5-17)

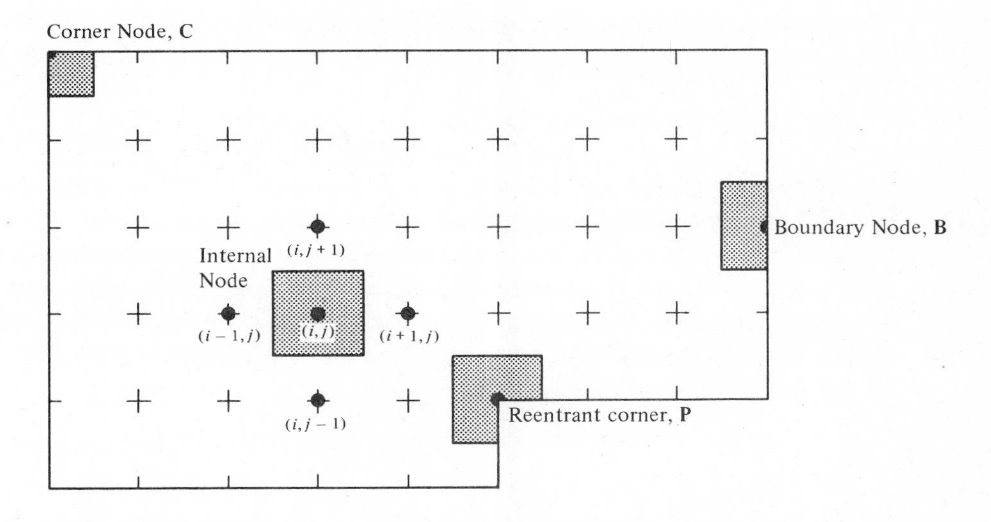

Figure 5-17 Possible Nodal Locations for a Two-Dimensional Problem.

$$\rho c\,(\Delta x \cdot \Delta y \cdot 1)\,\frac{T_{i,j}^{n+1} - T_{i,j}^{n}}{\Delta\tau} \quad \text{for an internal node} \tag{5-50}$$

$$\rho c\left(\frac{1}{2}\Delta x\cdot\Delta y\cdot 1\right)\frac{T_{\mathrm{B}}^{n+1}-T_{\mathrm{B}}^{n}}{\Delta\tau}\quad\text{for a boundary node}\tag{5-50a}$$

$$\rho c\left(\frac{1}{4}\Delta x\cdot\Delta y\cdot 1\right)\frac{T_{\mathrm{C}}^{n+1}-T_{\mathrm{C}}^{n}}{\Delta\tau}\quad\text{for a node at an external corner}\tag{5-50b}$$

and

$$\rho c\left(\frac{1}{2}\Delta x\Delta y\cdot 1+\frac{1}{4}\Delta x\Delta y\cdot 1\right)\frac{T_{\mathrm{P}}^{n+1}-T_{\mathrm{P}}^{n}}{\Delta\tau}$$

$$\text{for a node at a reentrant corner}\tag{5-50c}$$

In the above

$T_{i,j}^{n+1}$ is temperature at node (i, j) at time $(\tau + \Delta\tau)$

$T_{i,j}^{n}$ is temperature at node (i, j) at time τ

B is a point on the boundary, excluding a corner.

C is a point at an exterior corner.

P is a point at a reentrant corner.

Expression (5-50) and the left side of equation (5-23) must be equal in order to fulfill conservation of energy at an internal node. Hence

$$\rho c\Delta x\Delta y\frac{T_{i,j}^{n+1}-T_{i,j}^{n}}{\Delta\tau}=k\left(\Delta y\frac{T_{i-1,j}^{n}-T_{i,j}^{n}}{\Delta x}+\Delta y\frac{T_{i+1,j}^{n}-T_{i,j}^{n}}{\Delta x}\right.$$
$$\left.+\Delta x\frac{T_{i,j-1}^{n}-T_{i,j}^{n}}{\Delta y}+\Delta x\frac{T_{i,j+1}^{n}-T_{i,j}^{n}}{\Delta y}\right)$$

We observe that the above formulation is in the explicit form. Introducing $\lambda = \Delta y/\Delta x$ and simplifying, one obtains

$$T_{i,j}^{n+1}=\frac{\alpha\Delta\tau}{(\Delta x)^2}\left[T_{i-1,j}^{n}-2T_{i,j}^{n}+T_{i+1,j}^{n}\right.$$
$$\left.+\frac{1}{\lambda^2}\left(T_{i,j-1}^{n}-2T_{i,j}^{n}+T_{i,j+1}^{n}\right)\right]+T_{i,j}^{n}$$

or

$$T_{i,j}^{n+1} = \frac{\alpha \Delta \tau}{(\Delta x)^2} \left[T_{i-1,j}^{n} + T_{i+1,j}^{n} + \frac{1}{\lambda^2} \left(T_{i,j-1}^{n} \right) + \frac{1}{\lambda^2} \left(T_{i,j+1}^{n} \right) \right] + \left[1 - \frac{2\alpha \Delta \tau}{(\Delta x)^2} \left(1 + \frac{1}{\lambda^2} \right) \right] T_{i,j}^{n} \tag{5-51}$$

When $\Delta y = \Delta x$, or $\lambda = 1$, the above equation becomes

$$T_{i,j}^{n+1} = \frac{\alpha \Delta \tau}{(\Delta x)^2} \left(T_{i-1,j}^{n} + T_{i+1,j}^{n} + T_{i,j-1}^{n} + T_{i,j+1}^{n} \right) + \left(1 - \frac{4\alpha \Delta \tau}{(\Delta x)^2} \right) T_{i,j}^{n} \tag{5-51a}$$

Once again, we find from the forgoing equation that if the coefficient of $T_{i,j}^{n}$ is negative, there is a possibility that the laws of thermodynamics could be violated, resulting in an instability of the solution. Therefore, we require that for the numerical solution to be stable, the coefficient of $T_{i,j}^{n}$ must be positive. Hence, a simple condition for the stability of a numerical solution for a two-dimensional transient problem with $\Delta x = \Delta y$ is that

$$\left[\frac{4\alpha \Delta \tau}{(\Delta x)^2} \right] \leqslant 1 \quad \text{or} \quad \Delta \tau \leqslant \frac{(\Delta x)^2}{4\alpha} \tag{5-52}$$

It can likewise be shown that the condition for stability for a three-dimensional transient problem with $\Delta x = \Delta y = \Delta z$ is

$$\left[\frac{6\alpha \Delta \tau}{(\Delta x)^2} \right] \leqslant 1 \quad \text{or} \quad \Delta \tau \leqslant \frac{(\Delta x)^2}{6\alpha} \tag{5-52a}$$

For a one-dimensional system, it may be recalled that

$$\left[\frac{2\alpha \Delta \tau}{(\Delta x)^2} \right] \leqslant 1 \quad \text{or} \quad \Delta \tau \leqslant \frac{(\Delta x)^2}{2\alpha} \tag{5-52b}$$

was the criterion for stability.

When writing equations for the boundary nodes, we use the expressions on the left side of equations (5-25) through (5-30) to represent the net rate of inflow of energy into the boundary element. We equate these to the corresponding expressions from (5-50a, b, and c) for the change of internal energy. For example, if we had a convective boundary condition as in Figure 5-9(c), we would use expression (5-50a), substitute the subscripts (M, j) for **B**, and equate it to the left side of equation (5-27). Thus

$$\rho c\left(\frac{1}{2}\Delta x\Delta y\right)\frac{T_{M,j}^{n+1}-T_{M,j}^{n}}{\Delta\tau}=k\frac{1}{2}\Delta x\frac{T_{M,j+1}^{n}-T_{M,j}^{n}}{\Delta y}+k\frac{1}{2}\Delta x\cdot\frac{T_{M,j-1}^{n}-T_{M,j}^{n}}{\Delta y}+k\Delta y\frac{T_{M-1,j}^{n}-T_{M,j}^{n}}{\Delta x}+h\Delta y\left(T_{\infty}-T_{M,j}^{n}\right)$$

Solving for $T_{M,j}^{n+1}$, one has

$$T_{M,j}^{n+1}=\frac{2\alpha\Delta\tau}{(\Delta x)^2}\left[\frac{1}{2\lambda^2}\left(T_{M,j+1}^{n}-2T_{M,j}^{n}+T_{M,j-1}^{n}\right)+\left(T_{M-1,j}^{n}-T_{M,j}^{n}\right)+\frac{h\Delta x}{k\lambda^2}\left(T_{\infty}-T_{M,j}^{n}\right)\right] \tag{5-53}$$

For $\Delta x = \Delta y$, the above simplifies to

$$T_{M,j}^{n+1}=\frac{2\alpha\Delta\tau}{(\Delta x)^2}\left(\frac{1}{2}T_{M,j+1}^{n}+\frac{1}{2}T_{M,j-1}^{n}+T_{M-1,j}^{n}+\frac{h\Delta x}{k}T_{\infty}\right)+\left[1-\frac{2\alpha\Delta\tau}{(\Delta x)^2}\left(2+\frac{h\Delta x}{k}\right)\right]T_{M,j}^{n} \tag{5-53a}$$

The criterion for stability of the above equation is that the coefficient of $T_{M,j}^{n}$ on the right side of the equation must be nonnegative. If, for simplicity, Δx and Δy are assumed to be equal, one requires

$$\Delta\tau \leqslant \frac{(\Delta x)^2}{4\alpha\left(1+\dfrac{h\Delta x}{2k}\right)} \tag{5-54}$$

Condition (5-54) gives a value of $\Delta\tau$ that is smaller than that obtained from condition (5-52). For a corner node with a convective boundary, the requirement becomes

$$\Delta\tau \leqslant \frac{(\Delta x)^2}{4\alpha\left(1+\dfrac{h\Delta x}{k}\right)} \tag{5-54a}$$

It should be noted that if the dimensionless group $(h\Delta x/k)$ happens to be very large, due to a large value of h and/or a small value of k, a very small value of the time step must be used to ensure stability.

PROBLEMS (English Engineering System of Units)

5-1. For a certain one-dimensional steady-state conduction problem, the following equation governs the temperature distribution:

$$\frac{d}{dx}\left(k\frac{dT}{dx}\right) = 0$$

For a certain material, the thermal conductivity, k, is given by the equation

$$k = 120 + 0.05T + 0.0005T^2 \text{ (Btu/hr-ft°F)}$$

where T is the temperature in °F. Solve the governing differential equation analytically, and obtain the solution in the form $x = f(T)$ if at $x = 0$, $T =$ 400°F and at $x = 1$ ft, $T = 100$°F. Then, using a digital computer, determine T as a function of x, and hence determine the rate of heat flow.

5-2. A 3/8 inch diameter steel rod is 24 inches long and is maintained at 400°F at one end and at 250°F at the other end. The thermal conductivity of steel is 31 Btu/hr-ft°F and the heat transfer coefficient between the surface of the rod and the surroundings is 3.2 Btu/hr-ft²°F. Using a nodal spacing of 1 inch, determine (a) the temperature distribution in the rod, (b) the temperature gradient at each end of the rod, that is

$$\left(\frac{dT}{dx}\right)_{x=0\text{ in.}} \quad \text{and} \quad \left(\frac{dT}{dx}\right)_{x=24\text{ in.}}$$

graphically and numerically, and (c) heat lost by the rod per unit time. The surroundings are at 70°F.

5-3. A straight trapezoidal fin has a thickness of 0.25 inch at the base and of 0.05 inch at the tip. The base temperature is maintained at 600°F, and the tip may be considered to be insulated. The length of the fin is 1.5 inches. If the convective heat transfer coefficient is 4.5 Btu/hr-ft²°F, determine the temperature distribution in the fin and the rate of heat dissipation from the fin. The thermal conductivity may be taken as 50 Btu/hr-ft°F. The surrounding temperature is 100°F.

5-4. Various materials are being considered for a rectangular fin of 0.25-inch thickness and 2.25-inch length. Determine numerically the ratio (k/hL) for which the variation in temperature along the fin will be less than 10 percent of $(T_o - T_\infty)$ where T_o and T_∞ are the base temperature of the fin and the surrounding fluid temperature, respectively. *Hint:* Use a dimensionless temperature, θ defined by $\theta = (T - T_\infty)/(T_o - T_\infty)$

5-5. A rectangular aluminum fin of 0.25-inch thickness and having a conductivity of 95 Btu/hr-ft°F has its base temperature maintained at 300°F. The ambient temperature is 60°F and the convective heat transfer coefficient is 8 Btu/hr-ft^2°F. Construct a two-dimensional grid, and set up appropriate nodal equations. By varying the length of the fin from 0.5 inch to 2.5 inches in steps of 0.5 inch, compute the transverse temperature distribution for various values of the x coordinate, and hence, determine the limiting length below which the one-dimensional approach will cause more than 1 percent deviation in the temperature distribution at $x = L/2$. What is the heat dissipation rate in each case?

5-6. Hot gases at 450°F pass through a chimney made of common brick whose cross section is shown in EFigure 5-6. The outside surface is maintained at 105°F. The inside h value is 6.8 Btu/hr-ft^2°F. Determine the temperature distribution in the chimney. Heat conduction along the length of the chimney may be neglected. Assume k = 0.25 Btu/hr-ft°F.

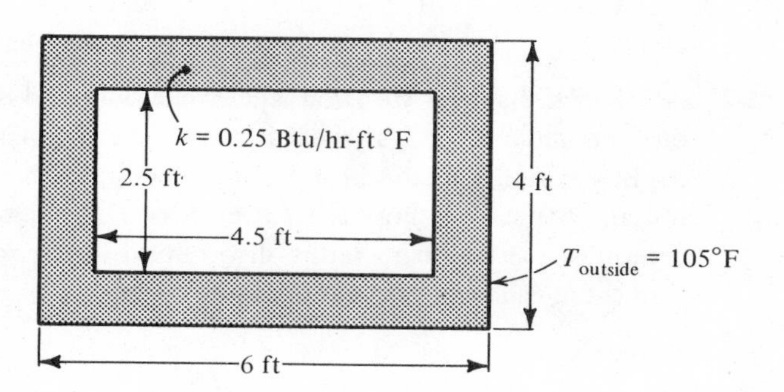

EFigure 5-6

5-7. Consider a two dimensional configuration as shown in EFigure 5-7.

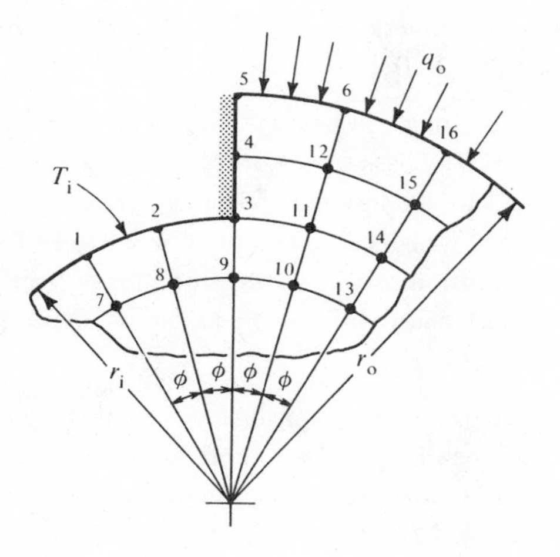

EFigure 5-7

The radial and the angular spacing is uniform. The boundary at $r = r_i$ is maintained at a uniform temperature of T_i. The boundary at $r = r_o$ receives a uniform heat flux of q_o Btu/hr-ft^2. The boundary on which the nodes 3, 4, and 5 are located is insulated. Perform an energy balance, and write the nodal equations for the nodes 11, 5, 3, and 2. Assume steady-state conditions.

5-8. An aluminum rod, 3/8 inch in diameter and 24 inches in length, is initially at 70°F. The surface of the rod is insulated. At time $\tau = 0$, the end temperatures of the rod are raised to 300°F. Determine numerically the time it will take for the temperature at $x = 12$ inches to reach within 2 percent of the steady-state temperature. $\alpha = 2.8$ ft^2/hr.

5-9. A brick wall is 9 inches thick. The convective heat transfer coefficients on the inside and the outside are 1.0 and 4.0 Btu/hr-ft^2°F, respectively. The outside ambient temperature varies as $T_\infty = 50 + 30 \sin (0.2618\tau)$, where T_∞ is in °F and τ is in hours. The air on the inside is maintained at a constant 68°F. Using numerical techniques, determine the temperature variation at $x = 1$ inch, $x =$ 4.5 inches, and at $x = 9$ inches for the first six hours. Assume $\alpha = 0.02$ ft^2/hr.

5-10. A current of 200 amperes passes through a steel wire of 0.1 inches in diameter and 1 ft in length. The resistivity of the steel is 65×10^{-6} ohm-cm. The outer surface is maintained at 350°F. Calculate analytically the temperature at the center of the wire and the heat transfer coefficient if $T_\infty = 70$°F and $k = 11$ Btu/hr-ft°F. Now, if the current is switched off, determine the time it will take for the surface temperature to drop to 100°F. Obtain a numerical solution if $\alpha = 0.204$ ft^2/hr. Assume h = 5.0 Btu/hr-ft^2°F.

5-11. Set up a nodal equation for a node on the common boundary between two materials constituting a composite material to determine the temperature at $(\tau + \Delta\tau)$ in terms of temperatures at time τ.

5-12. If the temperature at the surface of the wire of problem 5-10 is maintained at 350°F after the current is switched off, estimate the time required for the wire to attain steady state.

5-13. If the chimney wall of problem 5-6 is initially at 105°F and if the hot gases at 450°F start to flow through the chimney at $\tau = 0$, determine the largest time step that can be used without causing instability. Assume $\Delta x = 1$ inch, $\Delta y =$ 0.75 inch, $h_{inside} = 0.8$ Btu/hr-ft^2°F, $T_{outside} = 105$°F, and $\alpha = 0.02$ft^2/hr.

5-14. A 4000 pound vehicle is brought to rest from 60 mph in 300 ft by applying the brakes. The brakes are of the drum type, with brake-shoe dimensions of 3/8-inch thickness, 2.5-inch width, and 9-inch length. Assume that the entire kinetic energy is dissipated uniformly through the eight brake shoes. If the brake shoes are initially at 90°F and the film coefficient is 7 Btu/hr-ft^2°F, determine numerically the temperature-time history at $x = 0$, $x = 3/16$ inch, and $x = 3/8$ inch. Assume $k = 8$ Btu/hr-ft°F. Neglect the curvature of the brake shoes. Assume the ambient temperature to be 40°F and the thermal diffusivity to be 0.78 ft^2/hr. Also assume that the kinetic energy is dissipated at a uniform rate. Use a one-dimensional approach.

5-15. Consider the grid work in EFigure 5-15 to be used for a finite difference solution to a heat conduction problem. The following may be assumed.

$$k = 90.0 \text{ Btu/hr-ft}^\circ\text{F}$$

$$\text{Thickness} = 2 \text{ inches}$$

$$T_\infty = 210^\circ\text{F}$$

$$\rho = 250 \text{ lb}_m/\text{ft}^3$$

$$c_p = 0.1 \text{ Btu/lb}_m{}^\circ\text{F}$$

$$h = 20 \text{ Btu/hr-ft}^2{}^\circ\text{F}$$

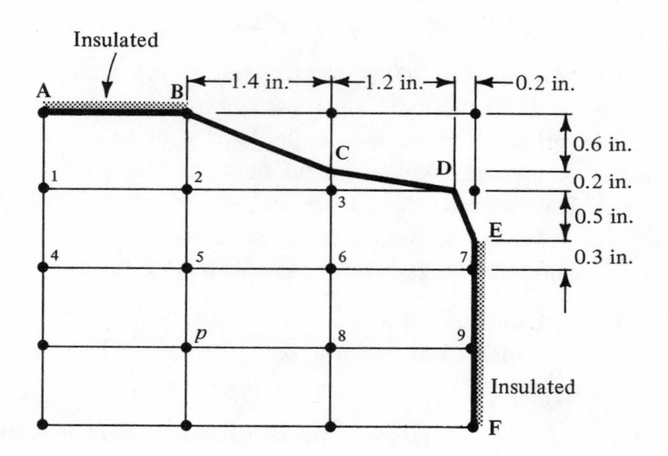

EFigure 5-15

There is no internal heat generation in the body. Estimate the maximum value of the time step, $\Delta\tau$, that can be used. Write equations for each of the following nodes on the basis of conservation of energy: (a) node 5, (b) node 2, (c) node 3, (d) node 7, by considering unsteady state.

PROBLEMS (SI System of Units)

5-1. The thermal conductivity of a plane wall of 25-cm thickness is given by

$$k = 73 + 0.08T + 0.0009T^2 \text{ (W/m}^\circ\text{K)}$$

where T is the temperature in °C. Assuming one-dimensional flow and following the procedure outlined in the text, write nodal equations for a numerical solution. If at $x = 0$, $T = 300^\circ$C and at $x = 25$ cm, $T = 50^\circ$C, using a digital computer, obtain the values of temperatures at various nodes. Also, determine the rate of heat flow.

5-2. An aluminum rod is 15 mm in diameter and is 40 cm long. The two ends of the rod are maintained at 200°C and 80°C, respectively. The convective heat transfer coefficient is 30 W/m2 °K and the thermal conductivity for aluminum may be taken as 204 W/m°K. Using a total of 20 nodal points, determine (a) the temperature distribution in the rod, and (b) the rate of heat loss from the rod. Assume T_∞ = 20°C.

5-3. A straight trapezoidal fin has a thickness of 6 mm at the base and of 1 mm at the tip. The length of the fin is 8 cm. The base temperature is maintained at 350°C, and the tip may be considered to be insulated. The thermal conductivity may be taken as 75 W/m°K. If the convective heat transfer coefficient is 25 W/m2 °K, determine the temperature distribution in the fin and the rate of heat dissipation from the fin. Assume the ambient temperature to be 20°C.

5-4. Determine numerically the ratio (k/hL) for which the variation in temperature along a rectangular fin will be less than 10 percent of $(T_o - T_\infty)$. T_o and T_∞ are the fin base temperature and the ambient temperature, respectively. Assume (L/t) = 10. [*Hint:* Use a dimensionless temperature, θ, defined by $\theta = (T - T_\infty)/(T_o - T_\infty)$]

5-5. A rectangular aluminum fin of 10-mm thickness and having conductivity of 150 W/m°K has its base temperature maintained at 100°C. The ambient temperature is 0°C and the convective heat transfer coefficient is 50 W/m2 °K. Construct a two-dimensional grid and set up appropriate nodal equations. By varying the length of the fin from 3 cm to 15 cm in steps of 2 cm, compute the transverse temperature distribution for various values of x coordinates and hence, determine the limiting length below which the one-dimensional approach will cause more than 1 percent deviation in the temperature distribution at $x = L/2$. What is the heat dissipation rate in each case?

5-6. The cross section of a chimney made of common brick is shown in SI Figure 5-6. The temperature of the hot gases passing through it is 225°C. The inside and the outside surfaces have h values of 60 W/m2 °K and 20 W/m2 °K, respectively. Determine the temperature distribution in the chimney. Heat conduction along the length of the chimney may be neglected. The ambient temperature on the outside is 25°C. The thermal conductivity of brick may be taken as 0.43 W/m°K.

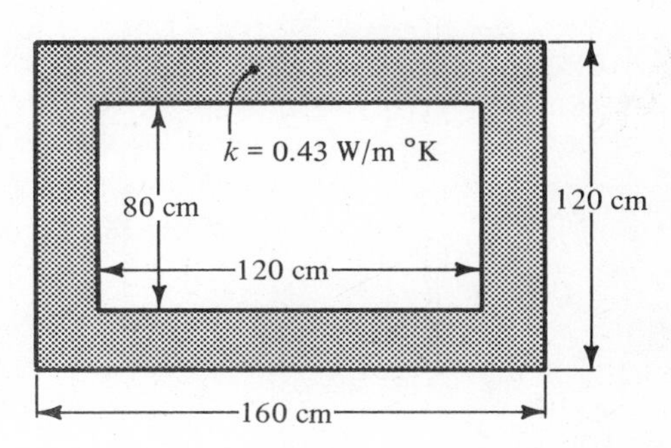

SIFigure 5-6

5-7. Consider a two-dimensional configuration as shown in SI Figure 5-7.

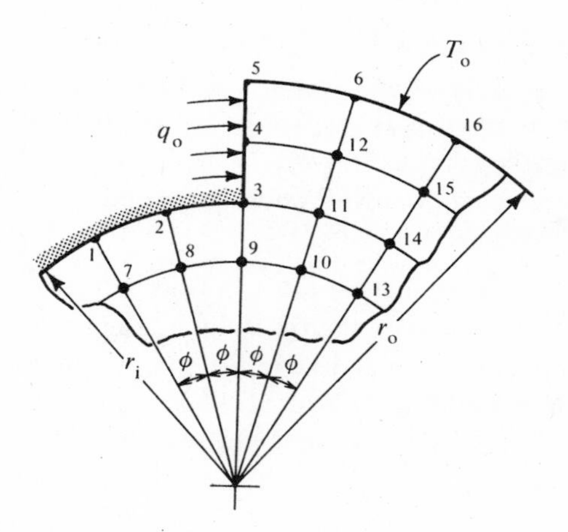

SIFigure 5-7

The radial and the angular spacing is uniform. The boundary at $r = r_o$ is maintained at a uniform temperature of T_o. The boundary on which the nodes 3, 4, and 5 are located receives a uniform heat flux of q_o W/m². The boundary at $r = r_i$ is insulated. Perform an energy balance, and write nodal equations for the nodes 11, 5, 3, and 2. Assume steady-state conditions.

5-8. An aluminum rod, 15mm in diameter and 40cm long, is initially at 15°C. At time $\tau = 0$, the end temperatures of the rod are raised to 150°C. Determine numerically the time it will take for the temperature at $x = 20$ cm to reach a temperature that is within 2 percent of the steady-state temperature distribution if $\alpha = 0.28$ m²/hr. Assume the surface of the rod to be insulated.

5-9. A brick wall is 20 cm thick. The convective heat transfer coefficients on the inside and the outside are 6.0 and 25.0 W/m² °K, respectively. The outside ambient temperature varies as $T_\infty = 5 + 10 \sin(0.2618\tau)$, where T_∞ is in °C and τ is in hours. The inside air temperature is maintained at a constant 22°C. Using numerical techniques, determine the temperature variation at $x = 3$ cm, $x = 11$ cm, and $x = 20$ cm for the first 6 hours. Assume $\alpha = 3.17 \times 10^{-3}$ m²/hr.

5-10. A current of 200 amperes passes through a steel wire 3 mm in diameter and 10 cm in length. The resistivity of the steel is 65×10^{-6} ohm-cm. The outer surface is maintained at 170°C. Calculate analytically the temperature at the center of the wire and the heat transfer coefficient if $T_\infty = 20$°C and if $k = 17$ W/m°K. Now, if the current is switched off, determine the time it will take for the surface temperature to drop to 70°C. Obtain a numerical solution. The thermal diffusity of steel may be taken as 0.036 m²/hr. Assume $h = 15$ W/m²°K.

5-11. Set up a nodal equation for a node on the common boundary between two materials constituting a composite material to determine the temperature at $(\tau + \Delta\tau)$ in terms of temperatures at time τ.

5-12. If the temperature at the surface of the wire of problem 5-10 is maintained at 70°C after the current is switched off, estimate the time required for the wire to attain steady state.

5-13. If the chimney wall of problem 5-6 is initially at 30°C and if the hot gases at 180°C start to flow through the chimney at $\tau = 0$, determine the largest time step that can be used without causing instability. Assume $\Delta x = 4$ cm, $\Delta y = 2$ cm, $T_{ambient} = 30°C$, and $\alpha = 4.0 \times 10^{-3} m^2/hr$.

5-14. Consider the grid work in SI Figure 5-14 for a finite difference solution to a heat conduction problem. The following may be assumed.

$$k = 500 \text{ W/m°K}$$

$$\text{Thickness} = 5 \text{ cm}$$

$$T_\infty = 95°C$$

$$\rho = 4000 \text{ kg/m}^3$$

$$c_p = 400 \text{ J/kg°K}$$

$$h = 25 \text{ W/m}^2\text{°K}$$

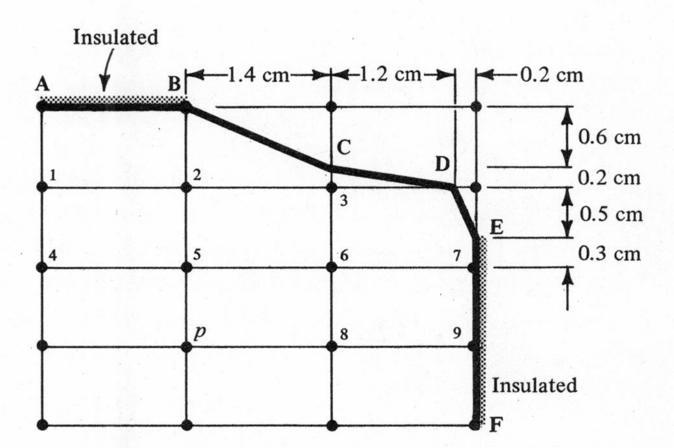

SIFigure 5-14

There is no internal heat generation in the body. Estimate the maximum value of the time step, $\Delta\tau$, that can be used. Write equations for each of the following nodes on the basis of conservation of energy (a) node 5, (b) node 2, (c) node 3, (d) node 7, by considering unsteady state.

REFERENCES

[1] Southwell, R. V., *Relaxation Methods in Engineering Science,* Oxford University Press, (1940).

[2] Dusinberre, G. M., *Numerical Analysis of Heat Flow*, McGraw-Hill Book Company, Inc., (1949).

[3] Kreith, F., *Principles of Heat Transfer*, 3rd ed., Intext Educational Publishers, New York, (1973).

[4] Leppert, G., "A Stable Numerical Solution for Transient Heat Flow Problems," *J. Am. Soc. Naval Engrs.,* **65**, (1973), p. 741.

[5] Fowler, C. M., "Analysis of Numerical Solutions of Transient Heat Flow Problems," *Quart. Appl. Math.,* 3, (1945), p. 361.

[6] Hidebrand, F. B., *Introduction to Numerical Analysis*, McGraw-Hill Book Company, Inc., (1956).

[7] Dusinberre, G. M., *Heat Transfer Calculations by Finite Differences*, International Textbook Company, Scranton, Pennsylvania (1961).

6

Thermal Radiation

6-1 INTRODUCTION

The transmission of heat by conduction and convection requires that matter be present to act as a transport vehicle for the process to occur. For conduction in solids, the lattice atom vibrations and the movement of free electrons account for the transfer of energy. For convection, a combination of conduction and fluid motion is responsible for the transfer of heat. However, an intermediary is not necessary for a surface to transmit heat to another surface by radiation. This is true since thermal radiation is electromagnetic radiation that is emitted solely as a result of the temperature of a surface. Consequently, its nature is the same as that of X-rays, visible light, and radio waves. The distinguishing feature of thermal radiation is that it possesses wavelengths between 0.1 and 100 microns (1 micron equals 10^{-6} meters). The wavelength distribution of the complete radiation spectrum is shown in Figure 6-1.

This chapter is divided into two distinct parts. Sections 6-2 to 6-7 consider a simplified model for the analysis of radiant energy exchange that enables the solution of a wide range of engineering problems. Sections 6-8 to 6-11 discuss various refinements that involve a deeper knowledge of the character of the radiation. For example, in the second part, we will answer such questions as why certain objects get warmer than others when placed in sunlight, and why a greenhouse stays warm on a cold, sunny, winter day.

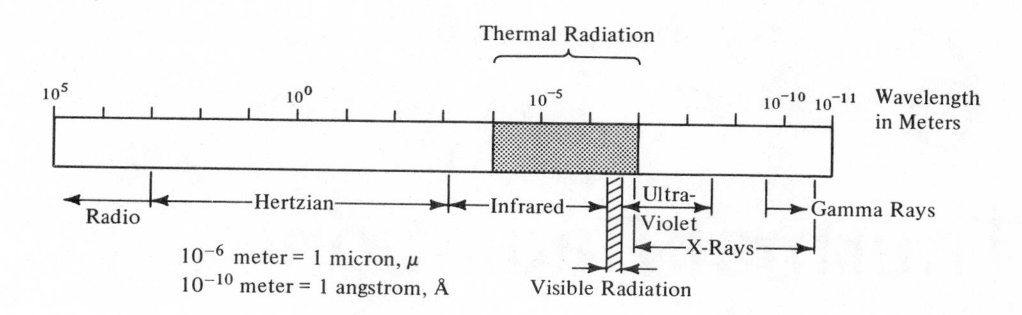

Figure 6-1 Wavelength distribution of electromagnetic radiation spectrum.

6-2 IDEALIZED RADIATION HEAT TRANSFER CALCULATIONS

There are certain fundamentals that we must know before beginning an analysis. They are:

(1) The Stefan-Boltzmann law and the Blackbody,
(2) Basic radiation properties of surfaces, and
(3) Radiation shape factors and their relationships.

The Stefan-Boltzmann law will permit us to calculate the energy radiated by a blackbody. This knowledge, when combined with information about the basic radiation properties of surfaces and shape factors, will allow us to calculate the net radiant heat transfer from the surfaces under somewhat idealized conditions. These concepts will now be discussed.

6-2.1 The Stefan-Boltzmann Law and the Blackbody

Solids, liquids, and some gases (especially water vapor and hydrocarbons) emit thermal radiation as a result of their temperature. An ideal emitter, called a *blackbody* or a *black surface*, emits thermal radiation according to the Stefan-Boltzmann equation

$$e_b = \sigma T^4 \tag{6-1}$$

In the above, e_b represents the *emissive power* of the blackbody, which is the total energy emitted per unit area and time, σ is the Stefan-Boltzmann constant, and T is the temperature in degrees absolute. The distinguishing feature of a blackbody is that at any temperature, T, its emissive power is the maximum emissive power of any body.

Values of the Stefan-Boltzmann constant in different systems of units are

$$\sigma = 0.1713 \times 10^{-8}\ \text{Btu/hr-ft}^2\ {}^\circ\text{R}^4$$
$$\sigma = 5.66 \times 10^{-8}\ \text{W/m}^2\ {}^\circ\text{K}^4$$

The two temperature scales that will be used are the Rankine (°R) and the Kelvin (°K) scales. We should note that

$$\left.\begin{aligned} {}^\circ\text{R} &= {}^\circ\text{F} + 459.7 \\ &\text{and} \\ {}^\circ\text{K} &= {}^\circ\text{C} + 273.17 \end{aligned}\right\} \tag{6-2}$$

where °F and °C represent degrees Fahrenheit and Celsius, respectively.

Although a true blackbody is rarely encountered in nature, the concept of a blackbody provides a very useful reference against which we can compare the radiant emission of real bodies. The opening of a peephole on the side of a large commercial boiler, usually used for visually inspecting the combustion process within the boiler, is a very good approximation of a black surface.

The emissive power, e, of a nonblack surface at a temperature, T, is the energy emitted by it per unit time and area. It is expressed in terms of the emissive power of a blackbody at the same temperature, T, and is given by

$$e = \epsilon e_b = \epsilon \sigma T^4 \tag{6-3}$$

where ϵ is the *emissivity* of the surface and ranges from zero to unity. The emissivity of a surface usually depends on the temperature and the nature of the surface.

6-2.2 Basic Radiation Properties

Consider radiant energy, G, impingent on a surface as shown in Figure 6-2.

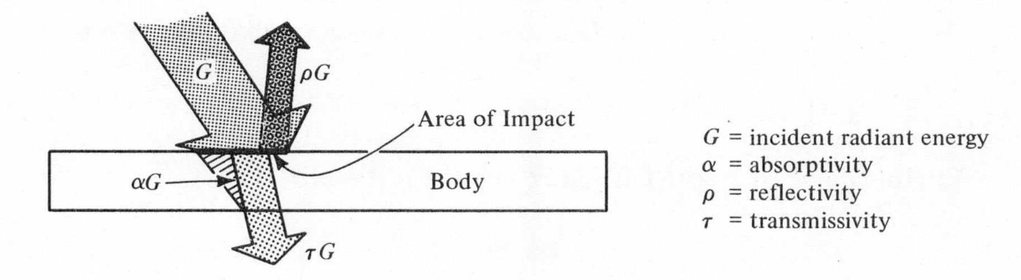

Figure 6-2 Radiant energy impingent upon a surface.

Let us define α, the *absorptivity*, as the fraction of the incident radiation that is absorbed by the material; ρ, the *reflectivity*, as the fraction of the incident radiation that is reflected by the material; and τ, the *transmissivity*, as the fraction of the incident radiation that is transmitted through the material.

The reflected radiation may be either diffuse, specular, or a combination of the two. If it is all diffuse, it will be uniformly distributed over a hemisphere above the area of impingement. If it is all specular, the angle made by the reflected beam with the normal to the surface will equal the angle made by the incident beam with the normal. A common household mirror reflects the incident beam of light in a nearly specular fashion, while a rough metal surface reflects in a more diffuse fashion. A perfect diffuse reflector will appear equally bright regardless of the direction from which it is viewed. In reality, most surfaces exhibit a mixture of specular and diffuse reflection.

From an energy balance (Figure 6-2), we can derive a relationship among basic radiation properties.

$$\text{energy coming in} = \text{energy leaving} + \text{energy absorbed}$$

$$G = (\rho G + \tau G) + \alpha G$$

or

$$\alpha + \rho + \tau = 1 \tag{6-4}$$

For opaque materials, $\tau = 0$, and $\alpha + \rho = 1$; whereas for most gases (other than water vapor, sulfur dioxide, ammonia, and hydrocarbons), $\alpha = 0$, $\rho = 0$, and $\tau = 1$. For a blackbody, $\tau = 0$, $\rho = 0$, and $\alpha = 1$. In general, the absorptivity, the reflectivity, and the transmissivity of a body are dependent on the temperatures of the source radiation and the nature of the surface.

Before beginning our analysis, it is necessary to relate the quantities α, τ, and ρ to the emissivity of a surface. To do this, we perform a thought experiment. Consider an evacuated enclosure whose walls are maintained at a fixed constant temperature and are black. The enclosure contains within it a body that we will refer to as body 1, which is held at the same temperature as the walls. Figure 6-3 is a sketch of the hypothetical system used in our thought experiment.

Let G_1 be the radiant energy incident on surface 1 per unit area and time from the black enclosure. If body 1 has an absorptivity α_1, then

$$\alpha_1 G_1 = \text{radiant energy } \textit{absorbed} \text{ by surface 1 per unit area and time.}$$

Furthermore, if body 1 is black ($\alpha_1 = 1$), it follows that

$$G_1 = \text{radiant energy } \textit{absorbed} \text{ by surface 1 per unit area and time.}$$

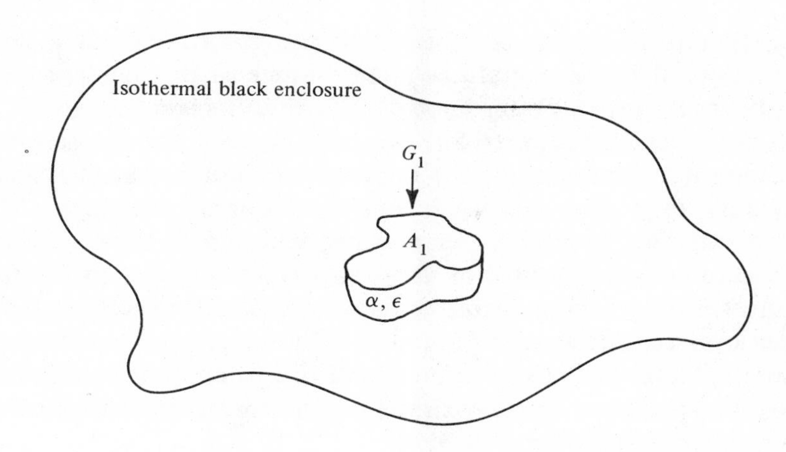

Figure 6-3 System used in thought experiment to relate α to ϵ.

Also, for the case where body 1 is black,

e_{b1} = radiant energy *emitted* by surface 1 per unit area and time.

If there are no sources or sinks within body 1 and if its temperature is steady, then the absorbed and emitted energies are in balance so that

$$G_1 = e_{b1} \tag{6-5}$$

If surface one were replaced by a nonblack body, then

$\alpha_1 G_1$ = radiant energy *absorbed* by surface 1 per unit area and time.

Since the emissive power of a real body is its emissivity times the emissive power of a blackbody at the same temperature, we have

$\epsilon_1 e_{b1}$ = radiant energy *emitted* by surface 1 per unit area and time.

For thermal equilibrium of body 1, we must have

$$\alpha_1 G_1 = \epsilon_1 e_{b1} \tag{6-5a}$$

Combining equations (6-5) and (6-5a), we conclude that

$$\alpha_1 = \epsilon_1 \tag{6-5b}$$

Therefore, it may be stated that the absorptivity of a body is equal to its emissivity if the temperatures of the source of the incident radiation and of the body are the same. This is known as Kirchhoff's law.

Although equation (6-5b) has been derived for the case where the emitting surface and the source of the incident radiation are at the same temperature, the equality of α and ϵ is used for more general situations. A body that obeys Kirchoff's law is referred to as a *graybody*. Such a body is one for which the radiation properties do not vary with wavelengths. Strictly speaking, there are almost no graybodies encountered in practice. However, a great many bodies are gray over certain wavelength ranges. If a body is gray to the major portion of the radiation that is incident upon it and that is emitted by it, then Kirchhoff's law may be applied as an approximation. The emissivities of various surfaces are given in the Appendix.

Although most bodies are not gray, many sound engineering calculations are made utilizing the gray body concept. That is, integrated averages are obtained for the radiation properties over all wavelengths, and the body is assumed to be gray. Thus, the body takes on the average values of α, τ, ρ, and ϵ. The radiation problem is then solved using these values in the manner presented in Section 6-4. Such an approach is similar to that of selecting an average thermal conductivity in a given temperature interval and working a conduction problem as though the material had a constant thermal conductivity.

When we want to apply the graybody analysis to a given body, it is necessary to compare its absorptivity in the wavelength band of the incident radiation with its emissivity in the wavelength band of the radiant energy which the body emits. If they are approximately equal, the graybody approach is applicable; if they are markedly different, it cannot be used. As an example, white paint has a very low absorptivity for incident solar energy, yet has a high emissivity in the infrared region, which is the portion of the spectrum in which it emits radiant energy. Consequently, it is not possible to apply the graybody assumption for such a condition.

6-2.3 Radiation Shape Factors and Their Relationships

Consider an enclosure of N surfaces, each maintained at a different temperature (Figure 6-4). The enclosure is filled with a transparent gas that does not participate in the radiation transfer process. We assume that the convection effects are negligible. It may be observed that each surface exchanges radiant energy with the remaining $(N - 1)$ surfaces. If we consider the radiant energy leaving any one of the surfaces, we conclude that different fractions of this energy strike various surfaces.

The radiation *shape factor* (also called angle factor or configuration factor) is designated as $F_{m\text{-}n}$ and is interpreted to mean the fraction of the energy leaving surface m and heading toward surface n. To take a specific example, $F_{1\text{-}2}$ represents the fraction of energy leaving surface 1 and heading toward surface 2.

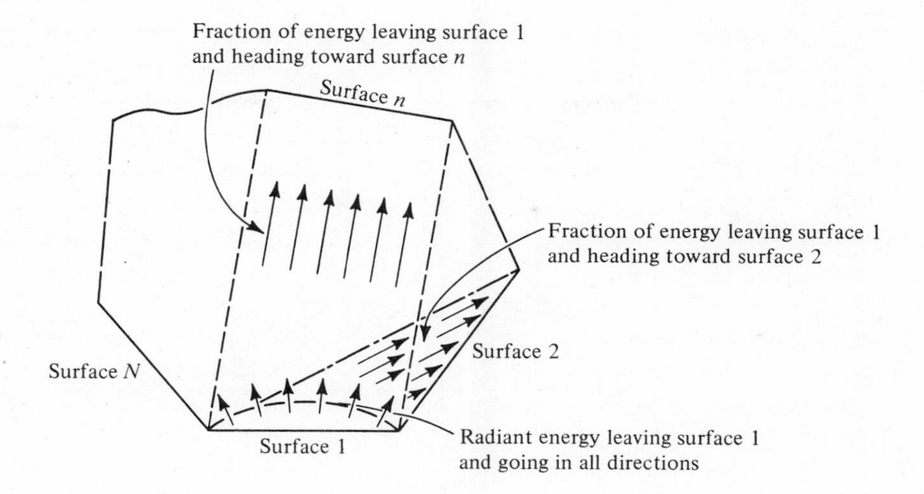

Figure 6-4 A schematic diagram of an enclosure for describing shape factors.

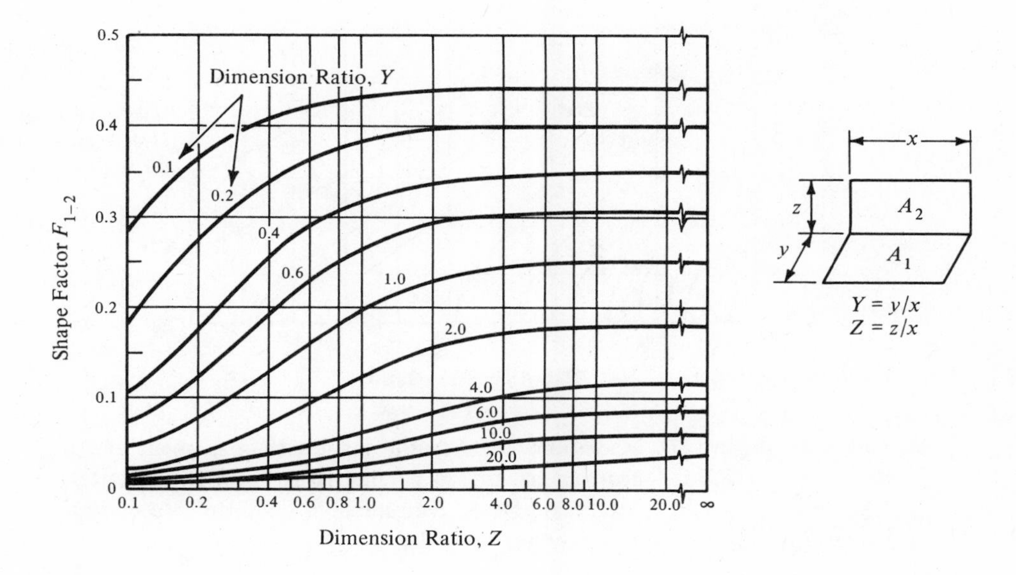

Figure 6-5 Radiation shape factor for adjacent rectangles in perpendicular planes.

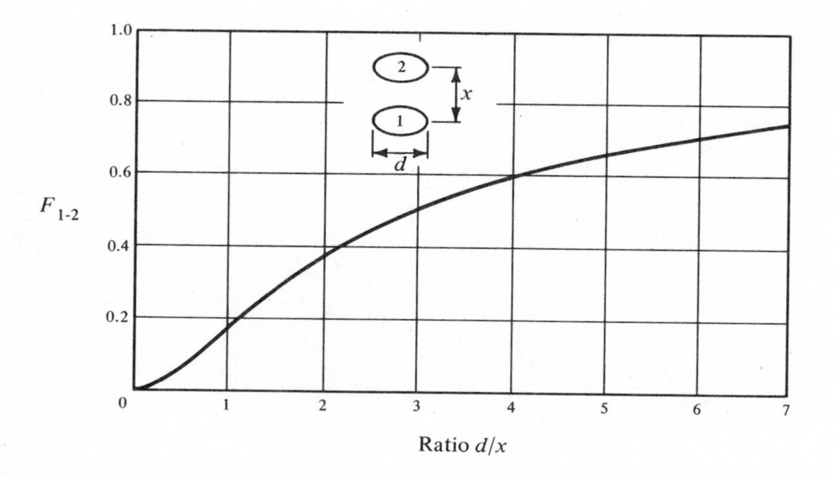

Figure 6-6 Radiation shape factor for radiation between parallel disks, from *Heat Transmission* by W. H. McAdams. Copyright 1954, by W. H. McAdams. Used with permission of McGraw-Hill Book Company.

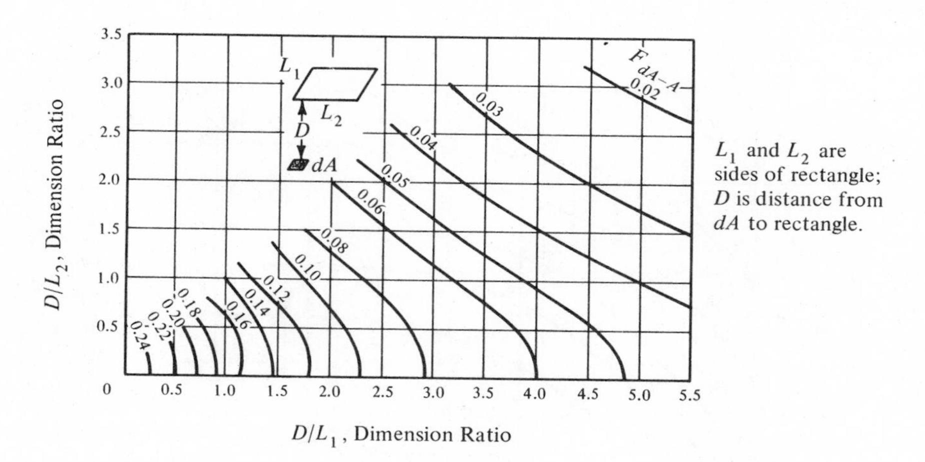

Figure 6-7 Radiation shape factor for a surface element and a rectangular surface parallel to it, from H. C. Hottel, "Radiant Heat Transmission," *ASME Mechanical Engineering*, 52, 1930.

The shape factor depends only on the relative orientation and the relative sizes of the two surfaces under consideration as long as the radiation is of a diffuse nature. Section 6-10 deals with the details of the mathematical relationships involved in the evaluation of shape factors. At the moment, however, we concern ourselves only with the values of shape factors for specific configurations. Figures 6-5, 6-6, 6-7, and 6-8 give the shape factors for some of the more commonly encountered geometries. Reference 1 gives a catalogue of information for shape factors.

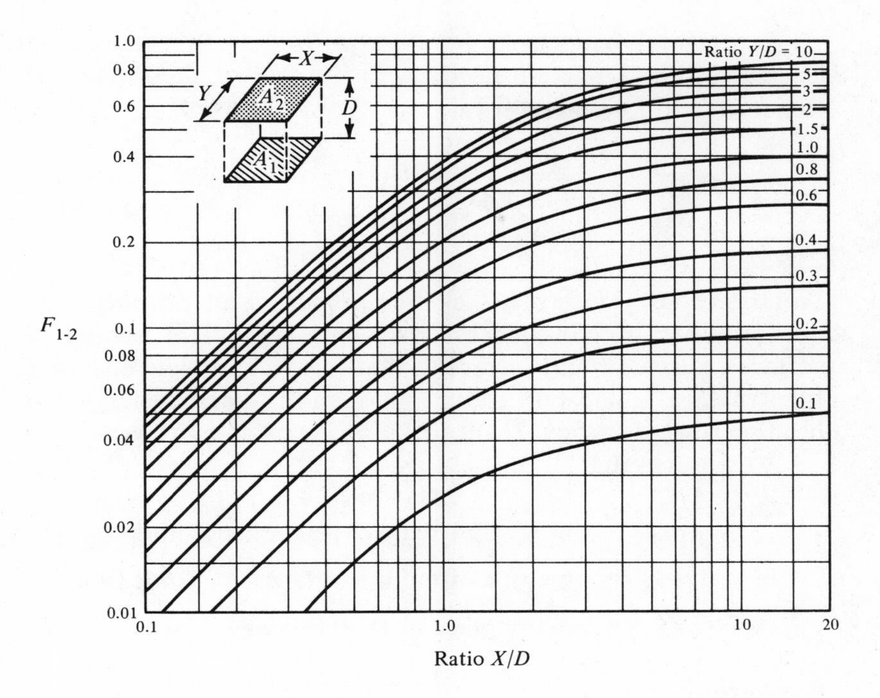

Figure 6-8 Radiation shape factor for radiation between parallel rectangles, from *Heat Transfer* by J. P. Holman. Copyright 1972 by McGraw-Hill, Inc. Used with permission.

We should note that if we have an enclosure of four flat surfaces, then body 1 *sees* a total of three other surfaces, and

$$F_{1\text{-}2} + F_{1\text{-}3} + F_{1\text{-}4} = 1$$

The above equation is also valid for convex surfaces. Also, surfaces 2, 3, and 4 can be concave.

The interpretation of the above equation is: The sum of the fraction of energy leaving surface 1 and heading toward surface 2, the fraction of energy leaving surface 1 and heading toward surface 3, and the fraction of energy leaving surface 1 and heading toward surface 4 represents the total energy leaving surface 1 and simply equals unity. In an enclosure of N surfaces, this means

$$\sum_{n=2}^{N} F_{1\text{-}n} = 1$$

This equation requires only that surface 1 be flat or convex. The others may be concave, convex, or flat.

Since surface 1 is assumed to be flat, it does not see itself, and $F_{1\text{-}1} = 0$.

If surface 1 happens to be concave, $F_{1\text{-}1} \neq 0$ because a part of the radiation leaving surface 1 falls on itself, as in the case of the inside surface of a cylinder. Hence, in a more general form

$$\sum_{n=1}^{N} F_{m\text{-}n} = 1 \quad \text{for } m = 1, 2, \ldots, n \tag{6-6}$$

Equation (6-6) is referred to as the *summation* relationship.

Next, let us consider the exchange of radiant energy between two black surfaces, surface 1 and surface 2, that form a complete enclosure. The enclosure is filled with a transparent gas. Consequently, radiant energy leaving surface 1 and heading toward surface 2 will strike surface 2 and vice versa.

We observe that

$$e_{b1} = \text{energy emitted by surface 1 per unit time and area}$$

$$A_1 e_{b1} = \text{energy emitted by surface 1 per unit time}$$

$$A_1 F_{1\text{-}2} e_{b1} = \text{energy per unit time leaving surface 1 and striking surface 2}$$

Similarly

$$A_2 F_{2\text{-}1} e_{b2} = \text{energy per unit time leaving surface 2 and striking surface 1}$$

Since blackbodies absorb all incident radiant energy, the energy absorbed by surface 1 is also $A_2 F_{2\text{-}1} e_{b2}$.

The net energy loss by surface 1, Q_1, is the difference between the energy leaving surface 1 and the energy absorbed by it. Thus

$$Q_1 = A_1 e_{b1} - A_1 e_{b1} F_{1\text{-}1} - A_2 F_{2\text{-}1} e_{b2}$$

$$= A_1 e_{b1} F_{1\text{-}2} - A_2 e_{b2} F_{2\text{-}1}$$

since

$$F_{1\text{-}1} = 1 - F_{1\text{-}2}$$

If we let $T_1 = T_2$, then $e_{b1} = e_{b2}$, and there can be no net loss or gain of energy by surface 1. Therefore

$$Q_1 = 0$$

resulting in

$$0 = A_1 F_{1\text{-}2} - A_2 F_{2\text{-}1}$$

or

$$A_1F_{1\text{-}2} = A_2F_{2\text{-}1}$$

and in general terms

$$A_mF_{m\text{-}n} = A_nF_{n\text{-}m} \tag{6-6a}$$

Equation (6-6a) is referred to as the *reciprocity* relationship. Although it is derived for blackbodies in thermal equilibrium, it holds in general since the areas and shape factors are only functions of the system geometry and are not functions of the thermodynamic state of the system provided we have diffuse radiation.

Sample Problem 6-1 Consider a 1-ft O.D. sphere located in a 2-ft I.D. sphere as shown in the following sketch. Determine $F_{2\text{-}1}$.

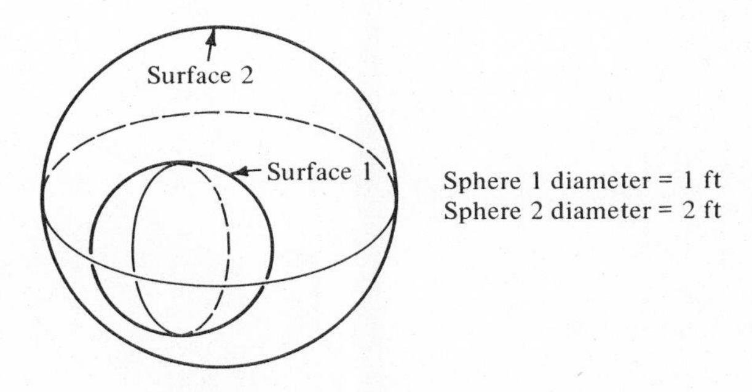

Solution: We know from the summation relationship that

$$F_{1\text{-}1} + F_{1\text{-}2} = 1$$

Also, none of the energy leaving surface 1 strikes itself, giving

$$F_{1\text{-}1} = 0$$

Therefore

$$F_{1\text{-}2} = 1$$

and we know from the reciprocity relationship that

$$A_1F_{1\text{-}2} = A_2F_{2\text{-}1}$$

so that

$$F_{2\text{-}1} = (A_1/A_2)F_{1\text{-}2}$$

and

$$F_{2\text{-}1} = \frac{4\pi r_1^2}{4\pi r_2^2}(1)$$

Since $r_1/r_2 = (0.5/1)$, it follows that

$$F_{2\text{-}1} = \frac{1}{4}$$

Sample Problem 6-2 Consider the *very long* triangular duct shown below. Determine $F_{1\text{-}2}$.

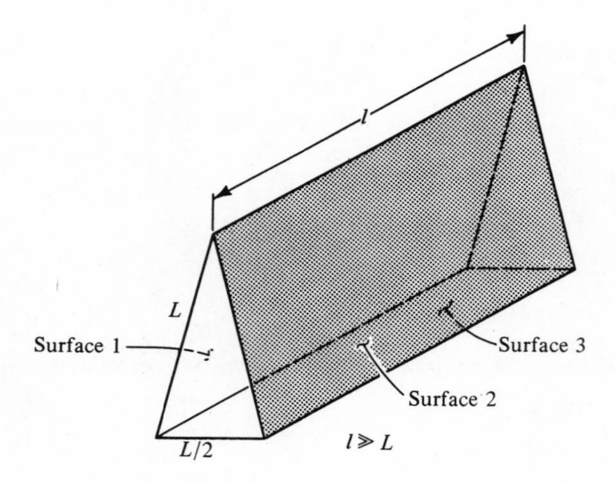

Solution: Ignore radiation lost out the end of the duct, since the duct is stated to be very long. We note that

$$F_{1\text{-}2} + F_{1\text{-}3} = 1$$

$$F_{3\text{-}1} + F_{3\text{-}2} = 1$$

By symmetry

$$F_{3\text{-}1} = F_{3\text{-}2}$$

From the last two equations we obtain

$$F_{3\text{-}1} = \frac{1}{2}$$

Now

$$A_3 F_{3\text{-}1} = A_1 F_{1\text{-}3}$$

and

$$F_{1\text{-}3} = \left(\frac{A_3}{A_1}\right) F_{3\text{-}1} = \frac{(L/2)(1)}{(L)(1)}\left(\frac{1}{2}\right) = \frac{1}{4}$$

Therefore

$$F_{1\text{-}2} = 1 - F_{1\text{-}3} = 1 - \frac{1}{4} = \frac{3}{4}$$

Sample Problem 6-3 Consider the figure below. Find $F_{1\text{-}2}$.

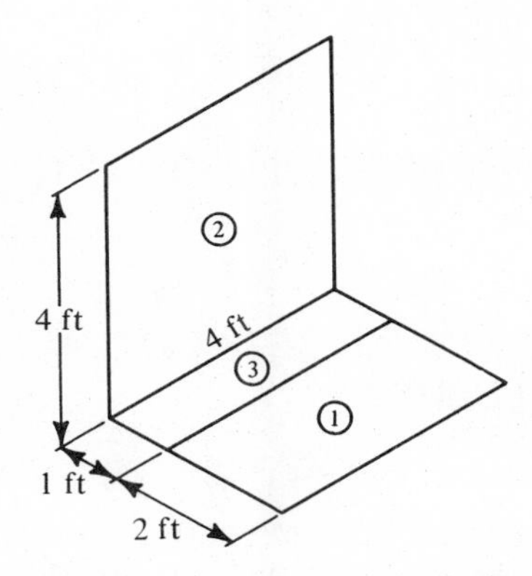

Solution: This shape factor cannot be determined directly from Figure 6-5. It can be determined by considering the energy arriving at surface 2 as follows:

$$\begin{bmatrix}\text{energy leaving surface 1}\\ \text{going to surface 2}\end{bmatrix} = \begin{bmatrix}\text{energy leaving surfaces}\\ \text{(1 + 3) going to surface 2}\end{bmatrix} - \begin{bmatrix}\text{energy leaving surface 3}\\ \text{going to surface 2}\end{bmatrix}$$

If the surfaces are assumed to be blackbodies, we obtain

$$A_1 F_{1\text{-}2}\, e_{b1} = A_{(1+3)} F_{(1+3)\text{-}2}\, e_{b(1+3)} - A_3 F_{3\text{-}2} e_{b3}$$

If we let $T_1 = T_2 = T_3$, then $e_{b1} = e_{b(1+3)} = e_{b3}$, and

$$A_1 F_{1\text{-}2} = A_{(1+3)} F_{(1+3)\text{-}2} - A_3 F_{3\text{-}2}$$

From Figure 6-5, for $F_{(1+3)\text{-}2}$: $Y = \frac{3}{4}$ and $Z = \frac{4}{4}$

Hence, $F_{(1+3)\text{-}2} = 0.24$. For $F_{3\text{-}2}$: $Y = \frac{1}{4}$ and $Z = 1$.

Hence, from Figure 6-5, $F_{3\text{-}2} = 0.37$

and

$$F_{1\text{-}2} = \left(\frac{1}{A_1}\right) [A_{(1+3)} F_{(1+3)\text{-}2} - A_3 F_{3\text{-}2}]$$

Substitution yields

$$F_{1\text{-}2} = \frac{1}{8} [(12)(0.24) - (4)(0.37)]$$

or

$$F_{1\text{-}2} = 0.175$$

The process used in this sample problem is known as shape factor algebra.

6-3 RADIANT HEAT TRANSFER BETWEEN TWO BLACKBODIES FORMING AN ENCLOSURE

Consider two black surfaces, exchanging radiant energy, forming an enclosure that is filled by a transparent gas. Such a situation exists whenever one black surface is completely enclosed by another black surface. A black sphere placed in an oven with black walls would be one such example. The amount of radiant energy leaving body 1 and arriving at and being absorbed by body 2 is $A_1 F_{1\text{-}2} e_{b1}$. In addition, if surface 2 is concave, some of its own emitted radiation is incident on itself. The amount of this radiant energy is $A_2 F_{2\text{-}2} e_{b2}$. The net rate of heat loss of surface 2 is the difference between the radiation emitted by 2 and the radiation absorbed by 2. That is

$$Q_2 = A_2 e_{b2} - (A_1 F_{1\text{-}2} e_{b1} + A_2 F_{2\text{-}2} e_{b2})$$

From the reciprocity relation

$$A_1 F_{1\text{-}2} = A_2 F_{2\text{-}1}$$

and, in addition

$$F_{2\text{-}1} + F_{2\text{-}2} = 1 \tag{6-7}$$

yielding

$$Q_2 = A_2 F_{2\text{-}1}(e_{b2} - e_{b1})$$

By similar reasoning it follows that

$$Q_1 = A_1 F_{1\text{-}2}(e_{b1} - e_{b2}) = -Q_2 \tag{6-7a}$$

6-4 RADIANT HEAT TRANSFER BETWEEN GRAYBODIES

Next, we will consider gray surfaces. A surface is said to be gray if $\alpha = \epsilon$, and ϵ is constant over the temperature range of the problem. The analysis is predicated on the assumptions that:

(1) All surfaces are gray.
(2) All reflections are diffuse.
(3) The temperature is uniform on each surface.
(4) All surfaces are opaque.
(5) The enclosure is filled with a transparent gas.
(6) All emissions are diffuse.

Two new quantities will be defined to aid the analysis. The first quantity is the *radiosity, J*, of a surface. By definition, the radiosity is the rate at which radiant energy leaves a surface per unit area and time. It is composed of both reflected and emitted energy. The second term is the *irradiation, G*, and is the radiant energy incident upon a surface per unit area and time. In view of the definition of J

$$J = \rho G + \epsilon e_b \tag{6-8}$$

where

ρ = the reflectivity of the surface
ϵ = the emissivity of the surface
e_b = the emissive power of the surface if it were a blackbody

The analysis presented here assumes that J is uniform on each surface.

Understanding of the concepts of radiosity and irradiation is facilitated if we visualize a control volume with one of its surfaces exchanging radiation energy with other surfaces located outside the control volume. We then consider the radiosity as the radiant energy leaving the control volume and the irradiation as the radiant energy entering the control volume. The quantity $(J - G)$ will then equal the net radiant heat loss from the surface per unit area and time, q. Or

$$Q = Aq = A(J - G) \tag{6-9}$$

If the surface is to be maintained at a constant temperature, an external source must supply energy at a rate of Q units per unit time to the surface.

Let us calculate the radiant energy loss, Q, from a surface that is one of two gray surfaces forming an enclosure. A gray sphere placed in an oven with gray walls would be one such example. The amount of radiant energy leaving surface 1 and striking surface 2 is $A_1 F_{1\text{-}2} J_1$. From surface 2, energy leaves at the rate of $A_2 J_2$, and if surface 2 is concave, an amount, $A_2 J_2 F_{2\text{-}2}$, is incident on surface 2 itself. Therefore

$$G_2 = (A_1 F_{1\text{-}2} J_1 + A_2 F_{2\text{-}2} J_2)/A_2$$

Then, in accordance with equation (6-9)

$$Q_2 = A_2 J_2 - (A_1 F_{1\text{-}2} J_1 + A_2 F_{2\text{-}2} J_2)$$

Next, the shape factor relations

$$A_1 F_{1\text{-}2} = A_2 F_{2\text{-}1} \quad \text{and} \quad F_{2\text{-}1} + F_{2\text{-}2} = 1$$

are used, giving

$$Q_2 = A_2 F_{2\text{-}1}(J_2 - J_1) \tag{6-10}$$

Similarly,

$$Q_1 = A_1 F_{1\text{-}2}(J_1 - J_2) = -Q_2 \tag{6-10a}$$

Note that equations (6-10) are very similar to equations (6-7), except that now we have radiosities J_1 and J_2 that are still unknown. The method of determining the Js will now be described.

When equation (6-8) is rewritten for surface 1 and rearranged, we obtain

$$G_1 = \frac{J_1 - \epsilon_1 e_{b1}}{\rho_1} \tag{6-11}$$

Next, equation (6-9) is rewritten for surface 1.

$$\frac{Q_1}{A_1} = J_1 - G_1 \tag{6-11a}$$

Combining equations (6-11) and (6-11a), we find

$$\frac{Q_1}{A_1} = J_1 - \frac{J_1 - \epsilon_1 e_{b1}}{\rho_1}$$

But $\rho_1 = 1 - \alpha_1 = 1 - \epsilon_1$ for a gray, opaque surface. Therefore

$$\frac{Q_1}{A_1} = \frac{(1 - \epsilon_1)J_1 - (J_1 - \epsilon_1 e_{b1})}{1 - \epsilon_1}$$

or

$$Q_1 = \frac{e_{b1} - J_1}{\dfrac{1 - \epsilon_1}{\epsilon_1 A_1}} \tag{6-12}$$

It can be likewise shown that the net loss of radiant energy by surface 2 per unit time is

$$Q_2 = \frac{e_{b2} - J_2}{\dfrac{1 - \epsilon_2}{\epsilon_2 A_2}} \tag{6-12a}$$

By substitution of Q_1 and Q_2 from equations (6-12) and (6-12a) into equation (6-10a), we have

$$\frac{e_{b1} - J_1}{\dfrac{1 - \epsilon_1}{\epsilon_1 A_1}} = \frac{J_1 - J_2}{\dfrac{1}{A_1 F_{1\text{-}2}}} = \frac{J_2 - e_{b2}}{\dfrac{1 - \epsilon_2}{\epsilon_2 A_2}} \tag{6-13}$$

This equation really contains a pair of algebraic equations for J_1 and J_2. These equations are easily solved, and expressions for J_1 and J_2 are then substituted back into equations (6-12) and (6-12a). The result is

$$Q_1 = -Q_2 = \frac{e_{b1} - e_{b2}}{\dfrac{1 - \epsilon_1}{\epsilon_1 A_1} + \dfrac{1}{A_1 F_{1\text{-}2}} + \dfrac{1 - \epsilon_2}{\epsilon_2 A_2}} \tag{6-13a}$$

6-4.1 Electrical Network for Radiative Exchange in an Enclosure of Two Graybodies

The structure of equation (6-13) is very similar to that of equation (1-4) for one-dimensional steady-state heat conduction problems for which electrical networks are drawn. We can look upon the quantities $(e_{b1} - J_1)$, $(J_1 - J_2)$, and $(J_2 - e_{b2})$ as driving potentials; the quantities $(1 - \epsilon_1)/(\epsilon_1 A_1)$, $1/A_1 F_{1\text{-}2}$, and $(1 - \epsilon_2)/(\epsilon_2 A_2)$ as resistances; and Q_1 and Q_2 as current flows. The resistance involving surface properties (ϵ_1 or ϵ_2) is termed a surface resistance, while the one involving the shape factor is termed a spatial resistance.

With such an analogy, we can draw circuit diagrams to represent each of the equations (6-12), (6-10), and (6-12a) as shown in Figure 6-9a, b, and c. These individual circuit diagrams can be combined into a single diagram representing equation (6-13a) as shown in Figure 6-9(d).

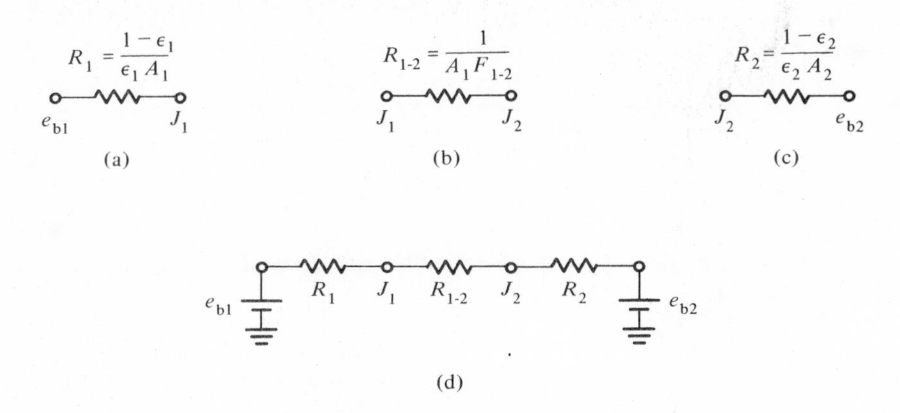

Figure 6-9 Electrical network for a radiative exchange in an enclosure of two bodies.

The analysis and solution of problems involving radiant exchange are greatly facilitated once an electrical network is drawn.

Sample Problem 6-4 Consider a 1-foot O.D. sphere situated in a 2-ft I.D. sphere as discussed in Sample Problem 6-1. If the emissivity of the outer surface of the inner sphere is unity (it is a black surface) and the emissivity of the inner surface of the outer sphere is 0.5, determine the radiant heat loss from the inner sphere if the temperatures of the inner and outer spheres are 1040°F and 540°F, respectively.

Solution: A schematic of the enclosure and the appropriate electrical network are shown on page 233.

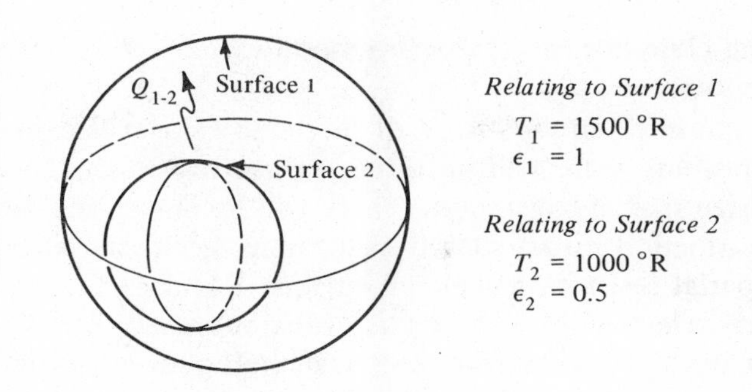

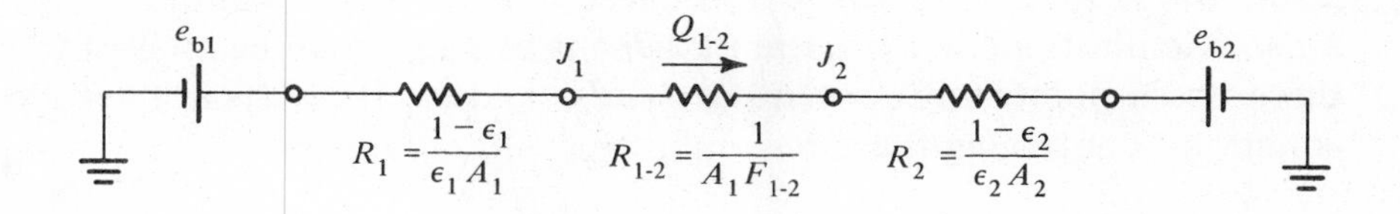

The blackbody emissive powers of the two surfaces are

$$e_{b1} = \sigma T_1^4 = (0.1713 \times 10^{-8})(1500)^4 = 8{,}672 \text{ Btu/hr-ft}^2$$
$$e_{b2} = \sigma T_2^4 = (0.1713 \times 10^{-8})(1000)^4 = 1{,}713 \text{ Btu/hr-ft}^2$$

Various resistances of the circuit are

$$R_1 = \frac{1-\epsilon_1}{\epsilon_1 A_1} = \frac{1-1}{(1)(4\pi 0.5^2)} = 0$$

$$R_2 = \frac{1-\epsilon_2}{\epsilon_2 A_2} = \frac{1-0.5}{(0.5)(4\pi 1^2)} = 0.079$$

$$R_{1\text{-}2} = 1/(A_1 F_{1\text{-}2}) = 1/(4\pi 0.5^2)(1) = 0.318$$

With these, the radiant heat loss from the inner sphere is given by

$$Q_1 = \frac{e_{b1} - e_{b2}}{R_1 + R_{1\text{-}2} + R_2} = \frac{8{,}672 - 1{,}713}{0 + 0.318 + 0.079} = 17{,}530 \text{ Btu/hr}$$

The rate of energy loss by surface 1 is equal to the rate of energy gain by surface 2.

6-4.2 Electrical Network for Three Graybodies

If we consider an enclosure of three gray surfaces exchanging radiant energy with one another, there will be three surface resistances (one for each surface present) and three spatial resistances. There will be one spatial resistance between surface 1 and surface 3, one spatial resistance between surface 2 and surface 3, and one spatial resistance between surface 1 and surface 2. In general, if there are N surfaces, there will be N surface resistances and $[N(N-1)]/2$ spatial resistances.

Figure 6-10 shows the electrical analog for a gray enclosure consisting of three surfaces. Examining Figure 6-10, we see that if the temperatures T_1, T_2, and T_3 are known, then e_{b1}, e_{b2}, and e_{b3} are also determined. Assuming the system geometry is given, we can determine the three shape factors, $F_{1\text{-}2}$, $F_{1\text{-}3}$, and $F_{2\text{-}3}$. Kirchhoff's law for currents arriving at a node can be applied to each of the three nodal points with potentials J_1, J_2, and J_3. It results in the following three equations, containing the three unknowns, J_1, J_2, and J_3.

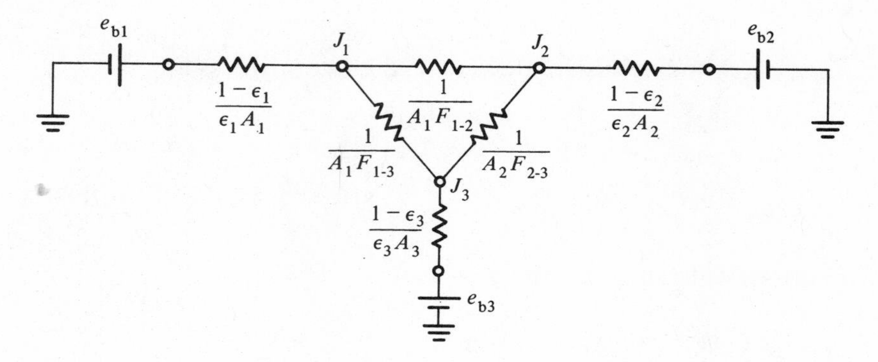

Figure 6-10 Electrical analog for a three-surface gray enclosure.

$$\left.\begin{aligned}
\frac{e_{b1}-J_1}{(1-\epsilon_1)/(\epsilon_1 A_1)}+\frac{J_2-J_1}{1/(A_1F_{1\text{-}2})}+\frac{J_3-J_1}{1/(A_1F_{1\text{-}3})}&=0\\
\frac{e_{b2}-J_2}{(1-\epsilon_2)/(\epsilon_2 A_2)}+\frac{J_1-J_2}{1/(A_1F_{1\text{-}2})}+\frac{J_3-J_2}{1/(A_2F_{2\text{-}3})}&=0\\
\frac{e_{b3}-J_3}{(1-\epsilon_3)/(\epsilon_3 A_3)}+\frac{J_1-J_3}{1/(A_1F_{1\text{-}3})}+\frac{J_2-J_3}{1/(A_2F_{2\text{-}3})}&=0
\end{aligned}\right\}\qquad(6\text{-}14)$$

There are some special situations for which the problem for the three-surface enclosure becomes simple and leads to two equations with two unknowns. These are:

(1) If any one of the surfaces is black, the emissivity of that surface is unity, and its surface resistance, $(1-\epsilon)/\epsilon A$, becomes zero. Consequently, the radiosity

for that surface becomes equal to its blackbody emissive power; i.e., if $\epsilon = 1$, $J = e_b$ (a fixed potential dependent upon the temperature of the surface).

(2) Consider the areas of two surfaces to be very small compared to the area of the third surface. If the area of the third surface is very large, the surface resistance, $(1 - \epsilon)/\epsilon A$, approaches zero, and the radiosity for the third surface becomes equal to its blackbody emissive power. That is, as $A \to \infty$, $(1 - \epsilon)/\epsilon A \to 0$, and $J = e_b$ (a fixed potential dependent upon the temperature of the third surface).

(3) If one of the surfaces has its back insulated, then $Q = 0$. In accordance with equation (6-9), the radiosity and irradiation for that surface must be equal, i.e., $J = G$. Furthermore, since $J = (1 - \epsilon)G + \epsilon e_b$ and if $G = J$

$$J = (1 - \epsilon)J + \epsilon e_b$$

from which it follows that

$J = e_b$ (a floating potential dependent upon the temperatures of the other two surfaces)

An adiabatic wall, which is thermally insulated, is called a *refractory* wall. A perfectly reflecting wall is also adiabatic.

Figure 6-11 shows the electrical analog for the simplified three-surface enclosure problems for the cases where one surface is black (a), where two surfaces are very small compared to the third surface (b), and where one surface is adiabatic (c).

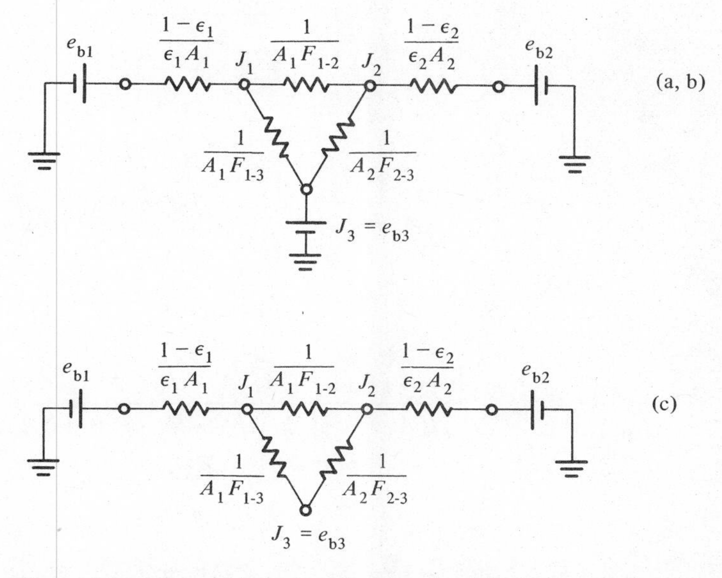

Figure 6-11 Electrical analog for the simplified three-surface enclosure problems. Body three is (a) the black surface, (b) the one with very large area, or (c) the adiabatic wall.

For cases (a) and (b), the potential $J_3 = e_{b3}$ is fixed by the temperature of the black surface or the temperature of the surface with the large area. To obtain a solution for these cases, currents (heat flows) are summed, at the junctions whose potentials are J_1 and J_2, according to Kirchhoff's law. The two resulting equations are then solved for J_1 and J_2 so that heat flow rates may be calculated. The two equations are:

$$\frac{e_{b1} - J_1}{(1 - \epsilon_1)/(\epsilon_1 A_1)} + \frac{J_2 - J_1}{1/A_1 F_{1\text{-}2}} + \frac{e_{b3} - J_1}{1/(A_1 F_{1\text{-}3})} = 0$$

$$\frac{e_{b2} - J_2}{(1 - \epsilon_2)/(\epsilon_2 A_2)} + \frac{J_1 - J_2}{1/(A_1 F_{1\text{-}2})} + \frac{e_{b3} - J_2}{1/(A_2 F_{2\text{-}3})} = 0 \tag{6-15}$$

where e_{b3} is known.

For case (c), the potential $J_3 = e_{b3}$ floats, and its magnitude depends on the temperatures of the other participating surfaces. There is no battery (source) of potential e_{b3} connected to that node. To obtain a solution to this problem, we note that since $J_3 = e_{b3}$ is a floating potential, the resistance $(1/A_1 F_{1\text{-}2})$ is in parallel with the resistances $(1/A_1 F_{1\text{-}3})$ and $(1/A_2 F_{2\text{-}3})$. If these resistances are combined into an equivalent resistance, then by adding the surface resistances and dividing the total resistance into the driving potential, $(e_{b1} - e_{b2})$, the heat loss Q_1 from surface 1 may be calculated as (See Figure 6-11c)

$$Q_1 = \frac{e_{b1} - e_{b2}}{\Sigma R}$$

Note that $\Sigma R = R_{eqv} + R_{surface}$ where

$$\frac{1}{R_{eqv}} = \frac{1}{1/(A_1 F_{1\text{-}2})} + \frac{1}{1/(A_1 F_{1\text{-}3}) + 1/(A_2 F_{2\text{-}3})}$$

or

$$R_{eqv} = \frac{[1/(A_1 F_{1\text{-}2})]\,[1/(A_1 F_{1\text{-}3}) + 1/(A_2 F_{2\text{-}3})]}{1/(A_1 F_{1\text{-}2}) + 1/(A_1 F_{1\text{-}3}) + 1/(A_2 F_{2\text{-}3})}$$

and

$$R_{surface} = [(1 - \epsilon_1)/(\epsilon_1 A_1)] + [(1 - \epsilon_2)/(\epsilon_2 A_2)]$$

Substitution gives

$$Q_1 = \frac{e_{b1} - e_{b2}}{\dfrac{1-\epsilon_1}{\epsilon_1 A_1} + \dfrac{[1/(A_1F_{1\text{-}2})]\,[1/(A_2F_{2\text{-}3}) + 1/(A_1F_{1\text{-}3})]}{1/(A_1F_{1\text{-}2}) + 1/(A_2F_{2\text{-}3}) + 1/(A_1F_{1\text{-}3})} + \dfrac{1-\epsilon_2}{\epsilon_2 A_2}} \tag{6-16}$$

Next, we note that

$$Q_1 = \frac{e_{b1} - J_1}{\dfrac{1-\epsilon_1}{\epsilon_1 A_1}}$$

and

$$Q_1 = -Q_2 = \frac{J_2 - e_{b2}}{\dfrac{1-\epsilon_2}{\epsilon_2 A_2}}$$

With Q_1 known from equation (6-16), the forgoing two equations can be used to determine J_1 and J_2. Once Q_1, J_1, J_2, and J_3 are determined, then T_3 may be evaluated from $J_3 = e_{b3} = \sigma T_3^4$.

Sample Problem 6-5 Two parallel discs 60 cm in diameter are spaced 30 cm apart with one disc located directly above the other disc. One disc is maintained at 500°C and the other at 227°C. The emissivities of the discs are 0.2 and 0.4, respectively. The discs are located in a very large space whose walls are maintained at 60°C. Determine the rate of heat loss by radiation from the inside surfaces of each disc

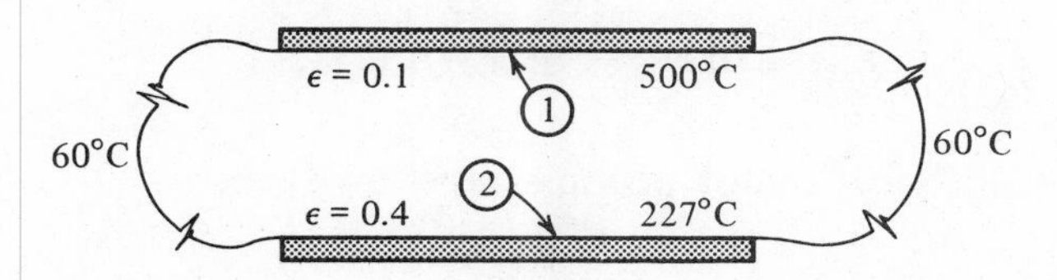

Solution: Figure 6-11a shows the appropriate electrical analog for this problem. We note that

hot disc:	$T_1 = 773°K$	$\epsilon_1 = 0.2$
cold disc:	$T_2 = 500°K$	$\epsilon_2 = 0.4$
room walls:	$T_3 = 333°K$	

From Figure 6-6

$$F_{1\text{-}2} = F_{2\text{-}1} = 0.38$$
$$F_{1\text{-}3} = 1 - F_{1\text{-}2} = 0.62$$
$$F_{2\text{-}3} = 1 - F_{2\text{-}1} = 0.62$$

Also

$$A_1 = A_2 = \pi r_1^2 = (\pi)(0.3)^2 = 0.283 \text{ m}^2$$

$$R_1 = \frac{1 - \epsilon_1}{\epsilon_1 A_1} = \frac{1 - 0.2}{(0.2)(0.283)} = 14.1 \text{ m}^{-2}$$

$$R_2 = \frac{1 - \epsilon_2}{\epsilon_2 A_2} = \frac{1 - 0.4}{(0.4)(0.283)} = 5.3 \text{ m}^{-2}$$

$$R_3 = \frac{1 - \epsilon_3}{\epsilon_3 A_3} = \frac{1 - \epsilon_3}{\epsilon_3(\infty)} \to 0$$

$$R_{1\text{-}2} = \frac{1}{A_1 F_{1\text{-}2}} = \frac{1}{(0.283)(0.38)} = 9.3 \text{ m}^{-2}$$

$$R_{1\text{-}3} = R_{2\text{-}3} = \frac{1}{A_1 F_{1\text{-}3}} = \frac{1}{A_2 F_{2\text{-}3}} = \frac{1}{(0.283)(0.62)} = 5.7 \text{ m}^{-2}$$

The emissive powers are

$$e_{b1} = \sigma T_1^4 = 20200 \text{ W/m}^2$$
$$e_{b2} = \sigma T_2^4 = 3540 \text{ W/m}^2$$
$$J_3 = e_{b3} = \sigma T_3^4 = 695 \text{ W/m}^2$$

The electrical analog becomes as shown below.

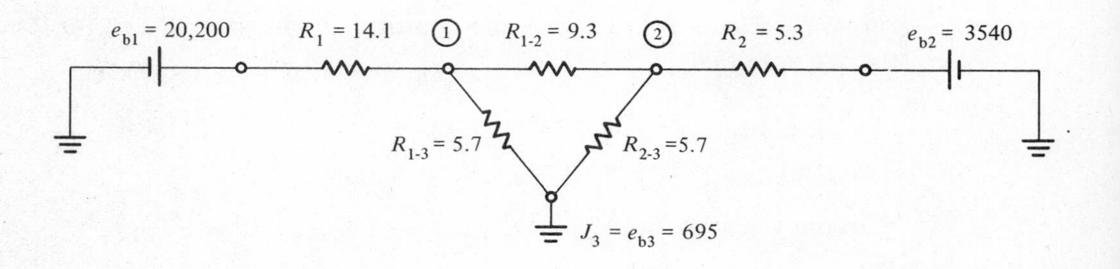

Summing currents around nodes 1 and 2, we write

$$\text{For } J_1: \quad \frac{20200 - J_1}{14.1} + \frac{J_2 - J_1}{9.3} + \frac{695 - J_1}{5.7} = 0$$

$$\text{For } J_2: \quad \frac{3540 - J_2}{5.3} + \frac{J_1 - J_2}{9.3} + \frac{695 - J_2}{5.7} = 0$$

Solving these simultaneous equations, we find

$$J_1 = 5266 \text{ W/m}^2$$

$$J_2 = 2876 \text{ W/m}^2$$

The rate of heat loss from the hot disc is given by

$$Q_1 = \frac{e_{b1} - J_1}{(1 - \epsilon_1)/\epsilon_1 A_1} = \frac{20200 - 5266}{14.1}$$

$$Q_1 = 1059 \text{ watt}$$

Also, the rate of heat loss by cool disc is

$$Q_2 = \frac{e_{b2} - J_2}{(1 - \epsilon_2)/\epsilon_2 A_2} = \frac{3540 - 2876}{5.3}$$

$$Q_2 = 125 \text{ watt}$$

To check our answers, the quanitity $(Q_1 + Q_2)$ should equal the heat flowing through the resistances between the potentials J_1 and J_3 and between the potentials J_2 and J_3. That is

$$Q = \frac{J_1 - J_3}{1/A_1 F_{1\text{-}3}} + \frac{J_2 - J_3}{1/A_2 F_{2\text{-}3}}$$

$$= \frac{5266 - 695}{5.7} + \frac{2876 - 695}{5.7} = 1184 \text{ watt}$$

Also

$$Q = Q_1 + Q_2 = 1059 + 125 = 1184 \text{ watt}$$

6-5 RADIANT ENERGY INTERCHANGE AMONG N GRAY SURFACES

One often encounters situations where more than three surfaces participate in the radiant energy interchange. We consider an enclosure formed by N gray surfaces, each of which is isothermal. We assume that all shape factors, $F_{i\text{-}j}$, are known.

The radiosity, J_i, of the ith surface, which is the total radiant energy leaving surface i per unit area and time, is given by

$$J_i = \rho_i G_i + \epsilon_i e_{bi} \tag{6-17}$$

or

$$J_i = (1 - \epsilon_i) G_i + \epsilon_i e_{bi} \tag{6-17a}$$

The irradiation, G_i, for the ith surface, which by definition is the amount of radiation energy arriving at the ith surface per unit area and time, is given by

$$G_i = \frac{1}{A_i} \sum_{j=1}^{N} A_j F_{j\text{-}i} J_j$$

In view of the reciprocity relation, $A_j F_{j\text{-}i} = A_i F_{i\text{-}j}$, the irradiation, G_i, can be rewritten as

$$G_i = \sum_{j=1}^{N} F_{i\text{-}j} J_j$$

Substituting for G_i from the above into equation (6-17a), we obtain

$$J_i = \left[(1 - \epsilon_i) \sum_{j=1}^{N} F_{i\text{-}j} J_j \right] + \epsilon_i e_{bi} \tag{6-18}$$

Equation (6-18) represents a set of N simultaneous algebraic equations, the solution of which yields the $J_1, J_2, \ldots, J_N$.

Equation (6-12), which was first developed for an enclosure of two surfaces, is actually valid for an enclosure of N gray surfaces. Therefore, we may write it in the following form

$$\frac{Q_i}{A_i} = q_i = \frac{\epsilon_i}{1 - \epsilon_i} (e_{bi} - J_i) \tag{6-19}$$

If the temperatures of all the surfaces of the enclosure are known, we know the values of e_{bi}. For such a situation, the only unknowns in equation (6-18) are the J_is. We can solve for the J_is by the same methods that were discussed in Chapter 5. Once the values of the radiosities are determined, we can calculate the heat flux from each surface by evaluating equation (6-19) for that surface.

If instead of the surface temperature, the heat flux, q_i, at each surface is known, then the objective is to determine the temperature of each surface. To this end, we rewrite equation (6-19) to read

$$e_{bi} = \frac{1 - \epsilon_i}{\epsilon_i} q_i + J_i \tag{6-20}$$

Substituting the above in equation (6-18) and simplifying, we obtain

$$J_i = q_i + \sum_{j=1}^{N} F_{i\text{-}j} J_j \tag{6-21}$$

Equation (6-21) represents a set of N simultaneous equations in the N unknown J_is, which again can be determined by the methods discussed in Chapter 5. Once the values of the J_is are determined, equation (6-20) can be used to determine the emissive power of each surface and hence its temperature.

A third possibility is that the temperatures of M surfaces and the heat fluxes for $(N - M)$ surfaces are prescribed. In such a situation, the objective is to determine the M unknown heat fluxes and the $(N - M)$ unknown temperatures. We may proceed as follows in such a case.

We generate M equations by employing equation (6-18) for those surfaces whose temperatures are known. We further generate $(N - M)$ equations by employing equation (6-21) for those surfaces for which heat fluxes are known. Thus, altogether we generate $[M + (N - M)]$ or N equations in N unknown J_is, which can be readily determined. Once the values of the J_is are determined, we can determine the unknown heat fluxes from equation (6-19) and the unknown temperatures from equation (6-20).

6-6 RADIATION SHIELDS

The smaller the emissivity of a surface, the less the radiative heat transfer from it, since the surface resistance $(1 - \epsilon)/\epsilon A$ approaches infinity as $\epsilon \rightarrow 0$. Since $\epsilon = (1 - \rho)$ for opaque materials, this indicates that the use of highly reflective surfaces $(\rho \rightarrow 1)$ is desirable when we wish to reduce radiant heat transfer. If this approach does not limit the heat transfer sufficiently, then radiation shields may be used.

Such shields placed between the heat transfer surfaces do not add or remove energy from the system. They add resistances and thereby reduce the net radiative heat transfer from a surface. Radiation shields are widely used when high temperatures are measured using a thermocouple.

To begin our analysis, let us consider the general two-surface enclosure for which the electric analog is shown in Figure 6-9d. The heat loss equation is readily derived as

$$Q_1 = \frac{e_{b1} - e_{b2}}{(1 - \epsilon_1)/(\epsilon_1 A_1) + (1/A_1 F_{1\text{-}2}) + (1 - \epsilon_2)/(\epsilon_2 A_2)} = -Q_2$$

Note that if surfaces 1 and 2 are infinite parallel planes, then

$$A_1 = A_2 \text{ and } F_{1\text{-}2} = 1$$

Therefore

$$\frac{Q_1}{A_1} = \frac{e_{b1} - e_{b2}}{(1/\epsilon_1) + (1/\epsilon_2) - 1} = -\frac{Q_2}{A_2} \tag{6-22}$$

If bodies 1 and 2 were two spheres as shown in Sample Problem 6-1, where 1 refers to the outer surface of the inner sphere and 2 refers to the inner surface of the outer sphere, we have (noting $F_{1\text{-}2} = 1$)

$$\frac{Q_1}{A_1} = \frac{e_{b1} - e_{b2}}{(1/\epsilon_1) + (A_1/A_2)\,[(1/\epsilon_2) - 1]} \tag{6-23}$$

Equation (6-23) also holds for two infinitely long cylindrical surfaces. If A_2 is much larger than A_1, then the ratio (A_1/A_2) approaches zero, and the rate of heat transfer from surface 1 is

$$Q_1 = A_1 \epsilon_1 (e_{b1} - e_{b2}) \tag{6-24}$$

Returning to our analysis of radiation shields, we shall analyze the system shown in Figure 6-12. The system depicted in Figure 6-12 is that of two surfaces, 1 and 2, with an adiabatic shield interposed between them. It will be assumed in our analysis that the emissivities of both surfaces of the shield are identical and are equal to ϵ as is the emissivity of surfaces 1 and 2.

We will compare the rate of heat transfer per unit area, (Q/A), from surface 1 with and without the shield. Observe that Figure 6-12c may be used to determine the overall resistance between surfaces 1 and 2 even if the emissivities are not all equal.

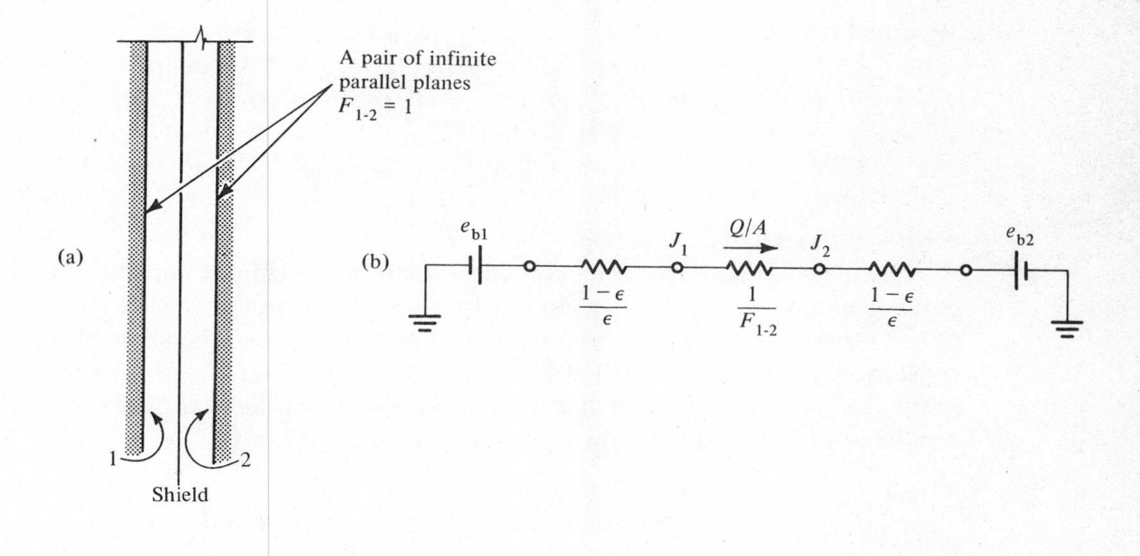

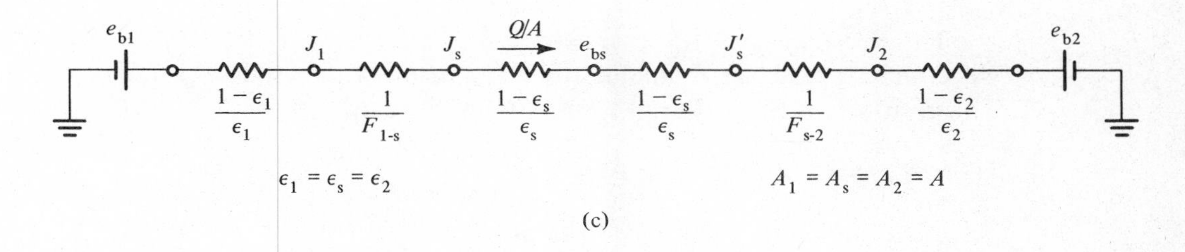

Figure 6-12 Radiation shield situated between infinite parallel planes and corresponding electrical network.

From Figure 6-12b, the heat flow in the absence of a shield is given by

$$\left(\frac{Q}{A}\right)_{\substack{\text{without}\\ \text{shield}}} = \frac{e_{b1} - e_{b2}}{(2/\epsilon) - 1}$$

The heat flow with a shield, noting that all shape factors are unity, is derived from Figure 6-12c to be

$$\left(\frac{Q}{A}\right)_{\substack{\text{with}\\ \text{shield}}} = \frac{e_{b1} - e_{b2}}{[4(1 - \epsilon)/\epsilon] + 2}$$

We conclude that

$$\left(\frac{Q}{A}\right)_{\substack{\text{with}\\ \text{shield}}} = \frac{1}{2}\left(\frac{Q}{A}\right)_{\substack{\text{without}\\ \text{shield}}}$$

provided all emissivities are equal.

Next, let us consider the case where there are n shields between the infinite parallel planes, all having an emissivity equal to ϵ. For this case, there will be $(n + 1)$ spaces created by a total of $(n + 2)$ parallel planes. The value of each space resistance is unity. There will also be $2n$ surface resistances for the n shields plus 2 more surface resistances for the original two surfaces. Each surface resistance has a value of $(1 - \epsilon)/\epsilon$. Therefore, the total resistance will be

$$\underbrace{(n + 1)(1)}_{\text{spatial}} + \underbrace{(2n + 2)\left(\frac{1 - \epsilon}{\epsilon}\right)}_{\text{surface}} = \underbrace{(n + 1)\left(\frac{2}{\epsilon} - 1\right)}_{\text{total}}$$

The resistance for the case where no shield is used is $[(2/\epsilon) - 1]$ and, hence, we conclude that for n shields the resistance is increased by a factor of $(n + 1)$. This results in a reduction of heat flow by the same factor.

$$\left(\frac{Q}{A}\right)_{\substack{\text{with}\\ \text{shields}}} = \frac{1}{n + 1}\left(\frac{Q}{A}\right)_{\substack{\text{without}\\ \text{shields}}} \tag{6-25}$$

It should be emphasized that equation (6-25) is valid only if all emissivities are equal. Sample Problem 6-6 shows that if the emissivity of the shield is much less than that of the original surfaces, then the heat flow may be reduced by a considerably larger amount than in the case where all emissivities are equal.

Sample Problem 6-6 Two large parallel planes at temperatures of 940°F and 440°F exchange heat by radiation and have emissivities of 0.4 and 0.9, respectively. Compare the original heat transfer between the planes to that which occurs when a single polished aluminum radiation shield having an emissivity of 0.05 is placed between them.

Solution:

$$e_{b1} = \sigma T_1^4 = 6580 \text{ Btu/hr-ft}^2;\ 1/\epsilon_1 = 2.5;\ F_{1\text{-}2} = 1$$

$$e_{b2} = \sigma T_2^4 = 1125 \text{ Btu/hr-ft}^2;\ 1/\epsilon_2 = 1.11;\ F_{2\text{-}1} = 1$$

$$\left(\frac{Q}{A}\right)_{\substack{\text{without}\\\text{shield}}} = \frac{e_{b1} - e_{b2}}{(1/\epsilon_1) + (1/\epsilon_2) - 1}$$

$$\left(\frac{Q}{A}\right)_{\substack{\text{without}\\\text{shield}}} = \frac{6580 - 1125}{2.5 + 1.11 - 1} = 2090 \text{ Btu/hr-ft}^2$$

Referring to Figure 6-12c, we can write

$$\left(\frac{Q}{A}\right)_{\substack{\text{with}\\\text{shield}}} = \frac{e_{b1} - e_{b2}}{(1 - \epsilon_1)/(\epsilon_1) + 2(1 - \epsilon_s)/\epsilon_s + (1 - \epsilon_2)/(\epsilon_2) + 2}$$

$$\left(\frac{Q}{A}\right)_{\substack{\text{with}\\\text{shield}}} = \frac{6580 - 1125}{1.5 + (2)(19) + 0.11 + 2} = 131 \text{ Btu/hr-ft}^2$$

The heat flow is cut from 2090 Btu/hr-ft^2 to 131 Btu/hr-ft^2 or by a factor of 16, considerably more than the 50% reduction that a shield of equal emissivity would have caused.

6-7 RADIATION-CONVECTION SYSTEMS

Whenever thermocouples are used to measure temperatures of high-temperature gases contained in a relatively low-temperature enclosure, it becomes difficult to obtain accurate results because of radiation losses. Since radiative losses are proportional to temperature to the fourth power; the higher the temperature, the more important the radiation effects in interpreting experimental data.

Let us consider the problem of measuring the temperature of a hot gas flowing in a tube whose walls are maintained at a relatively low temperature. Figure 6-13 depicts the situation to be analyzed.

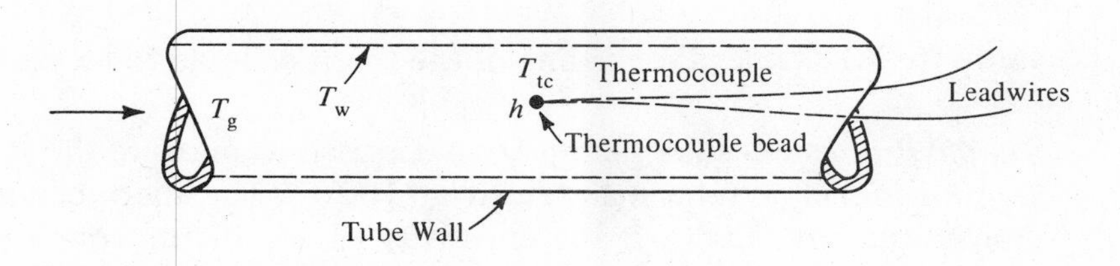

Figure 6-13 Measurement of temperature of a hot gas flowing in a relatively cool tube, $T_w << T_g$.

Let h be the convective heat transfer coefficient between the thermocouple junction and the gas, T_{tc} be the temperature of the thermocouple bead, T_g be the temperature of the gas, and T_w be the temperature of the tube wall. When steady-state conditions are reached, the temperature of the thermocouple does not change. However, it loses energy by radiation to the cool wall and must gain an equal amount of energy by convection from the hot gas. Hence, the thermocouple reading does not represent the true temperature of the gas, T_g, but represents a value that is less than T_g due to the combined convection-radiation process taking place. This is similar to the white frost problem discussed in Chapter 1.

When the thermocouple junction has attained thermal equilibrium and when conduction along the thermocouple lead wires is neglected, we have

$$Q_{conv} = Q_{rad}$$

If A_{tc} is the area of the thermocouple junction, then

$$Q_{conv} = hA_{tc} (T_g - T_{tc})$$

Since the area of the tube wall is much greater than the surface area of the thermocouple junction, equation (6-24) gives

$$Q_{rad} = A_{tc} \sigma \epsilon_{tc} (T_{tc}^4 - T_w^4)$$

where ϵ_{tc} is the emissivity of the thermocouple junction.

Substituting Q_{conv} and Q_{rad} in the energy balance equation yields

$$hA_{tc} (T_g - T_{tc}) = A_{tc} \sigma \epsilon_{tc} (T_{tc}^4 - T_w^4)$$

The actual gas temperature is then given by

$$T_g = T_{tc} + \frac{\sigma \epsilon_{tc} (T_{tc}^4 - T_w^4)}{h} \tag{6-26}$$

Equation (6-26) enables the reading of the thermocouple to be corrected.

Sample Problem 6-7 A thermocouple is used to measure the temperature of a gas flowing in a tube maintained at 100°C. The thermocouple indicates a temperature of 500°C. If the emissivity of the thermocouple junction is 0.5 and the convective heat transfer coefficient is 250W/m^2°K, determine the actual temperature of the gas.

Solution: In the steady state, equation (6-26) applies. Upon insertion of the given data, there follows that

$$T_g = 500 + \frac{(5.66 \times 10^{-8})(0.5)(773^4 - 373^4)}{250}$$

$$T_g = 500 + 38 = 538°\text{C}$$

We see that if we had ignored radiation effects, we would have had considerable error. Oftentimes radiation shields wrapped around the junction are used to cut down on such radiation losses.

6-8 WAVELENGTH-DEPENDENT CHARACTERISTICS

In the preceding sections, we discussed radiant energy transfer assuming graybodies for which radiation properties did not depend on wavelength. In this and the following sections, we will consider the variation of radiation properties with wavelength.

In all our previous analyses, we considered the total energy emitted by a body per unit area and time, e, not taking into account the fact that the energy emitted has a spectral distribution possessing various amounts of energy at different wavelengths. If a body is black, the *total* (i.e., integrated over all wavelengths) amount of energy emitted by that body when its temperature is T is given by the Stefan-Boltzmann equation.

$$e_b = \sigma T^4$$

In reality, the emissive power of the blackbody is distributed among different wavelengths as indicated in Figure 6-14. The *monochromatic emissive power* of a blackbody, $e_{b\lambda}$, is the amount of energy emitted per unit wavelength, area, and time at a given wavelength, λ, and at a given temperature, T. It has been shown by Max Planck (Reference 1) to be equal to

$$e_{b\lambda} = \frac{C_1}{\lambda^5 (e^{C_2/\lambda T} - 1)} \tag{6-27}$$

The units of $e_{b\lambda}$ are Btu/hr-ft$^2\mu$ or W/m$^2\mu$. Values of the constants C_1 and C_2 are

$$C_1 = 1.187 \times 10^8 \text{ Btu } \mu^4/\text{hr-ft}^2;\ C_1 = 3.740 \times 10^{-16} \text{ W/m}^2$$

$$C_2 = 2.5896 \times 10^4\ \mu°\text{R};\ C_2 = 0.014387 \text{ m}°\text{K}$$

If one wants to determine the total energy emitted by a blackbody, it is only necessary to add up the energy emitted over all wavelengths. That is, if e_b is the total energy emitted per unit area and time and if $e_{b\lambda}$ is the energy emitted at wavelength λ per unit area and time, then

$$e_b = \sum_{\substack{\text{all} \\ \text{wavelengths}}} e_{b\lambda} \Delta\lambda$$

Or in the integral form

$$e_b = \int_0^\infty e_{b\lambda} \, d\lambda = \sigma T^4 \tag{6-28}$$

Figure 6-14 shows a plot of $e_{b\lambda}$ versus wavelength for several temperatures. Equation (6-28) tells us that the areas under the respective curves in Figure 6-14 equal e_b. Note that as the temperature increases, there is a decrease in the wavelength, λ_{max}, at which the maximum energy is emitted. For a given temperature, one can determine λ_{max}, by differentiating equation (6-27), setting it equal to zero, and solving for λ. Doing this results in Wien's displacement law

$$\lambda_{max} = \frac{5215.6\ \mu°\text{R}}{T} \quad \text{or} \quad \lambda_{max} = \frac{2897.6\ \mu°\text{K}}{T} \tag{6-29}$$

where T is in degrees absolute and λ_{max} is expressed in microns.

If $(e_{b\lambda})_{max}$ is the maximum value of $e_{b\lambda}$, then a plot of $[e_{b\lambda}/(e_{b\lambda})_{max}]$ versus (λ/λ_{max}) can be drawn and is shown in Figure 6-15.

Recall that the area under the curve of $e_{b\lambda}$ versus λ is equal to e_b for a given temperature. On occasion, we may wish to know the amount of radiant energy emitted in a specific wavelength band, say between λ_1 and λ_2. Referring to Figure 6-16, we let $e_b(0 \rightarrow \lambda_2)$ be equal to the energy emitted between wavelengths $\lambda = 0$ and $\lambda = \lambda_2$. In general, the quantity $e_b(0 \rightarrow \lambda)$ denotes the emission in the range $\lambda = 0$ to $\lambda = \lambda$. Hence, for $e_b(0 \rightarrow \lambda)$, we write

$$e_b(0 \rightarrow \lambda) = \int_0^\lambda e_{b\lambda} \, d\lambda$$

To find $e_b(\lambda_1 \rightarrow \lambda_2)$, the integral between λ_1 and λ_2 can be written as a difference of integrals

$$e_b(\lambda_1 \rightarrow \lambda_2) = \int_{\lambda_1}^{\lambda_2} e_{b\lambda} \, d\lambda = \int_0^{\lambda_2} e_{b\lambda} \, d\lambda - \int_0^{\lambda_1} e_{b\lambda} \, d\lambda$$

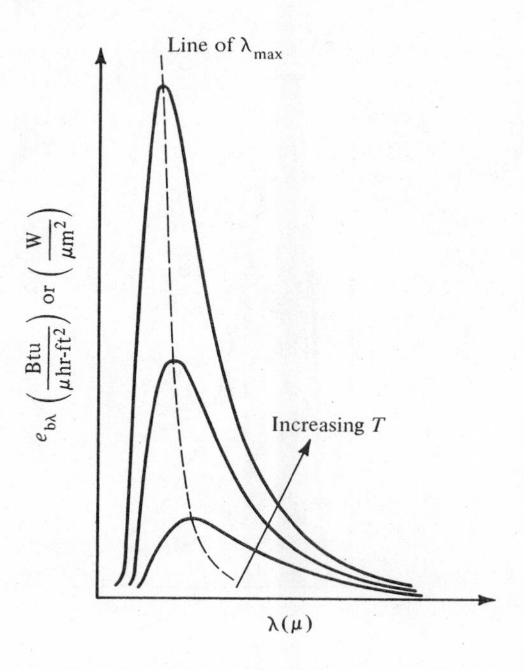

Figure 6-14 Blackbody monochromatic emissive power as a function of wavelength.

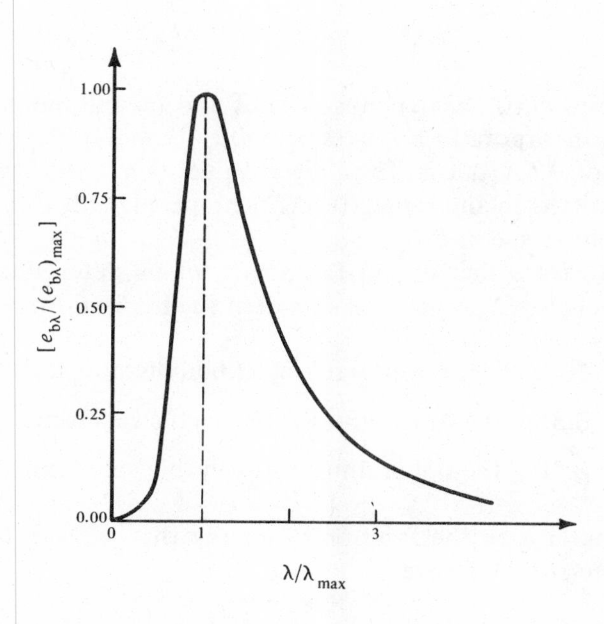

Figure 6-15 $(e_{b\lambda})/(e_{b\lambda})_{max}$ as a function of λ/λ_{max}.

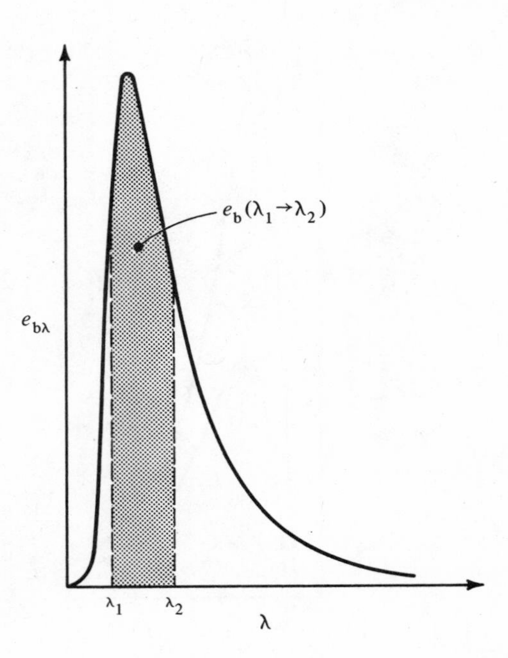

Figure 6-16 Nomenclature for determination of radiant energy emitted between wavelengths λ_1 and λ_2.

or

$$e_b(\lambda_1 \to \lambda_2) = e_b(0 \to \lambda_2) - e_b(0 \to \lambda_1)$$

The value of $e_b(0 \to \lambda)$ depends upon λ and on the temperature T of the surface. Table 6-1 incorporates both the variables, λ and T, by giving $[e_b(0 \to \lambda T)/e_b]$ as a function of λT. That is, for a given value of λT, the table gives the ratio of the energy emitted in the range $(0 \to \lambda T)$ compared with the total energy, e_b, emitted over the entire spectrum.

To illustrate the use of Table 6-1, let us determine the fraction of energy emitted between λ_1 and λ_2 as shown in Figure 6-16 for the following data:

$T = 10{,}000°\text{R}$, the effective blackbody temperature of the surface of the sun

$\lambda_1 = 0.35\mu$, the lower limit of the visible spectrum

$\lambda_2 = 0.70\mu$, the upper limit of the visible spectrum

We will determine the fraction of energy that the sun emits in the visible wavelength spectrum. We have

$$\lambda_1 T = 3500\ \mu°\text{R}$$

$$\lambda_2 T = 7000\ \mu°\text{R}$$

TABLE 6-1 Radiation Functions†

λT $\mu°R$	λT $\mu°K$	$\frac{e_{b\lambda} \times 10^5}{\sigma T^5}$	$\frac{e_b(0 \to \lambda T)}{\sigma T^4}$	λT $\mu°R$	λT $\mu°K$	$\frac{e_{b\lambda} \times 10^5}{\sigma T^5}$	$\frac{e_b(0 \to \lambda T)}{\sigma T^4}$	λT $\mu°R$	λT $\mu°K$	$\frac{e_{b\lambda} \times 10^5}{\sigma T^5}$	$\frac{e_b(0 \to \lambda T)}{\sigma T^4}$
1000	556	.0000394	0	7200	4000	10.089	.4809	13400	7445	2.714	.8317
1200	667	.001184	0	7400	4111	9.723	.5007	13600	7556	2.605	.8370
1400	778	.01194	0	7600	4222	9.357	.5199	13800	7667	2.502	.8421
1600	889	.0618	.0001	7800	4333	8.997	.5381	14000	7778	2.416	.8470
1800	1000	.2070	.0003	8000	4445	8.642	.5558	14200	7889	2.309	.8517
2000	1111	.5151	.0009	8200	4556	8.293	.5727	14400	8000	2.219	.8563
2200	1222	1.0384	.0025	8400	4667	7.954	.5890	14600	8111	2.134	.8606
2400	1333	1.791	.0053	8600	4778	7.624	.6045	14800	8222	2.052	.8648
2600	1445	2.753	.0098	8800	4889	7.304	.6194	15000	8333	1.972	.8688
2800	1556	3.872	.0164	9000	5000	6.995	.6337	16000	8889	1.633	.8868
3000	1667	5.081	.0254	9200	5111	6.697	.6474	17000	9445	1.360	.9017
3200	1778	6.312	.0368	9400	5222	6.411	.6606	18000	10000	1.140	.9142
3400	1889	7.506	.0506	9600	5333	6.136	.6731	19000	10556	.962	.9247
3600	2000	8.613	.0667	9800	5444	5.872	.6851	20000	11111	.817	.9335
3800	2111	9.601	.0850	10000	5556	5.619	.6966	21000	11667	.702	.9411
4000	2222	10.450	.1051	10200	5667	5.378	.7076	22000	12222	.599	.9475
4200	2333	11.151	.1267	10400	5778	5.146	.7181	23000	12778	.516	.9531
4400	2445	11.704	.1496	10600	5889	4.925	.7282	24000	13333	.448	.9589
4600	2556	12.114	.1734	10800	6000	4.714	.7378	25000	13889	.390	.9621
4800	2667	12.392	.1979	11000	6111	4.512	.7474	26000	14445	.341	.9657
5000	2778	12.556	.2229	11200	6222	4.320	.7559	27000	15000	.300	.9689
5200	2889	12.607	.2481	11400	6333	4.137	.7643	28000	15556	.265	.9718
5400	3000	12.571	.2733	11600	6445	3.962	.7724	29000	16111	.234	.9742
5600	3111	12.458	.2983	11800	6556	3.795	.7802	30000	16667	.208	.9765
5800	3222	12.282	.3230	12000	6667	3.637	.7876	40000	22222	.0741	.9881
6000	3333	12.053	.3474	12200	6778	3.485	.7947	50000	27778	.0326	.9941
6200	3445	11.783	.3712	12400	6889	3.341	.8015	60000	33333	.0165	.9963
6400	3556	11.480	.3945	12600	7000	3.203	.8081	70000	38887	.0092	.9981
6600	3667	11.152	.4171	12800	7111	3.071	.8144	80000	44445	.0055	.9987
6800	3778	10.808	.4391	13000	7222	2.947	.8204	90000	50000	.0035	.9990
7000	3889	10.451	.4604	13200	7333	2.827	.8262	100000	55556	.0023	.9992
								∞	∞	0	1.0000

†From R. V. Dunkle, *Trans. of ASME*, $e_b(0 - \lambda T)$ **76**, (1954), p. 549.

From Table 6-1

$$\text{for } \lambda_1 T, \quad e_b(0 \rightarrow \lambda_1 T)/e_b = 0.0587$$

$$\text{for } \lambda_2 T, \quad e_b(0 \rightarrow \lambda_2 T)/e_b = 0.4604$$

so that

$$e_b(\lambda_1 \rightarrow \lambda_2)/e_b = 0.4604 - 0.0587 = 0.4017$$

or 40.17 percent of the sun's energy is given off in the visible spectrum. We will compare this result with that for a tungsten filament light bulb later on in this section.

Sample Problem 6-7 A glass plate has a transmissivity of 0.94 for wavelengths between 0.40μ and 3μ and is opaque to all other wavelengths. Determine the percent of the incident solar energy transmitted through the glass.

Solution: The temperature of the sun is taken as 10,000°R.

$$\lambda_1 T = 4000\mu°\text{R} \quad \text{and} \quad \lambda_2 T = 30{,}000\mu°\text{R}$$

From Table 6-1

$$\text{for } \lambda_1 T, \quad e_b(0 \rightarrow \lambda_1 T)/e_b = 0.1051$$

$$\text{for } \lambda_2 T, \quad e_b(0 \rightarrow \lambda_2 T)/e_b = 0.9765$$

or

$$e_b(\lambda_1 \rightarrow \lambda_2)/e_b = 0.8714$$

This means that 87.14 percent of the solar energy that is incident on the glass is in the wavelength band from 0.4 to 3μ. Since the transmissivity is 0.94, the energy transmitted is (0.94) (0.8714) or 81.9 percent of that which is incident.

6-8.1 Wavelength-Dependent Properties of Real Surfaces

Up until now we have considered only gray surfaces for which the radiation properties are constant, and we have used the average emissivity, ϵ, for that surface. In reality, the emissivity may be different for different wavelengths, and we will call the emissivity at a given wavelength the *monochromatic emissivity* and call it ϵ_λ. At a given wavelength and temperature, $\epsilon_\lambda = \alpha_\lambda$. A plot of ϵ_λ can be made for various materials. Figure 6-17 shows the general trends of ϵ_λ as a function of λ for metals and for white or light-colored nonmetals (electrical conductors and nonconductors).

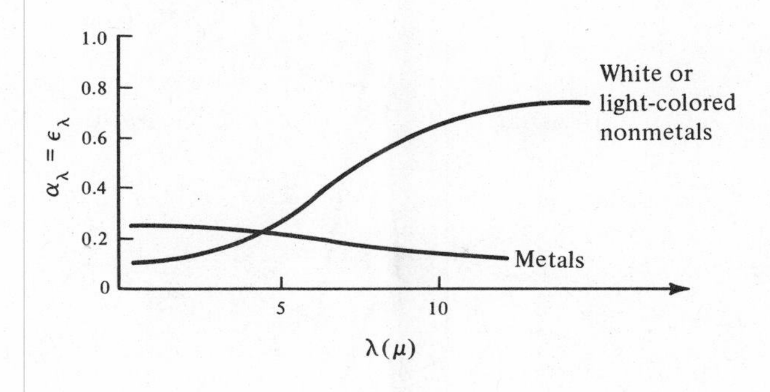

Figure 6-17 ϵ_λ as a function of λ for metals and nonmetals.

The variation of radiation properties with wavelength and temperature explains some everyday phenomena. As an example, consider a metal and nonmetal, such as a piece of aluminum and a piece of light-colored birch bark, exposed to bright sunlight. We know from experience that the metal gets hot and the nonmetal stays relatively cool. This is explained by considering the variation of radiation properties with temperature. For the incident solar energy, if we employ Wien's displacement law and use $T \approx 10{,}000°\text{R}$

$$\lambda_{max} = \frac{5215.6\mu°\text{R}}{T} = 0.52\mu$$

That is, the bulk of the radiation arrives with wavelengths around 0.5 microns. It is known that $\alpha_{0.5\mu} \approx 0.3$ for the aluminum and $\alpha_{0.5\mu} \approx 0.15$ for the bark. Consequently, the aluminum absorbs more of the incident radiation. Also, the temperature of the aluminum or the bark will be somewhere between 540°R and 660°R in the steady state, which is the temperature at which it emits radiant energy. At such temperatures, λ_{max} will be between 8μ-10μ. For the aluminum, $\epsilon_{8\mu\text{-}10\mu} \approx 0.1$ or less; for the bark, $\epsilon_{8\mu\text{-}10\mu} \approx 0.8$. This means that the wood not only absorbs less radiant energy than the aluminum, but it also emits more energy and thereby is cooler than the aluminum. One can perform an energy balance on the wood and on the aluminum and show quantitatively that the wood would be cooler than the aluminum.

Next, let us see how the variation of radiation properties with wavelength affects the energy emitted by a body. Figure 6-18 shows the nonchromatic emissive power of a real body and a graybody as a function of wavelength. The total energy emitted by the real body is the area under the e_λ versus λ curve. Note that for a graybody, $a_\lambda = \epsilon =$ constant, and the e_λ curve is the same as the $e_{b\lambda}$ curve but diminished in magnitude by the scale factor ϵ. Thus, for a gray surface, the same fraction of the total energy is emitted between two wavelengths, say λ_1 and λ_2, as for a blackbody at the same temperature. This is not true for a real surface.

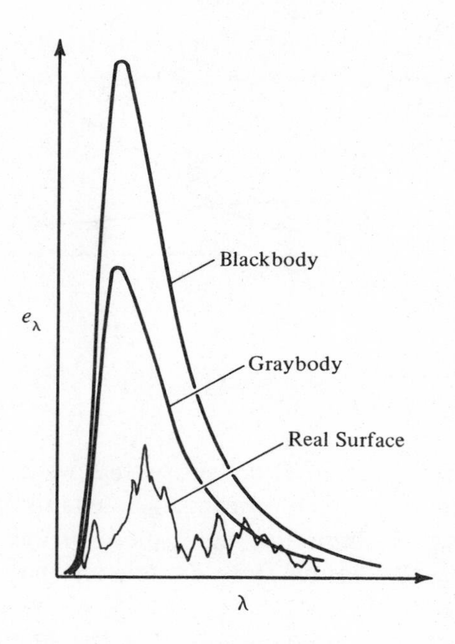

Figure 6-18 Monochromatic emissive power for real bodies and gray bodies.

Sample Problem 6-8 Consider a tungsten filament light bulb whose filament is at a temperature of 5000°R. If the filament is considered to be gray, what fraction of the total energy emitted by the bulb is in the visible wavelength spectrum from 0.35μ to 0.7μ? Comment on its efficiency as a light source.

Solution:

$$\lambda_1 T = 1750\ \mu°\text{R} \quad \text{and} \quad \lambda_2 T = 3500\ \mu°\text{R}$$

From Table 5-1

$$\text{for } \lambda_1 T, \quad e_b(0 \to \lambda_1 T)/e_b = 0.0002$$

$$\text{for } \lambda_2 T, \quad e_b(0 \to \lambda_2 T)/e_b = 0.0587$$

Therefore

$$\frac{e_b(\lambda_1 \to \lambda_2)}{e_b} = 0.0587 - 0.0002 = 0.0585$$

Only 5.85 percent of the energy is emitted in the visible wavelength range. This means 94.15 percent goes into heating the surrounding room. Thus, tungsten filament bulbs are highly inefficient as sources of light.

In view of the dependence of the radiation properties on wavelength, it is appropriate to rewrite the basic definitions for ϵ and α. We will denote surface temperature by T. *For the emissivity:*

$$\epsilon(T) = \frac{e}{e_b} = \frac{\int_0^\infty \epsilon_\lambda(T) e_{b\lambda}\, d\lambda}{\sigma T^4} \tag{6-30}$$

For the absorptivity: Let $G(T')$ be the monochromatic irradiation coming from a source at temperature T'.

$$\alpha(T, T') = \frac{\int_0^\infty \alpha_\lambda(T) G_\lambda(T')\, d\lambda}{\int_0^\infty G_\lambda(T')\, d\lambda} \tag{6-31}$$

Note that the emissivity, ϵ, is a function of the surface temperature, whereas the absorptivity, α, is a function of both the surface temperature and of the temperature of the incident radiation.

Materials such as glass and clear plastics exhibit a transmissivity. The transmissivity of these materials is highly dependent upon wavelength and the thickness of the glass or plastic. For example, common window glass is much more transparent to radiant energy in the visible spectrum than to radiant energy in the intermediate and far infrared regions. It is this dependency on wavelength that explains the greenhouse effect. That is, the windows behave like a check valve letting in radiant energy from the sun but stopping (acting opaque to) the radiant energy emitted by objects inside the greenhouse.

6-9 DIRECTIONAL ASPECTS OF EMITTED RADIATION

The energy emitted by a surface streams away in various directions from the emitting surface. The *intensity of radiation, i*, is the radiant energy leaving the surface per unit area normal to the direction of the rays, per unit solid angle, and per unit time. For black surfaces, the intensity, i_b, is uniform in all directions. The same is true for any surface that is a diffuse emitter. The definition of intensity applies both to monochromatic and total radiation. Here we discuss only total radiation.

We take an elemental black surface area, dA, at O (Figure 6-19) with OP as a normal to it. If we select an arbitrary direction OQ making an angle θ with the normal OP, then the "normal" area as seen from Q is not dA but $dA \cos\theta$. Also, let $d\omega$ represent a small solid angle around the line OQ. Then, from the definition of the intensity, the energy, $d\Phi$, passing through the solid angle is

$$d\Phi = i_b \, dA \cos\theta \, d\omega \tag{6-32}$$

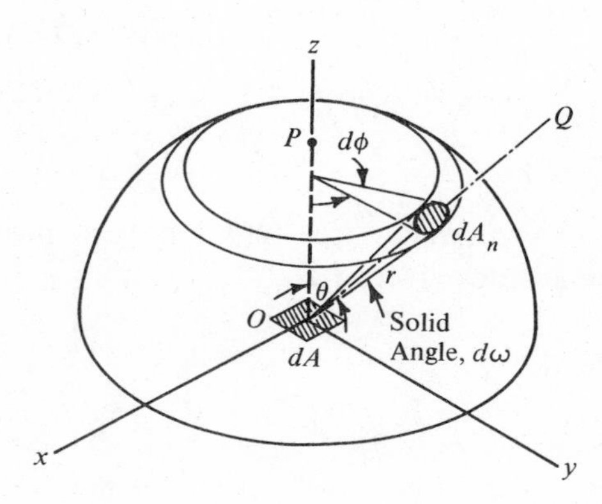

Figure 6-19 Nomenclature for intensity of radiation.

We can proceed to establish a relation between the intensity of the radiation, i_b, and the emissive power, e_b. If we sum up all the components of the energy arriving on the inside of the hemisphere above the surface, dA, we must have

$$\int d\Phi = e_b \, dA = \int i_b \, dA \cos\theta \, d\omega$$

The elemental solid angle $d\omega$ is given by

$$d\omega = \frac{dA_n}{r^2}$$

where $dA_n = (r \sin\theta \, d\phi)(r d\theta)$. Hence

$$e_b \, dA = \iint i_b \, dA \cos\theta \, (r^2 \sin\theta \, d\phi \, d\theta)/r^2$$

or

$$e_b = \int_{\theta=0}^{\pi/2} \int_{\phi=0}^{2\pi} i_b \sin\theta \cos\theta \, d\theta \, d\phi$$

Since i_b is a constant, it can be removed from the integral sign. Integration with respect to ϕ gives

$$e_b = 2\pi i_b \int_{\theta=0}^{\pi/2} \sin\theta \cos\theta \, d\theta$$

Integrating again with respect to θ results in

$$e_b = \pi i_b \tag{6-33}$$

For nondiffuse surfaces, a directional emissivity, ϵ_θ, can be defined as

$$\epsilon_\theta = i(\theta)/i_b$$

where $i(\theta)$ indicates that i is a function of θ. Figure 6-20 indicates the nomenclature for various emissivities that have been investigated by research workers.

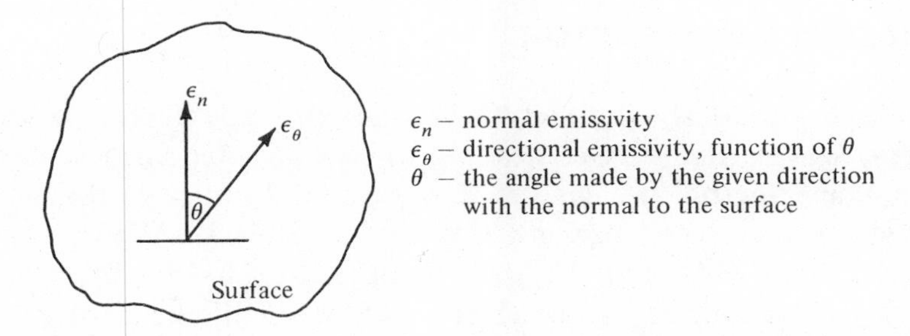

Figure 6-20 Nomenclature for various types of emissivities.

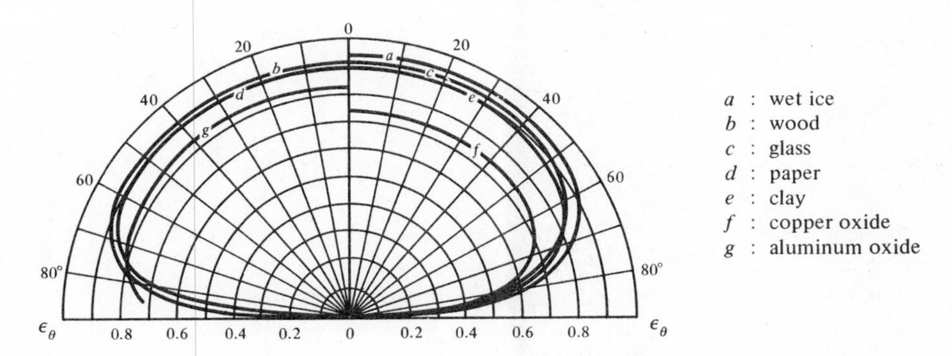

Figure 6-21 Directional variation of emissivity for several electrical nonconductors, from *Forschung im Ingenieurwesen*, **6**, pp. 175/1935–VDI–VERLAG GmbH-DÜSSELDORF.

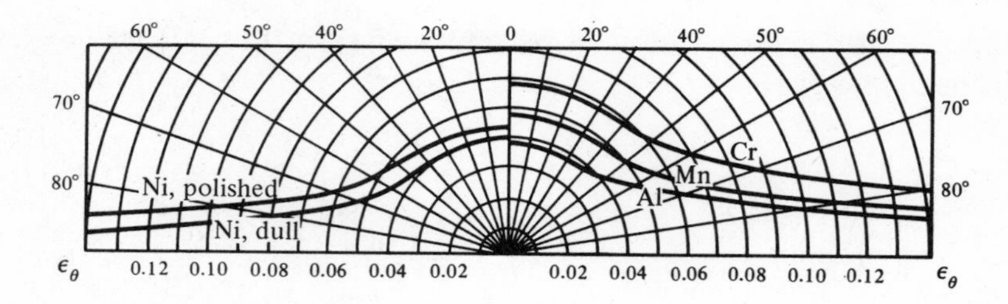

Figure 6-22 Directional variation of emissivity for several metals, from *Forschung im Ingenieurwesen*, **6**, pp. 175/1935–VDI– VERLAG GmbH–DÜSSELDORF.

The emissivity utilized previously is the weighted average emissivity through a hemispherical solid angle of 2π steradians. The quantity, ϵ_n, is the emissivity in the direction normal to the surface. Figures 6-21 and 6-22, respectively, give the variation of emissivity for several nonconductors and metals.

6-10 RADIATION SHAPE FACTOR

We have previously defined the radiation shape factor as the fraction of the energy leaving one surface heading toward another surface. Let us see how radiation shape factors are calculated. Figure 6-23 indicates the geometrical notation for shape factor calculations. Consider two black surfaces, A_1 and A_2, and elemental areas, dA_1 and dA_2, on them. Let i_b be the intensity of the radiation leaving a black surface. Since a black surface is a diffuse emitter, i_b is the same in all directions. We have already shown that for a blackbody, $i_b = e_b/\pi$.

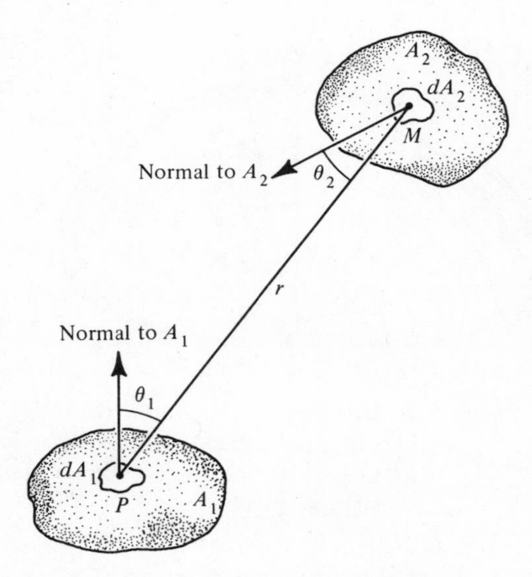

Figure 6-23 Geometrical notation for shape factor calculations.

The rate of energy, $d\Phi$, leaving surface, dA_1, and headed toward dA_2 along the direction PM will be proportional to the projection of the area, dA_1, along a plane normal to PM and to the solid angle subtended by dA_2 as viewed from P.

$$d\Phi = i_{b1} dA_1 \cos\theta_1 \, d\omega \tag{6-34}$$

where

i_{b1} = intensity of radiation from dA_1 = e_{b1}/π

$dA_1 \cos\theta_1$ = projection of area element, dA_1, along a plane normal to PM

$d\omega$ = solid angle subtended by receiving area dA_2

$$= \frac{dA_2 \cos\theta_2}{r^2}$$

Substituting for $d\omega$ and i_b from the preceding equations, we have

$$d\Phi = e_{b1} dA_1 \left(\frac{\cos\theta_1 \cos\theta_2 \, dA_2}{\pi r^2} \right)$$

The flux of energy leaving dA_1 and arriving at all the elements dA_2 that make up A_2 is obtained by integration over A_2.

$$\Phi_{dA_1-A_2} = e_{b1} dA_1 \int_{A_2} \frac{\cos\theta_1 \cos\theta_2 \, dA_2}{\pi r^2}$$

Then, the contribution of all elements dA_1 that make up A_1 may be summed up by integrating over A_1 giving

$$\Phi_{1\text{-}2} = e_{b1} \int_{A_1} \int_{A_2} \frac{\cos\theta_1 \cos\theta_2 \, dA_2 \, dA_1}{\pi r^2}$$

This is the radiant flux that leaves surface 1 and arrives at surface 2.

The radiation leaving area, A_1, in all directions is

$$\Phi_1 = e_{b1} A_1$$

The shape factor $F_{1\text{-}2}$ follows from its definition

$$F_{1\text{-}2} = \Phi_{1\text{-}2}/\Phi_1$$

so that

$$A_1 F_{1\text{-}2} = \int_{A_1} \int_{A_2} \frac{\cos \theta_1 \cos \theta_2 \, dA_1 \, dA_2}{\pi r^2} \tag{6-35}$$

Note that equation (6-35) is a quadruple integral. When performing the integration, it is necessary to express θ_1, θ_2, and r in terms of the variables contained in dA_1 and dA_2.

As an example of calculating shape factors, let us consider the shape factor of an elemental area, dA_1, with respect to a hemisphere placed above it as shown in Figure 6-24.

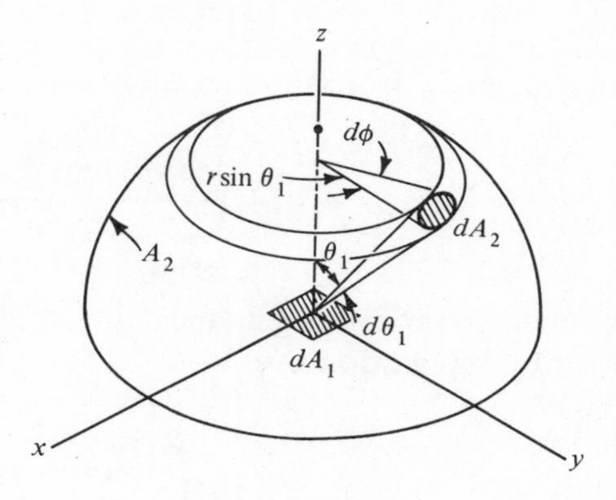

Figure 6-24 Shape factor between an elemental area and a hemispherical cap.

From equation (6-35)

$$A_1 F_{1\text{-}2} = \int_{A_1} \int_{A_2} \frac{\cos \theta_1 \cos \theta_2 \, dA_1 \, dA_2}{\pi r^2}$$

Since A_1 is infinitesimal, the above equation becomes

$$dA_1 F_{1\text{-}2} = dA_1 \int_{A_2} \frac{\cos \theta_1 \cos \theta_2 \, dA_2}{\pi r^2}$$

Note that $\cos\theta_2 = 1$ and that

$$dA_2 = (r \sin\theta_1 \, d\phi)(r \, d\theta_1)$$

Hence

$$F_{1\text{-}2} = \int_{\theta_1=0}^{\pi/2} \int_{\phi=0}^{2\pi} \frac{\cos\theta_1 \sin\theta_1 \, d\theta_1 \, d\phi}{\pi}$$

Noting that $\sin 2\theta = 2\sin\theta\cos\theta$

$$F_{1\text{-}2} = 2\pi \int_{\theta_1=0}^{\pi/2} \left(\frac{\sin 2\theta_1}{2\pi}\right) d\theta_1$$

$$F_{1\text{-}2} = \left[\frac{-\cos 2\theta_1}{2}\right]_0^{\pi/2} = 1$$

which means that all the energy leaving the elemental surface reaches the hemisphere, as would be expected.

6-11 RADIATION IN GASES

The analysis of radiant energy exchange between a gas and a heat transfer surface is considerably more complex than the situations discussed in the preceding sections. Although it is true that for most gases the transmissivity is unity, there are a number of cases of engineering importance in which absorption and emission from gases cannot be neglected. Most significantly, hydrocarbons, SO_2, CO_2, CO, ammonia, and water vapor have a transmissivity other than unity in various wavelength bands, which means that they may absorb, scatter, and emit radiant energy at these wavelengths. Because of the strong dependency on wavelength, it is unwise to model these gases as graybodies, thereby adding another dimension to the complexity of the problem. In addition, the radiation properties are also functions of the thickness of the gas layer. That is, the greater the thickness of the gaseous body, the greater is its absorptivity.

Consider a layer of gas L units thick as shown in Figure 6-25. If we let a beam of monochromatic radiation of intensity, $i_{\lambda,0}$, impinge on this body of gas, we find that the intensity coming out of the gas layer, $i_{\lambda,L}$, is less than $i_{\lambda,0}$. A quantity of energy is, therefore, absorbed in the layer of gas.

The decrease in intensity ($-di_{\lambda,x}$) over an infinitesimal thickness, dx, is found to be proportional to the intensity, $i_{\lambda,x}$, and to the thickness, dx. Thus

$$-di_{\lambda,x} = k_\lambda i_{\lambda,x} dx$$

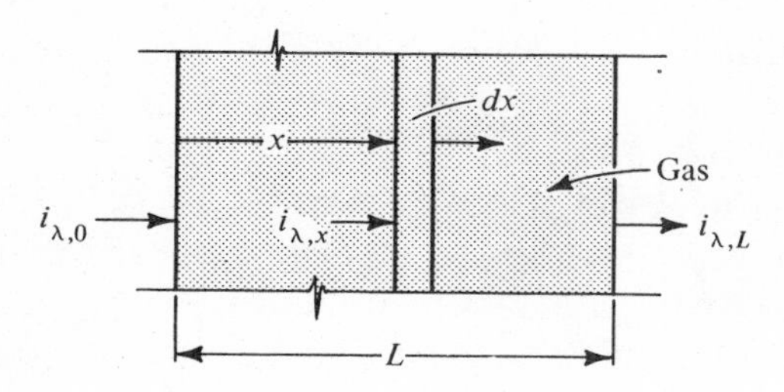

Figure 6-25 Absorption in gas.

where $i_{\lambda,x}$ is the intensity at $x = x$ and k_λ is a proportionality constant. The quantity, k_λ, is referred to as the monochromatic absorption coefficient, a property of the gas in question. The above equation, after separating the variables $i_{\lambda,x}$ and x, can be integrated between the limits $x = 0$ and $x = L$ to give

$$\Big[\ln i_{\lambda,x}\Big]_{i_{\lambda,0}}^{i_{\lambda,L}} = \Big[-k_\lambda x\Big]_0^L$$

or

$$i_{\lambda,L} = i_{\lambda,0} e^{-k_\lambda L}$$

The monochromatic absorptivity, α_λ, of the gas is given by

$$\alpha_\lambda = \frac{i_{\lambda,0} - i_{\lambda,L}}{i_{\lambda,0}} = \frac{i_{\lambda,0} - i_{\lambda,0} e^{-k_\lambda L}}{i_{\lambda,0}}$$

or

$$\alpha_\lambda = 1 - e^{-k_\lambda L} \tag{6-36}$$

The above expression is restricted to monochromatic radiation and takes into account a single direction. In a practical situation, it is necessary to account for the entire spectrum of wavelengths of the radiant energy and of all the possible directions in which it propagates through the gaseous body. Further complexities arise due to the fact that the monochromatic absorption coefficient, k_λ, is not a constant quantity but depends on the temperature and the pressure of the gas. For a thorough discussion of radiation through gases, the reader is referred to Reference 1.

Although we have indicated that it is unwise to model gases as graybodies, let us see what happens to the heat transfer rate between two infinite parallel planes when a gray gas is placed between them. We assume that a gas at a uniform temperature, T_g, is confined between two infinite parallel plates. The plates are at uniform temperatures, T_1 and T_2, respectively. We further assume that the nature of the gas is such that for the wavelength ranges relevant to the temperature, T_g, it behaves like a graybody. Figure 6-26 shows a sketch of the problem under investigation. The analysis, however crude, does give an idea of the effect of placing a participating gas in an enclosure.

In performing our analysis, we will assume that the gas is isothermal and nonreflecting, but the gas does absorb and reemit radiant energy as well as permit the direct transfer of radiant energy between surfaces 1 and 2 as a consequence of its transmissivity. A network approach will be used to reinforce that which was presented in the section on graybody radiant energy exchange.

We begin by noting that the radiant energy per unit time leaving surface 1 that is transmitted through the gas and reaches surface 2 is

$$A_1 F_{1\text{-}2} J_1 \tau_g$$

and that which leaves surface 2 and reaches surface 1 is

$$A_2 F_{2\text{-}1} J_2 \tau_g$$

The net radiant energy loss by direct transmission is then equal to

$$Q_1 = A_1 F_{1\text{-}2} \tau_g (J_1 - J_2) \qquad (6\text{-}37)$$

where the reciprocity relationship, equation (6-6a), has been applied. Since $\rho_g = 0$, we have

$$\tau_g + \alpha_g = 1$$

Since the gas is considered to be gray, $\alpha_g = \epsilon_g$, and

$$\tau_g + \epsilon_g = 1$$

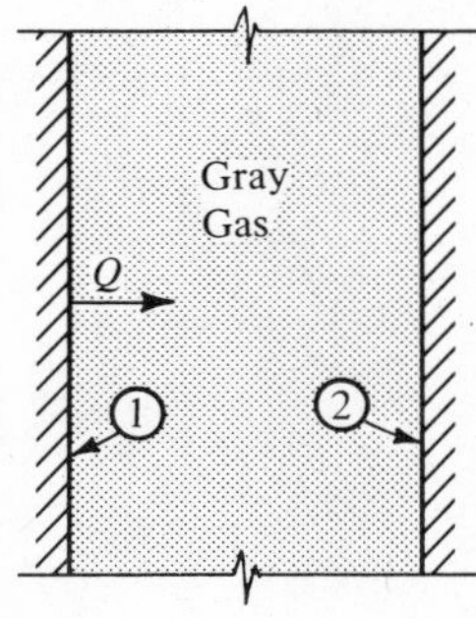

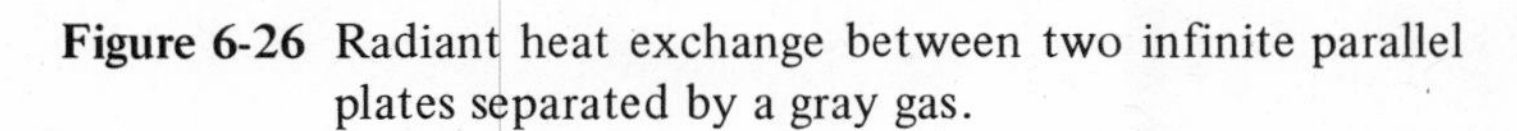
Figure 6-26 Radiant heat exchange between two infinite parallel plates separated by a gray gas.

We may therefore write

$$Q_1 = \frac{J_1 - J_2}{\dfrac{1}{A_1 F_{1\text{-}2}(1 - \epsilon_g)}} \tag{6-38}$$

Equation (6-38) accounts for the radiosities of the participating surfaces of the enclosure containing a gray gas. Note that radiosities are a function of surface properties alone. The appropriate network element used to describe equation (6-38) is shown in Figure 6-27a.

Next, let us consider the radiant energy, J_g, emitted by the isothermal gas.

$$J_g = \epsilon_g e_{bg}$$

The energy leaving the gas and reaching surface 1 is

$$A_g F_{g\text{-}1} \epsilon_g e_{bg}$$

In the above expression, A_g is the total surface area of the gas body and $F_{g\text{-}1}$ and $F_{g\text{-}2}$ are the appropriate shape factors between the gas body and surfaces 1 and 2. The interested reader may consult Reference 1 for the calculation of $F_{g\text{-}1}$ and $F_{g\text{-}2}$. The energy leaving the gas and reaching surface 2 is

$$A_g F_{g\text{-}2} \epsilon_g e_{bg}$$

Radiant energy that leaves surface 1 or 2 is said to have reached the gas if it is absorbed by the gas. That is, the radiant energy leaving surface 1 and reaching the gas is

$$\epsilon_g A_1 F_{1\text{-}g} J_1$$

and the radiant energy leaving surface 2 and reaching the gas is

$$\epsilon_g A_2 F_{2\text{-}g} J_2$$

Consequently, the net radiant energy loss by the gas becomes

$$Q_g = \underbrace{(A_g F_{g\text{-}1} \epsilon_g e_{bg} - A_1 F_{1\text{-}g} \epsilon_g J_1)}_{\text{net loss to surface 1}} + \underbrace{(A_g F_{g\text{-}2} \epsilon_g e_{bg} - A_2 F_{2\text{-}g} \epsilon_g J_2)}_{\text{net loss to surface 2}}$$

or, using reciprocity, we write

$$Q_g = \frac{e_{bg} - J_1}{1/(A_1 F_{1\text{-}g} \epsilon_g)} + \frac{e_{bg} - J_2}{1/(A_2 F_{2\text{-}g} \epsilon_g)} \tag{6-39}$$

Equation (6-39) can be represented by the network shown in Figure 6-27b.

In order to complete the network for the present problem, we must consider surface resistances. We recognize that the resistances will be the same as those developed in Section 6-4. Each surface resistance will have a value of $[(1 - \epsilon)/\epsilon A]$.

We can now construct Figure 6-27c to show the complete network for the exchange of radiant energy between two infinite parallel plates when the space between them is occupied by a gray transmitting, emitting gas. Note the similarity of the electrical networks in Figure 6-11c and Figure 6-27c. It is due to the fact that, under steady state, the gas does not lose or gain energy, or $Q_g = 0$. This results in a floating potential, e_{bg}, in Figure 6-27c.

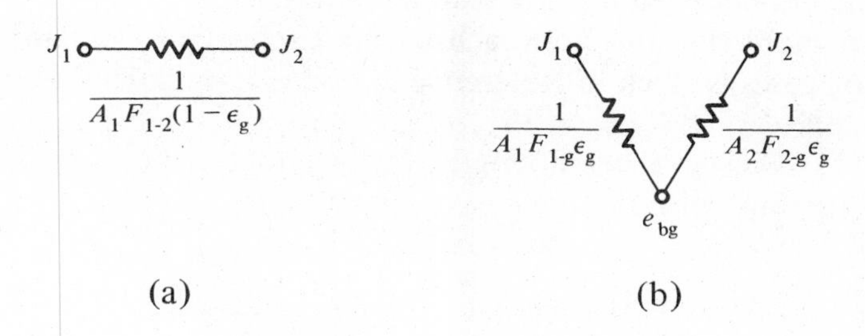

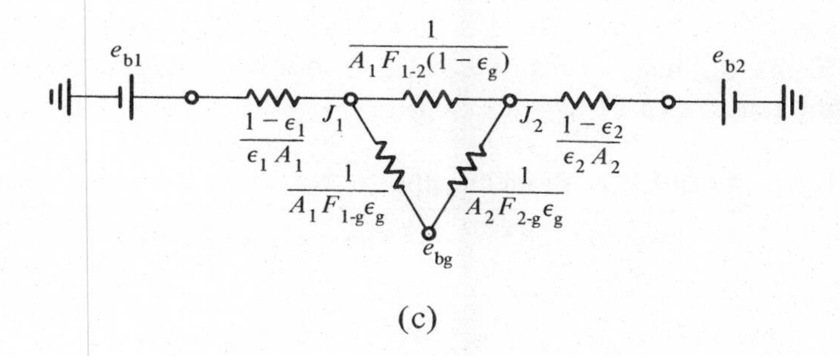

Figure 6-27 (a) Network element for radiant energy transmitted through a gray gas. (b) Network element for radiant energy lost by gray gas to two surfaces. (c) Network for radiant energy exchange between infinite parallel plates with participating gas between them.

With T_1 and T_2 fixed, the effect of introducing the gas has been to replace the spatial resistance, $(1/A_1F_{1-2})$, appearing between J_1 and J_2, with a new resistance having a value of

$$R = \frac{\dfrac{1}{A_1F_{1\text{-}2}(1-\epsilon_g)}\left(\dfrac{1}{A_1F_{1\text{-}g}\epsilon_g}+\dfrac{1}{A_2F_{2\text{-}g}\epsilon_g}\right)}{\dfrac{1}{A_1F_{1\text{-}2}(1-\epsilon_g)}+\dfrac{1}{A_1F_{1\text{-}g}\epsilon_g}+\dfrac{1}{A_2F_{2\text{-}g}\epsilon_g}} \tag{6-40}$$

The above is obtained from Figure 6-27c utilizing standard techniques for finding the equivalent resistance of a series-parallel circuit.

The introduction of a participating gas increases the resistance between nodes of potentials J_1 and J_2 and the gas, in effect, behaves like a radiation shield. This reduces heat flow between surfaces 1 and 2. We see that if $\epsilon_g = 0$, the network of Figure 6-27c will degenerate into that of Figure 6-9 for the case of a nonparticipating gas between infinite parallel plates.

A more rigorous approach to the exchange of radiant energy through participating gases is given in Reference 2, and is summarized below:

(1) Define a beam length, L, which determines the radiation characteristics of a gaseous volume

$$L = \frac{4V}{A_s} \tag{6-41}$$

where V is the volume of the gas, and A_s is the area of the bounding surface of the gas.

Table 6-2 lists values of L for commonly encountered geometries as calculated from equation (6-41).

TABLE 6-2 Beam Lengths

Shape	*Characteristic Dimension*	*Beam Length eq. (6-41)*
Sphere	Diameter, D	0.66D
Infinite Cylinder	Diameter, D	D
Cylinder height = diameter	Diameter, D	0.66D
Space between tubes on equilateral triangle, clearance = diameter	Diameter, D	3.45D

(2) Let p be the partial pressure (in atmospheres) of the participating gas. Then define a parameter, called the optical length, L_0, which is equal to $p \times L$

$$L_0 = pL$$

(3) Gas emissivities can be plotted as a function of L_0 and temperature. The two most commonly encountered gases of interest (because of combustion) are H_2O and CO_2. Figure 6-28 gives their emissivities as functions of L_0 and temperature.

(4) If both H_2O and CO_2 are present, the corrected emissivity is given by

$$\epsilon_g = \epsilon_{H_2O} + \epsilon_{CO_2} - \Delta\epsilon$$

where $(-\Delta\epsilon)$ takes into account the mutual absorption that takes place between the two gases. Figure 6-29 plots $\Delta\epsilon$ for $(H_2O - CO_2)$ mixtures as a function of $[p_w/(p_w + p_c)]$, where p_w and p_c are the partial pressures of the water vapor and CO_2, respectively. The parameters used in these figures are the summation of the optical lengths for H_2O and CO_2, and temperature.

(5) The absorptivity of the gases H_2O and CO_2 is found as follows:

(a) For CO_2, determine α_{CO_2} at T_s (temperature of the surface bounding the gas) from Figure 6-28a using $[L_0 (T_s/T_g)]$ instead of simply using L_0. Take this value of α_{CO_2} and multiply by $(T_g/T_s)^{0.65}$ to obtain the corrected value of α_{CO_2}.

(b) For H_20, determine α_{H_2O} at T_s, from Figure 6-28b using $[L_0(T_s/T_g)]$ instead of L_0. Take this value of α_{H_2O} and multiply it by $(T_g/T_s)^{0.45}$ to determine the corrected value of α_{H_2O}.

The corrections made above are necessary to account for the fact that the temperature of the gas itself as well as that of the surrounding surface has an effect on its absorptivity.

(6) A mutual absorptivity correction, $\Delta\alpha$, can be determined from Figure 6-29 in a fashion similar to that of determining $\Delta\epsilon$.

(7) If the bounding surface is black, the heat flow from the gas becomes

$$Q_g = \epsilon_g A_s \sigma T_g^4$$

(8) The radiant energy leaving the surrounding surface and absorbed by the gas is

$$Q_s = \alpha_g A_s \sigma T_s^4$$

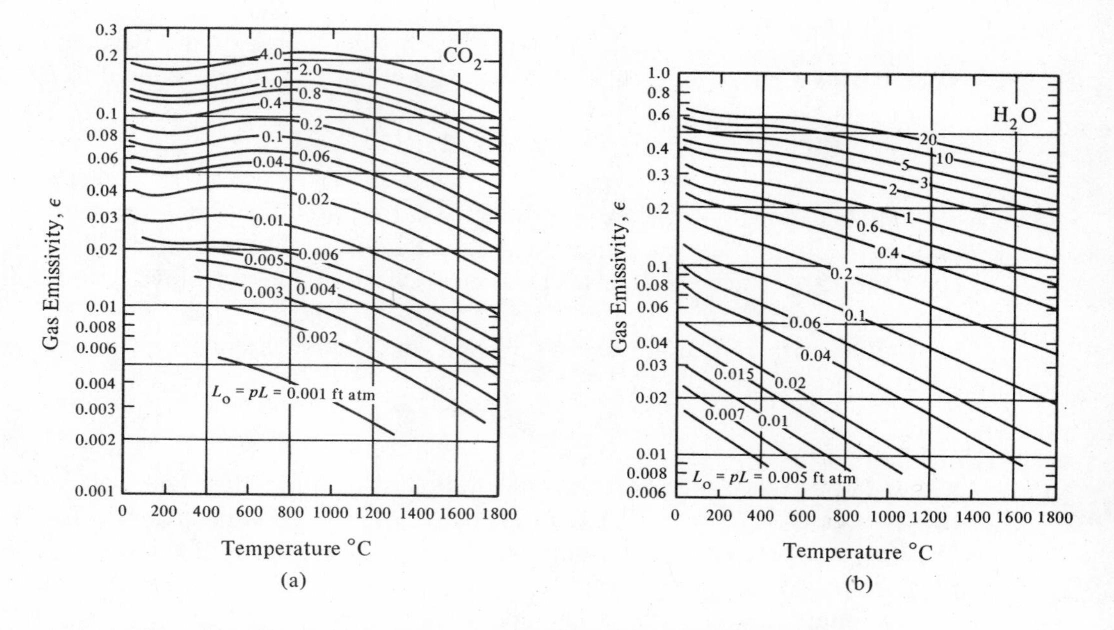

Figure 6-28 Emissivities of (a) carbon dioxide and (b) water vapor, from *Heat Transmission* by W. H. McAdams. Copyright 1954 by W. H. McAdams. Used with permission of McGraw-Hill Book Company.

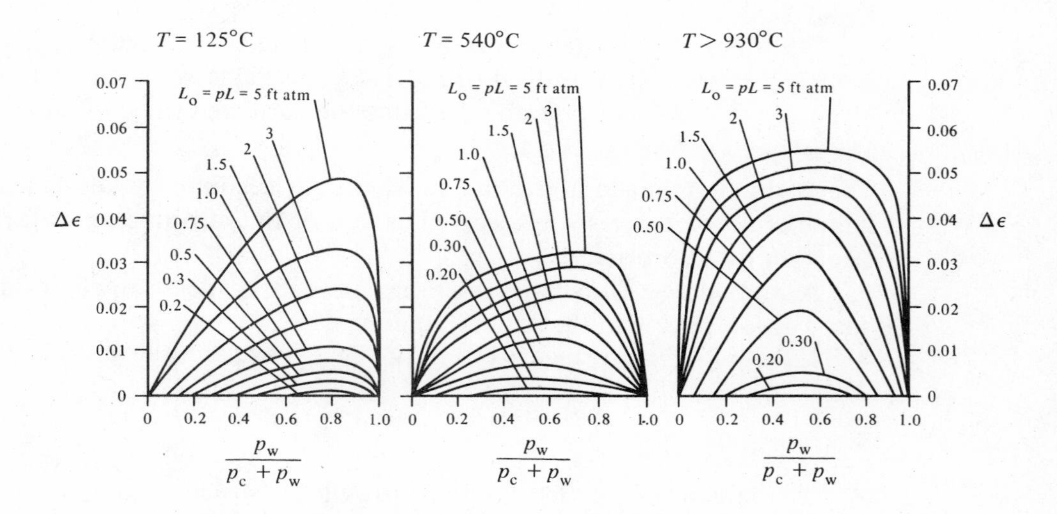

Figure 6-29 Correction for mutual absorptivity of water vapor and carbon dioxide, from *Heat Transmission* by W. H. McAdams. Copyright 1954 by W. H. McAdams. Used with permission of McGraw-Hill Book Company.

(9) Combining the above, we write the net radiant interchange between the gas and the surrounding surface as

$$Q = Q_g - Q_s = \sigma A_s(\epsilon_g T_g^4 - \alpha_g T_s^4)$$

The above procedure can best be illustrated by an example.

Sample Problem 6-11 A gas turbine combustion chamber is 1 ft in diameter and the walls are maintained at 940°F. The products of combustion are at 1840°F, a pressure of 1 atm, and contain 15 percent by volume of CO_2 and 15 percent by volume of H_2O. Assuming the combustion chamber to be very long (i.e., an infinite cylinder for mathematical purposes), determine the net radiant energy exchange between the gases and the combustion chamber wall.

Solution:

$$T_{CO_2} = T_{H_2O} = 1840°F = 1005°C \quad \text{and} \quad T_s = 940°F = 505°C$$

In view of Dalton's law of partial pressures, the partial pressure of a given gas in a mixture is equal to the product of the total pressure of the mixture and the volume fraction of the given gas. Thus

$$P_{CO_2} = P_{H_2O} = 0.15 \text{ atm}$$

From Table 6-2

$$L = D = 1 \text{ ft}$$

Determination of ϵ_{H_2O}:

$$L_0 = p_{H_2O} L = 0.15 \text{ ft atm}$$

From Figure 6-28b

$$\epsilon_{H_2O} = 0.06$$

Determination of ϵ_{CO_2}:

$$L_0 = p_{CO_2} L = 0.15 \text{ ft atm}$$

From Figure 6-28a

$$\epsilon_{CO_2} = 0.08$$

Determination of $\Delta\epsilon$:

$$p_{CO_2}L + p_{H_2O}L = 0.3 \text{ ft atm}$$

and

$$\frac{p_{H_2O}}{p_{H_2O} + p_{CO_2}} = 0.5$$

From Figure 6-29

$$\Delta\epsilon = 0.005$$

Consequently

$$\epsilon_g = \epsilon_{CO_2} + \epsilon_{H_2O} - \Delta\epsilon = 0.135$$

Determination of α_{CO_2}:

$$p_{CO_2}L\left(\frac{T_s}{T_g}\right) = (0.15)\left(\frac{778°K}{1278°K}\right) = 0.091 \text{ ft atm}$$

Referring to Figure 6-28a

$$\epsilon_{CO_2} = 0.07$$

It follows from step 5a that

$$\alpha_{CO_2} = (0.07)\left(\frac{T_g}{T_s}\right)^{0.65} = (0.07)\left(\frac{1278}{778}\right)^{0.65}$$

$$\alpha_{CO_2} = 0.097$$

Determination of α_{H_2O}:

$$p_{H_2O}L\left(\frac{T_s}{T_g}\right) = 0.091 \quad \text{at } T_s = 505°C, \epsilon_{H_2O} = 0.075$$

and

$$\alpha_{H_2O} = (0.075)\left(\frac{T_g}{T_s}\right)^{0.45} = (0.075)\left(\frac{1278}{778}\right)^{0.45}$$

$$\alpha_{H_2O} = 0.094$$

Proceeding according to step 6, we find that

$$\Delta\alpha_g \simeq 0$$

Therefore

$$\alpha_g = \alpha_{CO_2} + \alpha_{H_2O} - \Delta\alpha_g = 0.191$$

Now

$$\frac{Q_g}{A_s} = \epsilon_g \sigma T_g^4 = (0.135)(0.1713 \times 10^{-8})(2300)^4$$

$$\frac{Q_g}{A_s} = 6471 \text{ Btu/hr-ft}^2$$

and

$$\frac{Q_s}{A_s} = \alpha_g \sigma T_s^4 = (0.191)(0.1713 \times 10^{-8})(1400)^4 = 1257 \text{ Btu/hr-ft}^2$$

which gives

$$\frac{Q_g}{A_s} - \frac{Q_s}{A_s} = 5214 \text{ Btu/hr-ft}^2$$

PROBLEMS (English Engineering System of Units)

6-1. Calculate the emissive power of a blackbody at (a) 0°F, (b) 150°F, (c) 400°F, and (d) 10,000°F.

6-2. Calculate the emissive power at 400°F for (a) plaster, (b) polished aluminum, and (c) heavily oxidized aluminum.

6-3. Using Figures (6-5) through (6-8) determine $F_{1\text{-}2}$ for the geometries in EFigure 6.3.

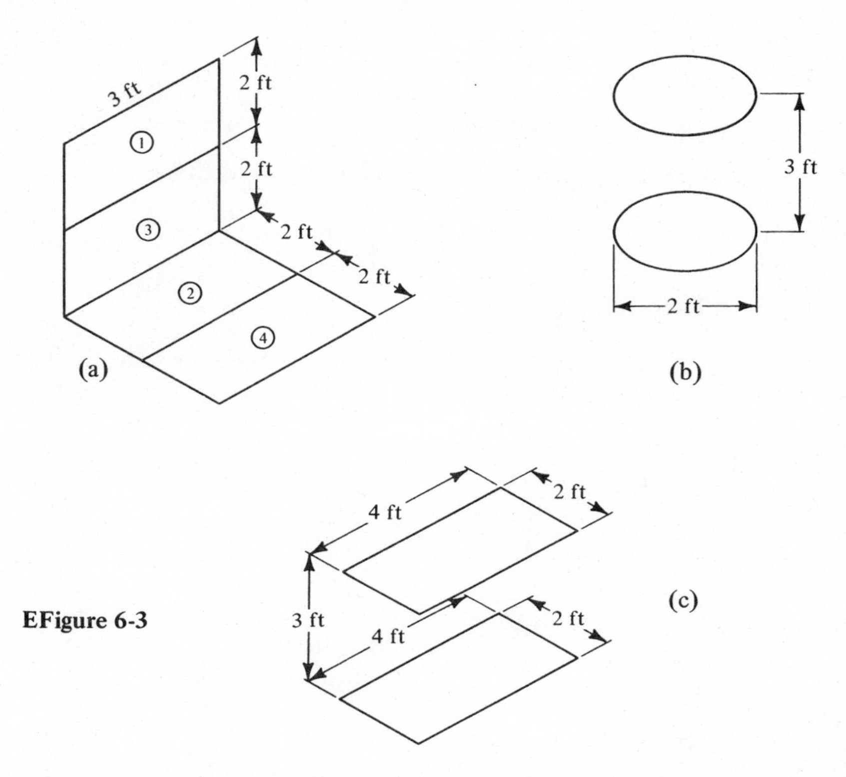

EFigure 6-3

6-4. A bottle of chilled martinis has an inside surface area of 100.5 square inches. When the top is removed, a hole 0.5 square inches in area results. Find the shape factor from the inside of the bottle to your thumb, which is placed across the hole when the bottle is empty.

6-5. Using charts and taking advantage of symmetry, determine $F_{1\text{-}2}$ for the geometry in EFigure 6-5.

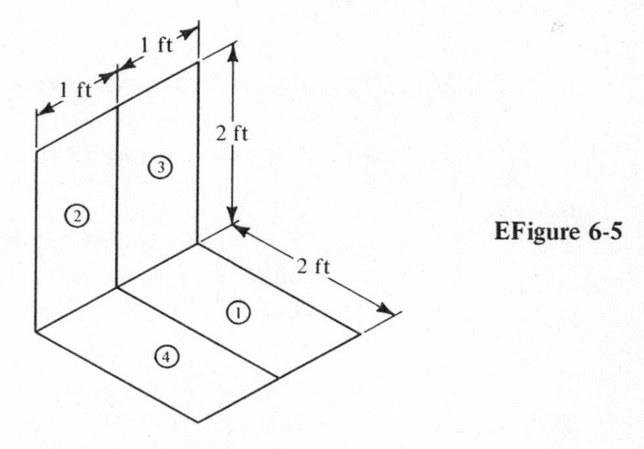

EFigure 6-5

6-6. A cube 1 foot on a side is located inside a second cube 2 feet on a side. If the cubes do not touch, determine the shape factor between the outer surface of the inner cube and the inner surface of the outer cube.

6-7. Determine the net radiant heat flow between two parallel black discs 2 feet in diameter separated by a distance of 3 feet if their surface temperatures are 1000°F and 500°F. Assume that the surroundings are at 0°R.

6-8. A 1-foot-square plate, which is thermally insulated on one side, is located in a vacuum. The other side of the plate receives incident radiant energy at the rate of 400 Btu/hr-ft^2. If the surface is gray, determine its steady-state temperature.

6-9. Consider N graybodies in an enclosure exchanging radiant energy with each other and nothing else. In an electrical analog, how many surface resistances would you have, and how many space resistances would you have?

6-10. Consider a groove cut in a black surface as sketched in EFigure 6-10. From the data given, determine the heat loss from the groove.

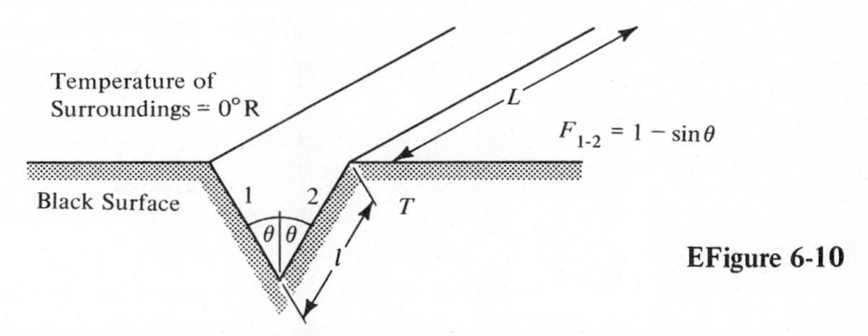

EFigure 6-10

6-11. A wire maintained at 1000°F having a diameter of 0.012 inches is located in a large enclosure whose walls are maintained at 200°F. If the emissivity of the wire is 0.3, determine the heat transfer between the wire and the enclosure.

6-12. Rework problem 6-7, only this time let the discs be connected by a reradiating but nonconducting wall.

6-13. Determine the temperature of the reradiating but nonconducting wall in problem 6-12.

6-14. Two parallel square plates 4 feet on a side placed 1 foot apart are placed in a large room whose walls are maintained at 70°F. The plates are maintained at 1040°F and 540°F and have emissivities of 0.5 and 0.6, respectively. Determine the heat lost by each plate and the net flow of radiant energy to the walls of the room.

6-15. If the two plates of problem 6-14 were located on the sides of spacecrafts and used to transfer heat from one vehicle to the other, how much energy would be transmitted? Assume that outer space is a blackbody at 0°R.

6-16. If the two plates of problem 6-14 were connected by a reradiating and nonconducting wall, what would be the net heat transfer between the plates.

6-17. A high pressure steam pipe 6 inches O.D. has an outer surface temperature of 350°F. If it is located in a large room whose walls are maintained at 80°F, determine the heat lost per linear foot of pipe. The emissivity of the outer surface of the pipe is 0.7.

6-18. Two large parallel plates have emissivities of 0.2 and 0.8 and are maintained at 1600°F and 500°F, respectively. If a radiation shield having an emissivity of 0.05 is placed between the plates, determine:
(a) the heat flow without the shield,
(b) the heat flow with the shield, and
(c) the temperature of the shield.

6-19. Consider three infinite parallel plates. Plate 1 is maintained at 2000°R, and plate 3 is maintained at 200°R. $\epsilon_1 = 0.8$, $\epsilon_2 = 0.5$, and $\epsilon_3 = 0.8$. Plate 2 is placed between plates 1 and 2 and receives no heat from external sources. What is the temperature of plate 2?

6-20. A thermocouple is placed in a large duct to measure the temperature of the air flowing through the duct. The duct wall is at 1000°F and the air is at 2000°F. Neglect any conduction in the thermocouple leads, and assume that the convective heat transfer coefficient between the thermocouple junction and the air is h.
(a) Will the thermocouple indicate an air temperature that is too high or too low?
(b) To cause the error in the temperature measurement to go to zero, what should be the value of h?
(c) If h were a fixed constant, how would you alter the absorptivity of the thermocouple junction to decrease error in the measurement?

6-21. A thermocouple is placed in the heating duct of a building where the circulated air has a temperature of 140°F. Will there be a significant error due to radiation losses? Why?

6-22. A thermocouple indicates a temperature of 1000°F when placed in a duct where a hot gas at 1100°F is flowing. If the convective heat transfer coefficient between the thermocouple and the gas is 20 Btu/hr-ft^2°F, estimate the duct wall temperature. The thermocouple has an emissivity of 0.5.

6-23. In problem 6-22, if the duct wall temperature were known to be 600°F, determine the value of the convective heat transfer coefficient.

6-24. On a clear night, the effective radiation temperature of the sky may be taken as −100°F. If you neglect the latent heat of vaporization for water, take the emissivity of water as unity, and assume the convective heat transfer coefficient between the dew and the air to be 5 Btu/hr-ft^2°R, estimate the temperature at which ice will begin to form.

6-25. A gas turbine combustion chamber is 18 inches in diameter and its walls are maintained at 1400°F. The products of combustion are at 2750°F, a pressure of one atmosphere, and contain 10 percent by volume of CO_2 and 20 percent by volume of H_2O. Assuming the combustion chamber to be a very long cylinder, determine the net radiant energy exchange between the gases and the combustion chamber walls.

PROBLEMS (SI System of Units)

6-1. Calculate the emissive power of a blackbody at (a) 0°C, (b) 70°C, (c) 200°C, and (d) 6000°C.

6-2. Calculate the emissive power of plaster, polished aluminum, and heavily oxidized aluminum at 300°C.

6-3. Using Figures (6-5 to 6-8), determine $F_{1\text{-}2}$ for the geometries in SIFigure 6-3.

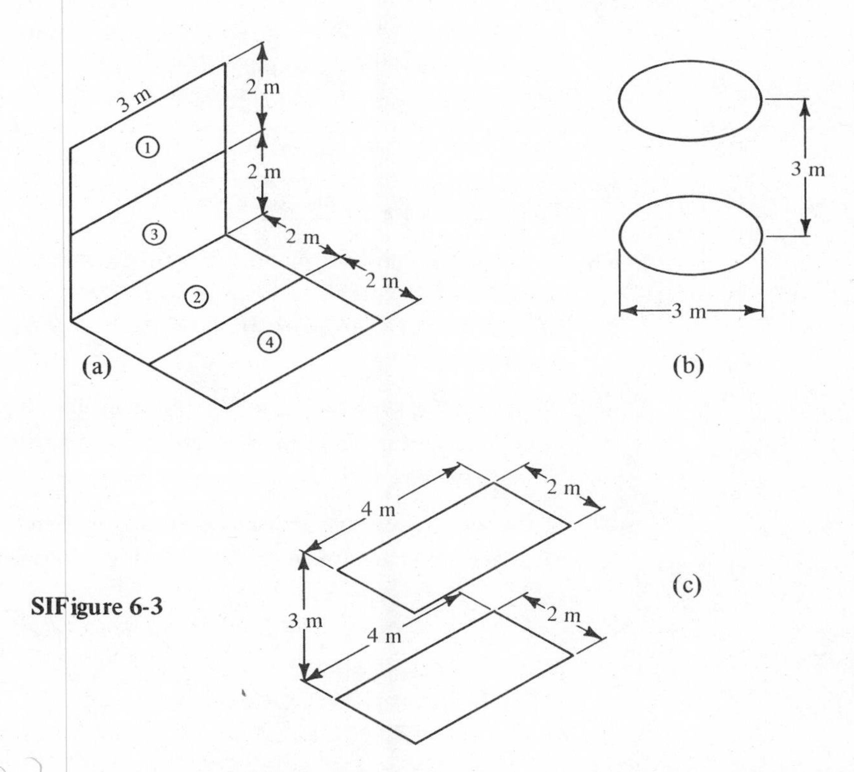

SIFigure 6-3

6-4. A bottle of chilled martinis has an inside surface area of 755 square cm. When the top is removed, a hole 5 square cm in area results. Find the shape factor from the inside of the bottle to your thumb, which is placed across the hole when the bottle is empty.

6-5. Using charts and taking advantage of symmetry, determine $F_{1\text{-}2}$ for the geometry in SIFigure 6-5.

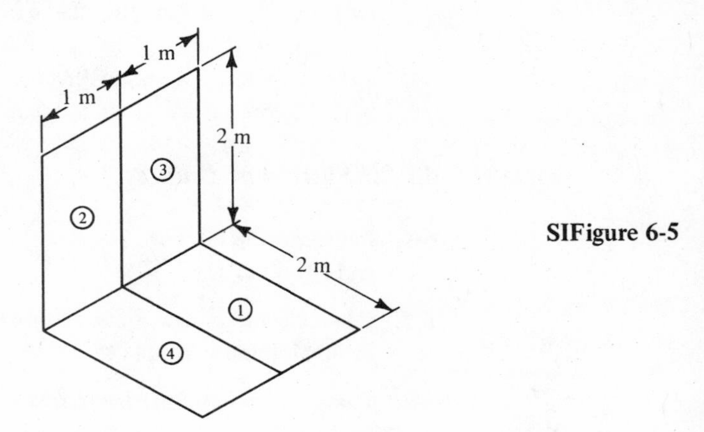

SIFigure 6-5

6-6. A cube 1 m on a side is located inside a second cube 2 m on a side. If the cubes do not touch, determine the shape factor between the outer surface of the inner cube and the inside surface of the outer cube.

6-7. Determine the net radiant heat flow between two parallel black rectangular plates 60 cm by 80 cm separated by a distance of 40 cm if their surface temperatures are 500°C and 200°C.

6-8. A 1-m-square circular plate, which is thermally insulated on one side, is located in a vacuum. The other side of the plate receives incident radiant energy at the rate of 1000 W/m². If the surface is gray, determine its steady-state temperature.

6-9. Consider N graybodies in an enclosure exchanging radiant energy with each other and nothing else. In an electrical analog, how many surface resistances would you have and how many space resistances would you have?

6-10. Consider a groove cut in a black surface as sketched in SIFigure 6-10. From the data given, determine the heat loss for the groove.

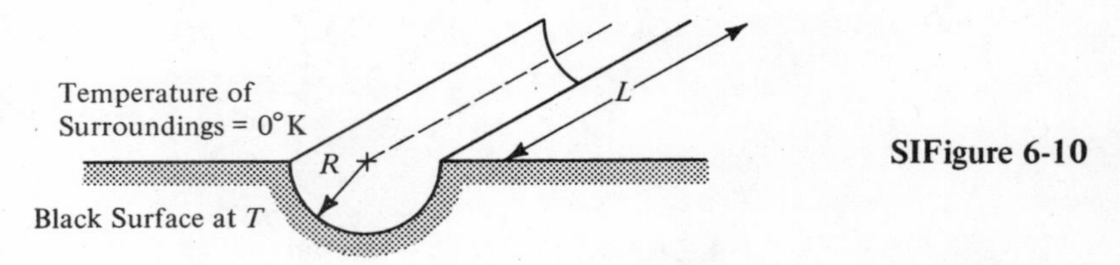

SIFigure 6-10

6-11. A wire maintained at 500°C and having a diameter of 0.3 mm is located in a large enclosure whose walls are maintained at 50°C. If the emissivity of the wire is 0.3, determine the heat transfer between the wire and the enclosure.

6-12. Rework problem 6-7, only this time let the plates be connected by a reradiating but nonconducting wall.

6-13. Determine the temperature of the reradiating but nonconducting wall in problem 6-12.

6-14. Two parallel circular discs 4 meters in diameter placed 1 meter apart are placed in a large room whose walls are maintained at 25°C. The plates are maintained at 500°C and 200°C and have emissivities of 0.5 and 0.6, respectively. Determine the heat lost by each plate and the net flow of radiant energy to the walls of the room.

6-15. If the two discs of problem 6-14 were located on the sides of spacecrafts and used to transfer heat from one vehicle to the other, how much energy would be transmitted? Assume that outer space is a blackbody at 0°K.

6-16. If the two discs of problem 6-14 were connected by a reradiating and nonconducting wall, what would be the net heat transfer between the plates.

6-17. A high pressure steam pipe 20-cm O.D. has an outer surface temperature of 150°C. If it is located in a large room whose walls are maintained at 30°C, determine the heat lost per linear meter of pipe. The emissivity of the outer surface of the pipe is 0.8.

6-18. Two large parallel plates have emissivities of 0.2 and 0.8 and are maintained at 100°C and 1000°C, respectively. If a radiation shield having an emissivity of 0.05 is placed between the plates, determine:

(a) the heat flow without the shield

(b) the heat flow with the shield, and

(c) the temperature of the shield.

6-19. Consider three infinite parallel plates. Plate 1 is maintained at 1500°K, and plate 3 is maintained at 100°K. $\epsilon_1 = 0.8$, $\epsilon_2 = 0.5$, and $\epsilon_3 = 0.8$. Plate 2 is placed between plates 1 and 3 and receives no heat from external sources. What is the temperature of plate 2?

6-20. A thermocouple is placed in a large duct to measure the temperature of the air flowing through the duct. The duct wall is at 700°C and the air is at 800°C. Neglect any conduction in the thermocouple leads, and assume that the convective heat transfer coefficient between the thermocouple junction and the air is h.

(a) Will the thermocouple indicate an air temperature that is too high or too low?

(b) To cause the error in the temperature measurement to go to zero, what should be the value of h?

(c) If h were a fixed constant, how would you alter the absorptivity of the thermocouple junction to decrease error in the measurement?

6-21. A thermocouple is placed in the heating duct of a building where the circulated air has a temperature of 60°C. Will there be a significant error due to radiation losses? Why?

6-22. A thermocouple indicates a temperature of 800°C when placed in a duct where a hot gas at 870°C is flowing. If the convective heat transfer coefficient between the thermocouple and the gas is 60 W/m²°K, estimate the duct wall temperature. The thermocouple has an emissivity of 0.5.

6-23. In problem 6-22, if the duct wall temperature were known to be 700°C, determine the value of the convective heat transfer coefficient.

6-24. On a clear night, the effective radiation temperature of the sky may be taken as −60°C. If you neglect the latent heat of vaporization for water, take the emissivity of water as unity, and assume the convective heat transfer coefficient between the dew and the air to be 35 W/m²°K, estimate the temperature at which ice will begin to form.

6-25. A gas turbine combustion chamber is 50 cms in diameter and its walls are maintained at 800°C. The products of combustion are at 1400°C, a pressure of one atmosphere, and contain 10 percent by volume of CO_2 and 20 percent by volume of H_2O. Assuming the combustion chamber to be a very long cylinder, determine the net radiant energy exchange between the gases and the combustion chamber walls.

6-26. A certain plastic has a transmissivity of 0.85 for wavelengths between 0.5μ and 3.5μ, and a transmissivity of zero for all other wavelengths. What fraction of incident solar energy will be transmitted through the plastic?

6-27. Estimate the surface temperature for two canoes placed in the hot summer sun. Assume that one canoe is made of birch bark and the other of aluminum. The incident solar energy is measured as 300 Btu/hr-ft² and the convective heat transfer coefficient between the canoes' surfaces and the ambient air is 5 Btu/hr-ft²°F. $T_\infty = 80°F = T_{surroundings}$.

6-28. Determine the amount of radiant energy emitted in the visible wavelength spectrum by a blackbody at (a) 100°F, (b) 500°F, (c) 500°C, (d) 1000°C.

REFERENCES

[1] Sparrow, E. M., and Cess, R. D., *Radiation Heat Transfer,* Brookes/Cole Publishing Company, (1966).

[2] Bayley, F. J., Owens, J. M., and Turner, A. B., *Heat Transfer,* Barnes and Noble, Publishers, New York, (1972).

7

Fluid Flow Background for Convective Heat Transfer

7-1 INTRODUCTION

Convection is the mechanism by which heat is transferred between a solid surface and a fluid moving adjacent to it. In general, the greater the mean velocity of a fluid, the greater is the rate of convective heat transfer for a prescribed temperature difference between the fluid and the solid boundary. In addition to the velocity magnitude, the pattern of the fluid flow motion also affects the heat transfer characteristics. It therefore becomes essential that we acquire some knowledge of the dynamics of fluid flow. This chapter is devoted to a discussion of a few selected fluid flow topics, while the following two chapters deal with the mechanism of convective heat transfer.

In studying different types of fluid flow that are relevant to convective heat transfer, we consider flow of viscous fluids through pipes or tubes and flow of fluids over objects. Such situations are encountered in the flow of water in the tubes of a boiler or of an automobile radiator, the flow of warm air in the heating ducts of a house, the flow of air over the wings of an airplane, etc.

Some of the pertinent variables in a typical fluid flow problem are density, viscosity, mean velocity of the fluid, and a characteristic dimension of the geometry of the flow. We are usually interested in determining the velocity distribution in a fluid flow, since it directly influences the transport of energy. Another quantity of interest is the pressure drop experienced by a fluid as it flows through a certain length. It is essential to know the pressure drop to compute the power requirement for pumping. The principles of conservation of mass and conservation of momentum form the bases for the analysis of fluid flows.

In the flow of real fluids, viscosity plays a very important role. It contributes to the variation of velocities and to the pressure drops. Therefore, our study of fluid flow will begin with the nature and effects of viscosity.

7-2 VISCOSITY

Consider a real fluid flowing between two parallel plates as shown in Figure 7-1. One plate is moving at a velocity, U, and the other plate is stationary. Such a situation is approximated by a thin film of a lubricant inside a journal bearing. It is found experimentally that the fluid adheres or sticks to the solid boundary regardless of whether the boundary is moving or is fixed. Thus, as the upper plate moves to the right at a velocity, U, the fluid particles adhering to it also move to the right at a velocity, U. Likewise, the fluid particles next to the bottom plate have zero velocity. Evidently, the fluid velocity must vary across the gap from 0 to U. It has been found that the velocity of the fluid in the x direction, denoted by u, varies linearly with the y coordinate* and is given by

$$u = \frac{y}{H} U \tag{7-1}$$

where H is the distance separating the two plates.

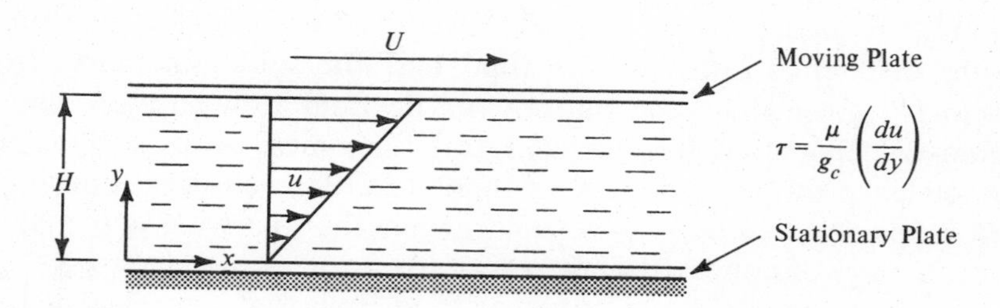

Figure 7-1 Fluid flow between two parallel plates.

The force, F, necessary to move the upper plate at a given velocity, U, depends on the type of fluid. This force causes a shear stress to be set up in all the fluid layers between the two plates. The shear stress in the layer near the top plate

*This is true for laminar flow conditions only. A discussion of laminar flow is presented in the following section.

is the ratio of the force, F, to the surface area of the plate. Newton postulated the following relation for the shear stress, τ, exerted by one fluid layer on its neighbor

$$\tau = \frac{\mu}{g_c}\left(\frac{du}{dy}\right) \tag{7-2}$$

where μ is a property of the fluid and is called the coefficient of dynamic viscosity or simply viscosity. Values of viscosity for different fluids are listed in the Appendix. The viscosity of a liquid usually decreases with an increase in its temperature, while that of a gas increases with an increase in its temperature. Fluids that obey equation (7-2) are called Newtonian fluids.

The units of viscosity in the English engineering system are lb_m/ft-sec or $lb_f sec/ft^2$. In the latter case, (μ/g_c) is replaced by μ. Inasmuch as values of viscosity are often given in the units of lb_m/ft-sec, we will adopt these units for use in this text. It requires that equation (7-2) contain the constant g_c, which equals 32.2 $lb_m\, ft/(lb_f sec^2)$. In the SI system, the constant g_c has a numerical value of unity, and μ has the unit of Ns/m^2 (Newton second/meter^2). The conversion factor is

$$1\ lb_m/\text{ft-sec} = 0.672\ Ns/m^2$$

The ratio of the viscosity, μ, to the density, ρ, is called the kinematic viscosity, ν, and has the units of ft^2/sec in the English system and m^2/sec in the SI system of units. Thus

$$\nu = \frac{\mu}{\rho} \tag{7-3}$$

7-3 LAMINAR AND TURBULENT FLOWS

If honey is poured out of a bottle, we see a very smooth flow with all the particles moving in well-organized paths. Such a flow is referred to as a laminar flow. It is a well-ordered pattern where fluid layers slide over one another. If a dye were injected into a laminar flow of water, a streak of colored water would be formed that would maintain its distinct identity throughout the entire flow field. In general, fluid particles in a laminar flow move in well-defined paths. The path described by a fluid particle as it moves through space is called a *pathline*. A line drawn through the fluid tangent to the velocity vectors at that instant is called a *streamline*. In a steady, laminar flow, the pathlines and the streamlines are identical. Such is not the case in a turbulent flow. If a faucet in a home is opened fully, the ensuing flow of water is usually turbulent. The fluid particles in a turbulent flow do not travel in a well-ordered fashion. Rather, there are components of velocity transverse to the principal direction of flow, and these components are constantly changing in magnitude. The distinction between laminar and turbulent flow can be demonstrated with the following experiment.

Consider a simple experiment (Figure 7-2) in which water flows through a transparent pipe and the flow rate is regulated by a valve on the downstream end. This is essentially the set-up of the classical experiment performed by Osborne Reynolds in 1889. Provision is made to insert a dye stream of the same density as that of water near the upstream end of the pipe. When the outlet valve is barely cracked open, the dye will form a very fine line through the flowing water (Figure 7-2a). As the valve is progressively opened more and more, the dye will continue to flow in a straight line until a flow rate is reached at which fluctuations in the dye filament become evident. This indicates the onset of transition from laminar flow to turbulent flow (Figure 7-2b). As the valve is further opened, the dye, soon after leaving the dye injector, disperses throughout the fluid, indicating a fully turbulent flow (Figure 7-2c).

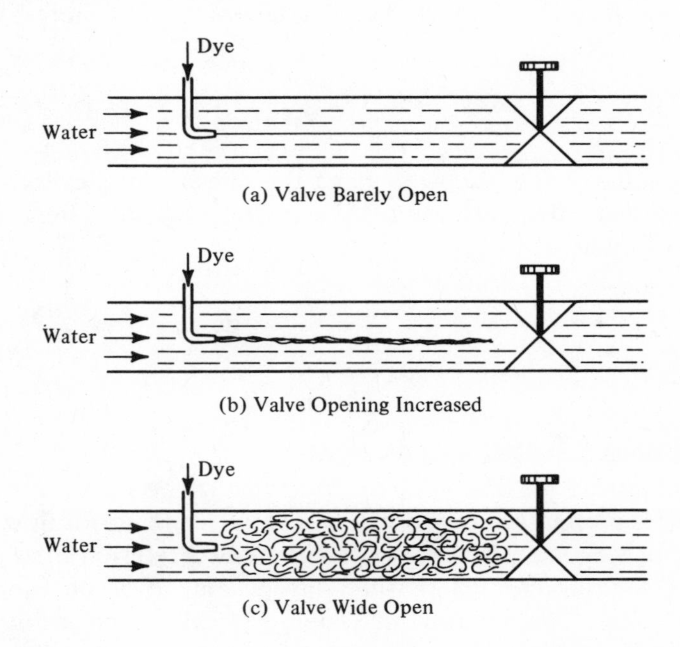

Figure 7-2 Reynolds experiment. (a) Laminar flow, (b) transitional flow, and (c) turbulent flow.

In performing an analysis of a fluid flow, it is necessary to know whether it is a laminar flow or a turbulent flow. Turbulent flows are much more complex to analyze than are laminar flows.

It is possible to predict the nature of a flow if we know its density, ρ; mean flow velocity, u_{av}; viscosity, μ; and a characteristic dimension, L_c, for the flow geometry. The characteristic dimension for a flow over a plate is the distance x from the leading edge of the plate. For the flow inside tubes, the characteristic length is the hydraulic diameter, D_H, given by

$$D_H = \frac{4\text{ (cross-sectional area)}}{\text{wetted perimeter}} \tag{7-4}$$

From equation (7-4), we see that for a circular tube, the hydraulic diameter is the same as the inner diameter of the tube.

The four quantities, ρ, u_{av}, L_c, and μ, are combined into a dimensionless number, called the Reynolds number

$$\text{Re} = \frac{\rho u_{av} L_c}{\mu} \tag{7-5}$$

The value of the Reynolds number determines the nature of the flow.

Flow inside circular tubes is always laminar for a Reynolds number less than 2300. Turbulent flow will commonly result if the Reynolds number is greater than 4000. When the Reynolds number ranges between 2300 and 4000, the flow in a tube is known as a *transitional flow*. The value of Reynolds number where transition from laminar flow to turbulent flow begins is known as the *critical Reynolds number*. Under very carefully controlled conditions, critical Reynolds numbers as high as 40,000 have been observed.

When the flow over a flat plate with a sharp leading edge is examined, a laminar flow is generally observed for Reynolds numbers less than 5×10^5. In the presence of a blunt leading edge or free stream turbulence, the critical Reynolds number can be lower than 5×10^5. Under controlled conditions, the critical Reynolds number can also be as high as 2×10^6. When a fluid flows over a flat plate of sufficient length, it is possible to observe a laminar flow near the leading edge of the plate followed by a transition region, which in turn is followed by a turbulent flow.

7-4 FLOW INSIDE CIRCULAR TUBES

Typically, a fluid, as it enters a tube from a large plenum chamber, has a uniform velocity distribution. For a circular tube, this means that the streamwise velocity, u, is the same for all values of radius at the entrance to the tube ($x = 0$). This is shown in Figure 7-3a. As the fluid flows down the tube, the velocity profile, which reflects the variation of u as a function of r, becomes more rounded. At some distance from the entrance of the tube, the velocity profile becomes established and does not change thereafter. This distance is called the *entrance length*, L_e. A theoretical expression for the laminar entrance length has been given by Langhaar (Reference 1), which agrees well with experimental results. It is

$$L_e = 0.058 \, \mathrm{Re} \, D \tag{7-6}$$

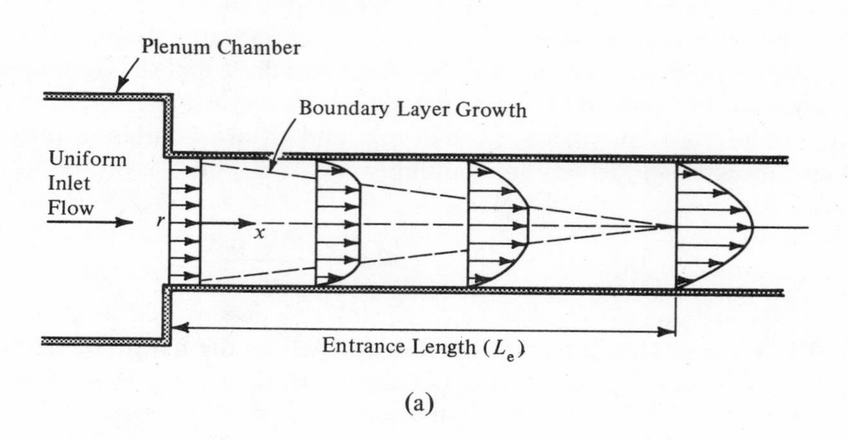

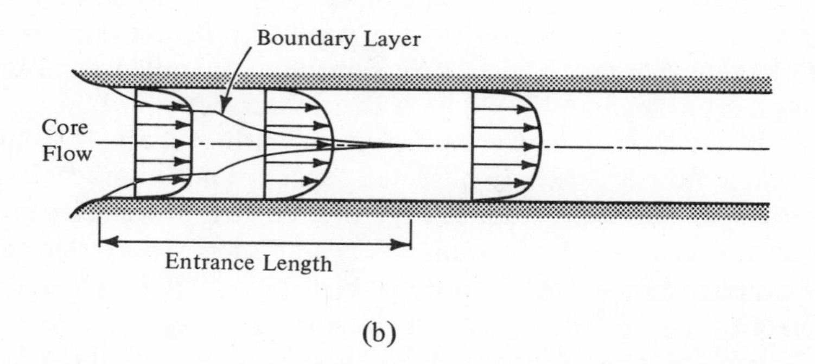

Figure 7-3 (a) Development of a laminar flow through a tube.
(b) Velocity profile for turbulent flow in a tube.

For locations beyond the entrance length, the flow is said to be fully developed. For laminar flow, the fully developed velocity profile is a parabola.

If conditions permit a turbulent flow, then at some distance from the entrance, a fully developed turbulent velocity profile will appear, as shown in Figure 7-3b. Latzko's (Reference 2) analysis for the prediction of the entrance length was improved upon by Holdhusen (Reference 3). The resulting pressure drop, shear stress at the wall, and energy loss in the inlet region agree with experiments.

It appears that there is no satisfactory general relationship available for the entrance length of a turbulent flow. Barbin and Jones (Reference 4) found that fully developed smooth pipe flow is not attained in an inlet length of 40.5 diameters for a Reynolds number of 388,000. They did find that the shear stress at the wall attained fully developed values in a length equal to 15 times the pipe diameter.

In a turbulent flow, chunks of fluid called *eddies* move in the transverse or radial direction in a random fashion, and they carry with them their momentum. Such a process tends to make the velocity profile more uniform in that region where the eddies exist. This region is called the *turbulent core* of the flow. Consequently, the velocity profile for a turbulent tube flow is almost flat in the core region, as seen in Figure 7-3b.

***Sample Problem** 7-1* Water flows through a tube of 3/4-inch I.D. at a rate of 0.4 gallon per minute. The viscosity of the water may be taken as 2.36 lb_m hr-ft. Determine if the flow is laminar or turbulent. If it is laminar, determine the entrance length.

Solution: The nature of the flow is determined by the magnitude of the Reynolds number, the evaluation of which requires values of the average velocity, the density, the viscosity of the water, and the inner diameter of the tube. The average velocity may be calculated from

$$u_{av} = \frac{\text{volume rate of flow}}{\text{cross-sectional area}} = \frac{0.4 \times (0.1337/60)}{\frac{\pi}{4}(3/4)^2 \times (1/144)}$$

or

$$u_{av} = 0.29 \text{ ft/sec}$$

Hence the Reynolds number is

$$\text{Re} = \frac{\rho u_{av} D}{\mu} = \frac{62.4 \times 0.29 \times (0.75/12)}{(2.36/3600)} = 1718$$

Since the Reynolds number is less than 2300, the flow is laminar. The entrance length, L_e, is given by

$$L_e = 0.058 \text{ Re } D = (0.058)(1718)(0.75/12)$$

or

$$L_e = 6.2 \text{ ft}$$

7-4.1 Laminar Flow in Tubes

Fully developed laminar flow in tubes is also known as *Hagen-Poiseuille flow*. Consider laminar flow of a fluid through a tube of inner radius, r_w. Let p be the pressure at any point in the fluid, and let u be the velocity of the fluid in the x direction at a radius, r. We note that for fully developed conditions, the pressure p, is a function of the x coordinate and gravity, and the velocity, u, is a function of the r coordinate in a cylindrical coordinate system (Figure 7-4). We wish to determine the velocity distribution as a function of the radial coordinate. The result will then be used to evaluate the shear stress at the wall, τ_w, which, in turn, can then be used to calculate the pressure drop over a streamwise distance, L.

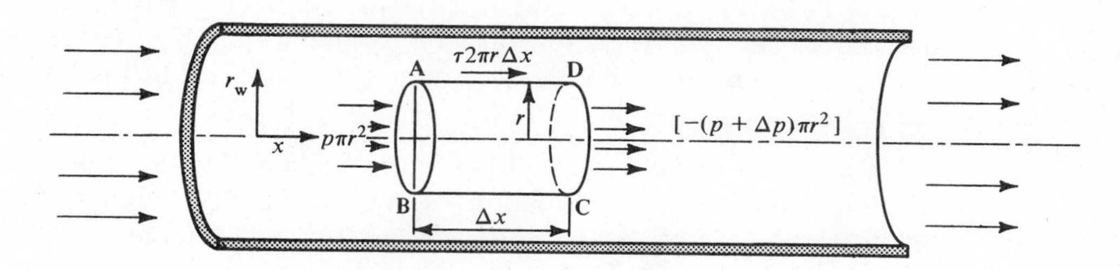

Figure 7-4 Control volume for laminar flow in a tube.

We begin by considering the conservation of momentum for a cylindrical control volume **ABCD** of radius, r, and length, Δx. For steady state, the momentum equation for the x direction is

$$\sum_{i=1}^{N} F_{i,x} = (\dot{Mom})_{\text{out},x} - (\dot{Mom})_{\text{in},x} \tag{7-7}$$

where the $F_{i,x}$s are the external forces acting on the control volume in the x direction, $(\dot{Mom})_{\text{out},x}$ is the rate of momentum leaving the control volume in the x direction, and $(\dot{Mom})_{\text{in},x}$ is the rate of momentum entering the control volume in the x direction.

The shapes of the velocity profiles at **AB** and **DC** are identical since the flow is fully developed. The net rate of outflow of momentum in the x direction is, therefore, equal to zero. Hence, the equation of conservation of momentum for the x direction reduces to

$$F_{AB} + F_{DC} + F_{ADBC} = 0$$

where

$$F_{\mathbf{AB}} = \text{External force acting on the face } \mathbf{AB} \text{ in the } x \text{ direction}$$
$$= p(\pi r^2)$$
$$F_{\mathbf{DC}} = \text{External force acting on the face } \mathbf{DC} \text{ in the } x \text{ direction}$$
$$= -(p + \Delta p)(\pi r^2)$$
$$F_{\mathbf{ADBC}} = \text{External force acting on the cylindrical face in the } x \text{ direction}$$
$$= \tau(2\pi r \Delta x)$$

Employing the expressions for all the forces, we obtain

$$p\pi r^2 - (p + \Delta p)\pi r^2 + \tau 2\pi r \Delta x = 0$$

Substituting for τ from equation (7-2)

$$\tau = \frac{\mu}{g_c}\left(\frac{du}{dr}\right)$$

and simplifying, we obtain

$$r\Delta p = \frac{2\mu}{g_c}\left(\frac{du}{dr}\right)\Delta x$$

As $\Delta x \to 0$, the forgoing equation becomes

$$\frac{2\mu}{g_c r}\left(\frac{du}{dr}\right) = \frac{dp}{dx} \tag{7-8}$$

If the variation in pressure due to hydrostatic head is neglected, then the pressure is a function of x alone. Thus, in equation (7-8), the left-hand side is a function of r alone, whereas the right-hand side is a function of x alone. The only way in which these functional dependences can be satisfied is when each side of equation (7-8) equals a constant. Let this constant be β. We then have

$$\beta = \frac{2\mu}{g_c r}\left(\frac{du}{dr}\right) \tag{7-8a}$$

and

$$\beta = \frac{dp}{dx} \tag{7-8b}$$

Upon separating the variables u and r in equation (7-8a), we obtain

$$\int_{u_o}^{u} du = \frac{g_c \beta}{2\mu} \int_{r=0}^{r} r\, dr$$

where u_0 is the velocity of the fluid at $r = 0$, that is, at the axis of the tube. Integration gives

$$u - u_o = \left(\frac{g_c \beta}{4\mu}\right) r^2 \tag{7-9}$$

At the inner wall of the tube, the fluid adheres to the surface, and, therefore, it has zero velocity; that is

$$\text{at } r = r_w, \qquad u = 0$$

Substitution of the above condition in equation (7-9) yields

$$u_o = \left(-\frac{g_c \beta}{4\mu}\right) r_w^2 \tag{7-9a}$$

Since the flow is in the positive x direction, u_o must also be positive. Therefore, we conclude from equation (7-9a) that β must be negative. Hence, equation (7-8b) demonstrates that the pressure decreases linearly with x. Substituting the expression for u_o from equation (7-9a) into equation (7-9), we obtain

$$u = -\frac{g_c \beta}{4\mu}(r_w^2 - r^2) \tag{7-9b}$$

Next, after substituting for β from equation (7-8b), we have an expression for the velocity distribution:

$$u = -\left(\frac{dp}{dx}\right)\frac{g_c r_w^2}{4\mu}\left[1 - \left(\frac{r}{r_w}\right)^2\right] \tag{7-10}$$

Also dividing equation (7-9b) by equation (7-9a), there results

$$\frac{u}{u_o} = 1 - \left(\frac{r}{r_w}\right)^2 \tag{7-10a}$$

Equation (7-10a) represents a parabola. Thus, the velocity profile for a fully developed laminar flow in a circular tube is parabolic, as was previously asserted. The average velocity of the flow is given by

$$u_{av} = \frac{1}{\pi r_w^2} \int_o^{r_w} u(2\pi r)dr$$

Since we already have an expression for u in equation (7-10a), substitution for u and integration yields

$$u_{av} = \frac{1}{2} u_o \tag{7-11}$$

The shear stress at the wall is given by

$$\tau_w = \frac{\mu}{g_c}\left(\frac{du}{dr}\right)_{r=r_w}$$

Differentiating equation (7-10), we get

$$\left(\frac{du}{dr}\right)_{r=r_w} = -\left(\frac{dp}{dx}\right)\frac{g_c r_w^2}{4\mu}\left[0 - \frac{2r}{r_w^2}\right]_{r=r_w} = \left(\frac{dp}{dx}\right)\frac{r_w}{2\mu}$$

from which it follows

$$\tau_w = \frac{1}{2}\left(\frac{dp}{dx}\right) r_w \tag{7-12}$$

It is common practice in fluid dynamics to express the pressure drop, $(p_1 - p_2)$, over a length, L, in terms of a friction factor, f, the ratio, (L/D), and the velocity head, $(\rho u_{av}^2/2g_c)$. The friction factor, f, is defined as

$$f = \frac{p_1 - p_2}{(\rho u_{av}^2/2g_c)(L/D)} \tag{7-13}$$

We observe that equations (7-9a) and (7-11) can be used to solve for β in terms of u_{av}. Thus

$$\beta = -\frac{8\mu u_{av}}{g_c r_w^2}$$

Also, employing equation (7-8b), we can write

$$\frac{dp}{dx} = -\frac{8\mu u_{av}}{g_c r_w^2}$$

or

$$\int_1^2 dp = -\int_1^2 \frac{32\mu u_{av}}{g_c D^2}\, dx$$

where r_w^2 has been replaced by $(D^2/4)$. Integration yields

$$p_1 - p_2 = \frac{32\mu}{g_c D^2} L u_{av}$$

where $L = x_2 - x_1$. Multiplying the numerator and the denominator of the right-hand side by $(2\rho u_{av})$ and regrouping terms, we obtain

$$p_1 - p_2 = 64 \left(\frac{\mu}{\rho u_{av} D}\right)\left(\frac{L}{D}\right)\left(\frac{\rho u_{av}^2}{2g_c}\right)$$

We recognize that $(\mu/\rho u_{av} D)$ is the reciprocal of the Reynolds number, (1/Re). A comparison of the forgoing equation with equation (7-13) indicates that the friction factor, f, for fully developed laminar flow in a circular tube is given by

$$f = \frac{64}{\text{Re}} \tag{7-13a}$$

For any fully developed laminar *duct* flow, the friction factor can be expressed by

$$f = C/\text{Re} \tag{7-13b}$$

where C is a constant. The value of C depends upon the shape of the duct cross section. Values of C are listed in Reference 5.

***Sample Problem** 7-2* If the 3/4-inch diameter tube of Sample Problem 7-1 transports water over a distance of 10,000 feet, determine the pressure drop and the corresponding pumping power necessary to maintain the flow.

Solution: The pressure drop is given by equations (7-13) and (7-13a). It is

$$p_1 - p_2 = \left(\frac{64}{\text{Re}}\right)\left(\frac{L}{D}\right)\left(\frac{\rho u_{av}^2}{2g_c}\right)$$

so that

$$\Delta p = \left(\frac{64}{1718}\right)\left(\frac{10{,}000}{0.75/12}\right)\left[\frac{62.4(0.29)^2}{2 \times 32.2}\right]$$

or

$$\Delta p = 486 \text{ lb}_f/\text{ft}^2 \quad \text{or} \quad 3.4 \text{ psi}$$

The pumping power in ft-lb_f/sec is calculated from

$$\text{power} = \dot{m}\,(\Delta p/\rho)$$

where $\dot{m}$ is the mass flow rate in lb_m/sec. Substituting for various terms, we obtain

$$\text{power} = \left(0.4 \times \frac{0.1337}{60} \times 62.4\right)\left(\frac{486}{62.4}\right)$$

or

$$\text{power} = 0.43 \text{ ft-lb}_f/\text{sec} \quad \text{or} \quad 0.78 \times 10^{-3} \text{ hp}$$

7-4.2 Turbulent Flow in Tubes

The analysis of turbulent flow in smooth tubes is much more complex than that for laminar flow. In a simplified analytical model, the region $0 \leqslant r \leqslant r_w$ is divided into three zones. Close to the wall, the flow is essentially laminar. This region is termed the *viscous sublayer.* For water flowing through a 1-inch smooth pipe at a Reynolds number of 10^6, the thickness of the viscous sublayer is of the order of

10^{-4} inch. In the region adjacent to the axis of the tube, there is a fully turbulent region which is called the *turbulent core.* A third region, called the *buffer zone,* separates the viscous sublayer and the turbulent core.

In order to describe the velocity distributions in the three regions, a quantity called the friction velocity is used. It is defined as

$$u^* = \sqrt{\frac{\tau_w g_c}{\rho}} \tag{7-14}$$

The velocity distribution for the three zones is expressed in terms of y, where $y = r_w - r$

(1) Viscous sublayer:

$$0 \leqslant \frac{\rho u^* y}{\mu} < 5 \qquad \frac{u}{u^*} = \frac{\rho u^* y}{\mu} \tag{7-15}$$

(2) Buffer layer:

$$5 \leqslant \frac{\rho u^* y}{\mu} < 30 \qquad \frac{u}{u^*} = -3.05 + 5 \ln\left(\frac{\rho u^* y}{\mu}\right) \tag{7-15a}$$

(3) Turbulent core:

$$\frac{\rho u^* y}{\mu} \geqslant 30 \qquad \frac{u}{u^*} = 5.5 + 2.5 \ln\left(\frac{\rho u^* y}{\mu}\right) \tag{7-15b}$$

The above equations describe the *universal* turbulent velocity distribution near a wall. This set of equations is often referred to as the *law of the wall.* They were obtained by Martinelli (Reference 6) by interpreting Nikuradse's (Reference 7) data. These equations result in discontinuities in the slope of u, but they do represent the momentum transport mechanism rather accurately. Deissler's (Reference 8) model eliminates the distinction between the buffer layer and the turbulent core, while Spalding (Reference 9) presented a single equation for the entire region.

A pipe is said to be a rough pipe if the height, ϵ, of the roughness elements exceeds the thickness, δ, of the viscous sublayer. The ratio (ϵ/D), is used to characterize the degree of roughness of a pipe. Velocity profiles for rough pipes are given by equations [similar to equation (7-15b)] in Reference 10. Such equations take the roughness of the pipe into consideration.

In turbulent flow, the pressure drop that occurs in either smooth or rough pipes is best calculated using equation (7-13) in conjunction with the friction factors given by the Moody diagram, Figure 7-5. The Moody diagram, in fact, gives friction factors for all values of the Reynolds numbers encountered in practice and for all practical degrees of roughness as measured by the dimensionless parameter, (ϵ/D).

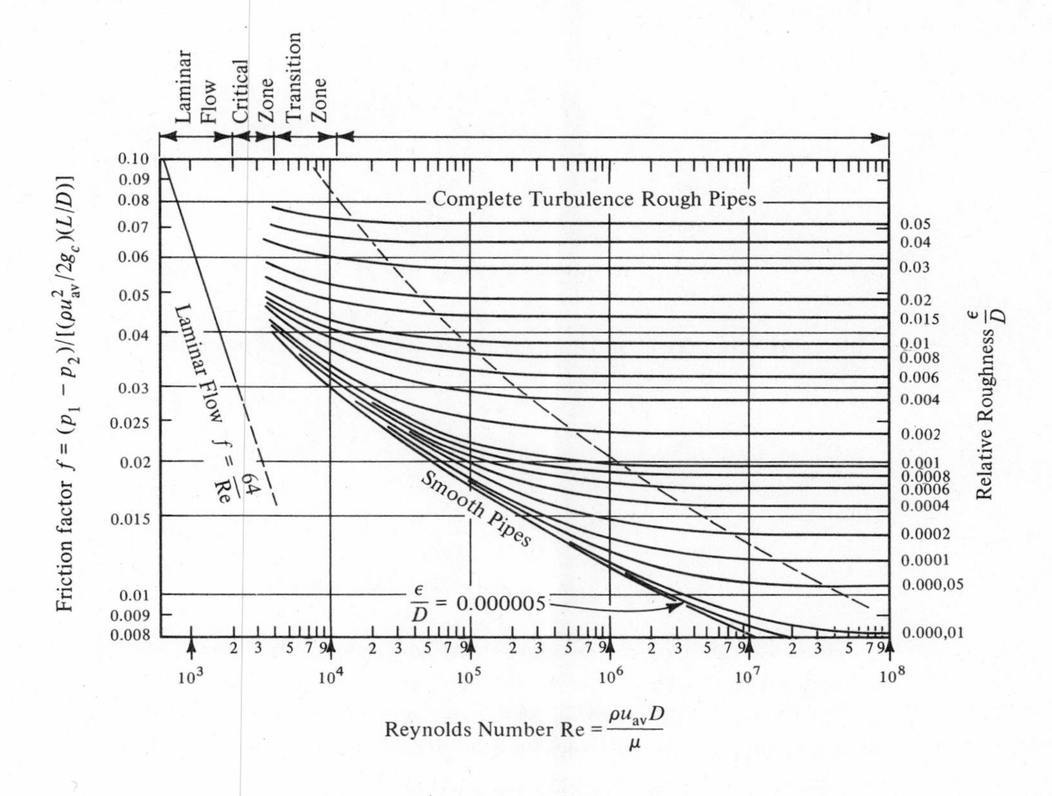

Figure 7-5 Moody diagram; friction factors for flow through pipes, from *ASME Trans.*, **68**, pp. 672, 1944.

Sample Problem 7-3 Determine the pressure drop in the tube of Sample Problem 7-1 if the flow rate were increased to 4 gallons per minute and if the tube were 10,000 feet long. Assume the tube to be smooth.

Solution: The effect of increasing the flow rate by a factor of 10 will be to increase u_{av} and Re by a factor of 10 since the other variables have not changed. Thus

$$u_{av} = 2.9 \text{ ft/sec} \quad \text{and} \quad Re = 17{,}180$$

The above value of the Reynolds number indicates a turbulent flow. Consequently, we must use the Moody diagram to determine the friction factor, f.

Referring to Figure 7-5, we obtain

$$f = 0.027$$

The pressure drop is given by equation (7-13). It is

$$p_1 - p_2 = f\left(\frac{L}{D}\right)\frac{\rho u_{\text{av}}^2}{2g_c} = 0.027\left(\frac{10{,}000}{0.75/12}\right)\frac{62.4\,(2.9)^2}{2 \times 32.2}$$

or

$$p_1 - p_2 = 34{,}900 \text{ lb}_f/\text{ft}^2 \quad \text{or} \quad 242 \text{ psi}$$

The pumping power is then given by (see Sample Problem 7-2)

$$\text{power} = \left(4 \times \frac{0.1337}{60} \times 62.4\right)\left(\frac{34{,}900}{62.4}\right) = 311 \text{ ft-lb}_f/\text{sec}$$

or

$$\text{power} = 0.57 \text{ hp}$$

It is interesting to observe that increasing the flow rate by a factor of 10 caused an increase in the pumping power by a factor of more than 700. The student can verify that if the pipe were extremely rough ($\epsilon/D = 0.05$), the friction factor for Sample Problem 7-3 would increase threefold, with a corresponding increase in pumping power.

7-5 FLOW OVER A FLAT PLATE

Convective heat transfer is also encountered in forced circulation of hot gases through a large rectangular furnace, in cold wind blowing over the exterior walls of a structure, etc. These are examples of external flows. The simplest type of external flow is that along a flat plate. Consider a flat plate of small thickness aligned parallel to an oncoming flow as shown in Figure 7-6. The effect of the plate is to disturb the otherwise uniform flow field. The flow upstream of the leading edge of the plate can be treated as if it were an ideal (nonviscous) fluid. Viscous effects, however, must be considered in studying the flow over the plate.

For the purpose of analysis, it is convenient to divide the flow field into two regions (Reference 11). We know that the velocity of the fluid at the plate surface is zero and that at a sufficient distance away from the plate the velocity is equal to u_∞. The region next to the surface of the plate, where 99 percent of the velocity variation takes place, is called the *boundary layer region.* The concept of the boundary layer was introduced by L. Prandtl with a view to simplifying the analysis. A hypothetical line separates the boundary layer region from the main stream. It is observed that as we move downstream from the leading edge, the thickness of the boundary layer increases. For air flowing at 50 ft/sec, the thickness of the boundary layer at a distance of one foot from the leading edge is of the order of 0.1 inch. The effects of viscosity must be taken into account when examining the flow inside the boundary layer. The flow in the region outside the boundary layer is called *free-stream flow*. The flow outside the boundary layer is treated as an ideal (nonviscous) flow.

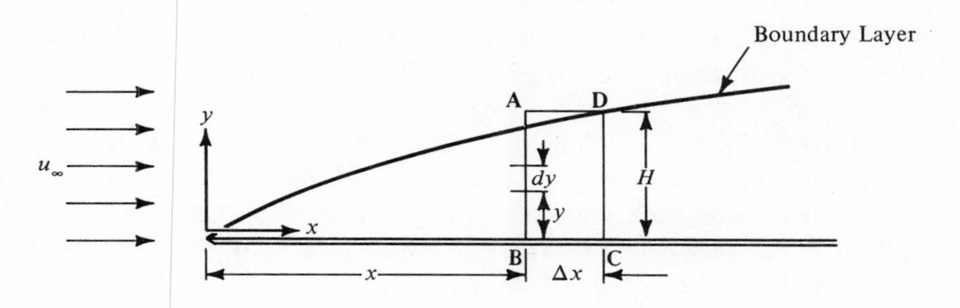

Figure 7-6 Control volume for an integral analysis of a laminar boundary layer.

The flow within the boundary layer can be laminar, transitional, or turbulent. The nature of the flow depends upon the local Reynolds number, where the value of the Reynolds number is based on the distance, x, from the leading edge. It has been found that if the free stream has very low turbulence, then the boundary layer is laminar if the local Reynolds number is less than 5×10^5. Transition from laminar flow to turbulent flow can begin at a Reynolds number of 5×10^5 or greater.

We will now present an approximate analysis of the laminar boundary layer.

7-5.1 Integral Analysis of the Laminar Boundary Layer

In the analysis of convective heat transfer from a flat plate, to be discussed in Chapter 8, we will need the velocity distribution inside the boundary layer. The velocity boundary layer is sometimes referred to as the hydrodynamic boundary layer. In the analysis of boundary layer flow, our objective is to determine the velocity profile, the boundary layer thickness, and the wall shear stress or drag. There are two methods of analysis, (a) the integral method, and (b) the differential method. We will discuss both methods. The first method, which is an approximate method, the credit for which is given to Von Karman (Reference 12) and Pohlhausen (Reference 13), consists of the following steps.

(1) Select a control volume that encompasses the entire thickness of the boundary layer and a small streamwise length, Δx. Such a control volume is illustrated in Figure 7-6.

(2) Apply principles of conservation of mass and momentum to generate equations in integral form.

(3) Assume that the velocity, u, can be represented in the form of a polynomial in y.

(4) Determine the coefficients in the polynomial from conditions that the velocity profile must satisfy at the plate surface and at the edge of the boundary layer.

(5) Substitute the polynomial velocity profile into the integral momentum equation and solve for the boundary layer thickness, δ.

(6) Calculate the wall shear stress by using the polynomial profile in conjunction with the expression for δ and equation (7-2).

Consider a flat plate aligned parallel to a fluid stream having constant thermophysical properties and a uniform free-stream velocity, u_∞. We assume a one-dimensional steady-state incompressible flow. Let the boundary layer on the plate be as shown in Figure 7-6. We select a control volume **ABCD**, which is of unit depth into the plane of the paper.

In order to apply the principle of conservation of mass, we examine the rates of fluid inflow and outflow at all the faces of the control volume. There is no flow across face **BC**, which is a solid boundary (the surface of the plate). The rate of flow into the control volume through an element of height dy on face **AB** is

$$\rho u\,(dy)\,(1)$$

where the depth into the plane of the page is taken as unity. The rate of flow into the control volume through **AB** is, therefore, given by

$$\dot{m}_{\mathbf{AB}} = \int_0^H \rho u\,dy$$

Using a Taylor series expansion, we can express the rate of outflow from the control volume through face **CD** in terms of $\dot{m}_{\mathbf{AB}}$ and its derivatives. Thus

$$\dot{m}_{\mathbf{CD}} \simeq \dot{m}_{\mathbf{AB}} + \frac{d}{dx}(\dot{m}_{\mathbf{AB}})\,\Delta x$$

Designating the rate of flow into the control volume through **AD** as $\dot{m}_{\mathbf{AD}}$, we can now write the equation for the conservation of mass under steady state as

$$\begin{pmatrix}\text{Rate of flow into the}\\ \text{control volume } \mathbf{ABCD}\end{pmatrix} = \begin{pmatrix}\text{Rate of flow out of the}\\ \text{control volume } \mathbf{ABCD}\end{pmatrix}$$

or

$$\dot{m}_{AB} + \dot{m}_{AD} = \dot{m}_{CD}$$

Then substituting for $\dot{m}_{AB}$ and $\dot{m}_{CD}$ and simplifying, we obtain

$$\dot{m}_{AD} = \frac{d}{dx}\left(\int_0^H u\,dy\right)\Delta x \tag{7-16}$$

It might appear that the quantity $\dot{m}_{AD}$ should be identically zero by virtue of the fact that the face **AD** is in the free stream where the velocity of the fluid is u_∞. It should be noted, however, that the division of the flow into two regions is artificial, and, in reality, the flow velocity at **AD** has components in the y as well as in the x direction. This gives rise to a nonzero value of $\dot{m}_{AD}$.
gives rise to a nonzero value of $\dot{m}_{AD}$.

Next, we apply the principle of conservation of momentum to the control volume **ABCD**. For steady-state conditions, it can be stated as

$$\begin{pmatrix}\text{Sum of the external forces}\\ \text{acting on the control volume}\\ \text{in the } x \text{ direction}\end{pmatrix} = \begin{pmatrix}\text{(Rate of outgoing momentum)}\\ \text{minus}\\ \text{(Rate of incoming momentum)}\end{pmatrix}$$

or

$$\sum_{i=1}^{N} F_{i,x} = (\dot{Mom})_{out,x} - (\dot{Mom})_{in,x} \tag{7-7}$$

The external forces acting on the control volume in the x direction are the pressure forces on faces **AB** and **CD** and the wall shear force at face **BC**. The pressure, p, is not expected to vary across the boundary layer, so that its value at any x coordinate is determined by the free-stream conditions via Bernoulli's equation, which is

$$\frac{p}{\rho} + \frac{u_\infty^2}{2g_c} + \frac{g}{g_c}z = \text{constant}$$

where z is the elevation.

Bernoulli's equation expresses the fact that for an ideal fluid flow, the sum of the flow work, (p/ρ), kinetic energy $(u_\infty^2/2g_c)$, and the potential energy, $(g/g_c)z$, is constant under steady-state conditions.

The pressure force on face **AB** is obtained by integrating the force on a small strip whose area is $(dy \cdot 1)$, with the result

$$F_{x,\,\mathbf{AB}} = \int_0^H p\, dy$$

The pressure force acting on face **CD** can be expressed in terms of $F_{x,\,\mathbf{AB}}$ and its derivative as

$$F_{x,\,\mathbf{CD}} \simeq -\left[F_{x,\,\mathbf{AB}} + \frac{d}{dx}(F_{x,\,\mathbf{AB}})\,\Delta x\right]$$

If we represent the wall shear stress acting along **BC** by τ_w we can write

$$F_{x,\,\mathbf{BC}} = -\tau_w\,(\Delta x \cdot 1)$$

Then, substituting the various expressions for the forces in the left-hand side of the momentum equation, simplifying, and interchanging the order of differentiation and integration, we obtain

$$\sum F_x = -\left[\int_0^H \left(\frac{dp}{dx}\right) dy + \tau_w\right]\Delta x \tag{7-17}$$

The right-hand side of the momentum equation consists of three momentum terms. They are the x-momenta entering the control volume through **AB** and **AD** and the x-momentum leaving the control volume through **CD**. The rate of x-momentum entering through a strip of dimension $(dy \cdot 1)$ on face **AB** is

$$\frac{1}{g_c}(\rho\, u\, dy \cdot 1)\, u$$

Therefore, the rate of x-momentum entering the control volume through **AB** is

$$(M\dot{o}m)_{\mathbf{AB},\,x} = \int_0^H \left(\frac{1}{g_c}\right)\rho u^2\, dy$$

The mass flow rate passing into the control volume through **AD** is given by equation (7-16). It possesses a velocity in the x direction, which is u_∞. Hence, the

x-momentum entering the control volume through **AD** is given by

$$(\dot{Mom})_{\mathbf{AD},\,x} = \frac{1}{g_c}\, u_\infty \frac{d}{dx}\left[\int_0^H \rho u \, dy\right]\Delta x$$

Finally, the x-momentum leaving the control volume through **CD** is determined by using a Taylor series expansion of $(\dot{Mom})_{\mathbf{AB}}$

$$(\dot{Mom})_{\mathbf{CD},\,x} \simeq (\dot{Mom})_{\mathbf{AB},\,x} + \frac{d}{dx}\left[(\dot{Mom})_{\mathbf{AB},\,x}\right]\Delta x$$

Substituting the three forgoing equations into the right-hand side of the momentum equation, we get

$$(\dot{Mom})_{\text{out},\,x} - (\dot{Mom})_{\text{in},\,x} = (\dot{Mom})_{\mathbf{CD},\,x} - \left[(\dot{Mom})_{\mathbf{AB},\,x} + (\dot{Mom})_{\mathbf{AD},\,x}\right]$$

or

$$(\dot{Mom})_{\text{out},\,x} - (\dot{Mom})_{\text{in},\,x} = \frac{1}{g_c}\frac{d}{dx}\left[\int_0^H \rho u^2 \, dy\right]\Delta x - \frac{1}{g_c}\, u_\infty\left[\frac{d}{dx}\int_0^H \rho u \, dy\right]\Delta x \qquad \text{(7-17a)}$$

Now, substitute equations (7-17) and (7-17a) in the momentum equation and simplify to yield

$$-\left[\int_0^H \left(\frac{dp}{dx}\right) dy + \tau_w\right]\Delta x = \frac{1}{g_c}\frac{d}{dx}\left[\int_0^H \rho u^2 \, dy\right]\Delta x - \frac{1}{g_c}\, u_\infty\left[\frac{d}{dx}\int_0^H \rho u \, dy\right]\Delta x \qquad \text{(7-17b)}$$

In order to simplify the forgoing equation, we differentiate Bernoulli's equation to obtain

$$\frac{1}{\rho}\frac{dp}{dx} + \frac{u_\infty}{g_c}\frac{du_\infty}{dx} = 0$$

since the density, ρ, and the elevation, z, are constant. Thus the integral in the first term in equation (7-17b) can be written as

$$\int_0^H \left(\frac{dp}{dx}\right) dy = -\frac{1}{g_c}\int_0^H \rho u_\infty \frac{du_\infty}{dx}\, dy = -\frac{1}{g_c}\rho\frac{du_\infty}{dx}\int_0^H u_\infty\, dy \qquad (7\text{-}17c)$$

Next, let

$$\int_0^H \rho u\, dy = f$$

Then

$$u_\infty \frac{d}{dx}\int_0^H \rho u\, dy = u_\infty \frac{df}{dx} = \frac{d}{dx}(u_\infty f) - f\frac{du_\infty}{dx}$$

where the product formula for differentiation has been used. Back-substitution for f gives

$$u_\infty \frac{d}{dx}\int_0^H \rho u\, dy = \frac{d}{dx}\int_0^H \rho u u_\infty\, dy - \frac{du_\infty}{dx}\int_0^H \rho u\, dy \qquad (7\text{-}17d)$$

When equations (7-17c) and (7-17d) are employed in equation (7-17b) there results

$$\frac{1}{g_c}\rho\frac{du_\infty}{dx}\int_0^H u_\infty\, dy - \tau_w = \frac{1}{g_c}\left[\frac{d}{dx}\left(\int_0^H \rho u^2\, dy\right) - \frac{d}{dx}\left(\int_0^H \rho u u_\infty\, dy\right) + \left(\frac{du_\infty}{dx}\right)\left(\int_0^H \rho u\, dy\right)\right]$$

A rearrangement gives

$$\tau_w = \frac{1}{g_c}\left\{\rho\left(\frac{du_\infty}{dx}\right)\int_0^H (u_\infty - u)\, dy + \frac{d}{dx}\left[\int_0^H \rho u(u_\infty - u)\, dy\right]\right\}$$

Since $(u_\infty - u) = 0$ for $y > \delta$, H in the above equation can be replaced by δ. Thus we have

$$\tau_w = \frac{1}{g_c}\left\{\rho\left(\frac{du_\infty}{dx}\right)\int_0^\delta \rho(u_\infty - u)\,dy + \frac{d}{dx}\left[\left[\int_0^\delta \rho u(u_\infty - u)\,dy\right]\right]\right\} \quad (7\text{-}18)$$

Equation (7-18) is known as the *integral momentum equation.* If the free stream has no pressure gradients, it has a uniform and constant velocity, and the quantity, (du_∞/dx), is identically zero. Consequently, the first term in the brace on the right-hand side of equation (7-18) vanishes. Also, for an incompressible fluid, the density, ρ, is a constant quantity. Therefore, equation (7-18) simplifies to

$$\frac{g_c \tau_w}{\rho} = \frac{d}{dx}\left[\int_0^\delta u(u_\infty - u)\,dy\right] \quad (7\text{-}19)$$

for an incompressible fluid with constant free-stream velocity.

Thus far, we have executed the first two steps outlined at the beginning of this section. Next, we assume a third-degree polynomial in y for the velocity profile. Let

$$\frac{u}{u_\infty} = a_0 + a_1\eta + a_2\eta^2 + a_3\eta^3 \quad (7\text{-}20)$$

where $\eta = y/\delta$. The four coefficients a_0, a_1, a_2, and a_3 can be determined by requiring that the velocity profile of equation (7-20) satisfy a set of boundary conditions. These conditions ensure that the actual profile is reasonably represented. The conditions are:

(1) At $y = 0$, $u = 0$. (No slip condition.)

(2) At $y = 0$, $(\partial^2 u/\partial y^2) = 0$. This condition follows from the governing partial differential equation for the boundary layer, which will be developed in the next section.

(3) At $y = \delta$, $u = u_\infty$.

(4) At $y = \delta$, $\partial u/\partial y = 0$. This condition ensures a smooth transition of the velocity profile from the boundary layer to the free-stream flow.

If we substitute the above four conditions into equation (7-20), we obtain the following four simultaneous equations, respectively.

$$a_0 = 0$$

$$2a_2 = 0$$

$$a_0 + a_1 + a_2 + a_3 = 1$$

$$a_1 + 2a_2 + 3a_3 = 0$$

The solution to the above equations is

$$a_0 = 0; \quad a_1 = 3/2; \quad a_2 = 0; \quad \text{and } a_3 = -1/2$$

Substitution of the values of these coefficients into equation (7-20) gives the following expression for the velocity profile

$$\frac{u}{u_\infty} = \frac{3}{2}\left(\frac{y}{\delta}\right) - \frac{1}{2}\left(\frac{y}{\delta}\right)^3 = \frac{3}{2}\eta - \frac{1}{2}\eta^3 \tag{7-21}$$

where δ is as yet unknown.

We can now determine an expression for the wall shear stress in equation (7-19) by employing equation (7-21). From the definition of τ [equation (7-2)]

$$\tau_w = \left(\frac{\mu}{g_c}\right)\left(\frac{\partial u}{\partial y}\right)_{y=0}$$

Taking the first derivative of equation (7-21) and substituting $y = 0$ in the resulting expression, we obtain

$$\tau_w = \left(\frac{\mu u_\infty}{g_c}\right)\left(\frac{3}{2\delta}\right) \tag{7-22}$$

Now, substitute τ_w from equation (7-22) and u from equation (7-21) into equation (7-19) to yield

$$\frac{3\mu u_\infty}{2\delta\rho} = \frac{d}{dx}\left\{\int_0^\delta (u_\infty)\left[\frac{3}{2}\left(\frac{y}{\delta}\right) - \frac{1}{2}\left(\frac{y}{\delta}\right)^3\right](u_\infty)\left[1 - \frac{3}{2}\left(\frac{y}{\delta}\right) + \frac{1}{2}\left(\frac{y}{\delta}\right)^3\right] dy\right\}$$

By performing the indicated multiplication in the braces and then taking u_∞^2 out of the integral sign, there results

$$\frac{3\mu u_\infty}{2\delta\rho} = u_\infty^2 \frac{d}{dx}\left\{\int_0^\delta \left[\frac{3}{2}\left(\frac{y}{\delta}\right) - \frac{1}{2}\left(\frac{y}{\delta}\right)^3 - \frac{9}{4}\left(\frac{y}{\delta}\right)^2 + \frac{3}{4}\left(\frac{y}{\delta}\right)^4 + \frac{3}{4}\left(\frac{y}{\delta}\right)^4 - \frac{1}{4}\left(\frac{y}{\delta}\right)^6\right] dy\right\}$$

Integration of the right-hand side gives

$$\frac{3\mu u_\infty}{2\delta\rho} = u_\infty^2 \frac{d}{dx}\left[\frac{3}{4}\delta - \frac{1}{8}\delta - \frac{9}{12}\delta + \frac{3}{20}\delta + \frac{3}{20}\delta - \frac{1}{28}\delta\right]$$

or

$$\frac{3\mu u_\infty}{2\delta\rho} = \left(\frac{39}{280}\right)u_\infty^2\left(\frac{d\delta}{dx}\right)$$

Separating the variables, x and δ, and rearranging, we obtain

$$\int_0^\delta \delta\, d\delta = \int_0^x \frac{140\mu}{13\rho u_\infty}\, dx$$

Integration yields

$$\frac{\delta^2}{2} = \frac{140}{13}\left(\frac{\mu x}{\rho u_\infty}\right) \tag{7-23}$$

or

$$\delta = 4.64\sqrt{\frac{\mu x}{\rho u_\infty}}$$

The above result, when stated in a dimensionless form, becomes

$$\frac{\delta}{x} = 4.64\sqrt{\frac{\mu}{\rho u_\infty x}} = \frac{4.64}{\sqrt{\mathrm{Re}_x}} \tag{7-23a}$$

where Re_x is the local Reynolds number and is given by

$$\mathrm{Re}_x = \frac{\rho u_\infty x}{\mu}$$

The local shear stress is given by equation (7-22). We evaluate it using δ from equation (7-23) to obtain

$$\tau_w = \left(\frac{3}{2g_c}\mu u_\infty\right)\left(\frac{1}{4.64}\sqrt{\frac{\rho u_\infty}{\mu x}}\right)$$

or

$$\tau_w = 0.646\left(\frac{1}{2g_c}\right)\rho u_\infty^2\sqrt{\frac{1}{\mathrm{Re}_x}} \tag{7-24}$$

Equation (7-24) expresses the local shear stress, τ_w, in terms of the velocity head, $[(1/2g_c)(\rho u_\infty^2)]$, and the local Reynolds number, Re_x. We can now introduce the local drag coefficient, C_f, also called the skin friction coefficient. It is defined as

$$C_f = \frac{\tau_w}{(1/2g_c)\,\rho u_\infty^2} \tag{7-25}$$

When equations (7-24) and (7-25) are combined, the final expression for C_f becomes

$$C_f = \frac{0.646}{\sqrt{\mathrm{Re}_x}} \tag{7-26}$$

The above expression agrees very well with the exact expression, which has a numerical constant of 0.664 in place of 0.646.

Equations (7-24) and (7-26) give expressions for the local wall shear and the local drag coefficient. If the entire plate has a laminar boundary layer over it, the average wall shear, $\tau_{w,\,av}$, and the average drag coefficient, $C_{f,\,av}$, can be determined as follows. They are defined as

$$\tau_{w,\,av} = \frac{1}{L}\int_0^L \tau_w\,dx \quad \text{and} \quad C_{f,\,av} = \frac{\tau_{w,\,av}}{(1/2g_c)\,\rho u_\infty^2}$$

Substitution for τ_w from equation (7-24) and integration gives

$$\tau_{w,\,av} = 1.292\left(\frac{1}{2g_c}\right)\rho u_\infty^2\left(\frac{1}{\sqrt{\mathrm{Re}_L}}\right) \tag{7-27}$$

where Re_L is the Reynolds number based on the length, L, of the plate. If the average wall shear, $\tau_{w,\,av}$, is multiplied by the surface area of the plate, it will give the force that must be exerted on the plate by a support to hold it in the stream. The expression for the average drag coefficient is readily obtained as

$$C_{f,\,av} = \frac{1.292}{\sqrt{\mathrm{Re}_L}} \tag{7-28}$$

Sample Problem 7-4 Air flows over a thin plate with a velocity of 2.5 m/sec. The plate is 1 meter long and 1 meter wide. Estimate the boundary layer thickness at the trailing edge of the plate and the force necessary to hold the plate in the stream of air. The air has a viscosity of 0.86×10^{-5} Ns/m^2 and a density of 1.12 kg/m^3.

Solution: The length, L_{cr}, for which the Reynolds number will be less than the critical value of 5×10^5 is given by

$$L_{cr} = \frac{\mu}{\rho u_\infty}(5 \times 10^5) = \frac{0.86 \times 10^{-5} \times 5 \times 10^5}{1.12 \times 2.5} = 1.53 \text{ m}$$

The actual length of the plate is 1.0 m, which is less than L_{cr}. Therefore, we have laminar flow over the plate, and equations (7-23a) and (7-27) can be used to determine the boundary layer thickness and the drag force, respectively. Equation (7-23a) is

$$\delta = \frac{4.64x}{\sqrt{\mathrm{Re}_x}} \tag{7-23a}$$

and

$$\mathrm{Re}_L = 5 \times 10^5 \left(\frac{1.00}{1.53}\right) = 3.27 \times 10^5$$

Hence

$$\delta = \frac{4.64 \times 1.0}{\sqrt{3.27 \times 10^5}} = 8.1 \times 10^{-3} \text{ m} = 8.1 \text{ mm}$$

The average shear stress is given by

$$\tau_{w,\,av} = 1.292\left(\frac{1}{2g_c}\right)\left(\rho u_\infty^2\right)\sqrt{\frac{1}{\mathrm{Re}_L}}$$

Substituting for g_c, ρ, u_∞, and Re_L, we obtain

$$\tau_{w,\,av} = 1.292\left(\frac{1}{2} \times 1.12 \times 2.5^2\right)\frac{1}{\sqrt{3.27 \times 10^5}}$$

$$= 0.0079 \text{ N/m}^2$$

The force, F, necessary to hold the plate in the stream has to overcome the drag on both faces of the plate. Thus

$$F = 2(\text{area of one face}) \times \tau_{\text{w, av}}$$
$$= 2 \times 1.0 \times 1.0 \times 0.0079$$
$$= 0.0158 \text{ Newton}$$

7-5.2 Differential Analysis of the Laminar Boundary Layer

In the integral analysis, we considered a control volume that spanned the entire thickness of the boundary layer. In the differential analysis, an elemental control volume of dimensions Δx by Δy by Δz, located inside the boundary layer, is considered as shown in Figure 7-7. We now present an approximate differential analysis of laminar boundary layers. Once again, we assume that the fluid has constant thermophysical properties; that the principal direction of flow is along the x axis, with u_∞ as the free-stream velocity; and that the flow is two-dimensional. The objective of the analysis is the same as before, that is, to determine the velocity distribution within the boundary layer and the wall shear stress.

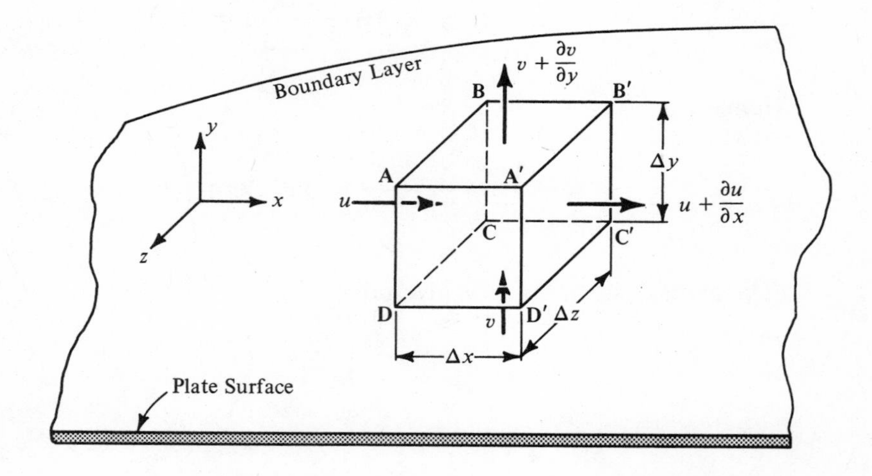

Figure 7-7 Control volume for differential analysis of a laminar boundary layer.

We now proceed to apply the principle of conservation of mass to the elemental control volume. We assume that the fluid flows into the control volume through faces **ABCD**, **DCC′D′**, and **BCC′B′** along the positive x, y, and z directions, respectively, and goes out through the remaining faces. The rate of inflow is given by the product of the density, the velocity normal to the area, and the area through which the fluid flows. Thus, if the x, y, and z components of velocity are represented by u, v, and w, we have

For face **ABCD**: $\dot{m}_{ABCD} = \rho u \Delta y \Delta z$

For face **DCC′D′**: $\dot{m}_{DCC'D'} = \rho v \Delta z \Delta x$

For face **BCC′B′**: $\dot{m}_{BCC'B'} = 0$

since w, the z component of velocity, is zero for a two-dimensional flow. The rate of outflows are obtained via a Taylor series expansion. They are

For face **A′B′C′D′**:

$$\dot{m}_{A'B'C'D'} \simeq \dot{m}_{ABCD} + \frac{\partial}{\partial x}(\dot{m}_{ABCD})\,\Delta x$$

$$\simeq \rho u \Delta y \Delta z + \left[\frac{\partial(\rho u)}{\partial x}\right] \Delta x \Delta y \Delta z$$

For face **ABB′A′**:

$$\dot{m}_{ABB'A'} \simeq \dot{m}_{DCC'D'} + \frac{\partial}{\partial y}(\dot{m}_{DCC'D'})\Delta y$$

$$\simeq \rho v \Delta z \Delta x + \left[\frac{\partial(\rho v)}{\partial y}\right] \Delta x \Delta y \Delta z$$

For face **AA′D′D**: $\dot{m}_{AA'D'D} = 0$

Since there are no sources or sinks of fluid present, the rate at which the fluid comes into the control volume must equal the rate at which the fluid leaves the control volume in order to maintain steady state. Hence

$$\dot{m}_{ABCD} + \dot{m}_{DCC'D'} + \dot{m}_{BCC'B'} = \dot{m}_{A'B'C'D'} + \dot{m}_{ABB'A'} + \dot{m}_{AA'D'D}$$

Substituting for different $\dot{m}$s and allowing Δx, Δy, and Δz, to approach zero, we obtain

$$\frac{\partial}{\partial x}(\rho u) + \frac{\partial}{\partial y}(\rho v) = 0$$

For an incompressible fluid (ρ = constant), the above equation becomes

$$\frac{\partial u}{\partial x} + \frac{\partial v}{\partial y} = 0 \tag{7-29}$$

This is known as the *continuity equation*. It is a partial differential equation in two unknowns, namely, u and v.

We next apply the principle of conservation of momentum for the x direction, equation (7-7). It is repeated here for convenience.

$$\sum_{i=1}^{N} F_{i,x} = (\dot{Mom})_{\text{out},x} - (\dot{Mom})_{\text{in},x} \tag{7-7}$$

The external forces acting on the control volume are due to the pressures and to the viscous shear stresses in the fluid. The pressure forces in the x direction are exerted on the faces **ABCD** and **A′B′C′D′**. The pressure force on face **ABCD** is

$$F_{p,x} = p\Delta y \Delta z \tag{7-30}$$

Using a Taylor series expansion, the pressure force on **A′B′C′D′** can be expressed as

$$F_{p,(x+\Delta x)} \simeq -\left[F_{p,x} + \left(\frac{\partial}{\partial x} F_{p,x}\right)\Delta x\right] \tag{7-30a}$$

The viscous shear stress as expressed earlier by equation (7-2) is valid when the flow has only one velocity component. When there are two (or three) components, it is given by

$$\tau_{yx} = \frac{\mu}{g_c}\left(\frac{\partial u}{\partial y} + \frac{\partial v}{\partial x}\right) \tag{7-31}$$

The first subscript in τ_{yx} specifies the direction of the normal to the surface on which τ_{yx} acts, and the second subscript gives the direction of the shear stress. Thus, τ_{yx} is the shear stress in the x direction acting on a surface that is normal to the y direction.

If the motion of the fluid is principally in the x direction, then the x-component of velocity, u, will be more significant than the y-component of velocity, v. Furthermore, because the boundary layer is very thin, $\partial/\partial y$ derivatives will generally be larger than $\partial/\partial x$ derivatives. Therefore, the term $(\partial v/\partial x)$ in equation (7-31) is negligible when compared to $(\partial u/\partial y)$, and we can write

$$\tau_{yx} \simeq \frac{\mu}{g_c}\left(\frac{\partial u}{\partial y}\right) \tag{7-31a}$$

For the same reasons, we neglect the viscous shear forces on faces **ABCD** and **A′B′C′D′** in Figure 7-7. Furthermore, because of the two-dimensional nature of the problem, there are no shear forces on faces **BCC′B′** and **AA′D′D**. Thus we are left with shear forces acting only on the faces **DCC′D′** and **ABB′A′**.

The viscous shear force acts in the negative x-direction on face **DCC′D′** and in the positive x-direction on the face **ABB′A′**. Recognizing that the shear force equals the product of the shear stress, τ_{yx}, and the area on which it acts, we have the shear force on face **DCC′D′**

$$F_{s,y} = -\tau_{yx}\Delta x \Delta z$$

where the subscript y in $F_{s,y}$ signifies a coordinate location and not a force direction (the force is in the x direction). Substituting for τ_{yx} from equation (7-31a), we obtain

$$F_{s,y} = -\frac{\mu}{g_c}\left(\frac{\partial u}{\partial y}\right)\Delta x \Delta z \qquad \text{(7-32)}$$

A Taylor series expansion can then be used to obtain the shear force on the face **ABB′A′** as

$$F_{s,(y+\Delta y)} \simeq -\left[F_{s,y} + \frac{\partial}{\partial y}(F_{s,y})\,\Delta y\right] \qquad \text{(7-32a)}$$

Now that we have obtained expressions for all the forces acting in the x-direction on the control volume, we can write the left-hand side of equation (7-7) as

$$\sum F_{i,x} = F_{p,x} + F_{p,(x+\Delta x)} + F_{s,y} + F_{s,(y+\Delta y)}$$

Substituting for the various terms on the right-hand side of the above equation from equations (7-30), (7-30a), (7-32), and (7-32a), and simplifying, we obtain

$$\sum F_{i,x} = \left[-\frac{\partial p}{\partial x} + \frac{\mu}{g_c}\left(\frac{\partial^2 u}{\partial y^2}\right)\right]\Delta x \Delta y \Delta z \qquad \text{(7-33)}$$

Next, we examine the net rate of momentum transfer through the control volume. Since we are interested in the x momentum only, we have to consider those faces of the control volume through which there is a flow and for which the fluid particles have an x-component of velocity. Such faces are **ABCD**, **A′B′C′D′** **DCC′D′**, and **ABB′A′**.

The mass flow rate into the control volume through face **ABCD** is $(\rho u \Delta y \Delta z)$. The x-component of velocity associated with this rate of mass flow is u. Hence, the rate at which x-momentum enters the control volume through **ABCD** is

$$(\dot{Mom})_x = \frac{1}{g_c}(\rho u \Delta y \Delta z)(u)$$

or

$$(\dot{Mom})_x = \frac{1}{g_c}(\rho u^2 \Delta y \Delta z) \tag{7-34}$$

Furthermore, the rate at which x-momentum leaves the control volume through **A′B′C′D′**, is given by

$$(\dot{Mom})_{x+\Delta x} \simeq (\dot{Mom})_x + \frac{\partial}{\partial x}[(\dot{Mom})_x]\,\Delta x \tag{7-34a}$$

Similarly, the quantity $(\rho v \Delta z \Delta x)$ represents the mass flow rate into the control volume through the face **DD′C′C**. As it enters the control volume, it carries with it an x-component of velocity,* u. Hence, the rate at which x-momentum enters the control volume through the face **DD′C′C** is

$$(\dot{Mom})_y = \frac{1}{g_c}(\rho v \Delta z \Delta x)u \tag{7-35}$$

The rate at which x-momentum leaves the control volume through face **ABB′A′** is given by

$$(\dot{Mom})_{y+\Delta y} \simeq (\dot{Mom})_y + \frac{\partial}{\partial y}[(\dot{Mom})_y]\Delta x \tag{7-35a}$$

The right-hand side of equation (7-7) can now be expressed as

$$(\dot{Mom})_{\text{out},x} - (\dot{Mom})_{\text{in},x} = [(\dot{Mom})_{x+\Delta x} + (\dot{Mom})_{y+\Delta y}] - [(\dot{Mom})_x + (\dot{Mom})_y]$$

*Strictly speaking, it is $(1/2)\{(u) + [u + (\partial u/\partial x)\Delta x]\}$. If the analysis is carried out using this expression, the final result is identical to that obtained by using u as the x-component of velocity when higher-order terms are neglected.

Substituting for the various terms on the right-hand side from equations (7-34), (7-34a), (7-35), and (7-35a), we obtain

$$(\dot{Mom})_{out,x} - (\dot{Mom})_{in,x} \simeq \frac{1}{g_c}\left[\frac{\partial}{\partial x}(\rho u^2) + \frac{\partial}{\partial y}(\rho uv)\right]\Delta x \Delta y \Delta z \quad (7\text{-}36)$$

The differential momentum equation for the laminar boundary layer is now obtained by introducing equations (7-33) and (7-36) into equation (7-7), giving

$$\left[-\frac{\partial p}{\partial x} + \frac{\mu}{g_c}\left(\frac{\partial^2 u}{\partial y^2}\right)\right]\Delta x \Delta y \Delta z = \frac{1}{g_c}\left[\frac{\partial}{\partial x}(\rho u^2) + \frac{\partial}{\partial y}(\rho uv)\right]\Delta x \Delta y \Delta z$$

As the control volume approaches zero, the above equation, for an incompressible fluid (ρ = constant) becomes

$$\frac{\partial}{\partial x}(u^2) + \frac{\partial}{\partial y}(uv) = -\frac{g_c}{\rho}\left(\frac{\partial p}{\partial x}\right) + \frac{\mu}{\rho}\left(\frac{\partial^2 u}{\partial y^2}\right)$$

Upon carrying out the differentiation, the left-hand side becomes

$$u\left(\frac{\partial u}{\partial x} + \frac{\partial v}{\partial y}\right) + \left[u\left(\frac{\partial u}{\partial x}\right) + v\left(\frac{\partial u}{\partial y}\right)\right]$$

The quantity in the first pair of parentheses is identically zero in view of the continuity equation, equation (7-29). The momentum equation for an incompressible fluid thus simplifies to

$$u\left(\frac{\partial u}{\partial x}\right) + v\left(\frac{\partial u}{\partial y}\right) = -\frac{g_c}{\rho}\left(\frac{\partial p}{\partial x}\right) + \nu\left(\frac{\partial^2 u}{\partial y^2}\right) \quad (7\text{-}37)$$

where $\nu = \mu/\rho$

If there are no pressure gradients in the flow outside the boundary layer, we expect that there will be no variations in pressure within the boundary layer. Therefore, the term, $(\partial p/\partial x)$, drops out, and the momentum equation becomes

$$u\left(\frac{\partial u}{\partial x}\right) + v\left(\frac{\partial u}{\partial y}\right) = \nu\left(\frac{\partial^2 u}{\partial y^2}\right) \quad (7\text{-}38)$$

7-5.3 Solution of the Laminar Boundary Layer Equations

The solution for the velocity distribution requires that the continuity and the momentum equations [equations (7-29) and (7-38)] be dealt with simultaneously. These are nonlinear, partial differential equations. Appropriate boundary conditions to solve these equations are

$$\text{at } y = 0, u = 0, \text{ and } v = 0 \tag{7-39}$$

and

$$\text{at } y \to \infty, u \to u_\infty, \text{ and } v = 0 \tag{7-39a}$$

The condition (7-39a) states that the velocity, u, approaches u_∞ in an asymptotic manner.

The forgoing equations were first solved by Blasius (Reference 14), a student of L. Prandtl, in 1908. He converted the two partial differential equations into a single, ordinary, nonlinear differential equation by employing an ingenious change of variables.

In the integral method for the analysis of the boundary layer (Section 7-5.1), the x-component of velocity was expressed as a function of η, where $\eta = (y/\delta)$. The boundary layer thickness, δ, was found to be proportional to $(x\mu/\rho u_\infty)^{1/2}$. If the variable η, now redefined as

$$\eta = y\sqrt{\frac{\rho u_\infty}{x\mu}} \sim \frac{y}{\delta} \tag{7-40}$$

is introduced in equation (7-38), it is possible to achieve some simplification. Also, to satisfy the continuity equation, equation (7-29), a function, Ψ, is introduced according to the following definitions

$$\frac{\partial \Psi}{\partial x} = v \quad \text{and} \quad \frac{\partial \Psi}{\partial y} = -u \tag{7-40a}$$

Direct substitution of equation (7-40a) into (7-29) validates that the latter is, indeed, satisfied. Further use of equation (7-40) and (7-40a) enables the momentum equation to be transformed into an ordinary differential equation. The details are beyond the scope of this text, but interested readers may find them in Reference 15. The end result of the transformation is

$$ff'' + f''' = 0 \tag{7-41}$$

where

$$f(\eta) = \frac{\Psi\eta}{yu_\infty} \quad \text{and} \quad f' \equiv \frac{df}{d\eta}$$

The boundary conditions become

$$\text{at } \eta = 0, f = f' = 0$$
$$\text{as } \eta \to \infty, f' \to 2$$

Solution of equations (7-41) can be obtained by numerical means only. Figure (7-8) shows the solution obtained and the experimental data of various workers.

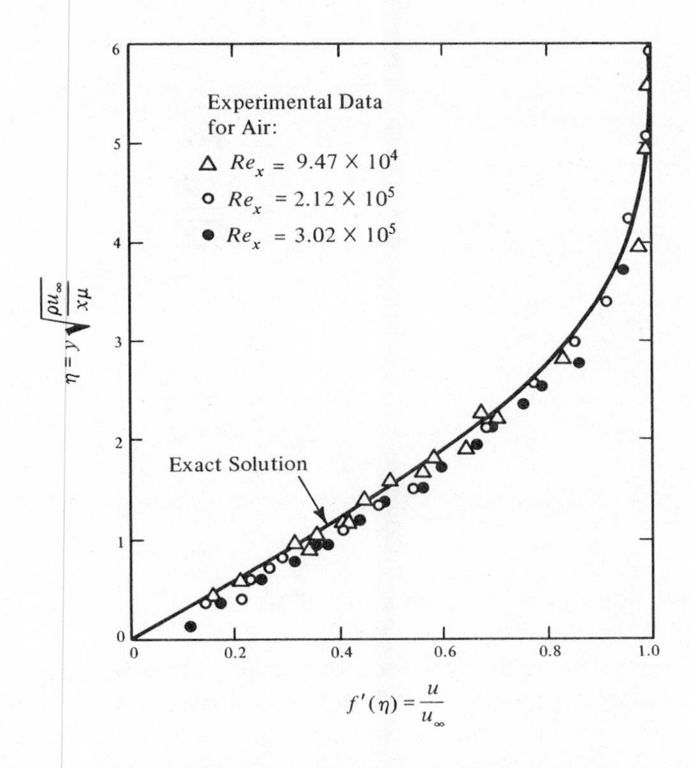

Figure 7-8 Dimensionless velocity profile in a laminar boundary layer over a flat plate, from "Direct Measurements of Skin Friction," by S. Dhawan, *NACA* T. N. No. 2567, Washington, D.C., 1952.

The boundary layer thickness, δ, is defined by some investigators as that y-distance from the plate surface where $u = 0.99u_\infty$. For this definition, the exact solution yields

$$\frac{\delta}{x} = \frac{5.0}{\sqrt{\text{Re}_x}} \tag{7-42}$$

Also, the local wall shear stress, τ_w, is obtained as

$$\tau_w = 0.664\left[\left(\frac{1}{2g_c}\right)\rho u_\infty^2\right]\sqrt{\frac{1}{\mathrm{Re}_x}}$$

and the average wall shear stress is given by

$$\tau_{w,\,av} = 1.328\left[\left(\frac{1}{2g_c}\right)\rho u_\infty^2\right]\sqrt{\frac{1}{\mathrm{Re}_L}}$$

Employing the definition of the drag coefficient, equation (7-25), the local and average drag coefficients are given by

$$C_f = \frac{0.664}{\sqrt{\mathrm{Re}_x}} \tag{7-44}$$

and

$$C_{f,\,av} = \frac{1.328}{\sqrt{\mathrm{Re}_L}} \tag{7-44a}$$

These results differ very little from those obtained from the approximate integral analysis.

7-5.4 Turbulent Boundary Layer on a Flat Plate

In a laminar flow, the diffusion of momentum is at a microscopic level and is due to a molecular property, the kinematic viscosity, ν. In a turbulent flow, a significant portion of the momentum diffusion is due to the macroscopic eddies. It is customary to assume an eddy diffusivity, ϵ_M, to account for such a contribution. The shear stress in a turbulent flow is then represented by

$$\tau = \left(\frac{1}{g_c}\right)\rho(\nu + \epsilon_M)\frac{du}{dy} \tag{7-45}$$

Analysis of turbulent flows is complicated by the fact that the eddy diffusivity, ϵ_M, is not a property of the fluid, but rather is dependent on the specific flow situation. Further complications arise due to the fact that the velocities are fluctuating with time, whereas one can deal effectively only with time-averaged values. Consequently, empirical relations and experimental data are extensively used to facilitate the analysis of turbulent flow. We give below a simplified method of dealing with the turbulent boundary layer. For a thorough discussion of turbulent flow, the reader is referred to Reference 16.

The analysis begins with the integral form of the momentum conservation principle, which was derived earlier [equation (7-19)] and is repeated here for convenience

$$\frac{\tau_w}{\rho} = \left(\frac{1}{g_c}\right)\frac{d}{dx}\int_0^\delta u(u_\infty - u)\,dy \qquad (7\text{-}19)$$

For the wall shear stress, one may use the empirical relation deduced by Blasius (Reference 17)

$$\tau_w = 0.0228\left(\frac{\rho u_\infty^2}{g_c}\right)\left(\frac{\nu}{u_\infty \delta}\right)^{1/4} \qquad 5 \times 10^5 < \text{Re}_x < 10^7 \qquad (7\text{-}46)$$

where δ is the thickness of the turbulent boundary layer.

The velocity distribution in a turbulent boundary layer in the absence of pressure gradients is represented fairly well by the equation

$$\frac{u}{u_\infty} = (y/\delta)^{1/7} \qquad (7\text{-}47)$$

as evidenced by Figure 7-9.

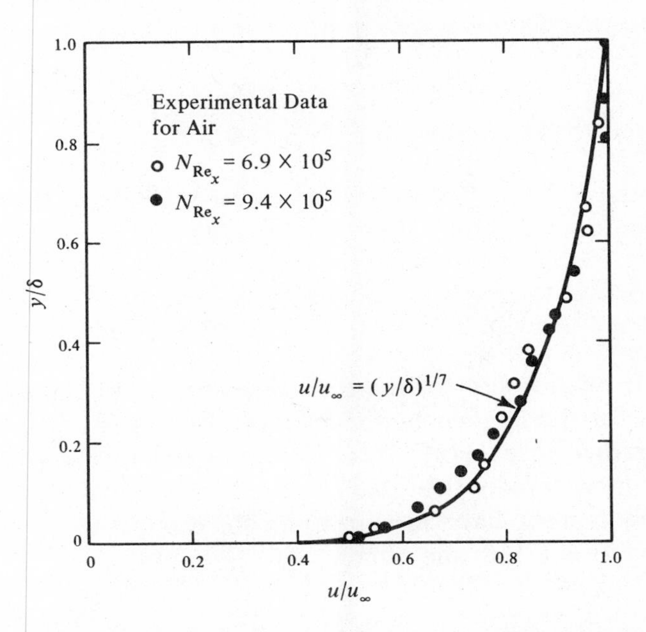

Figure 7-9 Dimensionless velocity profile in a turbulent boundary layer over a flat plate, from "Direct Measurements of Skin Friction," by S. Dhawan, *NACA* T. N. No. 2567, Washington, D.C., 1952.

It should be noted that equation (7-47) fails to give a meaningful value of the wall shear stress, since the derivative of u with respect to y contains $(y)^{6/7}$ in its denominator, it approaches infinity as $y \to 0$. However, when equation (7-47) is used in conjunction with the integral momentum equation [equation (7-19)] and with equation (7-46), it gives a satisfactory expression for the thickness of the turbulent boundary layer.

Substituting for τ_w and u from equations (7-46) and (7-47) into equation (7-19) and replacing the partial derivative by an ordinary derivative, we obtain from the integral momentum equation

$$0.0228\left(\frac{\nu}{u_\infty \delta}\right)^{1/4} = \frac{d}{dx}\int_0^\delta (y/\delta)^{1/7}\,[(1) - (y/\delta)^{1/7}]\,dy$$

When the integration is performed, there follows, after rearrangement

$$0.0228\left(\frac{72}{7}\right)\left(\frac{\nu}{u_\infty}\right)^{1/4} dx = \delta^{1/4}\,d\delta$$

Strictly speaking, when the above equation is to be integrated, the limits for integration cannot be from 0 to x and from 0 to δ, because we know that the equation is valid only when turbulence is established. As an approximation, however, we may still use the forgoing limits of integration (Reference 6) and integrate to yield

$$0.0228\left(\frac{72}{7}\right)\left(\frac{5}{4}\right)\left(\frac{\nu}{u_\infty}\right)^{1/4} x = \delta^{5/4}$$

or

$$\frac{\delta}{x} = \frac{0.375}{(\mathrm{Re}_x)^{1/5}} \tag{7-48}$$

This relationship for the thickness of the turbulent boundary layer is similar to that for the laminar boundary layer, except that the exponent of Re_x is now (1/5) rather than (1/2). Thus, the thickness of a turbulent boundary layer increases more rapidly than that of a laminar boundary layer.

Now that we have an expression for δ [equation (7-48)], we can introduce it into equation (7-46) and obtain

$$\tau_w = 0.0228\left(\frac{1}{g_c}\right)\left(\rho u_\infty^2\right)\left(\frac{\nu}{u_\infty}\right)^{1/4}\left[\frac{(\mathrm{Re}_x)^{1/5}}{0.375x}\right]^{1/4}$$

or

$$\tau_w = 0.02915\left(\frac{1}{g_c}\right)\left(\rho u_\infty^2\right)\left(\frac{\nu}{u_\infty x}\right)^{1/5} \tag{7-49}$$

The local drag coefficient is then given by

$$C_f = \frac{\tau_w}{\frac{1}{2}\rho u_\infty^2} = \frac{0.0583}{(\text{Re}_x)^{1/5}} \tag{7-49a}$$

***Sample Problem** 7-5* If the plate in Sample Problem 7-4 were 5 meters long, estimate the total drag force on the plate.

Solution: It was determined in the solution to Sample Problem 7-4 that the transition from laminar to turbulent flow occurred at $L_{cr} = 1.53$ m. We will compute the drag force on the first 1.53 m using equation (7-27) and on the remaining (5 − 1.53) or 3.47 m using equation (7-49). In other words, we assume that fully turbulent flow begins at $L = L_{cr}$.

Laminar Region:

$$\tau_{w,\,av} = 1.292\,[(1/2) \times 1.12 \times 2.5^2]\frac{1}{\sqrt{5 \times 10^5}} = 0.0064 \text{ N/m}^2$$

and

$$F_{\text{laminar}} = 2 \text{ (areas of one face)} \times \tau_{w,\,av}$$

$$= 2 \times 1.53 \times 1.0 \times 0.0064$$

$$= 0.0196 \text{ Newton}$$

Turbulent Region: The local shear stress, given by equation (7-49), is

$$\tau_w = 0.02915\,(1/g_c)\,(\rho u_\infty^2)\,(\nu/u_\infty)^{1/5}\,(x)^{-1/5}$$

The average shear stress for the region between $x = 1.53$ m and $x = 5$ m is, therefore, given by

$$\tau_{w,\,av} = \frac{\left[\int_{L_{cr}}^{L} \tau_w\, dx\right]}{\left[\int_{L_{cr}}^{L} dx\right]}$$

$$\tau_{w,\,av} = \frac{0.02915\,(1 \times 1.12 \times 2.5^2)}{(5 - 1.53)} \left(\frac{0.86 \times 10^{-5}}{1.12 \times 2.5}\right)^{1/5} \int_{1.53}^{5} x^{-1/5}\, dx$$

Integration gives

$$\tau_{w,\,av} = 4.643 \times 10^{-3} \left(\frac{x^{4/5}}{4/5}\right)\Bigg|_{1.53}^{5}$$

or

$$\tau_{w,\,av} = 4.643 \times 10^{-3} \times 5/4\,(5^{4/5} - 1.53^{4/5}) = 0.0129 \text{ N/m}^2$$

Hence

$$F_{turb} = 2 \times 1 \times (5 - 1.53) \times 0.0129$$
$$= 0.089 \text{ Newton}$$

Thus, the total force exerted on the plate is (0.0196 + 0.089) or 0.11 Newton.

7-6 FLOWS ACROSS A CYLINDER AND A SPHERE

After studying the simplest external flow, the flow over a flat plate, we now consider flows past some simple curved surfaces, namely, surfaces of a cylinder and a sphere.

In a manner analagous to the analysis of the flow over a flat plate, the flow past a curved surface can be divided into a boundary layer region near the surface and a nonviscous (inviscid) region away from the surface. The pressure gradient for flow about a cylinder or sphere is nonzero. This nonuniformity of pressure has a marked influence on the development of the boundary layer on a curved surface. In fact, the boundary layer invariably separates from the surface at some distance downstream of the point **P** in Figure 7-10. At point **P**, known as the *forward stagnation point*, the velocity head is completely converted into pressure head under idealized conditions. The phenomenon of separation significantly affects the drag force on a cylinder or a sphere. It is, therefore, worthwhile to look at the mechanism of separation in some detail.

One may reason that fluid particles tend to accelerate as they pass around the forward portion of the cylinder and then decelerate as they pass around the rear portion. This will result in a decreasing pressure on the forward portion and an increasing pressure on the rear portion. The relatively slow moving fluid particles situated adjacent to the cylinder wall are hard pressed to continue their forward motion in the face of the increasing pressure on the rear portion. At some point, they simply give up and begin to flow backward. This is called separation of the boundary layer. The condition for the separation to occur is (Reference 15)

$$\left(\frac{\partial u}{\partial y}\right)_{y=0} = 0 \tag{7-50}$$

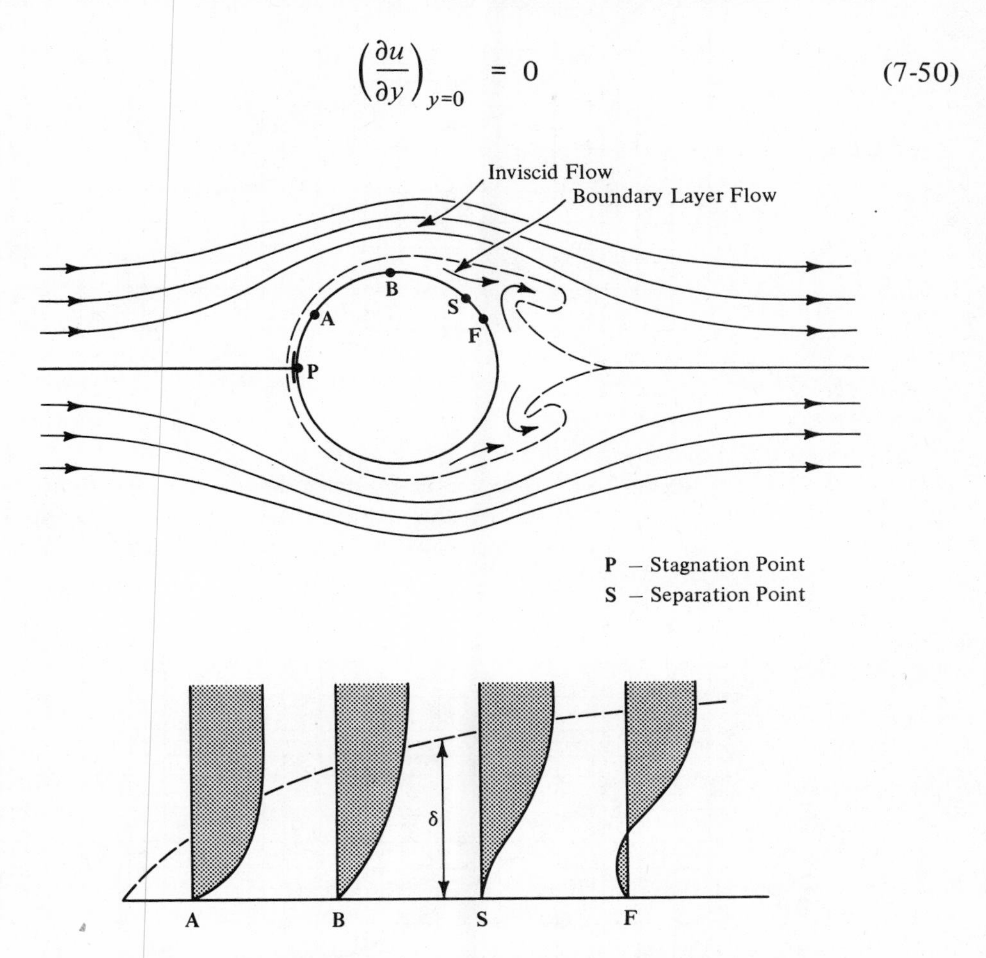

Figure 7-10 Flow past a cylinder and velocity profiles at different locations on the cylinder.

The separation is shown to occur at point **S** in Figure 7-10. The variation in the pressure gradient also causes changes in the velocity profile in the boundary layer as shown in Figure 7-10.

The separation of the boundary layer creates a localized low-pressure region on the rear portion of a cylinder or a sphere. In turn, this low pressure creates a drag called *form drag*. Thus, the total drag on a cylinder or a sphere is due to the wall shear (or fluid friction) and to the pressure forces or form drag. Figure 7-11 and 7-12 show the variation of the total drag coefficient, C_D, as a function of the Reynolds number. The Reynolds number is based on the outside diameter of the cylinder or the sphere.

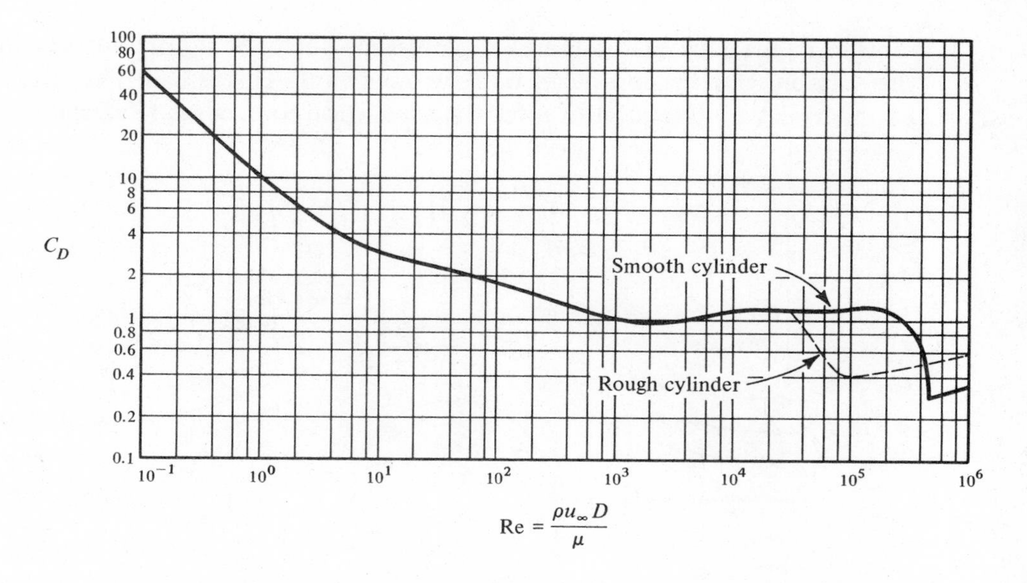

Figure 7-11 Drag coefficient for a circular cylinder, from *Boundary Layer Theory* by H. Schilchting. Copyright 1968 by McGraw-Hill, Inc.. Used by permission.

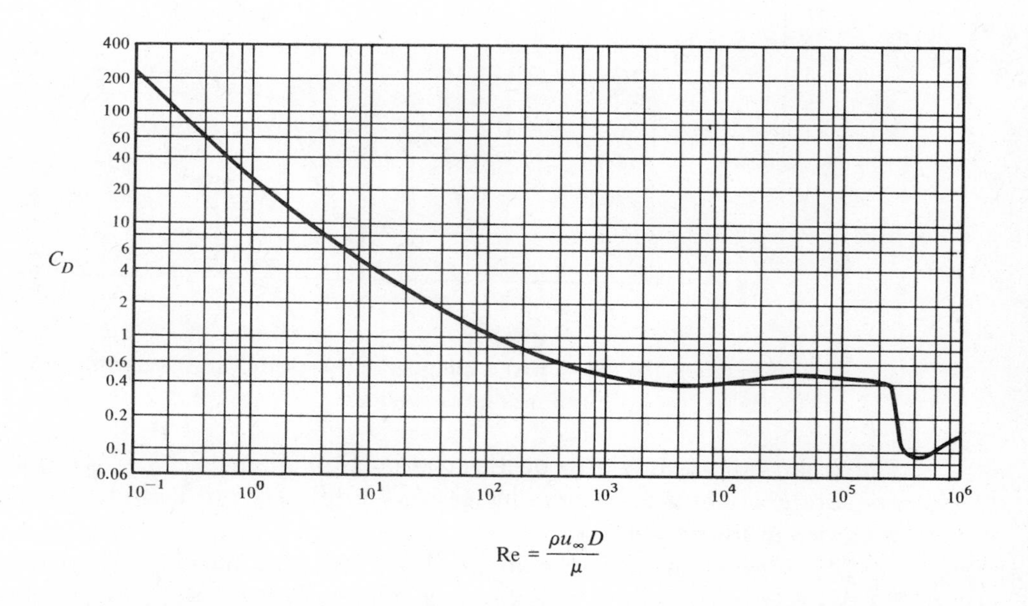

Figure 7-12 Drag coefficient for a sphere from *Boundary Layer Theory* by H. Schlichting. Copyright 1968 by McGraw-Hill, Inc. Used by permission.

If Figures 7-11 and 7-12 are examined closely, we can see four regions where the shape of the C_D versus Re curve changes. In the first zone ($Re \leqslant 2$), the slope of the curve is fairly steep. In this region, viscous effects are dominant, and the entire drag force is due to fluid friction. There is no flow separation in this region. In the second zone ($2 < Re \leqslant 2 \times 10^3$), the drag forces due to friction and to flow separation are of the same order of magnitude. The separated flow region becomes more and more turbulent as the Reynolds number exceeds 10^3. In the third zone ($2 \times 10^3 < Re < 3 \times 10^5$), the flow continues to be more and more turbulent, and form drag dominates. When the third zone terminates and the fourth zone begins ($Re \simeq 3 \times 10^5$), the boundary layer flow becomes fully turbulent. This causes a delayed separation of the boundary layer, thus reducing the contribution of the form drag and resulting in a sharp drop in the curve at about $Re = 3 \times 10^5$.

The behavior of the drag coefficient curve, discussed above, is qualitatively applicable to other bluff shapes such as elliptic cylinders and airfoils. Since the drag coefficient generally decreases with the Reynolds number, the reader should not think that the drag force decreases with an increase in the Reynolds number. In fact, the drag force will increase, since it is given by the product of the drag coefficient, C_D, the velocity head, $[(1/2g_c)\rho u_\infty^2]$, and the area of the object normal to the flow.

7-7 FLOW ACROSS BANKS OF TUBES

Banks of tubes are a common occurrence in large, commercial heat exchangers. In the design of these exchangers, it is necessary to calculate the pressure drop experienced by the fluid flowing over a bank of tubes. The pressure drop cannot be calculated using the relations for a single tube, since there is a substantial amount of interaction among the tubes, especially downstream of the first row.

The tube rows can be arranged either in-line or staggered as shown in Figure 7-13. The transverse pitch, S_t, and the longitudinal pitch, S_l, are measured between the tube centers as indicated in the figure. The outside tube diameter, D, is the characteristic dimension.

The pressure drop is given by

$$\Delta p = Nf\left(\frac{\rho u_{av}^2}{2g_c}\right) \tag{7-51}$$

where

Δp = pressure drop in lb_f/ft^2 or N/m^2

N = number of tube rows in flow direction

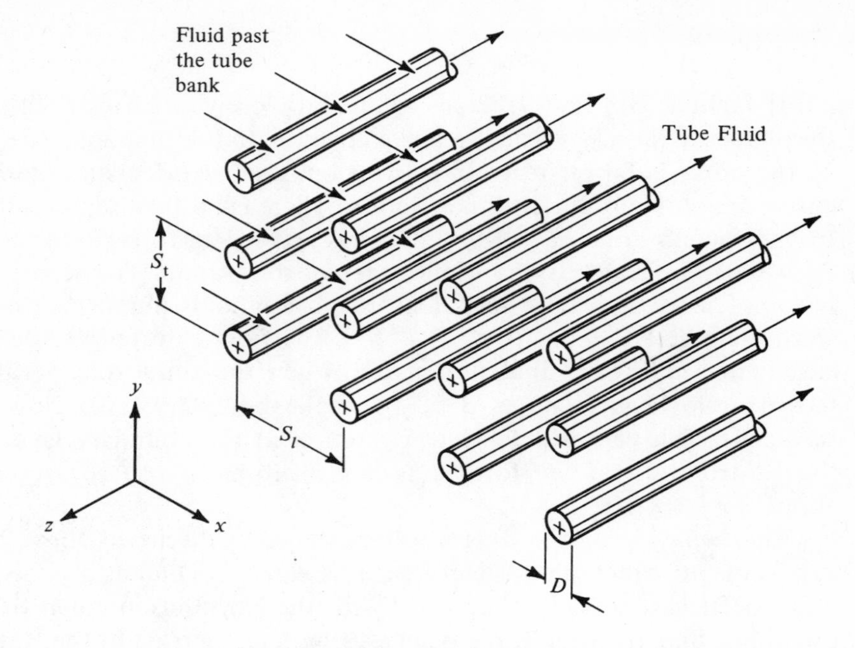

(a) In-line Arrangement

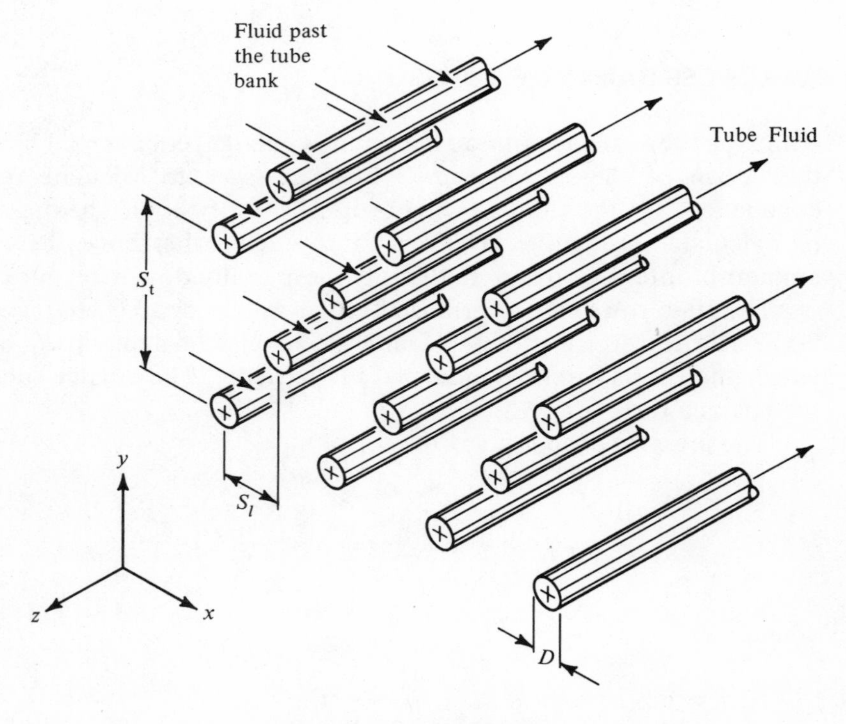

(b) Staggered Arrangement

Figure 7-13 Banks of tubes. (a) In-line arrangment, and (b) staggered arrangement.

ρ = density of the fluid flowing across the bank of tubes evaluated at the inlet temperature

u_{av} = average velocity of fluid through area between tubes

The friction factor, f, is plotted for in-line and staggered arrangements in Figure 7-14a and b. These figures are for a square arrangement and an arrangement of tubes in the form of isosceles triangles, respectively. For other arrangements, the value of the friction factor obtained from Figures 7-14a and b should be multiplied by the correction factor, g, plotted in the upper right-hand corners of Figures 7-14a and b.

Until the work of Zukauskas, the friction factor, f, was given by Reference 19 as

$$f = \left\{ 0.25 + \frac{0.1175}{[(S_t/D) - 1]^{1.08}} \right\} \mathrm{Re}_{max}^{-0.16} \tag{7-52}$$

for the staggered arrangement. For the in-line arrangement, it was given by

$$f = \left[0.044 + \frac{0.08\,(S_l/D)}{[(S_t - D)/D]^{[0.43 + (1.13D/S_l)]}} \right] \mathrm{Re}_{max}^{-0.15} \tag{7-52a}$$

In equations (7-52) $\mathrm{Re}_{max} = \rho u_{av} D/\mu$

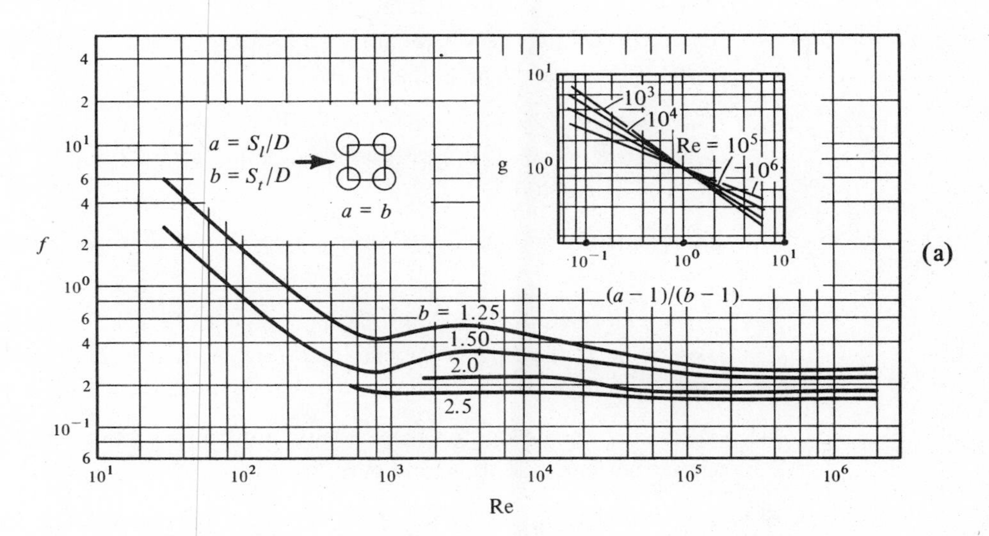

Figure 7-14 Friction factor, f, for banks of tubes: (a) In-line arrangement, and (b) staggered arrangement (g is the correction factor for arrangements other than square and isosceles triangular ones). Re is based on the velocity of the main flow. From *Heat Transfer in Banks of Tubes in Crossflow of Fluid* by A. Zukauskas, 'Mintis' Vilnius, pp. 124-125, 1968.

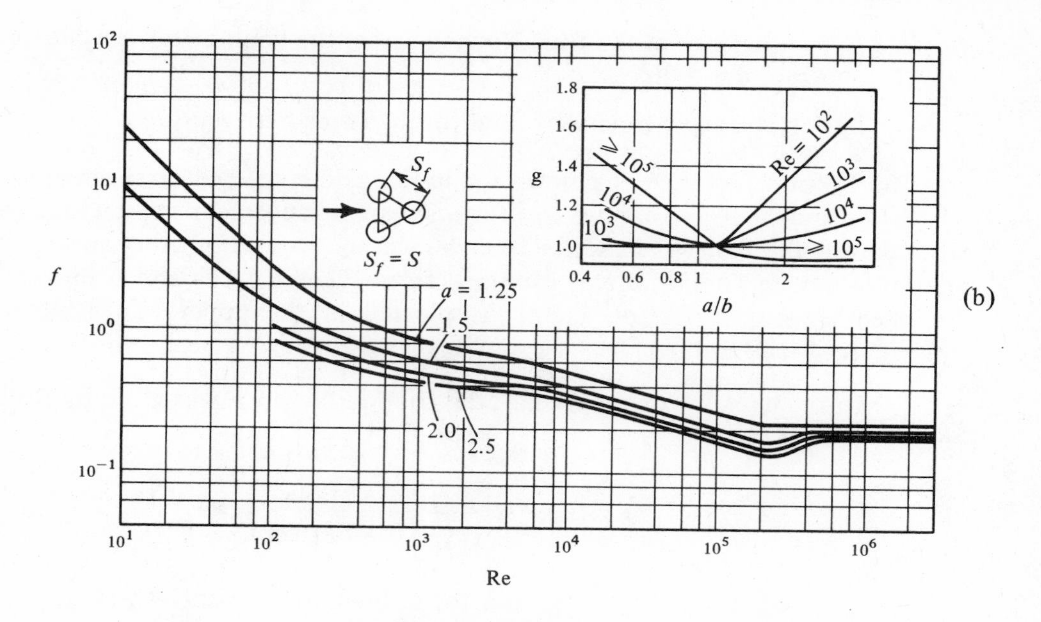

Figure 7-14 Continued

PROBLEMS (English Engineering System of Units)

7-1. The velocity distribution for a certain flow is given by $u = 60 \sin [(\pi/4)y]$ where y is the distance from the solid surface in inches and u is the velocity in ft/sec. Determine the viscous shear stress at $y = 0$ and $y = 0.2$ inches for (a) water, (b) engine oil, and (c) air at 60°F.

7-2. Calculate the Reynolds number for a flow rate of 0.1 cu ft/min through a tube having ½-inch I.D. for (a) saturated water at 400°F and 60°F, and (b) air at −100°F and 400°F.

7-3. Determine the entrance lengths for the flow of water and air using the data of problem 7-2.

7-4. Engine oil flows through a tube having 3/8-inch I.D. at a flow rate of 0.5 gpm. Determine if the flow is laminar or turbulent. If it is laminar, calculate the maximum flow velocity, and plot the velocity profile. $T_b = 68$°F.

7-5. Derive an expression for the velocity profile for laminar flow through an annulus of inner radius, r_i, and outer radius, r_o.

7-6. An experiment is to be designed to study the laminar flow of engine oil through a smooth tube of 3/4-inch I.D. Determine the maximum possible flow rate for laminar flow. Calculate the pressure gradient under this condition. $T_b = 68$°F.

7-7. Plot a graph of (u/u^*) versus $(\rho u^* y/\mu)$ for turbulent flow through a tube using equations (7-15).

7-8. Calculate the power requirement to deliver water through a 4-inch smooth pipe at a flow rate of 100 gpm over a distance of 5 miles.

7-9. Do problem 7-8 if the inner wall of the pipe has a roughness factor of 0.004.

7-10. Determine the boundary layer thickness for laminar flow over a flat plate by using a velocity profile $u/u_\infty = a_o + a_1(y/\delta)$. Follow the procedure of Section 7-5.1.

7-11. Air at 70°F flows over a thin plate with a velocity of 5 ft/sec. The plate is 8 feet long and 4 feet wide. Estimate the maximum boundary layer thickness and the force necessary to hold the plate in the air stream.

7-12. Oil at 300°F flows over a thin plate with a free-stream velocity of 10 ft/sec. Plot the velocity profile inside the boundary layer for $x = 0.5$ feet and $x = 5$ feet.

7-13. If the x-component of velocity in the laminar boundary layer is given by equation (7-21), obtain an expression for the y-component of velocity in the boundary layer. [*Hint:* Use equation (7-29).]

7-14. Plot the local skin friction coefficient as a function of length for a flow velocity of 8 ft/sec over a 4-foot long flat plate for (a) hydrogen, (b) air, (c) glycerine, and (d) engine oil. The temperature of the fluid in each case is 60°F. Use the results of the differential analysis of the laminar boundary layer.

7-15. Determine the values of the average skin friction coefficient for the fluids in problem 7-14, if $u_\infty = 100$ ft/sec and $T_\infty = 60°$F.

7-16. Calculate the drag force on a sphere of radius 6 inches held in a stream of air with $u_\infty = 100$ ft/sec and $T_\infty = 60°$F.

7-17. Estimate the drag force experienced by a cylinder with its axis normal to the flow of water. The cylinder is 4 inches in diameter and 4 feet in length. The free-stream velocity of water is 8 ft/sec. The temperature of the water is 70°F.

7-18. Calculate the pressure drop experienced by air flowing over a bank of tubes for the following conditions. Tubes are arranged in-line, 8 rows deep, and 12 rows high and have a diameter of ½ inch. The longitudinal and the transverse pitches are $S_l = 3/4$ inch and $S_t = 5/8$ inch. The air flow rate is 200 cu ft/min at 200°F. The tubes are 2 feet long.

7-19. If the tube arrangement in problem 7-18 is staggered, determine the pressure drop that the air experiences.

PROBLEMS (SI System of Units)

7-1. The velocity distribution for a certain flow is given by $u = 10 \sin (\pi/4)y$ where y is the distance from the solid surface in cm and u is the velocity in m/sec. Determine the viscous shear stress at $y = 0$ and $y = 5$ mm for (a) water, (b) engine oil, and (c) air. The fluid temperature is 20°C in all cases.

7-2. Calculate the Reynolds number for a flow rate of 50 cc/sec through a tube having 10-mm I.D. for (a) saturated water at 200°C and 20°C, and (b) air at −70°C and 200°C.

7-3. Determine the entrance length for the flow of water and air using the data of problem 7-2.

7-4. Water flows through a tube having an 8-mm I.D at a flow rate of 5 cc/sec. Determine if the flow is laminar or turbulent. If it is laminar, calculate the maximum flow velocity, and plot the velocity profile..

7-5. Derive an expression for the velocity profile for laminar flow through an annulus of inner radius, r_i, and outer radius, r_o.

7-6. An experiment is to be designed to study the laminar flow of engine oil through a smooth tube of 25-mm I.D. Determine the maximum possible flow rate for laminar flow. Calculate the pressure gradient under this condition.

7-7. Plot a graph (u/u^*) versus $(\rho u^* y/\mu)$ for turbulent flow through a tube using equations (7-15).

7-8. Calculate the power requirement to deliver water through a 140-cm I.D. smooth pipe at a flow rate of 2.5 m^3/min over a distance of 10km.

7-9. Do problem 7-8 if the inner wall of the pipe has a roughness factor of 0.004.

7-10. Determine the boundary layer thickness for laminar flow over a flat plate by using a velocity profile $(u/u_\infty) = a_o + a_1(y/\delta)$. Follow the procedure of Section 7-5.1.

7-11. Air at 22°C flows over a thin plate with a velocity of 2 m/sec. The plate is 3 m long and 1 m wide. Estimate the maximum boundary layer thickness and the force necessary to hold the plate in the air stream.

7-12. Engine oil at 150°C flows over a thin plate with a free-stream velocity of 3 m/sec. Plot the velocity profile inside the boundary layer for $x = 20$ cm and $x = 1$ m.

7-13. If the x-component of velocity in the laminar boundary layer is given by equation (7-21), obtain an expression for the y-component of velocity in the boundary layer. [*Hint:* Use equation (7-29).]

7-14. Plot the local skin friction coefficient as a function of length for a flow velocity of 3 m/sec over a 1.2-m long flat plate for (a) hydrogen, (b) air, (c) glycerine, and (d) engine oil. The temperature of the fluid in each case is 20°C. Use the results of the differential analysis of the laminar boundary layer.

7-15. Determine the values of the average skin friction coefficient for the fluids in problem 7-14, if $u_\infty = 20$ m/sec, $L = 4$ m, and $T_\infty = 20°C$.

7-16. Calculate the drag force on a sphere of radius 10 cm held in a stream of air with $u_\infty = 40$ m/sec and $T_\infty = 20°C$.

7-17. Estimate the drag force experienced by a cylinder with its axis normal to the flow of water. The cylinder is 7 cm in diameter and 1 m in length. The free-stream velocity of the water is 4 m/sec. The temperature of the water is 20°C.

7-18. Calculate the pressure drop experienced by air flowing over a bank of tubes for the following conditions. Tubes are arranged in-line, 8 rows deep, and 12 rows high and have a diameter of 10 mm. The longitudinal and the transverse pitches are $S_l = 16$ mm and $S_t = 13$ mm. The air flow rate is 1 m^3/min at 100°C. The tubes are 2.5 m long.

7-19. If the tube arrangement in problem 7-18 is staggered, determine the pressure drop that the air experiences.

REFERENCES

[1] Langhaar, H. L., "Steady Flow in the Transition Length of a Straight Tube," *Journal of Applied Mechanics*, **64**, (1942), p. 55.

[2] Latzko, H., "Heat Transfer in a Turbulent Liquid or Gas Stream," NACA TM 1068, (1944).

[3] Holdhusen, J. S., "The Turbulent Boundary Layer in the Inlet Region of Smooth Pipes," Ph.D. Thesis, University of Minnesota, (1952).

[4] Barbin, A. R., and Jones, J. B., "Turbulent Flow in the Inlet Region of a Smooth Pipe," *Journal of Basic Engineering*, **85**, (1963), p. 28.

[5] Olson, R. M., *Engineering Fluid Mechanics*, 2nd ed., International Textbook Company, (1966).

[6] Eckert, E. R. G., and Drake, R. M., Jr., *Heat and Mass Transfer*, McGraw-Hill Book Company, Inc., (1959).

[7] Nikuradse, J., *Stromungsgesetze in rauhen Rohren*, VDI-Forschungsheft, No. **361**, (1933). (English Translation in NACA TM 1292)

[8] Deissler, R. G., *Analytical and Experimental Investigation of Adiabatic Turbulent Flow in Smooth Tubes*, NACA TN 2138, Washington, D.C., (1950).

[9] Spalding, D. B., *A Single Formula for the Law of the Wall*, Journal of Applied Mechanics, Trans ASME, (1961), p. 455.

[10] Hansen, A. G., *Fluid Mechanics*, John Wiley and Sons, Inc., New York, (1967).

[11] Prandtl, L., "Uber Flussigkeitsbewegungen beisehr Kleiner Reibung," *Z. handlungen des III,* Internationalen Mathematiker Kongresses, Heidelberg, (1904).

[12] von Karman, T., "Uber Laminare und Turbulente Reibung," *Z. angew. Math. Mech.,* **I**, (1921), p. 233.

[13] Pohlhausen, K., "Zur naherungsweisen Integration der Differtialgleichung der laminaren Reibungsschicht," *Z. angnew. Math. Mech.,* **I**, (1921), p. 252.

[14] Blasius, H., "Grenzschichten in Flussigkeiten mit Kleiner Reibung," *Z. Math.-Physik,* **LVI**, (1908).

[15] Schlichting, H., *Boundary Layer Theory*, 4th ed., McGraw-Hill Book Company, (1968).

[16] Hinze, J. O., *Turbulence: An Introduction to its Mechanism and Theory*, 2nd ed., McGraw-Hill Book Company, (1975).

[17] Blasuis, H., "Das Ahnlichkeitsgesetz bei Reibungsvorgangen in Flussigkeiten," *Forschung a.d. Geb. d. Ingenieur wes, No.* **131**, (1913).

[18] Zhukauskas, A. A., *Heat Transfer in Banks of Tubes,* Mintis, Vilnius, Lithuania, (1968).

[19] Jakob, M., "Heat Transfer and Flow Resistance in cross flow of gases over Tuve Banks," Trans. ASME. Vol **60** p. 384, 1938.

8

Forced Convection

8-1 INTRODUCTION

The literal meaning of the word *convection* is the process or the action of carrying away. In the context of heat transfer, convection means the process of carrying heat energy away from a solid surface to an adjacent moving fluid in the presence of temperature differences, or vice versa. The convection process has two contributing mechanisms: (1) the conduction of heat from a solid surface to a thin layer of adjacent fluid, and (2) the movement of hot fluid particles away from the solid surface, their place in turn being taken by relatively cold fluid particles. The movement of the fluid particles can be attributed to pressure changes, to buoyancy, or to a combination of both. Thus, the study of convective heat transfer is intimately related to the study of fluid flow, which was discussed in Chapter 7.

Several examples of convective heat transfer can be found in our day to day life. On a cold winter morning with winds blowing at 15 miles per hour, a person waiting outside for a ride finds his nose and ears rapidly getting very cold. The blowing air (wind) literally carries heat away from his skin. In the case of an automobile radiator (which is really more of a convector than a radiator), the hot water from the engine block is pumped through the radiator tubes across which air is blown by a fan, thus cooling the water. Another example of convective heat transfer is that of a pipe carrying hot water and losing heat to the air in a room.

In the case of the automobile radiator, the flow of water was due to a pressure gradient, i.e., the water was pumped (forced) through the tubes by a water pump. This mode of convection is known as *forced convection.* In the case of the pipe losing heat to the room air, the motion of the air particles was due to buoyant forces caused by differences in the density of the air, that is, due to naturally occurring buoyant forces. Such a mode of heat transfer is called *natural* or *free convection.* The discussion of convective heat transfer can be conveniently divided into two areas, namely, forced convection and natural convection. Situations can occur where both modes of convection act simultaneously, and one has mixed convection. Usually, forced convection is of greater practical interest than natural convection because of its industrial applications. Most of the industrial heating or cooling devices such as heat exchangers, condensers, boilers, etc., involve heat transfer by forced convection. Natural convection, on the other hand, plays an important role in the heating of houses by baseboard heaters, in ecological problems, and in thermal pollution. We will discuss forced convection in this chapter, with the next chapter being devoted to natural convection.

We have seen in the preceding chapter that there are two types of flow, laminar and turbulent. The magnitude of the Reynolds number determines if the flow will be laminar or turbulent. The heat transfer analysis for convection is, in general, more complex than that for heat conduction, because the conservation of mass and momentum have to be satisfied in addition to the principle of energy conservation.

For turbulent flow, the analysis becomes extremely complex, owing to the transport of momentum and energy resulting from eddy motion. The majority of convective heat transfer problems encountered in practice involve turbulent flow. Therefore, it is customary to rely heavily on experimental data and empirical correlations. On the other hand, laminar heat transfer analyses are a useful vehicle for presenting the basic principles of convection and, at the same time, provide results of some practical importance. Therefore, we begin by analyzing heat transfer for laminar flow in tubes and in a laminar boundary layer over a flat plate. In the latter case, an approximate method and an exact method will be presented. Empirical correlations for turbulent flow heat transfer are then presented for many of the flow situations usually encountered.

8-2 THE CONVECTIVE HEAT TRANSFER COEFFICIENT

In 1701, more than one hundred years before Fourier formulated the basic law of conduction, Sir Isaac Newton proposed the following equation for predicting the rate of convective heat transfer, Q , from a solid surface to a surrounding fluid.

$$Q = hA(T_w - T_\infty) \tag{8-1}$$

where

h = the convective heat transfer coefficient, Btu/hr-ft^2 °F or W/m^2 °C

A = the surface area for convective heat transfer, ft^2 or m^2

T_w = the temperature of the solid surface, °F or °C

and

T_∞ = the temperature of the fluid sufficiently far from the solid surface so that it is not affected by the surface temperature, °F or °C

Representative values of the heat transfer coefficient are given in Chapter 1. Unlike the thermal conductivity of a material, the convective heat transfer coefficient is not a property. Its magnitude will change from one problem to another although the same solid and fluid may be involved in both problems. The value of the heat transfer coefficient depends upon a variety of factors, such as velocity, density, viscosity, thermal conductivity, and specific heat of the fluid; geometry of the surface; presence of buoyancy forces; etc. Such a broad dependence makes it difficult to arrive at an analytical expression for the heat transfer coefficient. There are a few simple cases that permit analytical solution. However, for a majority of problems of practical interest, one often relies on the experimental determination of the heat transfer coefficient employing dimensional analysis. Techniques for dimensional analysis will also be discussed later in this chapter.

8-2.1 The Nusselt Number

In the study of convective heat transfer, we are interested in determining the rate of heat transfer between a solid surface and an adjacent fluid, whenever a temperature difference exists. Consider a fluid flowing over a body. If the surface temperature is T_w and if the free-stream temperature is T_∞, the temperature of the fluid near the solid boundary will vary in some fashion as shown in Figure 8-1.

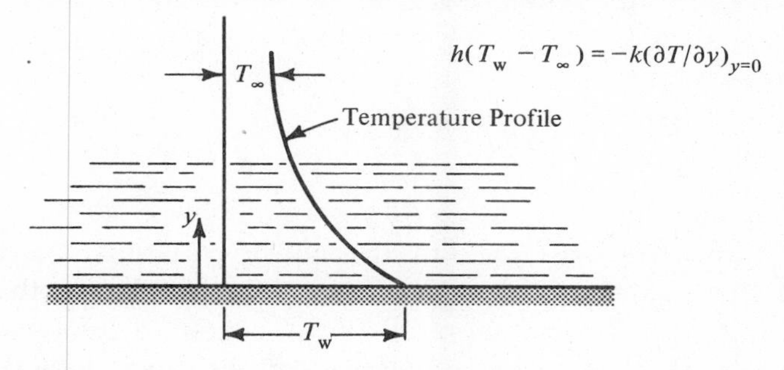

Figure 8-1 Temperature distribution in a flowing fluid near a solid boundary.

We can express the rate of heat transfer, Q, in the following manner.

$$Q = -kA\left(\frac{\partial T}{\partial y}\right)_{y=0} \tag{8-2}$$

where

k = the thermal conductivity of the fluid, Btu/hr-ft°F or W/m °C, evaluated at $(1/2)(T_W + T_\infty)$, and

$\left(\frac{\partial T}{\partial y}\right)_{y=0}$ = the value of the temperature gradient in the fluid at $y = 0$. The coordinate y is measured along the normal to the surface.

If equations (8-1) and (8-2) are combined, we obtain

$$hA(T_W - T_\infty) = -kA\left(\frac{\partial T}{\partial y}\right)_{y=0}$$

or

$$\frac{h}{k} = -\frac{1}{(T_W - T_\infty)}\left(\frac{\partial T}{\partial y}\right)_{y=0}$$

If a dimensionless distance η is defined as $\eta = (y/L_c)$ where L_c is a characteristic length, we obtain

$$\frac{h}{k} = -\frac{1}{(T_W - T_\infty)L_c}\left(\frac{\partial T}{\partial \eta}\right)_{\eta=0} \tag{8-3}$$

or

$$\text{Nu} = \frac{hL_c}{k} = -\frac{1}{(T_W - T_\infty)}\left(\frac{\partial T}{\partial \eta}\right)_{\eta=0}$$

The quantity, (hL_c/k), in the above equation is a dimensionless quantity called the *Nusselt number*. Its structure is similar to that of the Biot number encountered earlier in Chapter 4. The important difference is that k in the Nusselt number is the thermal conductivity of the fluid, while in the Biot number it represents that of the solid. It is customary to express the correlations for convective heat transfer, analytical or experimental, in terms of dimensionless quantities such as the Nusselt number and the Reynolds number.

In the following sections, we will analyze the transfer of heat for laminar flow in a circular tube and in the boundary layer over a flat plate. The results obtained in Chapter 7 for the velocity distribution will be used together with the principle of conservation of energy in arriving at the correlations for the convective heat transfer coefficient and the Nusselt number. We will also present empirical correlations for turbulent flows and a discussion of dimensional analysis.

8-3 HEAT TRANSFER FOR LAMINAR FLOW IN CIRCULAR TUBES

Consider laminar flow of a fluid through a tube of inner radius, r_w. We will assume that the flow is fully developed and that the velocity distribution is given by equation (7-10a), which is repeated here for convenience.

$$\frac{u}{u_o} = 1 - \left(\frac{r}{r_w}\right)^2 \tag{7-10a}$$

where

u = the velocity in the x direction at $r = r$, and

u_o = the velocity at the centerline of the tube given by equation (7-9a) as

$$u_o = -g_c\beta\left(\frac{r_w^2}{4\mu}\right) \tag{7-9a}$$

where $\beta = dp/dx$.

In a typical convective heat transfer problem for flow through a tube, we are interested in determining the heat transfer coefficient, h, along the wall of the tube. Equation (8-3) tells us that this can be accomplished if we can determine the temperature distribution in the flowing fluid. This necessitates application of the principle of conservation of energy to obtain an appropriate governing differential equation. When this equation is solved, subject to boundary conditions relevant to the problem, we will obtain the temperature distribution within the fluid and, in particular, the temperature gradient at the wall.

One possible thermal boundary condition for the present problem can be that the tube wall receives a uniform heat flux. Such a condition arises in radiant heating, electric resistance heating, and in counter-flow heat exchangers where the product of mass flow rate and specific heat is the same for the hot and the cold fluids. Another possible boundary condition is that in which the tube surface is maintained at a uniform temperature. The analysis for the latter condition is more complex than for the uniform heat flux condition. We will, therefore, confine ourselves to the uniform heat flux condition. Before we proceed with the analysis, it is convenient to define the bulk temperature of a fluid.

8-3.1 The Bulk Temperature

Recall that equation (8-1) involves the temperature, T_∞, of the fluid sufficiently far from the solid boundary such that it is unaffected by the temperature of the solid. In the case of flow through ducts, it is more convenient to use the *bulk temperature,* T_b. The bulk temperature, T_b, is the mean temperature of the fluid at a given cross section of the tube. If the fluid from a cross section were to be collected in a cup for a short period of time and thoroughly mixed without allowing it to exchange energy with the surroundings, the resulting temperature of the fluid would be the bulk temperature. It is also called the *mixing cup temperature* or the *mixed mean temperature*. In general, it will vary from one cross section of the tube to another. It can be expressed as

$$T_b = \frac{\text{(total thermal energy crossing a section of the tube in a unit time)}}{\text{(heat capacity of the fluid crossing the same section in a unit time)}} \tag{8-4}$$

If we consider a small elemental area, $(2\pi r\, dr)$, in a circular tube, then the rate of fluid flow through this area will be $[(\rho u)(2\pi r\, dr)]$. If the enthalpy, i, of the fluid is assumed to be equal to $c_p T$, then the thermal energy associated with this fluid flow rate is $[(c_p T)(\rho u)(2\pi r\, dr)]$. Thus, the total thermal energy crossing a section of a tube in a unit time is

$$\int_0^{r_w} (c_p T)(\rho u)(2\pi r\, dr)$$

The heat capacity of a fluid is the product of its specific heat (at constant pressure) and its mass. Therefore the heat capacity of a fluid crossing a small elemental area per unit time is $(c_p)(\rho u)(2\pi r\, dr)$. Integrating over a cross section of the tube, we have the following expression for the heat capacity of a fluid crossing a section of a tube in a unit time.

$$\int_0^{r_w} c_p(\rho u)(2\pi r\, dr)$$

Hence, the bulk temperature is given by

$$T_b = \frac{\int_0^{r_w} \rho(c_p T)(2\pi r u\, dr)}{\int_0^{r_w} \rho c_p\, 2\pi r u\, dr} \tag{8-5}$$

8-3.2 The Governing Differential Equation and its Solution

We now seek a temperature distribution in a fluid flowing through a tube. The resulting expression will then be used to determine the convective heat transfer coefficient and the bulk temperature. Consider a control volume in the shape of a cylindrical shell of radius, r, thickness, Δr, and length, Δx, as shown in Figure 8-2.

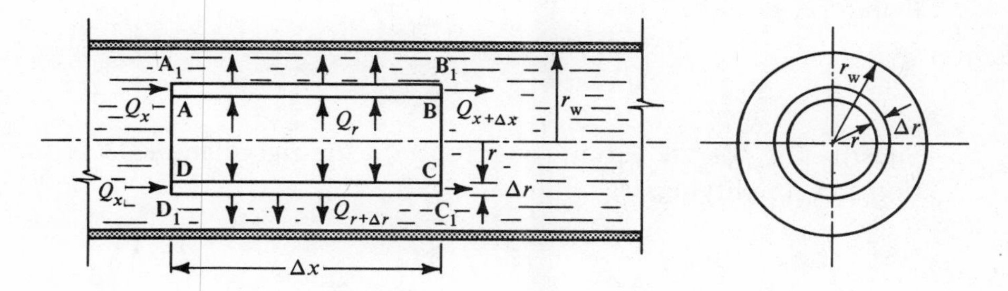

Figure 8-2 Control volume for energy balance for flow through a tube.

The principle of conservation of energy requires that under steady-state conditions, the *net* heat conducted into the control volume from the radial and axial directions must be equal to the net heat convected away in the x direction. There is no convection in the r direction in this problem, because the velocity is purely axial.

The rate at which heat is conducted into the control volume in the radial direction through the circumferential face **ABCD** is given by

$$Q_r = -k(2\pi r \Delta x)\frac{\partial T}{\partial r}$$

The rate of heat conduction out of the control volume through the circumferential face $\mathbf{A_1 B_1 C_1 D_1}$, in the radial direction, can be written by using a Taylor series expansion. It is

$$Q_{r+\Delta r} \simeq Q_r + \frac{\partial}{\partial r}(Q_r)\Delta r$$

In a similar fashion, the rates of axial heat conduction into and out of the control volume, Q_x and $Q_{x+\Delta x}$, can be expressed as

$$Q_x = -k(2\pi r \Delta r)\frac{\partial T}{\partial x}$$

and

$$Q_{x+\Delta x} \simeq Q_x + \frac{\partial}{\partial x}(Q_x)\Delta x$$

The *net* rate at which heat is conducted into the elemental control volume is, therefore, given by

$$Q_{\text{cond, net}} = Q_r - Q_{r+\Delta r} + Q_x - Q_{x+\Delta x}$$

Substituting for the various quantities on the right-hand side of the above equation and simplifying, we obtain

$$Q_{\text{cond, net}} = 2\pi k \frac{\partial}{\partial r}\left(r\frac{\partial T}{\partial r}\right)\Delta r\,\Delta x + 2\pi k r\left(\frac{\partial^2 T}{\partial x^2}\right)\Delta r\,\Delta x \tag{8-6}$$

The rate at which energy is convected or carried into the control volume can be expressed in terms of the mass flow rate of the fluid entering the control volume and energy associated with it. As noted earlier, fluid enters the control volume only in the axial direction, since the radial component of the velocity is zero for fully developed flow. The mass flow rate into the control volume is

$$\rho u(2\pi r\,\Delta r)$$

The energy associated with this mass flow rate, which is convected into the control volume, is given by

$$Q_{\text{conv},\,x} = i(\rho u 2\pi r\,\Delta r)$$

where i is the enthalpy of the fluid per unit mass. The energy convected out of the control volume is found by using a Taylor series expansion and is

$$Q_{\text{conv},\,x+\Delta x} \simeq Q_{\text{conv},\,x} + \frac{\partial}{\partial x}(Q_{\text{conv},\,x})\,\Delta x$$

Hence, the net rate of convection out of the elemental control volume is

$$Q_{\text{conv, net}} = Q_{\text{conv},\,x+\Delta x} - Q_{\text{conv},\,x}$$

or

$$Q_{\text{conv, net}} = 2\pi\rho r u\left(\frac{\partial i}{\partial x}\right)\Delta r\Delta x \tag{8-6a}$$

Observe that u does not depend on the x coordinate and, therefore, $(\partial/\partial x)(ui)$ has become $(u)(\partial i/\partial x)$. If we assume that

$$di = c_p dT$$

equation (8-6a) can be written as

$$Q_{\text{conv, net}} = 2\pi\rho c_p r u\left(\frac{\partial T}{\partial x}\right)\Delta r \Delta x \tag{8-6b}$$

Conservation of energy requires that

$$Q_{\text{cond, net}} = Q_{\text{conv, net}}$$

Substitution from equations (8-6) and (8-6b) gives

$$2\pi k\frac{\partial}{\partial r}\left(r\frac{\partial T}{\partial r}\right)\Delta r \Delta x + 2\pi k r\left(\frac{\partial^2 T}{\partial x^2}\right)\Delta r \Delta x = 2\pi\rho c_p r u\left(\frac{\partial T}{\partial x}\right)\Delta r \Delta x$$

which becomes

$$\frac{1}{r}\frac{\partial}{\partial r}\left(r\frac{\partial T}{\partial r}\right) + \left(\frac{\partial^2 T}{\partial x^2}\right) = u\left(\frac{\rho c_p}{k}\right)\frac{\partial T}{\partial x}$$

or

$$\frac{1}{r}\frac{\partial}{\partial r}\left(r\frac{\partial T}{\partial r}\right) + \frac{\partial^2 T}{\partial x^2} = \frac{u}{\alpha}\left(\frac{\partial T}{\partial x}\right) \tag{8-7}$$

where $\alpha = (k/\rho c_p)$, the thermal diffusivity of the fluid.

Equation (8-7) is the governing differential equation obtained from the principle of conservation of energy for fully developed flow through a circular tube. The boundary condition for the present problem is that at the tube wall ($r = r_w$), the heat flux coming radially inward, designated as q_w, equals the rate of heat conduction in the fluid. It can be stated as

$$\text{at } r = r_w, \; -\left(-k\frac{\partial T}{\partial r}\right) = q_w \tag{8-8}$$

The minus sign in front of the $(-k\ \partial T/\partial r)$ indicates radially inward flow of heat. In equation (8-8), k is the thermal conductivity of the fluid.

Since the fluid receives a uniform heat flux, q_w, it is logical to expect its bulk temperature to increase linearly along the direction of the fluid flow. That is, (dT_b/dx) is a constant quantity. It is shown in Reference 1 that at locations sufficiently downstream from the inlet of the tube

$$\frac{\partial T}{\partial x} = \frac{dT_b}{dx} = \text{constant} \tag{8-8a}$$

for the uniform heat flux case. Since $(\partial T/\partial x)$ is a constant, then $(\partial^2 T/\partial x^2)$ is identically zero. Equation (8-7) then takes the form

$$\frac{1}{ur}\frac{\partial}{\partial r}\left(r\frac{\partial T}{\partial r}\right) = \lambda = \frac{1}{\alpha}\left(\frac{\partial T}{\partial x}\right)$$

where λ is a constant. If we now replace $(\partial/\partial r)$ by (d/dr), which is permissible since x does not explicitly appear on the left-hand side, the above equation becomes

$$\frac{1}{ur}\frac{d}{dr}\left(r\frac{dT}{dr}\right) = \lambda$$

Substituting for u from equation (7-10a) and separating variables, it follows that

$$\int d\left(r\frac{dT}{dr}\right) = \int \lambda u_o\left(1 - \frac{r^2}{r_w^2}\right) r\,dr$$

Integration gives

$$r\frac{dT}{dr} = \lambda u_o\left[\frac{r^2}{2} - \frac{r^4}{4r_w^2}\right] + C_1$$

where C_1 is a constant of integration. Once again, separating variables and integrating, we obtain

$$T = \lambda u_o\left[\frac{r^2}{4} - \frac{r^4}{16r_w^2}\right] + C_1 \ln(r) + C_2$$

where C_2 is another constant of integration.

At $r = 0$, the temperature of the fluid must be a finite quantity. Since $\ln(r)$ approaches minus infinity as r approaches zero, it follows that C_1 must be zero. If we now substitute $r = 0$, the above solution for T indicates that C_2 is equal to the temperature, T_o, at the centerline of the tube. Thus, the solution becomes

$$T - T_o = \lambda u_o \left(\frac{r^2}{4} - \frac{r^4}{16 r_w^2} \right)$$

Recalling that $\lambda = (1/\alpha)(\partial T/\partial x)$, the above becomes

$$T - T_o = \frac{u_o r_w^2}{4\alpha} \left(\frac{\partial T}{\partial x} \right) \left[\left(\frac{r}{r_w} \right)^2 - \frac{1}{4} \left(\frac{r}{r_w} \right)^4 \right] \tag{8-9}$$

Note that T_o in equation (8-9) is still an unknown and will remain so, because the boundary condition of constant heat flux prescribes the temperature gradient but not the temperature. Hence, the final solution can be obtained within a constant term only. The boundary condition of constant heat flux is used in Reference 1 to show that $(\partial T/\partial x)$ is constant as stated in equation (8-8a).

We now proceed to solve for $(\partial T/\partial x)$ in terms of the heat flux, q_w. This is done by first differentiating equation (8-9) with respect to r and solving for $(\partial T/\partial r)_{r=r_w}$. Thus

$$\frac{\partial T}{\partial r} = \frac{u_o r_w^2}{4\alpha} \left(\frac{\partial T}{\partial x} \right) \left[\frac{2r}{r_w^2} - \frac{r^3}{r_w^4} \right]$$

For $r = r_w$, we obtain from the above equation

$$\left(\frac{\partial T}{\partial r} \right)_{r=r_w} = \frac{u_o r_w}{4\alpha} \left(\frac{\partial T}{\partial x} \right)$$

However, from equation (8-8), we have

$$\left(\frac{\partial T}{\partial r} \right)_{r=r_w} = \frac{q_w}{k} \tag{8-10}$$

Combining the two forgoing equations, we obtain

$$\left(\frac{\partial T}{\partial x} \right) = \frac{4 q_w \alpha}{u_o r_w k} \tag{8-10a}$$

The above expression for $(\partial T/\partial x)$ can now be substituted into equation (8-9), with the result

$$T - T_o = \frac{q_w r_w}{k}\left[\left(\frac{r}{r_w}\right)^2 - \frac{1}{4}\left(\frac{r}{r_w}\right)^4\right] \tag{8-11}$$

The bulk temperature may now be evaluated as follows. Substituting for T from equation (8-11) into equation (8-5), there results

$$T_b = \frac{\int_0^{r_w} \rho c_p\left\{T_o + \frac{q_w r_w}{k}\left[\left(\frac{r}{r_w}\right)^2 - \frac{1}{4}\left(\frac{r}{r_w}\right)^4\right]\right\} 2\pi r u\, dr}{\int_0^{r_w} \rho c_p 2\pi r u\, dr}$$

or

$$T_b = \frac{\int_0^{r_w} \rho c_p T_o 2\pi r u\, dr}{\int_0^{r_w} \rho c_p 2\pi r u\, dr} + \frac{\int_0^{r_w} \rho c_p\left(\frac{q_w r_w}{k}\right)\left[\left(\frac{r}{r_w}\right)^2 - \frac{1}{4}\left(\frac{r}{r_w}\right)^4\right]2\pi r u\, dr}{\int_0^{r_w} \rho c_p 2\pi r u\, dr}$$

We observe that the first term on the right-hand side is T_o, since temperature, T_o, at the centerline of the tube does not depend on the radius, r. In order to integrate the second term on the right-hand side, we substitute for u from equation (7-10a) to obtain

$$T_b = T_o + \frac{\int_0^{r_w} \rho c_p \frac{q_w r_w}{k}\left[\left(\frac{r}{r_w}\right)^2 - \frac{1}{4}\left(\frac{r}{r_w}\right)^4\right]2\pi r u_o\left[1 - \left(\frac{r}{r_w}\right)^2\right]dr}{\int_0^{r_w} \rho c_p 2\pi r u_o\left[1 - \left(\frac{r}{r_w}\right)^2\right]dr}$$

Carrying out the multiplication inside the integral sign and simplifying yields

$$T_b = T_o + \frac{\frac{q_w r_w}{k}\int_0^{r_w}\left[\frac{r^3}{r_w^2} - \frac{1}{4}\left(\frac{r^5}{r_w^2}\right) - \frac{r^5}{r_w^4} + \frac{1}{4}\left(\frac{r^7}{r_w^6}\right)\right]dr}{\int_0^{r_w}\left(r - \frac{r^3}{r_w^2}\right)dr}$$

Integration gives

$$T_b = T_o + \frac{\dfrac{q_w r_w}{k}\left[\dfrac{r^4}{4r_w^2} - \dfrac{1}{24}\left(\dfrac{r^6}{r_w^4}\right) - \dfrac{r^6}{6r_w^4} + \dfrac{r^8}{32r_w^6}\right]_0^{r_w}}{\left[\dfrac{r^2}{2} - \dfrac{r^4}{4r_w^2}\right]_0^{r_w}}$$

or

$$T_b = T_o + \frac{7}{24}\left(\frac{q_w r_w}{k}\right) \tag{8-12}$$

The Nusselt number for the present problem is (hD/k), where D is the inner diameter of the tube. As already noted, the heat transfer coefficient for tube flow is usually based on the wall-to-bulk temperature difference, $(T_w - T_b)$. Upon writing $q_w = h(T_w - T_b)$, it follows from equation (8-8) that

$$k\left(\frac{\partial T}{\partial r}\right)_{r=r_w} = h(T_w - T_b)$$

or

$$\frac{h}{k} = \frac{1}{(T_w - T_b)}\left(\frac{\partial T}{\partial r}\right)_{r=r_w}$$

We can solve for T_w from equation (8-11) after substituting $r = r_w$

$$T_w = T_o + \frac{3}{4}\left(\frac{q_w r_w}{k}\right) \tag{8-12a}$$

Equations (8-12) and (8-12a) can be combined to give

$$T_w - T_b = \frac{11}{24}\left(\frac{q_w r_w}{k}\right)$$

Also, from equation (8-10), we have

$$\left(\frac{\partial T}{\partial r}\right)_{r=r_w} = \frac{q_w}{k}$$

Hence

$$\frac{h}{k} = \frac{(q_w/k)}{\frac{11}{24}\left(\frac{q_w r_w}{k}\right)}$$

or

$$\frac{h}{k} = \frac{24}{11}\left(\frac{1}{r_w}\right)$$

The Nusselt number is, therefore, given by

$$\text{Nu} = \frac{hD}{k} = \frac{2hr_w}{k} = \frac{48}{11} = 4.364 \tag{8-13}$$

The above result is in agreement with the work of Sellers et al (Reference 2). The Nusselt number for this case is independent of the x coordinate. In general, when the Nusselt number is independent of the axial coordinate, it corresponds to a thermally developed condition. Near the inlet of a tube, where the heat transfer coefficient varies with x, thermal development takes place.

The value of the Nusselt number for the case where the tube wall temperature is maintained constant is 3.658 (Reference 1), 16 percent less than the solution for the case where the tube wall receives a constant heat flux.

Sample Problem 8-1 Water flows in a tube of 3/4-inch I.D. at a rate of 0.4 gallon per minute. If it receives a uniform heat flux of 400 Btu/hr-ft^2 at the wall, determine (1) the wall-to-centerline temperature difference, (2) the bulk temperature, (3) the rate at which the temperature changes from one section to another, and (4) the heat transfer coefficient. For water, μ = 2.36 lb$_m$/ft-hr, k = 0.35 Btu/hr-ft°F, c_p = 0.998 Btu/lb$_m$ °F, and ρ = 62.25 lb$_m$/ft^3.

Solution: The flow data for this problem are the same as those for Sample Problem 7-1. The average flow velocity, u_{av}, and the Reynolds number, Re, were found to be 0.29 ft/sec and 1718, respectively, indicating a laminar flow. We can, therefore, use the equations developed in Section 8-2.

(1) Equation (8-11) can be evaluated at $r = r_w$ to yield $(T_w - T_o)$, thus

$$(T_w - T_o) = \frac{3}{4}\left(\frac{q_w r_w}{k}\right)$$

Substituting for q_w, r_w, and k, we obtain

$$(T_w - T_o) = (3/4)\left[\frac{400 \times (1/2)(3/4)(1/12)}{0.35}\right] = 26.8°F$$

(2) The bulk temperature, T_b, is expressed by equation (8-12)

$$T_b = T_o + \frac{7}{24}\left(\frac{q_w r_w}{k}\right)$$

or

$$T_b = T_o + (7/24)\left[\frac{400 \times (1/2)(3/4)(1/12)}{0.35}\right] = (T_o + 10.4)\ °F$$

(3) The rate at which the temperature changes axially is given by equation (8-10a)

$$\frac{\partial T}{\partial x} = \frac{4 q_w \alpha}{u_o r_w k}$$

Also

$$\alpha = \frac{k}{\rho c_p} = \frac{0.35}{(0.998)(62.25)} = 5.63 \times 10^{-3}\ \text{ft}^2/\text{hr}$$

and

$$u_o = 2u_{av} = 2(0.29)$$

$$= 0.58\ \text{ft/sec} = 2088\ \text{ft/hr}$$

Hence

$$\frac{\partial T}{\partial x} = \frac{4 \times 400 \times 5.63 \times 10^{-3}}{2088 \times \frac{1}{2} \times \frac{3}{4} \times \frac{1}{12} \times 0.35} = 0.4\ °F/ft$$

(4) Equation (8-13) gives the heat transfer coefficient, h

$$h = 4.364 \frac{k}{D} = 4.364 \frac{0.35}{(3/4) \times (1/12)}$$

or

$$h = 24.44 \text{ Btu/hr-ft}^2 \text{ °F}$$

8-3.3 Nusselt Number for the Entrance Region of a Circular Tube

In the preceding section, we assumed that the temperature and the velocity profiles were fully developed. On the other hand, in Chapter 7, we saw that it takes a certain distance called the entrance length, L_e, for the velocity profile to be developed. Similarly, when a fluid with a uniform temperature enters a tube that receives a uniform heat flux, q_w, at its wall, a development length is required before the temperature profile, equation (8-11), becomes established. This distance is called the **thermal entry length**. The value of the Nusselt number given by equation (8-13) is applicable at points that are beyond the thermal entry length. Tubes in a heat exchanger are very often operated under conditions where entrance effects should be reflected in the value of the Nusselt number used.

The Prandtl number, Pr, is an important parameter in the analysis of convection problems. It is defined in terms of the kinematic viscosity, ν, and the thermal diffusivity, α, as

$$\text{Pr} = \frac{\mu c_p}{k} = \left(\frac{\rho c_p}{k}\right)\frac{\mu}{\rho}$$

or

$$\text{Pr} = \left(\frac{1}{\alpha}\right)\nu = \frac{\nu}{\alpha} \tag{8-14}$$

Physically, the Prandtl number is a measure of how rapidly momentum is dissipated compared to the rate of diffusion of heat through a fluid. Most gases have a Prandtl number of the order of unity. In particular, the Prandtl number for air at atmospheric pressure has a value in the range of 0.718 to 0.690 corresponding to a temperature range of 0°F to 1000°F. Saturated liquid water has a Prandtl number of 13.35 at 32°F, which drops to 1.00 at 400°F. The value of the Prandtl number for fresh engine oil decreases by a factor of more than 500 as the temperature is increased from 32°F to 320°F. At 68°F, the Prandtl number for fresh engine oil is 10,400. Liquid metals, owing to their large thermal conductivities, have Prandtl numbers of the order of 10^{-2}.

Table 8-1 summarizes the values of the local Nusselt number, Nu, for a uniform wall *heat flux* for the thermal entrance region. They are tabulated as a function of the dimensionless coordinate, x^*, defined by

$$x^* = \frac{(x/r_w)}{\text{Re Pr}}$$

where

x = the distance from the entrance of the tube

r_w = the inner radius of the tube

Re = the Reynolds number, $\left(\frac{\rho u_{av} D}{\mu}\right)$

and

Pr = the Prandtl number, $\left(\frac{\mu c_p}{k}\right)$

TABLE 8-1 Local Laminar Nusselt Numbers for a Circular Tube with Constant Wall Heat Flux

x^*	*Nu*
0.002	12.00
0.004	9.93
0.010	7.49
0.020	6.14
0.040	5.19
0.100	4.51
∞	4.36

Note that as x^* becomes very large, the value of the Nusselt number approaches the value obtained for fully developed flow.

In general, the average heat transfer coefficient, h_{av}, for flow through a tube of length, L, is defined by

$$h_{av} = \frac{\frac{1}{A}\int q\,dA}{\frac{1}{L}\int_0^L (T_w - T_b)\,dx} \tag{8-15}$$

where the quantity, (qdA), is the rate of heat transfer across an elemental area dA of the tube. In the above equation, the integral in the numerator represents the total quantity of heat transferred across surface area, A, of the tube, and the denominator represents the mean temperature difference between the tube wall and the fluid. If the heat flux is uniform and is designated as q_w, then the expression for h_{av} is simplified to

$$h_{av} = \frac{q_w L}{\int_0^L (T_w - T_b)\, dx} = \frac{q_w L}{(T_w - T_b)_{av}} \tag{8-15a}$$

It should be noted that the purpose of the heat transfer solution is to relate the temperature at the wall of the tube, T_w, to the bulk temperature, T_b, through the definition of the heat transfer coefficient. That is

$$T_w - T_b = \frac{q_w}{h} = \frac{q_w D}{\text{Nu} k}$$

The values of Nu for the case of uniform heat flux given in Table 8-1 can be used to calculate the quantity $(T_w - T_b)$ for different values of x. Next, an integration may be performed to determine the value of $(T_w - T_b)_{av}$, which then can be used to evaluate h_{av} from equation (8-15a).

Listed in Table 8-2 are the values of the local Nusselt number, and the average Nusselt number for the distance 0 to x^* for flow through a circular tube with *constant wall temperature*. At the entrance the flow has a fully developed profile and a uniform fluid temperature. This is a tabulated solution to the classical Graetz problem. The value of h_{av} obtained by using Table 8-2 should be used with the logarithmic mean temperature difference, LMTD. The general definition of the logarithmic mean temperature difference is given by equation (10-19) in the chapter on fins and heat exchangers. For a fluid with inlet temperature, T_1, and outlet temperature, T_2, flowing through a tube with a wall temperature, T_w (constant), the expression for the logarithmic mean temperature difference, LMTD, is

$$\text{LMTD} = \frac{(T_w - T_1) - (T_w - T_2)}{\ln \dfrac{(T_w - T_1)}{(T_w - T_2)}} \tag{8-15b}$$

The values of the average Nusselt Number, Nu_{av}, in Table 8-2 are based on the following definition (Reference 1).

$$Nu_{av} = \frac{h_{av}D}{k} = \frac{1}{x^*}\int_0^{x^*} Nu\,dx^* \tag{8-15c}$$

Results for combined thermal and hydrodynamic entry length problems, that is, neither the temperature profile nor the velocity profile is fully developed, are presented in Reference 2.

TABLE 8-2 Local and Average Laminar Nusselt Numbers for a Circular Tube with Constant Wall Temperature

x^*	Nu	Nu_{av}
0.001	12.86	22.96
0.004	7.91	12.59
0.010	5.99	8.99
0.040	4.18	5.87
0.080	3.79	4.89
0.100	3.71	4.66
0.200	3.66	4.16
∞	3.66	3.66

Sample Problem 8-2 Water flows at a rate of 10 cm^3/sec through a tube of 15 mm I.D. The tube wall is maintained at a constant temperature of 60°C. Calculate the local heat transfer coefficient at a distance of 1.0 m from the entrance of the tube. Also, calculate the average value of the heat transfer coefficient between $x = 0$ and $x = 1.0$ m. If the water enters the tube at a uniform temperature of 20°C, determine the amount of heat transferred from the tube wall to the water over the first 1-m length of the tube.

Solution: To determine the local and the average values of h, we need u_{av}, Re, Pr, and x^*. The appropriate values of the Nusselt number can then be determined from Table 8-2.

$$u_{av} = \frac{\text{volume flow rate}}{(\pi/4)\,D^2}$$

or

$$u_{av} = \frac{10 \times 10^{-6}}{(\pi/4)\,(0.015)^2} = 0.057 \text{ m/sec}$$

We need the thermophysical properties of water to evaluate Re and Pr. They change significantly with temperature. On the other hand, values in Table 8-2 are based on constant properties. We will, therefore, use the arithmetic mean of the temperature of the water at the inlet and the temperature of the tube wall, that is, at $T = (1/2)\ (20 + 60) = 40°\text{C}$. From the Appendix, we find the following values for the properties of saturated water at 40°C.

$$c_p = 4178.4 \text{ J/kg °K}; \qquad \nu = 0.658 \times 10^{-6} \text{ m}^2/\text{sec}$$

$$k = 0.628 \text{ W/m °K}; \qquad \rho = 994.59 \text{ kg/m}^3$$

The Reynolds number and the Prandtl number are then obtained as

$$\text{Re} = \frac{u_{av} D}{\nu} = \frac{(0.057)\ (0.015)}{0.658 \times 10^{-6}} = 1299$$

$$\text{Pr} = \frac{c_p \mu}{k} = \frac{(4178.4)\ (0.658 \times 10^{-6})\ (994.59)}{0.628} = 4.35$$

Hence

$$x^* = \frac{(x/r_w)}{\text{Re Pr}} = \frac{1.0/(0.015/2)}{1299 \times 4.35} = 0.0236$$

A graphical interpolation of the values in Table 8-2 gives Nu = 5.2 and Nu_{av} = 7.2. Thus

$$h = 5.2\left(\frac{k}{D}\right) = 217.7 \text{ W/m}^2 \text{ °K}$$

and

$$h_{av} = 7.2\left(\frac{k}{D}\right) = 301.4 \text{ W/m}^2 \text{ °K}$$

We now proceed to determine the amount of heat transferred to the water in the first one meter length of the tube. The h_{av} obtained above is to be used with the logarithmic mean temperature difference, LMTD, equation (10-19), which for the present problem is

$$\text{LMTD} = \frac{(60 - T_{x=0}) - (60 - T_{x=1})}{\ln\left(\dfrac{60 - T_{x=0}}{60 - T_{x=1}}\right)}$$

where $T_{x=0}$ and $T_{x=1}$ are the bulk temperatures at the entrance to the tube and at $x = 1$ meter.

Also, the amount of heat energy gained by the water over a distance of one meter equals the amount of heat transferred from the tube wall. Thus

$$(\rho c_p)\,(\text{volume flow rate})\,(T_{x=1} - T_{x=0})$$
$$= h_{\text{av}}\,(\text{surface area of the tube wall})\,(\text{LMTD})$$

Employing equation (8-15b) for the LMTD and substituting values for the various quantities we obtain

$$(994.59 \times 4178.4)\,(10 \times 10^{-6})\,(T_{x=1} - 20)$$
$$= 301.4\,(\pi 0.015 \times 1)\left\{\frac{(60 - 20) - (60 - T_{x=1})}{\ln\left[\dfrac{(60 - 20)}{(60 - T_{x=1})}\right]}\right\}$$

Simplification results in

$$\ln\left(\frac{40}{60 - T_{x=1}}\right) = 0.342$$

or

$$T_{x=1} = 31.6\ ^{\circ}\text{C}$$

Hence the rate of heat transferred, Q, over the first one meter section of the tube is

$$Q = \rho c_p\,(\text{volume flow rate})\,(T_{x=1} - T_{x=0})$$
$$= 994.59 \times 4178.4 \times 10 \times 10^{-6}\,(31.6 - 20)$$
$$= 482 \text{ watts}$$

We used a temperature of 40°C to read the property values from the Appendix. The problem should now be reworked by selecting property values corresponding to a mean temperature of (1/2) (20 + 30.3) or 25°C. This is done in Sample Problem 8-3.

8-4 HEAT TRANSFER FOR TURBULENT FLOW IN CIRCULAR TUBES—ANALYTICAL RESULTS

In Sections 7-4.2 and 7-5.4, the concept of the eddy diffusivity was introduced to account for the macroscopic transfer of momentum. The equation for shear stress in the fluid was then written as

$$\tau = \frac{\rho}{g_c}(\epsilon_M + \nu)\frac{\partial u}{\partial y} \tag{8-16}$$

where the subscript M denotes momentum transfer. When convective heat transfer takes place in a turbulent flow, a significant contribution to the heat diffusion in the fluid is made by the macroscopic transport owing to the eddy motion. This contribution is characterized by the eddy diffusivity for heat, ϵ_H, which is defined by

$$q = \rho c_p(\epsilon_H + \alpha)\frac{\partial T}{\partial y} \tag{8-16a}$$

The quantity $(\rho c_p \epsilon_H)$ may be considered as an increase in the conductivity of the fluid due to turbulence. The ratio, (ϵ_M/ϵ_H), is referred to as the turbulent Prandtl number, $\mathrm{Pr_t}$.

The similarity of equations (8-16) and (8-16a) is striking. If the Prandtl number equals unity ($\alpha = \nu$) and if the eddy diffusivities for heat and momentum are assumed equal ($\mathrm{Pr_t} = 1$), then the velocity profile and the temperature profile will be similar. The similarity of the velocity profile and the temperature profile results in a heat flux distribution, $q(y)$, which is identical to the shear stress distribution, $\tau(y)$. This is known as the *Reynolds analogy.* We will now look at some of the analytical results for turbulent heat transfer.

Fully Developed Flow Deissler (Reference 3) and Sparrow et al (Reference 4) have obtained analytical solutions to the problem of heat transfer in a fully developed turbulent flow in a circular tube with **uniform heat flux** at the wall. They considered a fluid with constant properties and with the turbulent Prandtl number, $\mathrm{Pr_t}$, equal to unity. Deissler found good agreement with the data of ten investigators. Webb (Reference 5) has recently shown that the Deissler results can be expressed in the following form.

$$\mathrm{Nu} = \frac{(f/8)\,\mathrm{Re}\,\mathrm{Pr}}{1.07 + 9\sqrt{f/8}\,(\mathrm{Pr} - 1)\,\mathrm{Pr}^{-1/4}} \tag{8-17}$$

where f is the friction factor defined in equation (7-13). Its value is obtained from the equation

$$\sqrt{\frac{8}{f}} = 2.78 \ln\left(\mathrm{Re}\sqrt{\frac{f}{8}}\right) - 0.39 \tag{8-17a}$$

Webb has also shown that the Nusselt numbers computed from equation (8-17) are within ±2.0 percent of the analytical results of Sparrow et al.

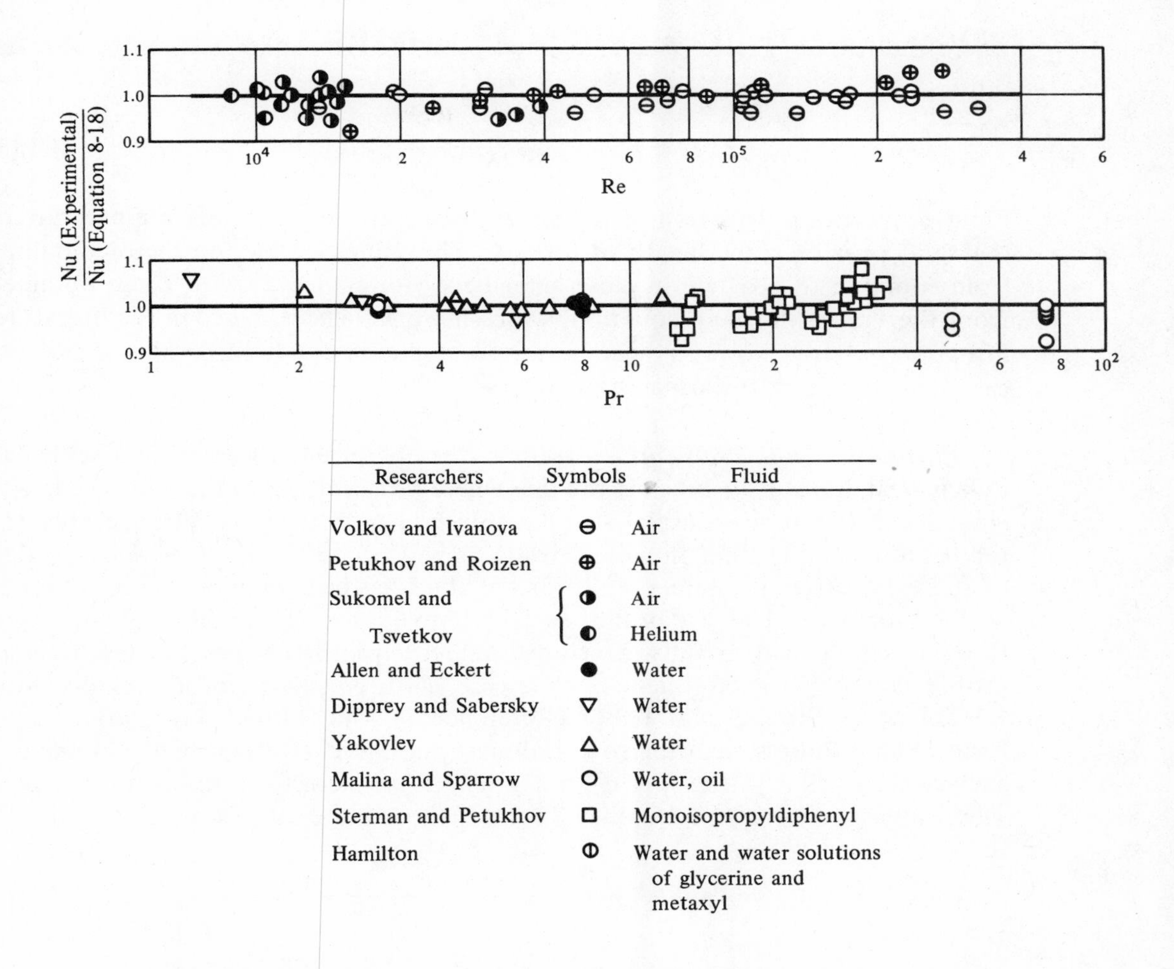

Figure 8-3 Ratio of the experimental Nusselt number to the Nusselt number from Equation (8-18), with permission from Dr. B. S. Petukhov [14], High Temperature Institute, Academy of Science of the U.S.S.R., Moscow, U.S.S.R.

Petukhov and Popov (Reference 6) obtained an analytical solution to the problem of turbulent heat transfer in a circular tube **with uniform heat flux.** They demonstrated that the results of their analyses are within ±6 percent of the experimental data reported in nine different papers (Figure 8-3). The result due to Petukhov and Popov is preferred to equation (8-17) and is given below.

$$\mathrm{Nu} = \frac{(f/8)\,\mathrm{Re}\,\mathrm{Pr}}{1.07 + 12.7\sqrt{f/8}\,(\mathrm{Pr}^{2/3} - 1)} \tag{8-18}$$

where the friction factor, f, is calculated from the Filonenko (Reference 5) equation

$$f = [1.82 \log_{10}(\text{Re}) - 1.64]^{-2} \tag{8-18a}$$

Fluid properties in equation (8-18) are evaluated at the mean bulk temperature of the fluid over the tube length of interest. The values of friction factor obtained from equation (8-18a) are in close agreement (Reference 5) with those obtained from the Prandtl-Karman equation, which has a wide acceptance in the literature. Equation (8-18a), however, can be solved explicitly for f, while such is not the case with the Prandtl-Karman equation.

Entrance Length Problems Another class of problems associated with turbulent heat transfer in tubes is the one where heat transfer in the entrance length is involved. If a tube is insulated for sufficient length from its entrance, then the temperature of the fluid flowing through it will become uniform at some section of the tube. After the temperature of the fluid has become uniform, let the tube wall be maintained at a constant surface temperature. This will result in a flow through a tube with a fully developed velocity profile, where the temperature profile is not yet established. Such a case, **with constant surface temperature,** was solved by Sleicher and Tribus (Reference 7). They obtained a solution in the form of an infinite series. Sparrow, Hallman, and Siegel (Reference 4) obtained an analytical solution to the thermal entry length problem for a **uniform heat flux**. The reader is referred to Reference 8 for details of these solutions.

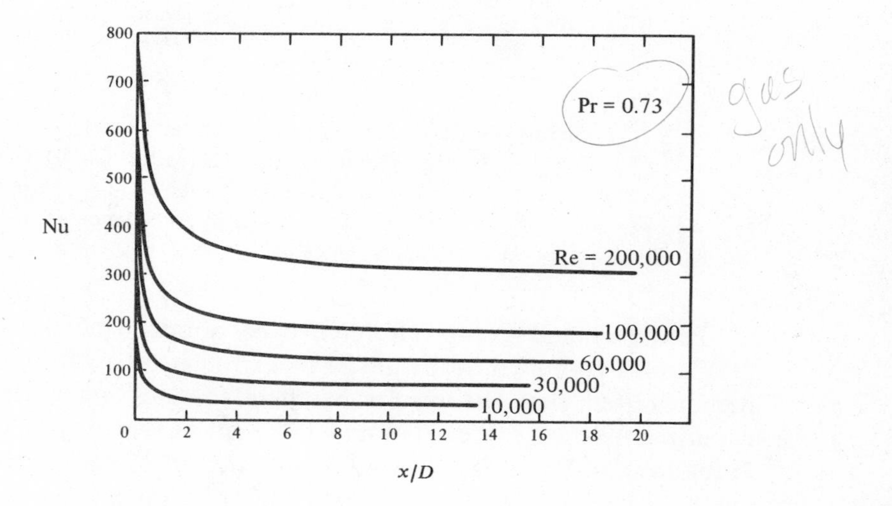

Figure 8-4 Nusselt number in the combined thermal and hydrodynamic entry length of a circular tube with uniform heat flux at wall, from *Handbook of Heat Transfer*, by W. M. Rohsenow and J. P. Hartnett. Copyright 1973, by McGraw-Hill, Inc. Used with permission.

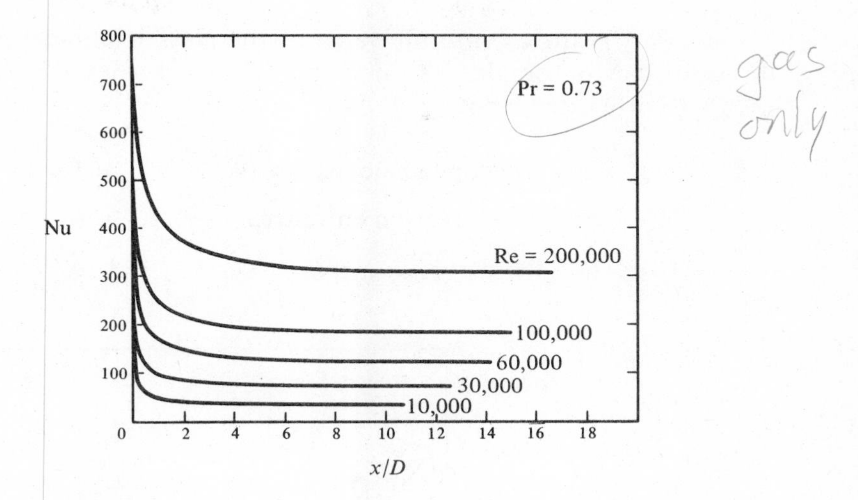

Figure 8-5 Nusselt number in the combined thermal and hydrodynamic entry length of a circular tube with constant wall temperature. From *Handbook of Heat Transfer*, by W. M. Rohsenow and J. P. Hartnett. Copyright 1973, by McGraw-Hill, Inc. Used with permission.

If neither the velocity profile nor the temperature profile is established, we have the case of a combined thermal and hydrodynamic entry length. If the free-stream turbulence is at a low level and if there is a bellmouth or a nozzle entrance, a laminar boundary layer may ensue for a Reynolds number not much greater than 2300. If under such low Reynolds number conditions a disturbance is created at the entrance, then turbulent flow will result. Analytical results (Reference 9) and experimental results (Reference 10) for the local Nusselt number for air are presented in Figures 8-4, 8-5, and 8-6. The analytical results for the local Nusselt number for the case of uniform heat flux are plotted as a function of (x/D) in Figure 8-4, where x is the distance from the entrance of the tube. Figure 8-5 shows the predicted variation in the local Nusselt number for a circular tube whose surface temperature is maintained constant. The experimental results for a constant wall temperature are presented in Figure 8-6.

The mean Nusselt number, $\mathrm{Nu_m}$, which is useful in the design of heat exchangers, may be expressed by the following relation for $(L/D) > 20$ for any gas (Reference 8) experiencing the combined entry length problem.

$$\frac{\mathrm{Nu_m}}{\mathrm{Nu_\infty}} = 1 + \frac{C}{(L/D)} \tag{8-19}$$

where

$$\mathrm{Nu_m} = \frac{1}{L}\int_0^L \mathrm{Nu}\,dx$$

and Nu_∞ is the asymptotic value of the Nusselt number as L tends to infinity. Its value can be obtained from Figure 8-4 or Figure 8-5. Values of C in equation (8-19) are listed below.

(1) Fully developed velocity profile: $C = 1.4$

(2) Abrupt contraction entrance: $C = 6$

(3) 180° Bend: $C = 6$

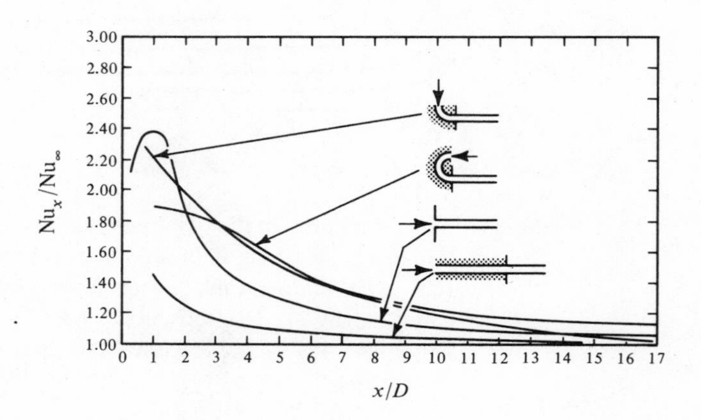

Figure 8-6 Ratio of local Nusselt number to the Nusselt number for fully developed flow, in the entry region of a circular tube for various entry configurations for air with constant surface temperature. From *Handbook of Heat Transfer*, by W. M. Rohsenow and J. P. Hartnett. Copyright 1973, by McGraw-Hill, Inc. Used with permission.

8-5 EMPIRICAL RELATIONS FOR HEAT TRANSFER IN FLOW THROUGH TUBES

Although analytical solutions for laminar flow through tubes are available, it is the turbulent flow case that is encountered most frequently in practice. Solutions for heat transfer in turbulent flow were restricted to relatively simple situations until recent innovations enabled a broadening of the range of problems that could be handled. Therefore, considerable effort has been expended in obtaining correlations for the Nusselt number. For turbulent duct flow, major attention has been given to devising correlations for the fully developed Nusselt number, owing to the fact that the thermal entrance region is short. The correlations are generally of the form

$$\mathrm{Nu} = C\,\mathrm{Re}^m\,\mathrm{Pr}^n$$

where the constant C and the exponents m and n are experimentally determined. Both the Nusselt number and the Reynolds number are based on the inside diameter of the tube. Correlations expressed in the forgoing form are convenient to use. They usually represent the data to within ±20 percent. Some of the accepted correlations are presented below. The Reynolds number appearing in these corelations is based on the average flow velocity.

Turbulent flows Dittus and Boelter (Reference 11) recommended the following correlation for fully developed turbulent flow inside tubes.

$$\text{Nu} = 0.023\ \text{Re}^{0.8}\ \text{Pr}^{0.4} \quad \text{for heating} \tag{8-20}$$

$$\text{Nu} = 0.023\ \text{Re}^{0.8}\ \text{Pr}^{0.3} \quad \text{for cooling} \tag{8-20a}$$

Another correlation due to Colburn (Reference 12) is also quite popular. It is

$$\text{Nu} = 0.023\ \text{Re}^{0.8}\text{Pr}^{1/3} \tag{8-20b}$$

The properties for equations (8-20) are evaluated at the bulk temperature of the fluid.

Webb (Reference 5) has demonstrated the superiority of the Petukhov-Popov equation, equation (8-18), over the Dittus-Boelter and the Colburn equations and has therefore recommended that the Petukhov-Popov equation be used for intermediate Prandtl numbers in the range of 1 to 50. For $0.5 \leqslant \text{Pr} \leqslant 1$, Reference 8 suggests the use of the following correlations.

$$\text{Nu} = 0.022\ \text{Re}^{0.8}\text{Pr}^{0.6} \quad \text{for uniform heat flux} \tag{8-21}$$

$$\text{Nu} = 0.021\ \text{Re}^{0.8}\text{Pr}^{0.6} \quad \text{for constant wall temperature} \tag{8-21a}$$

where the fluid properties are evaluated at the bulk temperature.

Whenever temperature of a fluid changes, its viscosity also changes. Changes in viscosity due to temperature are significant in the case of liquids. Seider and Tate (Reference 13) recommended the following correlation to account for variation in viscosity values in a fully developed turbulent flow.

$$\text{Nu} = 0.027\ \text{Re}^{0.8}\text{Pr}^{1/3}\left(\frac{\mu}{\mu_w}\right)^{0.14} \tag{8-22}$$

In the above, μ_w is the viscosity at the tube wall temperature, while all other properties are evaluated at the bulk temperature. A more recent correlation for heat transfer in fully developed turbulent flow with large differences in temperature has been obtained by Petukhov (Reference 14). It is

$$\mathrm{Nu} = \frac{(f/8)\,\mathrm{Re}\,\mathrm{Pr}}{1.07 + 12.7\sqrt{f/8}\,(\mathrm{Pr}^{2/3} - 1)}\left(\frac{\mu}{\mu_w}\right)^{0.11} \quad \text{for heating} \tag{8-23}$$

$$\mathrm{Nu} = \frac{(f/8)\,\mathrm{Re}\,\mathrm{Pr}}{1.07 + 12.7\sqrt{f/8}\,(\mathrm{Pr}^{2/3} - 1)}\left(\frac{\mu}{\mu_w}\right)^{0.25} \quad \text{for cooling} \tag{8-23a}$$

$$0.08 \leqslant \frac{\mu}{\mu_w} \leqslant 40$$

$$5 \times 10^3 \leqslant \mathrm{Re} \leqslant 1.25 \times 10^5$$

$$2 \leqslant \mathrm{Pr} \leqslant 140$$

All the properties to be used in the above equations, except μ_w, are to be evaluated the mean bulk temperature. The friction factor, f, in equations (8-23) is determined from equation (8-18a). The quantity, μ_w, is to be evaluated at the tube wall temperature. Equations (8-23) fit recent experimental data rather well. Therefore it is recommended that equations (8-23) be used for calculating heat transfer involving large temperature differences.

All the forgoing correlations are for smooth tubes only. Rough tubes have higher friction factors and heat transfer coefficients than those for smooth tubes. Dipprey and Sabersky (Reference 15) have correlated their data for flow through tubes with artificial roughness receiving uniform heat flux. The artificial roughness consisted of a close-packed, granular type of surface with roughness-height to diameter ratios, (ϵ/D), of 0.0024 to 0.049. Their correlation is

$$\mathrm{Nu} = \left(\frac{f}{8}\right)\mathrm{Re}\,\mathrm{Pr}\,\{1 + [5.19\,\mathrm{Pr}^{0.44}(e^*)^{0.2} - 8.48]\sqrt{f/8}\} \tag{8-24}$$

$$1.2 \leqslant \mathrm{Pr} \leqslant 5.9$$

$$6 \times 10^4 < \mathrm{Re} < 5 \times 10^5$$

where

$$e^* = \mathrm{Re}\,(\sqrt{f/8})\,\frac{\epsilon}{D} \tag{8-24a}$$

and the fluid properties are evaluated at the mean bulk temperature. The quantity, f, in the above equations is the friction factor defined by equation (7-13), and the diameter, D, is defined as

$$D = \sqrt{4 \text{ volume}/(\pi L)}$$

Another correlation for turbulent flow through rough tubes with constant wall temperature is due to Nunner (Reference 16). It is

$$\text{Nu} = \frac{(f/8) \text{ Re Pr}}{1 + 1.5 \text{ Re}^{-1/8} \text{ Pr}^{-1/6} [(f/f_s) \text{ Pr} - 1]} \tag{8-25}$$

where f_s is the friction factor for a smooth tube. Fluid properties in the above equation are evaluated at the mean bulk temperature. This equation is recommended for $\text{Pr} < 1$, and for Reynolds number in the range of 500 to 80,000.

Laminar Flows For laminar flow in tubes at **constant wall temperature**, Seider and Tate (Reference 13) suggested the following relation

$$\text{Nu}_{av} = 1.86 (\text{Re Pr})^{1/3} \left(\frac{D}{L}\right)^{1/3} \left(\frac{\mu}{\mu_w}\right)^{0.14} \tag{8-26}$$

All properties are evaluated at the bulk temperature except μ_w, which is evaluated at the wall temperature. The average heat transfer coefficient in the above equation is to be used with the arithmetic mean of the inlet and outlet wall-to-bulk temperature differences. Equation (8-26) cannot be used for extremely long tubes, since it will yield a zero value for the Nusselt number. The equation is valid for

$$\text{Re Pr} (D/L) > 10$$

A correlation-type representation of an analytical solution for laminar flow in tubes possessing **constant wall temperature** is due to Hausen (Reference 17). It gives the average value of the Nusselt number as a function of the length, L, of the tube. It is

$$\text{Nu}_{av} = 3.66 + \frac{0.0668 (D/L) \text{ Re Pr}}{1 + 0.04 [(D/L) \text{ Re Pr}]^{2/3}} \tag{8-27}$$

When using the value of h_{av}, as obtained from the above equation, the logarithmic mean temperature difference, LMTD, should be used to calculate the heat transfer rate. The LMTD is defined in equation (10-19).

Sample Problem 8-3 Calculate the average Nusselt numbers for the data in Sample Problem 8-2 by employing Seider-Tate and Hausen correlations. Compare the results obtained with the value from the analytical result.

Solution: The Seider-Tate correlation as given by equation (8-26) is

$$\mathrm{Nu}_{av} = 1.86\,(\mathrm{Re\,Pr})^{1/3}\left(\frac{D}{L}\right)^{1/3}\left(\frac{\mu}{\mu_w}\right)^{0.14} \tag{8-26}$$

From the solution to Sample Problem 8-2, we observe that the approximate mean temperature of the water as it flows through the first one meter length of the tube is (1/2) (20 + 30) or 25°C or 77°F. Properties of saturated water at 25°C are

$$c_p = 0.998\ \mathrm{Btu/lb}_m\,°\mathrm{F}, \qquad \mu \text{ at } 77°\mathrm{F} = 2.16\ \mathrm{lb}_m/\mathrm{ft\text{-}hr}$$

$$k = 0.353\ \mathrm{Btu/hr\text{-}ft}\,°\mathrm{F}, \qquad \mu_w \text{ at } 60°\mathrm{C} \text{ or } 122°\mathrm{F} = 1.14\ \mathrm{lb}_m/\mathrm{ft\text{-}hr}$$

$$\rho = 62.2\ \mathrm{lb}_m/\mathrm{ft}^3, \qquad \mathrm{Pr} = 6.13$$

Also

$$D = 15 \text{ mm or } 0.0492 \text{ ft}; \quad L = 1 \text{ m or } 3.28 \text{ ft}$$

$$u_{av} = 0.057 \text{ m/sec or } 0.187 \text{ ft/sec}$$

giving

$$\mathrm{Re} = \frac{\rho u_{av} D}{\mu} = \frac{62.2 \times 0.187 \times 3600 \times 0.0492}{2.16} = 958$$

Thus

$$\mathrm{Nu}_{av} = 1.86\,(958 \times 6.13)^{1/3}\left(\frac{0.0492}{3.28}\right)^{1/3}\left(\frac{2.16}{1.14}\right)^{0.14}$$

$$= 1.86\,(18.04)\,(0.2466)\,(1.094)$$

$$= 9.05$$

The Hausen equation is

$$\mathrm{Nu_{av}} = 3.66 + \frac{0.0668\,(D/L)\,\mathrm{Re\,Pr}}{1 + 0.04\,[(D/L)\,\mathrm{Re\,Pr}]^{2/3}}$$

$$= 3.66 + \frac{0.0668\,(0.0492/3.28)\,(958 \times 6.13)}{1 + 0.04\,[(0.0492/3.28)\,(958 \times 6.13)]^{2/3}}$$

$$= 6.94$$

If the procedure used in the solution to Sample Problem 8-2 is repeated with new values of Re and Pr, we find

$$x^* = \frac{1.0/(0.015/2)}{958 \times 6.13} = 0.0227$$

and interpolation of the values in Table 8-2 gives

$$\mathrm{Nu_{av}} = 7.5$$

We therefore conclude that the Hausen correlation gives a result that is closer to the analytical result than the one obtained by the Seider-Tate correlation.

Sample Problem 8-4 Water flows through a 1-inch I.D. copper tube at a rate of 1.5 gallons per minute. The inlet temperature of the water is 50°F. The tube wall receives a uniform heat flux of 4000 Btu/hr-ft. Determine the length of the tube necessary to heat the water to a bulk temperature of 130°F and the average value of $(T_w - T_b)$.

Solution: The mean temperature of the water is (1/2) (50 + 130) or 90°F. The relevant thermophysical properties of water at 90°F are

$$c_p = 0.997\ \mathrm{Btu/lb_m\,°F} \qquad \rho = 62.11\ \mathrm{lb_m/ft^3}$$

$$\mu = 1.85\ \mathrm{lb_m/ft\text{-}hr} \qquad k = 0.36\ \mathrm{Btu/hr\text{-}ft\,°F}$$

$$\mathrm{Pr} = 5.12$$

We will now determine if the flow is laminar or turbulent

$$u_{av} = \frac{\text{volume flow rate}}{\text{cross-sectional area}}$$

or

$$u_{av} = \frac{1.5 \times (0.1337/60)}{(\pi/4)\,(1)^2 \times (1/144)} = 0.613 \text{ ft/sec}$$

Hence, the Reynolds number is

$$\text{Re} = \frac{\rho u_{av} D}{\mu} = \frac{62.11 \times 0.613 \times (1) \times (1/12)}{(1.85/3600)} = 6174 > 2300$$

which indicates a turbulent flow.

We will use the Filonenko equation for the friction factor, f, and the Petukhov-Popov equation for the Nusselt number, Nu.

$$f = [1.82 \log_{10}(\text{Re}) - 1.64]^{-2} \tag{8-18a}$$

$$= [1.82 \log_{10}(6174) - 1.64]^{-2} = 0.03616$$

and

$$\text{Nu} = \frac{(f/8)\,\text{Re Pr}}{1.07 + 12.7\sqrt{f/8}\,(\text{Pr}^{2/3} - 1)} \tag{8-18b}$$

$$= \frac{(0.03616/8)\,(6174 \times 5.12)}{1.07 + 12.7\sqrt{0.03616/8}\,(5.12^{2/3} - 1)}$$

$$= 51.9$$

Also

$$h = \frac{k\text{Nu}}{D} = \frac{0.36 \times 51.9}{(1/12)}$$

$$= 224 \text{ Btu/hr-ft}^2\ {}^\circ\text{F}$$

The change in the enthalpy of the water flowing through the tube, Δi, over length, L, of the tube is given by

$$\Delta i = (\text{mass flow rate})\,(\text{specific heat})\,(T_{out} - T_{in})$$

$$= (1.5 \times 8.337)\,0.997\,(130 - 50)$$

$$= 997 \text{ Btu/min}$$

With $q_w = 4000$ Btu/hr-ft, the length, L, of the tube is given by

$$L = \left(997 \frac{\text{Btu}}{\text{min}}\right) \times \left(\frac{1}{4000} \frac{\text{hr-ft}}{\text{Btu}}\right)\left(\frac{60 \text{ min}}{\text{hr}}\right)$$
$$= 14.96 \text{ ft}$$

Also

$$h_{av} (T_w - T_b)_{av} \pi D L = q_w L$$

or

$$(T_w - T_b)_{av} = \frac{4000}{\pi (1)(1/12)} \times \frac{1}{224}$$
$$= 68°\text{F}$$

Since the mean bulk temperature of the water is 90°F, the approximate mean temperature of the inside wall of the tube will be (68 + 90) or 158°F.

8-6 DIMENSIONAL ANALYSIS

In the development of the relations for convective heat transfer in tubes, we encountered several dimensionless groups such as the Nusselt number (Nu), the Prandtl number (Pr), and the Reynolds number (Re). These are dimensionless because no matter what system of units (English, SI, or CGS) is used for the different variables involved in the groups, the final results have no units or dimensions associated with them. The principal advantage of using dimensionless groupings is to reduce the number of independent variables in a problem.

Consider a typical problem in convection where the following seven variables are involved:

$$h, k, L, \rho, u_\infty, \mu, c_p$$

Let us suppose that we are interested in performing an experiment to determine the effect of each variable on the heat transfer coefficient, h. It would be necessary to vary each of the remaining six variables, k, L, ρ, u_∞, μ, and c_p, one at a time. Let us assume that we vary each variable only four times. We would then have a total of 4^6 or 4096 data points. It would, indeed, be a formidable undertaking to obtain and interpret so many data points. In the results presented earlier in this chapter, there were three dimensionless groups (Nu, Re, and Pr), which included all seven variables. If the Reynolds number and the Prandtl number are varied 4 times each, the resulting 16 points will provide as much information as the previous 4096 points. The use of dimensionless parameters thus proves to be very advantageous.

There are four primary dimensions, namely, mass $[M]$, length $[L]$, time $[\tau]$, and temperature $[T]$. Sometimes force is also considered to be a basic dimension. The dimensions of all other physical variables can be obtained from the combination of the four basic dimensions $[M, L, \tau, T]$. For example

$$\text{force } F = (\text{constant})\ (\text{mass})\ (\text{acceleration})$$

or

$$[F] = [M]\ [L/\tau^2]$$

or

$$[F] = [ML\tau^{-2}]$$

Note that the constant itself does not enter into the dimensions of $[F]$. Thus, force is said to have the dimensions, $[ML\tau^{-2}]$. Similarly

$$\text{power} = \text{work/time}$$

or

$$\text{power} = \text{force} \times \text{displacement/time}$$

Substitution of dimensions for force, displacement, and time gives

$$[P] = [ML\tau^{-2}]\ [L]/[\tau]$$

Thus

$$[P] = [ML^2\tau^{-3}]$$

Likewise

$$\text{Viscosity} = \frac{\text{shear stress}}{(du/dy)}$$

$$= \frac{[\text{force/area}]}{[\text{velocity/length}]}$$

$$= \frac{[ML\tau^{-2}]/[L^2]}{[L/\tau]/[L]}$$

or

$$[\mu] = [ML^{-1}\tau^{-1}]$$

Dimensions for energy and work are identical, since work is a specific form of energy. Some of the commonly encountered variables, their symbols, and dimensions in the $[ML\tau T]$ form are given in Table 8-3. A more complete table is given in the Appendix.

Buckingham's* π *Theorem If we have a number of dimensionless groups, they are said to be *independent dimensionless groups* if none of the groups can be derived by combining the rest of the groups *in any manner*. It is important in correlating results for any problem that we have all the possible independent dimensionless groups.

Buckingham's π theorem tells us how many independent dimensionless groups can be obtained from a set of variables. It states that the number of independent dimensionless groups that can be formed from a set of N physical variables that govern a problem equals $(N - D)$, where D is the number of basic dimensions needed to define the dimensions of all the N variables.

Consider the following seven variables

h = convective heat transfer coefficient $[M\tau^{-3}T^{-1}]$

k = thermal conductivity $[ML\tau^{-3}T^{-1}]$

L = characteristic length $[L]$

ρ = density $[ML^{-3}]$

u_∞ = velocity $[L\tau^{-1}]$

μ = viscosity $[ML^{-1}\tau^{-1}]$

c_p = specific heat $[L^2\tau^{-2}T^{-1}]$

These have four basic dimensions, namely $[M]$, $[L]$, $[\tau]$, and $[T]$. According to Buckingham's theorem, the number of independent dimensionless groups that are pertinent to the problem is $(7 - 4)$ or 3. We already know that these are the Nusselt number (hL/k), the Reynolds number $(\rho u_\infty L/\mu)$, and the Prandtl number $(\mu c_p/k)$.

Any equation governing a physical phenomenon and derived from the basic principles of physics is always dimensionally correct. That is, if we substituted the M, L, τ, T dimensions for each variable on each side of such an equation and simplified the equation, we would find that the respective exponents of M, L, τ, and T on either side of the equation are equal. This principle forms the basis of one method of obtaining dimensionless groups, as presented below.

TABLE 8-3 Some Physical Quantities and Their Dimensions

Quantity	*Symbol*	*Dimensions*
Mass	m	M
Length	x, L	L
Time	t or τ	τ
Temperature	T	T
Velocity	u_∞	$L\tau^{-1}$
Acceleration	a	$L\tau^{-2}$
Force	F	$ML\tau^{-2}$
Work	W	$ML^2\tau^{-2}$
Energy, Heat	E, Q	$ML^2\tau^{-2}$
Power	P	$ML^2\tau^{-3}$
Pressure or Stress	p	$ML^{-1}\tau^{-2}$
Density	ρ	ML^{-3}
Specific Heat	c	$L^2\tau^{-2}T^{-1}$
Viscosity	μ	$ML^{-1}\tau^{-1}$
Kinematic Viscosity	ν	$L^2\tau^{-1}$
Thermal Conductivity	k	$ML\tau^{-3}T^{-1}$
Heat Transfer Coefficient	h	$M\tau^{-3}T^{-1}$
Thermal Diffusivity	α	$L^2\tau^{-1}$
Coefficient of Thermal Expansion	β	T^{-1}

When the actual relationship among different variables of a problem is not known, for the purpose of obtaining pertinent dimensionless groups, we may assume a relation of the following form.

$$V_1^a V_2^b V_3^c V_4^d V_5^e = A, \text{ a dimensionless quantity} \tag{8-28}$$

In the above equation, $V_1, V_2, \ldots, V_5$ are the relevant variables in the problem, and a, b, c, d, and e are unknown exponents (five variables are selected for concreteness). If we substitute the dimensions in terms of M, L, τ, and T for each of the variables in equation (8-28), then the sum of the exponents of each of the principal dimensions M, L, τ, and T, must be zero. This leads to a set of four simultaneous equations containing a, b, c, d, and e as the unknowns. The number of unknowns in the four simultaneous equations is five. Any four of these five unknowns can be solved for in terms of the remaining one. Then, back-substitution in equation (8-28) will lead to one independent dimensionless group. The forgoing procedure will now be illustrated.

Consider a problem where the variables are velocity, V, characteristic length, L_c, and gravitational acceleration, g. We then have

$$V^a L_c^b g^c = A, \text{ a dimensionless quantity}$$

Substituting the dimensions for V, L, and g, we find

$$[L\tau^{-1}]^a [L]^b [L\tau^{-2}]^c = A$$

The basic dimensions are L and τ, since M and T are not contained in the given variables.

$$\text{Exponent of } L\text{:} \quad a + b + c = 0$$

$$\text{Exponent of } \tau\text{:} \quad -a - 2c = 0$$

Solving the above two equations in terms of c, we obtain

$$a = -2c \quad \text{and} \quad b = c$$

Back-substitution for a and b gives

$$V^{-2c} L_c^c g^c = \text{A}$$

or

$$\frac{V^2}{L_c g} = A$$

thus yielding $(V^2/L_c g)$ as the dimensionless group. In this example, there were three variables (V, L_c, g) and two basic dimensions (L and τ). Therefore, there is only one independent dimensionless group, $(V^2/L_c g)$, known as the Froude number. It plays an important role in flows that are influenced by the gravitational field, most commonly appearing in the analysis of open channel flow.

In general, when the above method is used for deducing the dimensionless groups, one is very likely to come up with groups that are not the conventional ones for the problem. The form of the dimensionless groups depends upon how the method is employed. In the preceding example, had we solved for b and c in terms of a, we would have obtained $(V/\sqrt{gL_c})$ as the dimensionless group. Although it is correct, it is not a standard one.

Another way to derive the dimensionless groups is to work with the available governing differential equation and to make it dimensionless. This procedure will be illustrated through an example. The differential equation for the thermal boundary layer adjacent to a flat plate, equation (8-40), is

$$u\left(\frac{\partial T}{\partial x}\right) + v\left(\frac{\partial T}{\partial y}\right) = \alpha\left(\frac{\partial^2 T}{\partial y^2}\right)$$

Let $u = u_\infty \bar{u}$, $v = u_\infty \bar{v}$, $x = L_c \bar{x}$, $y = L_c \bar{y}$, and $T = T_o \bar{T}$, where the symbols with bars on them are dimensionless variables and u_∞, L_c, and T_o are a reference velocity, a characteristic length, and a reference temperature, respectively. Then, $dx = L_c d\bar{x}$,

$dy = L_c d\bar{y}$, and $dT = T_o d\bar{T}$. Substituting the above in the differential equation, we obtain

$$\left(\frac{u_\infty T_o}{L_c}\right)\bar{u}\left(\frac{\partial \bar{T}}{\partial \bar{x}}\right) + \left(\frac{u_\infty T_o}{L_c}\right)\bar{v}\left(\frac{\partial \bar{T}}{\partial \bar{y}}\right) = \left(\alpha \frac{T_o}{L_c^2}\right)\left(\frac{\partial^2 \bar{T}}{\partial \bar{y}^2}\right)$$

TABLE 8-4 Physical Interpretation of Some Dimensionless Groups

Name of the Dimensionless Group	*Group*	*Physical Interpretation*
Euler (Eu)	$\dfrac{p}{\rho u_\infty^2}$	$\dfrac{\text{pressure forces}}{\text{inertia forces}}$
Froude (Fr)	$\dfrac{u_\infty^2}{L_c g}$	$\dfrac{\text{inertia forces}}{\text{gravity forces}}$
Grashof (Gr)	$\dfrac{g L_c^3 \beta (T_w - T_\infty)}{\nu^2}$	$\dfrac{\text{(buoyancy forces) (inertia forces)}}{\text{(viscous forces)}^2}$
Mach (M)	$\dfrac{u_\infty}{u_c}$	$\dfrac{\text{velocity}}{\text{sonic velocity}}$
Nusselt (Nu)	$\dfrac{h L_c}{k}$	ratio of temperature gradients
Peclet (Pe)	$\dfrac{c_p \rho u_\infty L_c}{k}$	$\dfrac{\text{convective heat transfer}}{\text{conductive heat transfer}}$
Prandtl (Pr)	$\dfrac{\mu c_p}{k}$	$\dfrac{\text{momentum diffusivity}}{\text{thermal diffusivity}}$
Rayleigh (Ra)	$\dfrac{g L_c^3 \beta (T_w - T_\infty)}{\nu \alpha}$	$\dfrac{\text{forces due to buoyancy and inertia}}{\text{forces due to viscosity and thermal diffusion}}$
Reynolds (Re)	$\dfrac{\rho u_\infty L_c}{\mu}$	$\dfrac{\text{inertia forces}}{\text{viscous forces}}$
Stanton (St)	$\dfrac{h}{c_p \rho u_\infty}$	$\dfrac{\text{heat transfer at wall}}{\text{convective heat transfer}}$

A division by $[u_\infty(T_o/L_c)]$ yields

$$\bar{u}\left(\frac{\partial \bar{T}}{\partial \bar{x}}\right) + \bar{v}\left(\frac{\partial \bar{T}}{\partial \bar{y}}\right) = \left(\frac{\alpha}{L_c u_\infty}\right)\left(\frac{\partial^2 \bar{T}}{\partial \bar{y}^2}\right)$$

Since the left-hand side of the above equation contains only dimensionless quantities and the equation has to be dimensionally correct, it follows that the quantity in the parentheses on the right-hand side must be dimensionless. Thus

$$\frac{\alpha}{L_c u_\infty} = \frac{\mu}{\rho L_c u_\infty} \cdot \frac{k}{\mu c_p} = \left(\frac{1}{\text{Re}} \cdot \frac{1}{\text{Pr}}\right)$$

gives the dimensionless groups for the problem. Table 8-4 gives some of the dimensionless groups that are relevant to convection problems along with their interpretations.

8-7 CONVECTION FROM A FLAT PLATE—LAMINAR FLOW IN THE THERMAL BOUNDARY LAYER

The prior sections of this chapter have dealt primarily with tube flows. Now, we turn to boundary layers on plates.

In the discussion of flow over a flat plate (Section 7-5), the concept of a hydrodynamic boundary layer was introduced. If consideration is given to the analysis of heat transfer between a plate and a fluid flowing over it, the concept of a thermal boundary layer proves very useful. As illustrated in Figure 8-1, the temperature of the fluid changes from T_w (temperature of the plate surface) to the free-stream temperature, T_∞, as we move away from the surface of the plate. We can envision a thin layer of thickness, δ_t, called the *thermal boundary layer*, across which 99 percent of the temperature change takes place. Similar to the hydrodynamic boundary layer, the thickness of the thermal boundary layer is zero at the leading edge of the plate and increases as one goes downstream on the plate. The thicknesses of these two layers are not necessarily equal.

Among the possible boundary conditions, the plate surface could be maintained at a uniform temperature or it could supply a uniform heat flux to the fluid flowing over it. As usual, we seek an expression for the convective heat transfer coefficient.

The analytical approach is the same as that used for the analysis of convective heat transfer in a tube. That is, we employ the principle of energy conservation to obtain the governing equation for the temperature distribution. Expressions for the velocity distribution, which will appear in the energy equation, have already been developed in Chapter 7. We will use both the approximate integral technique and the differential formulation to analyze the problem. The integral approach will be described first.

Application of the principle of conservation of energy to a control volume for an analysis of the boundary layer requires consideration of

(1) kinetic energy transported into and out of the control volume
(2) potential energy transported into and out of the control volume
(3) work done by viscous forces
(4) internal energy convected into and out of the control volume
(5) work done by pressure forces (the flow work)
(6) rate of change of internal energy of fluid within the control volume
(7) heat conducted into and out of the control volume

We now examine the above seven items in some detail. The changes in the kinetic and the potential energies experienced by a fluid passing through a control volume are exactly balanced by a *portion* of the work done by viscous forces. The rest of the viscous work is dissipative and is often negligible. The internal energy, u, and the work done by the pressure forces, pv, can be combined into the enthalpy, i, by the relation

$$i = u + pv$$

Also, under steady-state conditions, temperature does not change with time, and therefore, the rate of change of internal energy of the fluid within the control volume is zero. We now proceed to analyze heat transfer in a boundary layer over a flat plate by using the integral approach.

8-7.1 Integral Analysis of Thermal Boundary Layer

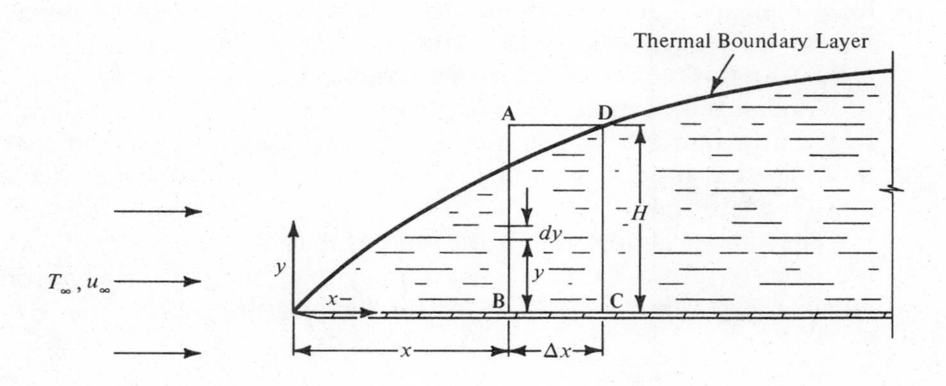

Figure 8-7 Energy balance on a control volume for the integral approach.

Consider a control volume of streamwise width, Δx, and height, H, which includes the entire thickness of the thermal boundary layer (Figure 8-7). The control volume is situated in a region where the flow is laminar. Energy flows into the control volume by convection and by conduction from the various sides of the control volume. The considerations for the energy balance discussed in the preceding section will be employed for the analysis.

Convection The energy associated with fluid entering the control volume through an area of height, dy, and depth unity into the plane of the paper consists of the internal energy, u, and the work due to pressure forces, pv. Combining u and pv into the enthalpy, i, we can write the following expression for the energy associated with the fluid.

$$\rho u (dy \cdot 1) i$$

where $\rho u(dy \cdot 1)$ is the mass flow rate and i is the enthalpy per unit mass. Then, the energy entering the control volume through face **AB** by convection is obtained by adding up the contributions of the elements, dy, that make up **AB**, so that

$$Q_{\mathbf{AB},\,\mathrm{conv}} = \int_0^H \rho u i \, dy$$

The convected energy leaving the control volume through the face **DC**, located a distance, Δx, away from the face **AB** and parallel to it, is

$$Q_{\mathbf{DC},\,\mathrm{conv}} \simeq Q_{\mathbf{AB},\,\mathrm{conv}} + \frac{d}{dx}(Q_{\mathbf{AB},\,\mathrm{conv}})\,\Delta x$$

The rate of mass flow into the control volume from the face **AD** is given by equation (7-16) as

$$\dot{m}_{\mathbf{AD}} \simeq \frac{d}{dx}\left(\int_0^H \rho u \, dy\right)\Delta x$$

The internal energy associated with the fluid entering through the face **AD** and the work done by the pressure forces on the face **AD** can be combined into the enthalpy, i_∞, measured on a unit mass basis. Thus the energy brought into the control volume through the face **AD** is

$$Q_{\mathbf{AD},\,\mathrm{conv}} = \dot{m}_{\mathbf{AD}}\, i_\infty$$

or

$$Q_{\mathbf{AD},\,\text{conv}} \simeq i_\infty \frac{d}{dx}\left(\int_0^H \rho u \, dy\right)\Delta x$$

Conduction Next, we consider the rate of conduction of heat through the four faces of the control volume. For any face, its magnitude is given by the product of the thermal conductivity of the fluid, the area of the face, and the temperature gradient normal to the face. Since the temperature of the fluid in the boundary layer is expected to vary principally in the y direction alone, little change of temperature is expected across the faces **AB** and **DC**.

$$\left(\frac{\partial T}{\partial x}\right)_{\mathbf{AB}} \simeq 0 \quad \text{and} \quad \left(\frac{\partial T}{\partial x}\right)_{\mathbf{DC}} \simeq 0$$

Also, the temperature of the fluid outside the boundary layer, that is, at **AD**, is a constant, T_∞. Hence

$$\left(\frac{\partial T}{\partial y}\right)_{\mathbf{AD}} = 0$$

Thus we conclude that the rate of heat transfer due to conduction across the faces **AB**, **DC**, and **AD** is negligible.

The only face of the control volume for which conduction is not accounted for, as yet, is face **BC**. Heat energy flows into the control volume across the face **BC** by conduction from the plate. The rate of heat conduction into the control volume across **BC** is given by

$$Q_{\mathbf{BC},\,\text{cond}} = -k\left(\frac{\partial T}{\partial y}\right)_{y=0} \Delta x$$

where k is the thermal conductivity of the fluid.

For conservation of energy, we require that under steady state, the inflow of energy to the control volume must equal the outflow of energy from the control volume on a unit time basis.
Thus we have

$$Q_{\mathbf{AB},\,\text{conv}} + Q_{\mathbf{AD},\,\text{conv}} + Q_{\mathbf{BC},\,\text{cond}} = Q_{\mathbf{DC},\,\text{conv}}$$

Substituting expressions for the various Qs, we obtain

$$\int_0^H \rho u i\, dy + i_\infty\left(\frac{d}{dx}\int_0^H \rho u\, dy\right)\Delta x - k\left(\frac{\partial T}{\partial y}\right)_{y=0}\Delta x$$

$$= \int_0^H \rho u i\, dy + \left(\frac{d}{dx}\int_0^H \rho u i\, dy\right)\Delta x$$

Observing that i_∞ is constant and simplifying, there results

$$\frac{d}{dx}\left[\int_0^H \rho u(i_\infty - i)\, dy\right] = k\left(\frac{\partial T}{\partial y}\right)_{y=0}$$

Since $(i_\infty - i) = 0$ for $y > \delta_t$, the above equation takes the form

$$\frac{d}{dx}\int_0^{\delta_t} \rho u(i_\infty - i)\, dy = k\left(\frac{\partial T}{\partial y}\right)_{y=0}$$

For an incompressible fluid with constant properties, and with

$$(i_\infty - i) = c_p(T_\infty - T)$$

The above equation becomes

$$\frac{d}{dx}\int_0^{\delta_t} u(T_\infty - T)\, dy = \alpha\left(\frac{\partial T}{\partial y}\right)_{y=0} \tag{8-29}$$

Equation (8-29) is the integral energy equation for the boundary layer over a flat plate. The next step in the integral method is to assume a temperature profile, which can be of the following form

$$\frac{T - T_w}{T_\infty - T_w} = b_0 + b_1\eta_t + b_2\eta_t^2 + b_3\eta_t^3 \tag{8-30}$$

where

$$\eta_t = \frac{y}{\delta_t}$$

The coefficients in the above equation can be evaluated with the aid of the following boundary conditions

(1) at $y = 0$, the temperature of the fluid must be the same as that of the plate, or $T = T_w$
(2) at $y = 0$, $(\partial^2 T/\partial y^2) = 0$, [see equation (8-40)]
(3) at $y = \delta_t$, $T = T_\infty$, the free-stream temperature
(4) at $y = \delta_t$, $(\partial T/\partial y) = 0$, which ensures a smooth transition from T to T_∞

The above conditions, when applied to equation (8-30), lead to a set of simultaneous equations, given below

$$b_0 = 0$$

$$2b_2 = 0$$

$$b_0 + b_1 + b_2 + b_3 = 1$$

$$b_1 + 2b_2 + 3b_3 = 0$$

When these equations are solved, we obtain

$$b_0 = 0; \quad b_1 = 3/2; \quad b_2 = 0; \quad b_3 = -1/2$$

so that equation (8-30) can be rewritten as

$$\frac{T - T_w}{T_\infty - T_w} = \frac{3}{2}\eta_t - \frac{1}{2}\eta_t^3 = \frac{3}{2}\left(\frac{y}{\delta_t}\right) - \frac{1}{2}\left(\frac{y}{\delta_t}\right)^3 \tag{8-31}$$

If we now differentiate the above equation with respect to y and evaluate the derivative at $y = 0$, there follows

$$\left(\frac{\partial T}{\partial y}\right)_{y=0} = (T_\infty - T_w)\left(\frac{3}{2\delta_t}\right) \tag{8-32}$$

Note that δ_t is as yet an unknown quantity.

At this point, if we examine equation (8-29), we find that we already have an expression for u in terms of (y/δ) in equation (7-21) and for T in equation

(8-31). Also $(\partial T/\partial y)_{y=0}$ has just been evaluated. Putting all these together in equation (8-29), there results

$$\frac{d}{dx}\int_0^{\delta_t}\left\{\frac{3}{2}\left(\frac{y}{\delta}\right)-\frac{1}{2}\left(\frac{y}{\delta}\right)^3\right\}\left\{T_\infty-(T_\infty-T_w)\left[\frac{3}{2}\left(\frac{y}{\delta_t}\right)-\frac{1}{2}\left(\frac{y}{\delta_t}\right)^3\right]-T_w\right\}dy$$

$$=(T_\infty-T_w)\frac{3\alpha}{2\delta_t}\cdot\frac{1}{u_\infty}$$

Dividing the above equation by $(T_\infty - T_w)$ and carrying out the multiplication inside the integral sign gives

$$\frac{d}{dx}\int_0^{\delta_t}\left[\frac{3}{2}\frac{y}{\delta}-\frac{y^3}{2\delta^3}-\frac{9y^2}{4\delta\delta_t}+\frac{3}{4}\frac{y^4}{\delta^3\delta_t}+\frac{3}{4}\frac{y^4}{\delta\delta_t^3}-\frac{1}{4}\frac{y^6}{\delta^3\delta_t^3}\right]dy=\frac{3\alpha}{2\delta_t u_\infty}$$

or

$$\frac{d}{dx}\left[\frac{3}{4}\frac{\delta_t^2}{\delta}-\frac{\delta_t^4}{8\delta^3}-\frac{9\delta_t^3}{12\delta\delta_t}+\frac{3\delta_t^5}{20\delta^3\delta_t}+\frac{3\delta_t^5}{20\delta\delta_t^3}-\frac{1}{28}\frac{\delta_t^7}{\delta^3\delta_t^3}\right]=\frac{3\alpha}{2\delta_t u_\infty}$$

Next, let $\zeta = \delta_t/\delta$, so that the above equation becomes

$$\frac{d}{dx}\left\{\delta\left[\frac{3}{4}(\zeta^2)-\frac{1}{8}(\zeta^4)-\frac{3}{4}(\zeta^2)+\frac{3}{20}(\zeta^4)+\frac{3}{20}(\zeta^2)-\frac{1}{28}(\zeta^4)\right]\right\}=\frac{3\alpha}{2\delta_t u_\infty}$$

or

$$\frac{d}{dx}\left\{\delta\left[\frac{3}{20}(\zeta^2)-\frac{3}{280}(\zeta^4)\right]\right\}=\frac{3\alpha}{2\delta_t u_\infty}$$

The coefficient of ζ^4 in the above equation is less than the coefficient of ζ^2 by a factor of 14. Also, for most gases, $\zeta < 1$. The term containing ζ^4 can, therefore, be neglected when compared with the ζ^2 term. The above equation then simplifies to

$$\frac{d}{dx}\left\{\delta\left[\frac{3}{20}(\zeta^2)\right]\right\}=\frac{3\alpha}{2\delta_t u_\infty}$$

Carrying out the differentiation on the left side, we have

$$\frac{1}{10}\left[\frac{d\delta}{dx}(\zeta^2) + \delta 2\zeta\left(\frac{d\zeta}{dx}\right)\right] = \frac{\alpha}{\delta_t u_\infty} \tag{8-33}$$

We recall from equation (7-23) that the hydrodynamic boundary layer thickness, δ, is given by

$$\delta^2 = \frac{280}{13}\left(\frac{\mu x}{\rho u_\infty}\right) \tag{8-34}$$

Differentiation yields

$$2\delta\left(\frac{d\delta}{dx}\right) = \frac{280}{13}\left(\frac{\mu}{\rho u_\infty}\right)$$

Hence, equation (8-33) becomes

$$\frac{1}{10}\left[\frac{140}{13}\left(\frac{\mu}{\rho u_\infty}\right)\left(\frac{\delta_t}{\delta}\right)\zeta^2 + 2\zeta\delta\delta_t\left(\frac{d\zeta}{dx}\right)\right] = \frac{\alpha}{u_\infty}$$

Recognizing that δ_t equals $\zeta\delta$, and using equation (8-34) for eliminating δ^2, we have

$$\frac{1}{10}\left[\frac{140}{13}\left(\frac{\mu}{\rho u_\infty}\right)\zeta^3 + 2\zeta^2\left(\frac{280}{13}\right)\left(\frac{\mu x}{\rho u_\infty}\right)\left(\frac{d\zeta}{dx}\right)\right] = \frac{\alpha}{u_\infty}$$

Simplifying further, we obtain

$$\zeta^3 + 4x\zeta^2\left(\frac{d\zeta}{dx}\right) = \frac{13}{14}\left(\frac{\alpha}{\nu}\right)$$

Noting that $d(\zeta^3) = 3\zeta^2 d\zeta$, and that α/ν is the reciprocal Prandtl number, Pr^{-1}, the above becomes

$$\zeta^3 + \left[\frac{4}{3}(x)\frac{d}{dx}(\zeta^3)\right] = \frac{13}{14}\mathrm{Pr}^{-1}$$

This is an ordinary, nonhomogeneous, linear differential equation in ζ^3, the solution to which is

$$\zeta^3 = Ax^{-3/4} + (13/14)\,\mathrm{Pr}^{-1} \tag{8-35}$$

where A is a constant of integration.

If the entire plate is at a uniform temperature so that both the thermal boundary layer and the hydrodynamic boundary layer start to develop at the leading edge, we expect ζ to be finite at $x = 0$. However, from equation (8-29) we see that ζ approaches ∞ as $x \to 0$. Therefore, we must have $A = 0$ in equation (8-35). The ratio, ζ, then takes the form

$$\zeta = \frac{\delta_t}{\delta} = 0.976 \, \mathrm{Pr}^{-1/3} \tag{8-36}$$

Substituting for δ from equation (8-34), we obtain the following expression for the thermal boundary layer thickness, δ_t.

$$\delta_t = \left[\frac{280}{13}\left(\frac{\mu x}{\rho u_\infty}\right)\right]^{1/2} 0.976 \, \mathrm{Pr}^{-1/3}$$

or

$$\delta_t = 4.53 x \, \mathrm{Re}_x^{-1/2} \, \mathrm{Pr}^{-1/3} \tag{8-36a}$$

where Re_x is the local Reynolds number, $(\rho u_\infty x/\mu)$

The Heat Transfer Coefficient and the Nusselt Number We can now proceed to determine the local convective heat transfer coefficient. It is given by

$$h(T_w - T_\infty) = -k\left(\frac{\partial T}{\partial y}\right)_{y=0}$$

An expression for $(\partial T/\partial y)_{y=0}$ was developed in equation (8-32), which, after substituting for δ_t from equation (8-36a), becomes

$$\left(\frac{\partial T}{\partial y}\right)_{y=0} = -(T_w - T_\infty)\frac{3}{2}\left(\frac{1}{4.53 x \mathrm{Re}_x^{-1/2} \mathrm{Pr}^{-1/3}}\right)$$

where Re_x is the local Reynolds number. Combining the two forgoing equations and simplifying, there results

$$h = 0.331\left(\frac{k}{x}\right)\mathrm{Re}_x^{1/2}\mathrm{Pr}^{1/3} \tag{8-37}$$

The local Nusselt number can now be obtained by multiplying both sides of the equation by (x/k). Thus

$$\mathrm{Nu} = \frac{hx}{k} = 0.331 \, \mathrm{Re}_x^{1/2} \, \mathrm{Pr}^{1/3} \tag{8-37a}$$

The average Nusselt number for a plate of length, L, over which laminar flow exists is then given by

$$\mathrm{Nu}_{av} = \frac{h_{av} L}{k}$$

where

$$h_{av} = \frac{\dfrac{1}{L}\displaystyle\int_0^L q\,dx}{\dfrac{1}{L}\displaystyle\int_0^L (T_w - T_\infty)\,dx} = \frac{q_{av}}{(T_w - T_\infty)_{av}}$$

For a uniform wall temperature, the local heat flux, q, equals $h(T_w - T_\infty)$ and the average heat transfer coefficient is given by

$$h_{av} = \frac{1}{L}\int_0^L h\,dx \tag{8-37b}$$

Consequently, the expression for the average heat transfer coefficient, h_{av}, for laminar boundary layer heat transfer from a flat plate maintained at a uniform temperature, T_w, is

$$h_{av} = \frac{1}{L}\int_0^L 0.331\left(\frac{k}{x}\right)\left(\frac{\rho u_\infty x}{\mu}\right)^{1/2} \mathrm{Pr}^{1/3}\,dx$$

or

$$h_{av} = 0.662\left(\frac{k}{L}\right)\mathrm{Re}_L^{1/2}\mathrm{Pr}^{1/3} \tag{8-37c}$$

and

$$\mathrm{Nu}_{av} = \frac{h_{av} L}{k} = 0.662\,\mathrm{Re}_L^{1/2}\,\mathrm{Pr}^{1/3} \tag{8-37d}$$

where $\mathrm{Re}_L = (\rho u_\infty L/\mu)$ is the Reynolds number based on the length, L, of the plate. It should be noted that equations (8-37c) and (8-37d) are valid only if the laminar boundary layer extends over the full length of the plate.

Sample Problem 8-5 Air flows over a thin flat plate, 1.0 m wide and 1.5 m long, at a velocity of 1.0 m/sec. The free stream temperature is 4°C. Calculate the amount of heat that must be supplied to the plate in order to maintain it at a uniform temperature of 50°C.

Solution: We need to know the properties of air so that the Reynolds number and the Prandtl number can be determined. Since the properties change with temperature, it is customary to determine properties at the film temperature, T_f, defined by

$$T_f = (1/2)(T_w + T_\infty) \tag{8-38}$$

For the present problem, $T_f = (1/2)(50 + 4) = 27°C$ for which we obtain the following property values from the Appendix.

$$c_p = 1005.7 \text{ J/kg°K}; \quad \nu = 15.68 \times 10^{-6} \text{ m}^2\text{/sec}$$
$$k = 0.02624 \text{ W/m°K}; \quad \text{Pr} = 0.708$$

We will first determine the Reynolds number and thereby ascertain if the flow is laminar.

$$\text{Re}_L = \frac{u_\infty L}{\nu} = \frac{(1)(1.5)}{15.68 \times 10^{-6}} = 9.55 \times 10^4 < 5 \times 10^5$$

Since the Reynolds number is less than the critical Reynolds number, 5×10^5, the flow in the boundary layer is laminar over the entire length of the plate. We now use equation (8-37c) to calculate the average heat transfer coefficient.

$$h_{av} = 0.662\left(\frac{k}{L}\right)\text{Re}_L^{1/2}\text{Pr}^{1/3}$$

or

$$h_{av} = (0.662)\frac{0.02624}{1.5}(9.55 \times 10^4)^{1/2}(0.708)^{1/3}$$

$$= 3.2 \text{ W/m}^2 \text{ °K}$$

Since the plate has two surfaces from which heat is lost by convection, the rate of heat supplied to the plate is given by

$$Q = 2hA(T_w - T_\infty)$$

$$= 2(3.2)(1.0 \times 1.5)(50 - 4)$$

$$= 441.6 \text{ watts}$$

We now proceed to analyze the thermal boundary layer over a flat plate, maintained at a uniform temperature, by employing differential analysis.

8-7.2 Differential Analysis of Thermal Boundary Layer

Consider a control volume (Figure 8-8) situated within the thermal boundary layer. The component of velocity in the z direction is assumed to be zero. Also, there is assumed to be no variation of temperature in the z direction. Therefore, energy will be transported in and out through the faces **ABCD**, **A′B′C′D′**, **DD′C′C**, and **AA′B′B**. We now proceed to develop expressions for energy transported across the boundaries of the control volume so that the principle of conservation of energy may be applied.

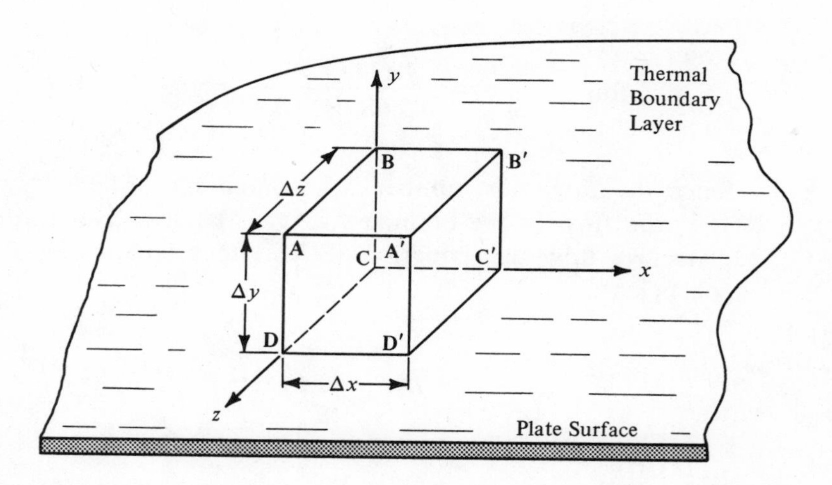

Figure 8-8 Energy balance on a control volume for the differential analysis.

Convection The mass flow rate into the control volume through the face ABCD is

$$\rho u \Delta y \Delta z$$

The energy associated with this fluid per unit mass is the enthalpy, the kinetic energy, and the potential energy. The enthalpy, i, includes the internal energy, u, and the work done by the pressure forces, pv. The changes in the kinetic energy and the potential energy experienced by the fluid, as it goes through the control volume, are exactly balanced by a part of the shear work due to viscous forces. The other part of the shear work is viscous dissipation. Hence, the rate of energy flow into the control volume through the face **ABCD** is

$$Q_{x,\,\text{conv}} = \rho u i \Delta y \Delta z$$

The rate of energy flow out of the control volume through the face **A′B′C′D′**, located Δx away from the face **ABCD** and parallel to it, is given by using a Taylor series expansion.

$$Q_{x+\Delta x,\,\text{conv}} \simeq Q_{x,\,\text{conv}} + \frac{\partial}{\partial x}(Q_{x,\,\text{conv}})\,\Delta x$$

The rate of mass inflow through face **DD′C′C** of the control volume is

$$\rho v \Delta z \Delta x$$

and the corresponding rate of energy inflow is

$$Q_{y,\,\text{conv}} = \rho v i \Delta z \Delta x$$

In addition, the rate of energy flow out of the control volume through **AA′B′B**, located Δy away from the face **DD′C′C**, is

$$Q_{y+\Delta y,\,\text{conv}} \simeq Q_{y,\,\text{conv}} + \frac{\partial}{\partial y}(Q_{y,\,\text{conv}})\,\Delta y$$

Conduction The rate of heat conduction in the x direction is usually very small compared to that in the y direction, since the primary temperature gradient is normal to the plate. Therefore, expressions for heat conduction will be developed only for the y direction.

The rate at which heat is conducted in the y direction at face **DD′C′C** into the control volume is

$$Q_{y,\,\text{cond}} = -k\left(\frac{\partial T}{\partial y}\right)\Delta z\,\Delta x$$

The rate of heat conduction away from the control volume across the face **AA′B′B** is given by

$$Q_{y+\Delta y,\,\text{cond}} \simeq Q_{y,\,\text{cond}} + \frac{\partial}{\partial y}(Q_{y,\,\text{cond}})\,\Delta y$$

Work Due to Viscous Forces We have seen in Section 7-5.2 that for a two- or three-dimensional flow, the viscous shear stress, τ_{yx}, is given by

$$\tau_{yx} = \frac{\mu}{g_c}\left(\frac{\partial u}{\partial y} + \frac{\partial v}{\partial x}\right)$$

The first subscript in τ_{yx} specifies the direction of the normal to the plane on which τ_{yx} acts, and the second subscript gives the direction of the shear stress. Thus, τ_{yx} is the shear stress on the face **DD′C′C** and τ_{xy} is the shear stress on the face **ABCD**.

The rate of work due to viscous forces on a face equals the product of the area of the face, the viscous shear stress acting on it, and the component of velocity along the direction of the shear stress. For most fluids $\tau_{xy} = \tau_{yx}$. Therefore

the viscous work on the face ABCD $= (\Delta y \Delta z)\left[\dfrac{\mu}{g_c}\left(\dfrac{\partial u}{\partial y} + \dfrac{\partial v}{\partial x}\right)\right](v)$, and

the viscous work on the face **DD′C′C** $= (\Delta x \Delta z)\left[\dfrac{\mu}{g_c}\left(\dfrac{\partial u}{\partial y} + \dfrac{\partial v}{\partial x}\right)\right](u)$

Since the x component of velocity, u, is much greater than the y component of velocity, v, a comparison of the above two expressions indicates that the viscous work on the face **ABCD** is negligible when compared to that on the face **DD′C′C**. Further simplification results when we observe that

$$\frac{\partial v}{\partial x} \ll \frac{\partial u}{\partial y}$$

giving the rate of viscous work (W_y) done on the face **DD′C′C** as

$$W_y = \underbrace{\frac{\mu}{g_c}\left(\frac{\partial u}{\partial y}\right)}_{\substack{\text{viscous}\\ \text{shear}\\ \text{stress}}}\underbrace{(\Delta x \Delta z)}_{\text{area}}\underbrace{u}_{\text{velocity}}$$

Next we want to determine if the above equation represents the rate of work done on the control volume or by the control volume. According to the standard sign convection of solid body mechanics, which is applicable here, the face **DD′C′C** is considered a negative face, since an outward normal to it points in the negative y direction. By the same convention, face **AA′B′B** is considered to be a positive face. The shear stresses acting on these two faces must be opposite in direction. If an expression representing the shear stress is positive, then its direction on the negative face is in the $(-x)$ direction and on the positive face it is in the $(+x)$ direction. We therefore conclude that work, W_y, is done ***by the control volume*** on the surrounding fluid at the face **DD′C′C** since the external viscous force is in the negative x direction and the x component of fluid velocity is in the positive x direction.

By similar reasoning, we conclude that work is done by the external fluid ***on the control volume*** at face **AA′B′B**, since the external viscous force and the x component of fluid velocity are both in the positive x direction. On a unit time basis, it is

$$W_{y+\Delta y} \simeq W_y + \frac{\partial}{\partial y}(W_y)\Delta y$$

The principle of conservation of energy for steady-state conditions may now be stated in words as:

Rate of ***outflow*** of the convected energy *from* the control volume
\+ Rate of heat conduction ***away from*** the control volume
\+ Rate of work done ***by*** the control volume on the external fluid
= Rate of ***inflow*** of the convected energy *to* the control volume
\+ Rate of heat conduction ***into*** the control volume
\+ Rate of work done ***on*** the control volume by the external fluid

or

$$Q_{x+\Delta x,\,\text{conv}} + Q_{y+\Delta y,\,\text{conv}} + Q_{y+\Delta y,\,\text{cond}} + W_y = Q_{x,\,\text{conv}} + Q_{y,\,\text{conv}} + Q_{y,\,\text{cond}} + W_{y+\Delta y}$$

Substituting appropriate expressions for the various Qs and Ws and simplifying, we obtain

$$\frac{\partial}{\partial x}(\rho u i) + \frac{\partial}{\partial y}(\rho v i) + \frac{\partial}{\partial y}(-k)\left(\frac{\partial T}{\partial y}\right) - \frac{\partial}{\partial y}\left(\frac{\mu}{g_c}\right)u\left(\frac{\partial u}{\partial y}\right) = 0$$

For an incompressible fluid with constant thermophysical properties, the above equation simplifies to

$$\rho \frac{\partial}{\partial x}(ui) + \rho \frac{\partial}{\partial y}(vi) = k \frac{\partial^2 T}{\partial y^2} + \frac{\mu}{g_c} \frac{\partial}{\partial y}\left(u \frac{\partial u}{\partial y}\right) \tag{8-39}$$

The last term on the right-hand side is the only term that does not contain the temperature. It represents the net work done per unit time by the frictional forces on the control volume. It can be expanded as

$$\frac{\mu}{g_c} \frac{\partial}{\partial y}\left(u \frac{\partial u}{\partial y}\right) = \frac{\mu}{g_c}\left(\frac{\partial u}{\partial y}\right)^2 + \frac{\mu}{g_c} u \frac{\partial^2 u}{\partial y^2}$$

The first term on the right-hand side is the viscous dissipation and is negligible for moderately low velocities. The second term on the right-hand side balances the changes in the kinetic energy and the potential energy, which were so far not included in our analysis. Thus, for low velocities, we can neglect the last term in equation (8-39). Dividing the resulting equation by the density, ρ, we obtain

$$\frac{\partial}{\partial x}(ui) + \frac{\partial}{\partial y}(vi) = \frac{k}{\rho}\left(\frac{\partial^2 T}{\partial y^2}\right)$$

or

$$u\left(\frac{\partial i}{\partial x}\right) + i\left(\frac{\partial u}{\partial x}\right) + v\left(\frac{\partial i}{\partial y}\right) + i\left(\frac{\partial v}{\partial y}\right) = \frac{k}{\rho}\left(\frac{\partial^2 T}{\partial y^2}\right)$$

or

$$u\left(\frac{\partial i}{\partial x}\right) + v\left(\frac{\partial i}{\partial y}\right) + i\left(\frac{\partial u}{\partial x} + \frac{\partial v}{\partial y}\right) = \frac{k}{\rho}\left(\frac{\partial^2 T}{\partial y^2}\right)$$

Since $(\partial u/\partial x) + (\partial v/\partial y) = 0$ [continuity equation (7-29)], and $di = c_p dT$, the above equation becomes

$$u\left(\frac{\partial T}{\partial x}\right) + v\left(\frac{\partial T}{\partial y}\right) = \alpha\left(\frac{\partial^2 T}{\partial y^2}\right) \tag{8-40}$$

Equation (8-40) is the energy equation for an incompressible constant-property boundary layer on a flat plate. For boundary conditions, consi-

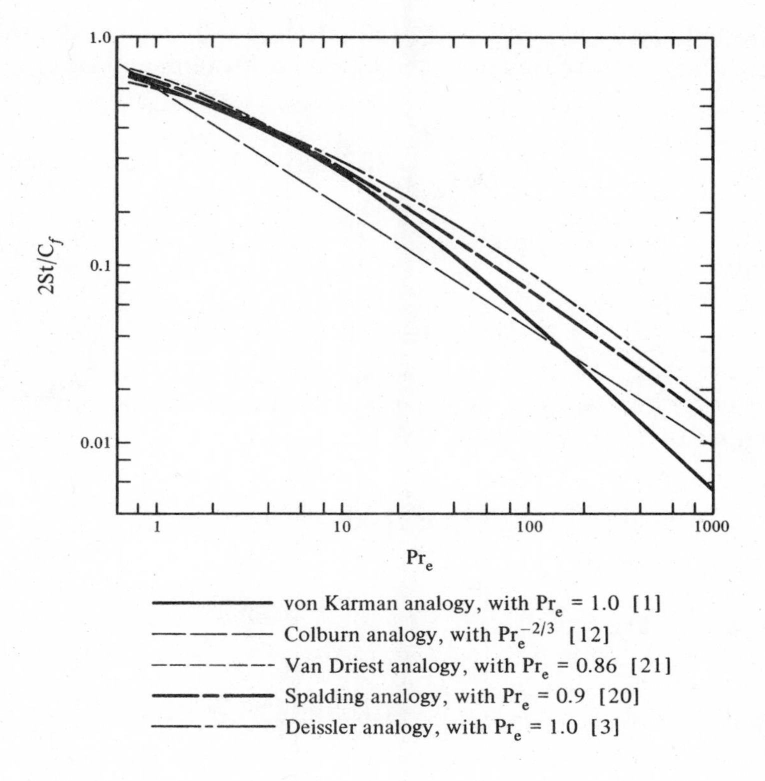

Figure 8-10 Influence of the effective Prandtl number on the Stanton number from *Handbook of Heat Transfer*, by W. M. Rohsenow, and J. P. Hartnett. Copyright 1973, by McGraw-Hill, Inc. Used with permission.

The local friction coefficient as determined by experimental measurements for turbulent flow is

$$C_f = 0.0576 \, \mathrm{Re}_x^{-1/5} \quad \text{for } 5 \times 10^5 < \mathrm{Re}_x < 10^7 \tag{8-51}$$

Applying Colburn's analogy, equation (8-48), we obtain

$$\mathrm{St} \, \mathrm{Pr}^{2/3} = C_f/2 = 0.0288 \, \mathrm{Re}_x^{-1/5} \tag{8-51a}$$

or

$$h = 0.0288 \rho c_p u_\infty \mathrm{Re}_x^{-1/5} \mathrm{Pr}^{-2/3} \tag{8-51b}$$

We note that there is normally a laminar flow up to the critical distance, x_{cr}, corresponding to $\mathrm{Re}_{cr} = 5 \times 10^5$. If we assume that the transitional zone can be neglected, the equation for h_{av} [equation (8-37b)] takes the following form for a plate whose surface is maintained at a uniform temperature.

$$h_{av} = \frac{1}{L}\left[\int_0^{x_{cr}} h_{\text{laminar}}\,dx + \int_{x_{cr}}^{L} h_{\text{turbulent}}\,dx\right]$$

where

$$x_{cr} = 5 \times 10^5 (\mu/\rho u_\infty)$$

Noting that h_{laminar} is given by equation (8-37) and that $h_{\text{turbulent}}$ is given by equation (8-51b), we obtain

$$h_{av} = \frac{1}{L}\left[\int_0^{x_{cr}} 0.331\left(\frac{k}{x}\right)\text{Re}_x^{1/2}\text{Pr}^{1/3}dx + \int_{x_{cr}}^{L} 0.0288\rho c_p u_\infty \text{Re}_x^{-1/5}\text{Pr}^{-2/3}\,dx\right]$$

Integration gives

$$h_{av} = \frac{k}{L}(0.036\text{Re}_L^{4/5} - 836)\text{Pr}^{1/3} \quad (8\text{-}52)$$

or

$$\text{Nu}_{av} = (0.036\text{Re}_L^{4/5} - 836)\text{Pr}^{1/3} \quad (8\text{-}52a)$$

and

$$\text{St}_{av} = (0.036\text{Re}_L^{-1/5} - 836\text{Re}_L^{-1})\text{Pr}^{-2/3} \quad (8\text{-}52b)$$

It is emphasized that the value of h_{av} obtained by using equation (8-52) is approximate; for better accuracy, one should use Figure 8-10 and carry out the integration in accordance with equation (8-15).

Based on the experimental work of Zhukauskas and Ambrazyavichyus (Reference 23) and the Colburn analogy, Whitaker (Reference 24) recommends the following equation for the local Nusselt number.

$$\text{Nu}_x = 0.029\,\text{Re}_x^{0.8}\text{Pr}^{0.43} \quad (8\text{-}53)$$

Whitaker assumes that transition takes place at $\text{Re}_{cr} = 2 \times 10^5$ and gives the following correlation for the average Nusselt number.

$$\mathrm{Nu_{av}} = 0.036\,[\mathrm{Re}_L^{0.8}\mathrm{Pr}^{0.43} - 17{,}400] + 297\,\mathrm{Pr}^{1/3} \tag{8-53a}$$

$$\text{for } 10^5 < \mathrm{Re}_L < 5.5 \times 10^6,$$

$$0.7 < \mathrm{Pr} < 380$$

$$\text{and } 0.26 < \frac{\mu_\infty}{\mu_w} < 3.5$$

When large temperature differences are present, Whitaker recommends the correlation

$$\mathrm{Nu_{av}} = 0.036\,\mathrm{Pr}^{0.42}\,(\mathrm{Re}_L^{0.8} - 9200)\,(\mu_\infty/\mu_w)^{1/4} \tag{8-53b}$$

The ranges of Re, Pr, and (μ_∞/μ_w) are the same as those for equation (8-53a). If the Reynolds number becomes large in relation to the critical Reynolds number, he recommends that the numerical constant, (9200), in the above equation be dropped. The quantities, μ_∞ and μ_w, are to be evaluated at the free-stream temperature and the wall temperature, respectively.

Sample Problem 8-8 A flat plate, 1.0 m wide and 1.5 m long, is to be maintained at 90°C in air with a free-stream temperature of 10°C. Determine the velocity at which the air must flow over the flat plate so that the rate of energy dissipation from the plate is 3.75 kW.

Solution: We will assume that the flow over the plate is turbulent and that it experiences transition from laminar flow to fully turbulent flow at

$$\mathrm{Re_{cr}} = 5 \times 10^5$$

The properties of air at a film temperature of (1/2) (90 + 10) or 50°C are found from the Appendix to be

$$\rho = 1.0877\ \mathrm{kg/m^3} \qquad k = 0.02813\ \mathrm{W/m^\circ K}$$

$$c_p = 1007.3\ \mathrm{J/kg^\circ K} \qquad \mathrm{Pr} = 0.703$$

$$\mu = 2.029 \times 10^{-5}\ \mathrm{kg/ms}$$

Also

$$Q = 3.75\ \mathrm{kW},\ T_w = 90^\circ\mathrm{C},\ T_\infty = 10^\circ\mathrm{C}$$

The average heat transfer coefficient is obtained from

$$Q = h_{av}A\,(T_w - T_\infty)$$

Hence

$$h_{av} = \frac{3.75 \times 10^3}{(90 - 10)(1.0 \times 1.5)} = 31.25 \text{ W/m}^2 \text{ °C}$$

Also, h_{av} is given by equation (8-52) as

$$h_{av} = (k/L)[0.036 \text{ Re}_L^{4/5} - 836] \text{Pr}^{1/3}$$

When we solve the above equation for Re_L, we obtain

$$\text{Re}_L = \left\{ \frac{1}{0.036} \left[\left(\frac{h_{av} L}{k} \right) \text{Pr}^{-1/3} + 836 \right] \right\}^{5/4}$$

Substitution for various quantities yields

$$\text{Re}_L = \left\{ \frac{1}{0.036} \left[\frac{31.25 \times 1.5}{0.02813} (0.703)^{-1/3} + 836 \right] \right\}^{5/4}$$

$$= 1.247 \times 10^6 > \text{Re}_{cr}$$

Hence, the free-stream velocity u_∞ is given by

$$u_\infty = \frac{\mu \text{Re}_L}{\rho L} = \frac{(2.029 \times 10^{-5}) \times 1.247 \times 10^6}{(1.0877)(1.5)}$$

or

$$u_\infty = 15.5 \text{ m/sec}$$

8-10 HEAT TRANSFER IN LIQUID METALS

When high heat transfer coefficients are desired, liquid metals, such as sodium, potassium, or mercury, are often given consideration. Being metals, they possess very high thermal conductivities when compared with nonmetallic liquids. Also, their boiling points are much higher than conventional fluids like water, glycol, methanol, etc. This makes it possible to operate a power plant, with a liquid metal as a working fluid, over a much wider temperature range than a conventional one and yet use only moderate pressures. The use of liquid metals is often considered for the transfer of the large amounts of heat released in a nuclear reactor. Liquid metals, on the other hand, are highly corrosive and are difficult to handle, which often makes it impractical to use them extensively. We will now briefly discuss some correlations for heat transfer in liquid metals flowing through tubes and over a flat plate.

Flow Through Tubes Some of the empirical relations for heat transfer for flow of liquid metals through tubes are given below. Seban and Shimazaki (Reference 25) investigated turbulent flow of a liquid metal in a smooth circular tube with a fully developed velocity profile and a temperature profile that is invariant in the x coordinate. Under such conditions, the Nusselt number does not vary along the length of the tube. They recommend the following correlation for the Nusselt number for tubes with *constant wall temperature.*

$$\text{Nu} = 5.0 + 0.025\,(\text{Re Pr})^{0.8} \qquad \text{for } (\text{Re Pr}) > 1000 \tag{8-54}$$

In the above equation, all the properties are evaluated at the bulk temperature.

For turbulent flow of liquid metals through tubes receiving a *constant heat flux* and with fully developed velocity and temperature profiles, Skupinski et al (Reference 26) recommend the following correlation.

$$\text{Nu} = 4.82 + 0.0185\,(\text{Re Pr})^{0.827} \qquad \text{for } 3.6 \times 10^3 < \text{Re} < 9.05 \times 10^5,$$

$$\text{and } 10^2 < (\text{Re Pr}) < 10^4$$

Flow Over a Flat Plate It is possible to obtain an expression for the Nusselt number for flow over a flat plate by employing the integral energy formulation. We consider a flat plate maintained at a temperature, T_w, along which there is a laminar flow of a liquid metal having a free-stream temperature equal to T_∞. We now proceed to determine the thermal boundary layer thickness and the heat transfer coefficient.

The Prandtl number (ν/α) for liquid metals is of the order of 0.01, which means that the diffusion of momentum under laminar flow conditions will be two orders of magnitude slower than the diffusion of heat energy. This results in a very small thickness of the hydrodynamic boundary layer, δ, compared to the thickness of the thermal boundary layer, δ_t.

The integral thermal boundary layer equation, equation (8-29), is repeated here.

$$\frac{d}{dx}\int_0^{\delta_t} u(T_\infty - T)dy = \alpha\left(\frac{\partial T}{\partial y}\right)_{y=0} \tag{8-29}$$

Since $\delta \ll \delta_t$, we can assume that, for all practical purposes, $u = u_\infty$ within the thermal boundary layer $(0 \leqslant y \leqslant \delta_t)$. This assumption is supported by the

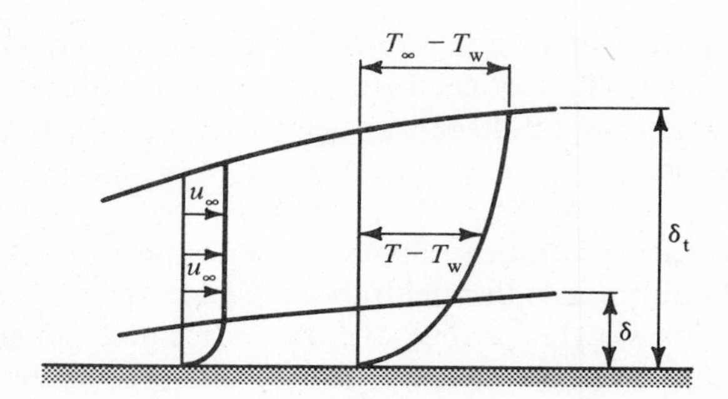

Figure 8-11 Thermal and Hydrodrynamic boundary layers in laminar flow of a liquid metal over a flat plate.

illustrative profiles shown in Figure 8-11. We now employ the third-degree polynomial for the temperature profile as expressed by equation (8-31).

$$\frac{T - T_w}{T_\infty - T_w} = \frac{3}{2}\left(\frac{y}{\delta_t}\right) - \frac{1}{2}\left(\frac{y}{\delta_t}\right)^3 \qquad (8\text{-}31)$$

Using equation (8-31), the following integral form for the thermal boundary layer equation for liquid metal flow is obtained.

$$\frac{d}{dx}\left[\int_0^{\delta_t}\left\{1 - \frac{3}{2}\left(\frac{y}{\delta_t}\right) + \frac{1}{2}\left(\frac{y}{\delta_t}\right)^3\right\}dy\right] = \frac{3\alpha}{2\delta_t u_\infty} \qquad (8\text{-}55)$$

Integration leads to

$$\frac{3}{8}\frac{d}{dx}(\delta_t) = \frac{3}{2}\left(\frac{\alpha}{\delta_t u_\infty}\right)$$

Separation of variables (δ_t and x) and integration gives

$$\delta_t = \sqrt{\frac{8\alpha x}{u_\infty}} \quad \text{or} \quad \frac{\delta_t}{x} = \sqrt{8}\,(\text{Re}_x \text{Pr})^{-1/2} \qquad (8\text{-}56)$$

The local convective heat transfer coefficient may be obtained from

$$h\,(T_w - T_\infty) = -k\,(\partial T/\partial y)_{y=0}$$

with the result

$$h = \frac{3k}{2\delta_t} = \frac{3\sqrt{2}}{8} k \sqrt{\frac{u_\infty}{\alpha x}} \tag{8-57}$$

The local Nusselt number then follows as

$$\text{Nu} = \frac{hx}{k} = 0.530\,(\text{Re}_x \text{Pr})^{1/2} \tag{8-57a}$$

Now, we can check on our assumption, $\delta \ll \delta_t$. Recall that the expression for δ was found earlier to be

$$\frac{\delta}{x} = \frac{4.64}{\text{Re}_x^{1/2}}$$

Therefore

$$\frac{\delta}{\delta_t} = \frac{4.64x}{\text{Re}_x^{1/2}} \frac{1}{\sqrt{8}} (\text{Re}_x \text{Pr})^{1/2} = 1.64\sqrt{\text{Pr}}$$

For Pr = 0.01, $\delta = 0.16\,\delta_t$, which is in reasonable agreement with our assumption. A student interested in further information on liquid metal heat transfer may see Reference 27.

8-11 FLOW ACROSS A TUBE AND ACROSS TUBE BANKS

The heat transfer characteristics of a cylinder of circular cross section situated in a cross flow have been studied extensively by Giedt (Reference 28). The results that he obtained are shown in Figure 8-12. It can be seen from this figure that there is a considerable circumferential variation in the local value of the Nusselt number, Nu_θ, which is substantially influenced by the value of the Reynolds number. For the values of the Reynolds number reported, values of Nu_θ decrease until θ equals, approximately, 80°. For Re < 100,000, separation of the laminar boundary layer occurs at this value of θ and the value of Nu_θ gradually increases for $\theta > 80°$.

For Re > 100,000, the increase in Nu_θ is more pronounced as θ exceeds 80° and Nu_θ reaches a peak value at about $\theta = 110°$. The local Nusselt number then exhibits another minimum value at about $\theta = 145°$. For such large values of the Reynolds number, the first minimum value of Nu_θ corresponds to the transition from the laminar boundary layer to the turbulent boundary layer, and the second minimum value corresponds to the separation of the turbulent boundary layer from the cylinder.

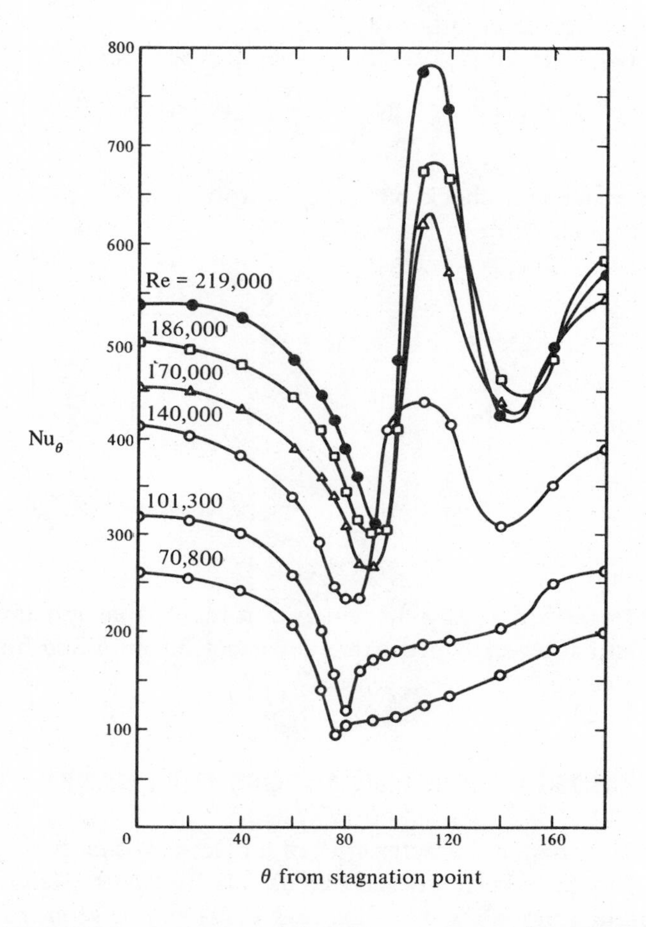

Figure 8-12 Variation in the local Nusselt number for flow past a cylinder, from W. H. Giedt, [28].

For most practical applications, we are interested in the average value of the Nusselt number. It is not possible to predict the Nusselt number for flow across a cylinder because of the complex nature of the process of separation. Correlations based on the experimental work of Hilpert (Reference 29) for gases and of Knudsen and Katz (Reference 30) for liquids are, therefore, widely used. These are given in Table 8-6. The average Nusselt number for air is plotted as a function of the Reynolds number in Figure 8-13. Subsequent work of Fand (Reference

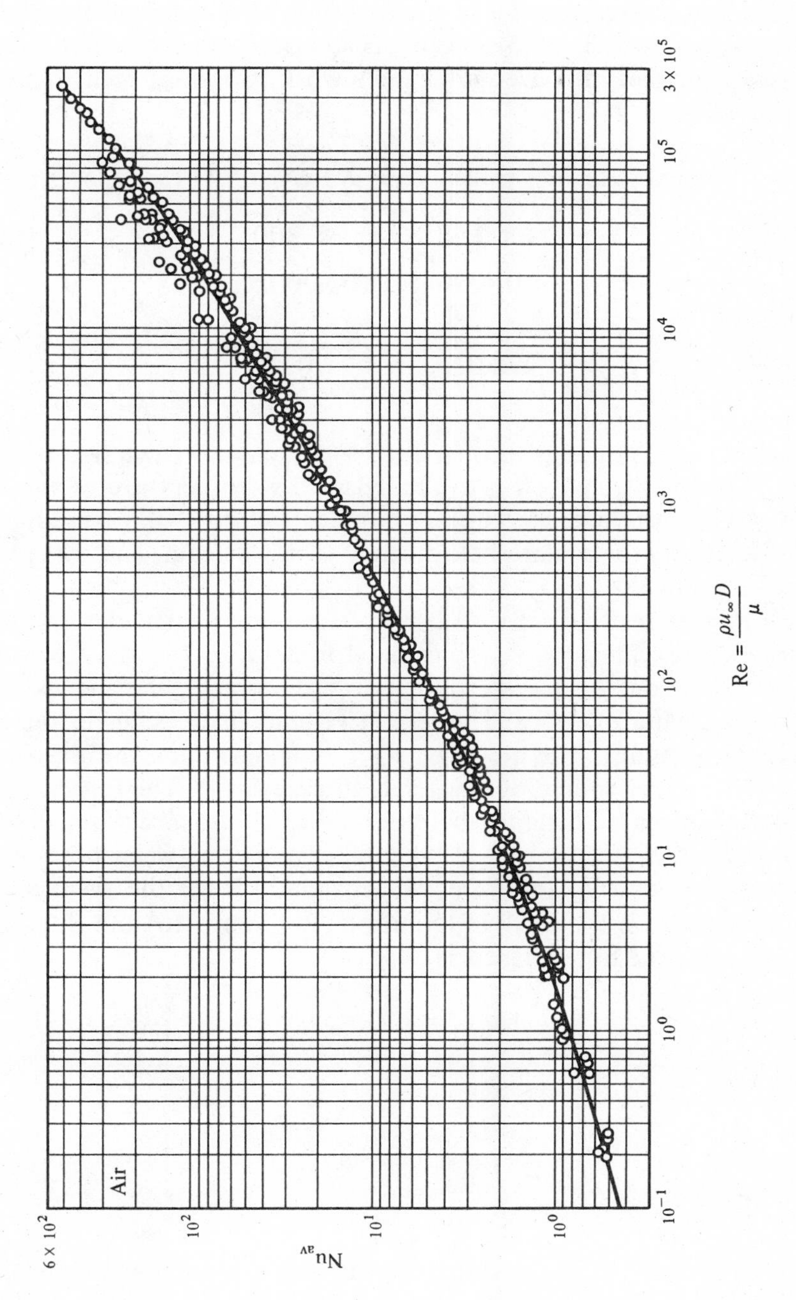

Figure 8-13 Experimental data for flow of air past a cylinder. From *Heat Transmission* by W. H. McAdams. Copyright 1954, by W. H. McAdams. Used with permission of McGraw-Hill Book Company.

31), which correlates data for $10^{-1} < \mathrm{Re} < 10^5$ for liquids, indicates a better correlation for the Nusselt number, which is also included in the table. Whitaker (Reference 24) has proposed a new correlation that is in agreement with experimental results to within ± 25 percent. His correlation is given in Table 8-6. Since the correlation takes into account Fand's work, it is suggested that the Whitaker correlation be used. It is

$$\mathrm{Nu_{av}} = (0.4\,\mathrm{Re}^{1/2} + 0.06\,\mathrm{Re}^{2/3})\mathrm{Pr}^{0.4}(\mu_\infty/\mu_w)^{1/4}$$

$$\text{for } 0.67 < \mathrm{Pr} < 300$$

$$10 < \mathrm{Re} < 100{,}000$$

$$\text{and } 0.25 < \left(\frac{\mu_\infty}{\mu_w}\right) < 5.2 \qquad (8\text{-}58)$$

In the above equations, all properties are evaluated at the free-stream temperature, T_∞, except, μ_w, which is evaluated at the temperature of the surface, T_w. Correlations for the average Nusselt number for flow of gases and liquids past noncircular tubes, for a plate held normal to the stream, and for spheres are also included in Table 8-6.

Heat transfer in flow across a bank of tubes is of particular importance in the design of heat exchangers. As mentioned in Section 7-7, the Reynolds number, Re_{max}, is based on the largest velocity, u_{max}, which the fluid experiences as it flows across a bank of tubes. The largest velocity may occur in the transverse or the diagonal openings. The quantity, u_{max}, depends upon the arrangement of the tubes, which can be in-line or staggered (Figure 8-14). The longitudinal pitch, S_l, is the center-to-center distance between tubes in *the direction of flow*, and the transverse pitch, S_t, is the center-to-center distance between tubes along a direction that is *transverse to the flow*. The correlations for the average Nusselt number for air, based on the work of Grimison (Reference 36) and Kays and Lo (Reference 37), have the following form

$$\mathrm{Nu_{av}} = C(\mathrm{Re}_{max})^n\,\mathrm{Pr}^{1/3}$$

where

$$\mathrm{Re}_{max} = \frac{\rho u_{max} D}{\mu}$$

All fluid properties in the various dimensionless groups appearing in the above correlation are evaluated at the film temperature, T_f, which is the mean of the free-stream temperature, T_∞, and the surface temperature, T_w. The values of C and n are presented in Table 8-7 for tube banks of 10 rows or more. In other

TABLE 8-6 Correlations for Flow across Bodies*

Geometry	*Range of Reynolds Number*	*Correlation for Nu_{av}*	*Reference*
Circular Tube ○ ($L_c = D$) (gases and liquids)	0.4 to 4	$0.989\ Re^{0.33}\ Pr^{1/3}$	Hilpert [29] and Knudsen & Katz [30]
	4 to 40	$0.911\ Re^{0.385}\ Pr^{1/3}$	
	40 to 4000	$0.683\ Re^{0.466}\ Pr^{1/3}$	
	4000 to 40,000	$0.193\ Re^{0.618}\ Pr^{1/3}$	
	40,000 to 400,000	$0.0266\ Re^{0.805}\ Pr^{1/3}$	
Circular Tube ○ ($L_c = D$) liquids $0.67 < Pr < 300$	0.1 to 100,000	$(0.35 + 0.56\ Re^{0.52})Pr^{0.3}$	Fand [31]
	10 to 100,000	$(0.4\ Re^{1/2} + 0.06\ Re^{2/3})\ Pr^{0.4}(\mu_b/\mu_w)^{1/4}$	Whitaker [24]
Square Tube gas → ◇ L_c	5000 to 100,000	$0.246\ Re^{0.588} Pr^{1/3}$	Jakob [32]
gas → □ L_c	5000 to 100,000	$0.102\ Re^{0.675} Pr^{1/3}$	
Hexagonal Tube gas → ⬡ L_c	5000 to 100,000	$0.153\ Re^{0.638} Pr^{1/3}$	
gas → ⬡ L_c	5000 to 19,500	$0.16\ Re^{0.638} Pr^{1/3}$	
	19,500 to 100,000	$0.0385\ Re^{0.782} Pr^{1/3}$	
Vertical Plate gas → I L_c	4000 to 15,000	$0.228\ Re^{0.731} Pr^{1/3}$	
Sphere ($L_c = D$) gas → ○	17 to 70,000	$0.37\ Re^{0.6}$	McAdams [33]
liquid → ○	1 to 2000	$(0.97 + 0.68\ Re^{0.5})\ Pr^{0.3}$	Kramers [34]
{oil, water} → ○	1 to 200,000	$(1.2 + 0.53\ Re^{0.54})\ Pr^{0.3}[\mu_f/\mu_w]^{0.25}$	Vliet & Leppert [35]
{gas, liquids $0.71 < Pr < 380$} → ○	3.5 to 76,000	$2 + (0.4\ Re^{1/2} + 0.06\ Re^{2/3})\ Pr^{0.4}\ (\mu_b/\mu_w)^{1/4}$	Whitaker [24]

*All fluid properties with subscript f are evaluated at the film temperature, $T_f = (1/2)\ (T_w + T_\infty)$. The Reynolds number and the Prandtl number are evaluated at film temperature. Subscript b refers to bulk temperature or free-stream temperature.

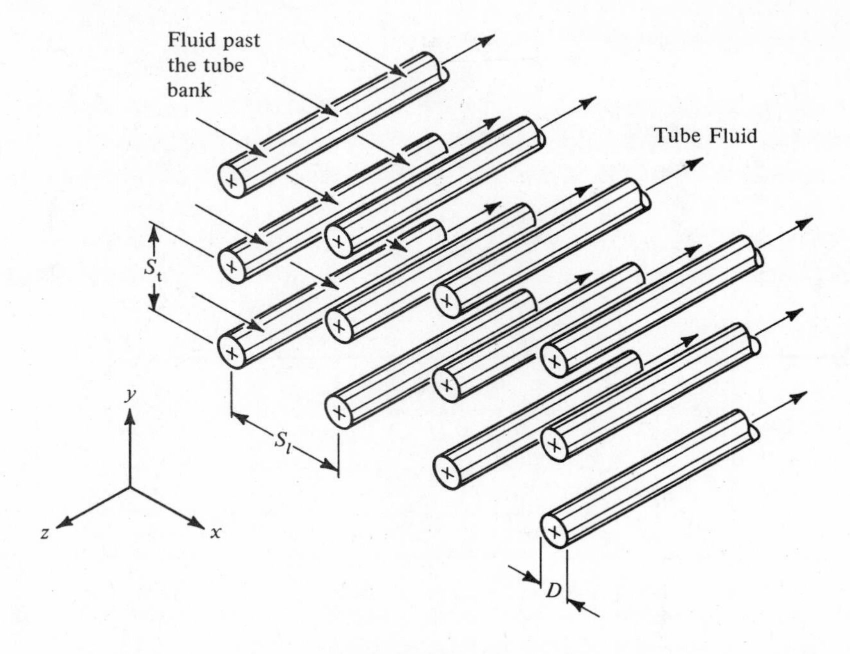

(a) In-line Arrangement

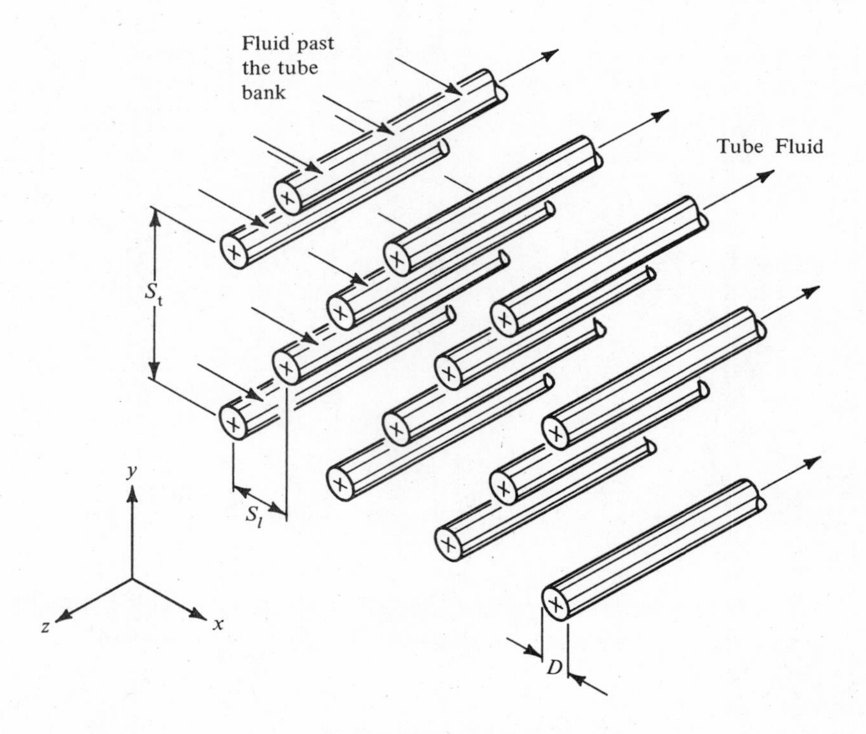

(b) Staggered Arrangement

Figure 8-14 Nomenclature for flow across tube banks.

words, once the number of rows in a bank of tubes exceeds 10, the value of Nu does not change. If the number of rows is less than 10, then the Nu value obtained by using Table 8-7 is multiplied by the appropriate factor from Table 8-8.

TABLE 8-7 Values of C and n in the Correlation for Heat Transfer in Flow of Air across Tube Banks of 10 Rows or More

$$\mathrm{Nu} = C\,\mathrm{Re}^n$$

$$(2000 < \mathrm{Re}_{max} < 40{,}000)$$

		S_l/D							
	S_t/D	1.25		1.5		2.0		3.0	
		C	n	C	n	C	n	C	n
In-line:	1.250	0.348	0.592	0.275	0.608	0.100	0.704	0.0633	0.752
	1.500	0.367	0.586	0.250	0.620	0.101	0.702	0.0678	0.744
	2.000	0.418	0.570	0.299	0.602	0.229	0.632	0.198	0.648
	3.000	0.290	0.601	0.357	0.584	0.374	0.581	0.286	0.608
Staggered:	0.600	–	–	–	–	–	–	0.213	0.636
	0.900	–	–	–	–	0.446	0.571	0.401	0.581
	1.000	–	–	0.497	0.558	–	–	–	–
	1.125	–	–	–	–	0.478	0.565	0.518	0.560
	1.250	0.518	0.556	0.505	0.554	0.519	0.556	0.522	0.562
	1.500	0.451	0.568	0.460	0.562	0.452	0.568	0.488	0.568
	2.000	0.404	0.572	0.416	0.568	0.482	0.556	0.449	0.570
	3.000	0.310	0.592	0.356	0.580	0.440	0.562	0.421	0.574

TABLE 8-8 Ratio of (Nu_{av}) for N Rows Deep to (Nu_{av}) for 10 Rows Deep, Based on Table 8-7

N	1	2	3	4	5	6	7	8	9	10
Ratio for in-line tubes	0.64	0.80	0.87	0.90	0.92	0.94	0.96	0.98	0.99	1.0
Ratio for staggered tubes	0.68	0.75	0.83	0.89	0.92	0.95	0.97	0.98	0.99	1.0

A more recent correlation for predicting the average Nusselt number for staggered tube banks consisting of ten or more rows, proposed by Whitaker (Reference 24), is

$$\mathrm{Nu}_{av} = (0.5\,\mathrm{Re}_B^{1/2} + 0.2\,\mathrm{Re}_B^{2/3})\mathrm{Pr}^{1/3}\left(\frac{\mu_b}{\mu_w}\right)^{0.14} \tag{8-59}$$

$$\text{for } 100 \leqslant \text{Re}_B < 10^5, 0.7 < \text{Pr} < 760$$

$$0.18 < \left(\frac{\mu_b}{\mu_w}\right) < 4.3, \quad \epsilon \leqslant 0.65$$

where

$$\text{Re}_B = \frac{3\rho D u_{av}}{2\mu_b(1 - \epsilon)}$$

and

$$\epsilon = \text{the void fraction} = \frac{\text{area for flow with the tubes}}{\text{total area for flow without the tubes}}$$

The fluid properties in equation (8-59) are evaluated at the mean bulk temperature.

For $\text{Re}_B < 100$, Whitaker recommends that the following correlation be used.

$$\text{Nu}_{av} = 2\text{Re}_B^{1/3}\text{Pr}^{1/3}(\mu_b/\mu_w)^{0.14} \tag{8-59a}$$

Sample Problem 8-9 Air flows at a rate of 900 ft^3/min over a cross flow heat exchanger consisting of seven tubes in the direction of the flow and eight tubes in the direction transverse to the flow. The length of each tube is four feet. The outer diameter of the tubes is 3/4 in, the longitudinal pitch, S_l, is 1.50 in, and the transverse pitch, S_t, is 1.125 in. The temperature of the air entering the heat exchanger is 400 °F and the temperature of the surface of the tubes may be taken as 200°F. The air has the following properties at 300°F: $\mu = 0.0574$ lb_m/hr-ft, $k = 0.0203$ Btu/hr-ft°F, $\rho = 0.052$ lb_m/ft^3, c_p = 0.243 Btu/lb_m °F, and Pr = 0.686. If the arrangement of the tubes is in-line, estimate the convective heat transfer coefficient between the air and the tubes.

Solution: We will first calculate the Nusselt number for flow across 10 tubes using Table 8-7 and then use Table 8-8 to account for the fact that there are actually only seven rows. From the data given in the problem, we have

$$u_{max} = \frac{\text{flow rate}}{\text{minimum area for flow}}$$

$$= \frac{(900/60)}{\underbrace{\left[\left(\frac{1.125 \times 8}{12}\right) \times 4\right]}_{\text{Area for flow ahead of the tubes}} - \underbrace{\left[\left(\frac{0.75 \times 8}{12}\right) \times 4\right]}_{\text{Projected area of the tubes blocking the flow}}} = 15 \text{ ft/sec}$$

$$\text{Re}_{max} = \frac{\rho u_{max} D}{\mu} = \frac{0.052 \times 15 \times (0.75/12)}{(0.0574/3600)} = 3057$$

$$\frac{S_t}{D} = \frac{1.125}{0.75} = 1.5 \text{ and } \frac{S_l}{D} = \frac{1.5}{0.75} = 2.0$$

For the in-line arrangement, with the forgoing values of the parameters, we obtain the following values of C and n from Table 8-7.

$$C = 0.101 \quad \text{and} \quad n = 0.702$$

so that

$$\text{Nu}_{av} = 0.101 \, \text{Re}_{max}^{0.702}$$

$$= 0.101 \, (3057)^{0.702}$$

$$= 28.25$$

From Table 8-8

$$\frac{\text{Nu}_{av} \text{ for seven rows deep}}{\text{Nu}_{av} \text{ for 10 rows deep}} = 0.96$$

Therefore, for the present problem

$$h_{av} = 0.96 \times 28.25 \times \left[\frac{0.0203}{(0.75/12)}\right]$$

$$= 8.81 \text{ Btu/hr-ft}^2 \, ^\circ\text{F}$$

PROBLEMS (English Engineering System of Units)

8-1. The average value of the Nusselt number for flow through a tube of 1.5-in. I.D. is 4.6. Determine the heat transfer coefficient if the fluid in the tube is (a) air, (b) water, and (c) engine oil. The temperature for evaluating fluid properties may be taken as 100°F.

8-2. The average value of the Nusselt number for flow over a surface is 11.0. The characteristic dimension is 4 ft. The temperature of the surface is 300°F, and the free-stream temperature is 60°F. Determine the rate of heat transfer from the plate to the fluid if the fluid is (a) air, (b) water, and (c) engine oil.

8-3. The temperature and the velocity distributions for flow through a tube are given by

$$T = c\left[1 - \left(\frac{r}{r_w}\right)^2\right] \quad \text{and} \quad u = u_o$$

where r_w is the inner radius of the tube, and c and u_o are constants. Obtain an expression for the bulk temperature of the fluid.

8-4. Water flows through a tube of 2 in. diameter. Determine the flow rate that will result in a Reynolds number of 1600. If the tube receives heat at a rate of 500 Btu/hr-ft length of the tube, determine the average heat transfer coefficient between the water and the tube wall. Assume that the velocity and the temperature profiles are fully established. Use properties at 60°F.

8-5. If the bulk temperature of the water in problem 8-4 is 75°F at a point where the temperature profile is fully established, estimate the distance from this point for the bulk temperature to become 125°F.

8-6. Plot a graph of $(T - T_o)$ versus radius for the water in problem 8-5.

8-7. If the diameter of the tube in problem 8-5 is doubled, what is the length of the tube necessary to increase the bulk temperature from 75°F to 125°F?

8-8. Air flows through a tube of 0.5-in. I.D. at an average velocity of 5 ft/sec. The bulk temperature of the air is 65°F at the entrance section of the tube. The tube wall is maintained at 120°F. Estimate the length of the tube necessary to heat the air to a bulk temperature of 100°F.

8-9. Oil flows through a tube of 3-in. diameter at a velocity of 20 ft/sec. Using equation (8-18), determine the length of the tube necessary to raise the bulk temperature of the oil from 162°F to 190°F if the tube wall receives a constant heat flux of 2×10^4 Btu/hr-ft^2 on the surface of the tube, and the mean T_w.

8-10. Water flows through a 3/8-in. tube at a rate of 5 gpm. If the temperature of the water at the entrance is 60°F, estimate the length of the tube required for heating the water to a temperature of 150°F when the tube wall is maintained at 500°F. Use equation (8-22).

8-11. Plot a graph of Nu_{av} versus (D/L) on log-log paper for (Re Pr) = 40,000, using equations (8-26) and (8-27). Determine the value of (D/L) for which the difference in the values of Nu_{av} from the two equations exceeds 10 percent. Assume $\mu = \mu_w$.

8-12. Determine if the following quantities are dimensionless or not:

(a) $\dfrac{L_c x}{c_p \mu}$

(b) $\dfrac{p}{\rho L_c g}$

(c) $\dfrac{L_c^3 \rho^2 \beta(\Delta T)}{\mu^2}$

(d) $\dfrac{\rho u_\infty c_p L_c^2}{h}$

8-13. For the integral analysis of the thermal boundary layer over a flat plate, assume

$$\frac{T - T_w}{T_\infty - T_w} = c_o + c_1 \eta_t$$

Proceeding in the same manner as in Section 8-7.1, obtain an expression for the thermal boundary layer thickness.

8-14. If a flat plate is maintained at $T = T_w$ from $x = x_o$ onwards, determine an expression for the thermal boundary layer thickness, δ_t. [*Hint:* Start with equation (8-35)].

8-15. Oil at 80°F flows over a thin plate at a free-stream velocity of 20 ft/sec. The plate is 3 ft long and is maintained at 200°F. Estimate the rate of heat transfer from the plate to the oil.

8-16. A thin plate is maintained at 180°F. Determine the critical length of the plate as a function of the free-stream velocity, u_∞, to ensure laminar flow if the fluid is (a) air, (b) glycerine, and (c) water. The temperature of the fluid in each case is 60°F.

8-17. Estimate the rate of heat transfer for the lengths of the plate obtained in problem 8-16. Assume the widths of the plates to be 1 foot.

8-18. Liquid bismuth flows through a 2-in. diameter stainless steel tube at a rate of 20 lb_m/sec. It enters at 600°F and is heated to 700°F as it flows through the tube. The tube wall receives a constant heat flux, and the tube wall is maintained at a temperature 80°F higher than the bulk temperature of the bismuth. Calculate the length of the tube necessary to bring about the desired heat transfer.

8-19. Two lb_m/sec of liquid bismuth enters a 1-in. diameter stainless steel pipe at 600°F. The tube wall temperature is maintained constant at 875°F. Calculate the bismuth exit temperature if the tube is 4 ft long.

8-20. Air at atmospheric pressure and at a temperature of 60°F flows over a heated cylinder of 3-in. diameter whose surface is maintained at 280°F. Determine the loss of heat from the cylinder if the air velocity is (a) 50 ft/sec, (b) 100 ft/sec, and (c) 250 ft/sec.

8-21. Assuming that a man can be approximated by a cylinder 14 in. in diameter and 5 ft 8 in. high with a surface temperature of 75°F, calculate the heat he would lose while standing in a 30 mph wind at 10°F.

8-22. Air at 70 ft/sec flows across a tube. The tube could be square with a side of 2 inches or circular with a diameter of 2 inches. Compare the rate of heat transfer in each case. Consider both cases of the square tube as shown in Table 8-6. Assume T_f = 80°F.

8-23. Air at atmospheric pressure and 70°F flows across a bank of tubes 12 rows high and 6 rows deep at a velocity of 25 ft/sec measured upstream of the tubes. The surfaces of the tubes are maintained at 200°F. The diameter of the tubes is 1¼ inch, and the tubes are arranged in a staggered fashion. The centers of the tubes form corners of an equilateral triangle of sides 2 inches. Calculate the total heat transfer rate if the tubes are 8 ft long.

PROBLEMS (SI System of Units)

8-1. The average value of the Nusselt number for flow through a tube of 25-mm I.D. is 4.6. Determine the heat transfer coefficient if the fluid in the tube is (a) air, (b) water, and (c) engine oil. The temperature for evaluating fluid properties may be taken as 40°C.

8-2. The average value of the Nusselt number for flow over a surface is 10.0. The characteristic dimension is 1 m. The temperature of the surface is 160°C, and the free-stream temperature is 10°C. Determine the rate of heat transfer from the plate to the fluid if the fluid is (a) air, (b) water, and (c) engine oil.

8-3. The temperature and the velocity distributions for flow through a tube are given by

$$T = c\left[1 - \left(\frac{r}{r_w}\right)^2\right] \quad \text{and} \quad u = u_o$$

where r_w is the inner radius of the tube, and c and u_o are constants. Obtain an expression for the bulk temperature of the fluid.

8-4. Water flows through a tube of 50 mm diameter. Determine the flow rate that will result in a Reynolds number of 1600. If the tube receives heat at a rate of 800 W/m length of the tube, determine the average heat transfer coefficient between the water and the tube wall. Assume that the velocity and the temperature profiles are fully established.

8-5. If the bulk temperature of the water in problem 8-4 is 25°C at a point where the temperature profile is fully established, estimate the distance from this point for the bulk temperature to become 50°C.

8-6. Plot a graph of $(T - T_o)$ versus radius for the water in problem 8-5.

8-7. If the diameter of the tube in problem 8-5 is doubled, what is the length of the tube necessary to increase the bulk temperature from 25°C to 50°C?

8-8. Air flows through a tube of 10-mm I.D. at an average velocity of 2 m/sec. The bulk temperature of the air is 15°C at the entrance section of the tube. The tube wall is maintained at 140°C. Estimate the length of the tube necessary to heat the air to a bulk temperature of 80°C.

8-9. Oil flows through a tube of 80 mm diameter at a velocity of 5 m/sec. Using equation (8-18), determine the length of the tube necessary to raise the bulk temperature of the oil from 150°C to 160°C if the tube wall receives a constant heat flux of 60 kW/m2 on the surface of the tube, and the mean T_w.

8-10. Water flows through a 10-mm tube at a rate of 400 cc/sec. If the temperature of the water at the entrance is 8°C, estimate the length of the tube required for heating the water to a temperature of 50°C when the tube wall is maintained at 250°C. Use equation (8-22).

8-11. Plot a graph of Nu_{av} versus (D/L) on log-log paper for (Re Pr) = 40,000 using equations (8-26) and (8-27). Determine the value of (D/L) for which the difference in the values of Nu_{av} from the two equations exceeds 10 percent. Assume $\mu = \mu_w$.

8-12. Determine if the following quantities are dimensionless or not:

(a) $\dfrac{L_c h}{c_p \rho v}$ (c) $\dfrac{L_c^3 \rho^2 \beta (\Delta T) \nu}{\mu^2 \alpha}$

(b) $\dfrac{g L_c \rho}{p}$ (d) $\dfrac{\rho u_\infty c_p L_c^2}{h}$

8-13. For the integral analysis of the thermal boundary layer over a flat plate, assume

$$\frac{T - T_w}{T_\infty - T_w} = c_o + c_1 \eta_t$$

Proceeding in the same manner as in Section 8-7.1, obtain an expression for the thermal boundary layer thickness.

8-14. If a flat plate is maintained at $T = T_w$ from $x = x_o$ onward, determine an expression for the thermal boundary layer thickness, δ_t. [*Hint:* Start with equation (8-35)].

8-15. Oil at 100°C flows over a thin plate at a free-stream velocity of 10 m/sec. The plate is 1 m long and is maintained at 140°C. Estimate the rate of heat transfer from the plate to the oil.

8-16. A thin plate is maintained at 90°C. Determine the critical length of the plate as a function of the free-stream velocity, u_∞, to ensure laminar flow if the fluid is (a) air, (b) glycerine, and (c) water. The temperature of the fluid in each case is 20°C.

8-17. Estimate the rate of heat transfer for the lengths of plate obtained in problem 8-16. Assume the widths of the plates to be 1 m.

8-18. Liquid bismuth flows through a 50 cm diameter stainless steel tube at a rate of 10 kg/sec. It enters at 300°C and is heated to 400°C as it flows through the tube. The tube wall receives a constant heat flux, and the tube wall is maintained at a temperature of 50°C higher than the bulk temperature of the bismuth. Calculate the length of the tube necessary to bring about the desired heat transfer.

8-19. One kg/sec of liquid sodium enters a 20-mm diameter stainless steel pipe at 300°C. The tube wall temperature is maintained constant at 375°C. Calculate the sodium exit temperature if the tube is 2 m long.

8-20. Air at atmospheric pressure and at a temperature of 17°C flows over a heated cylinder of 75 mm diameter whose surface is maintained at 200°C. Determine the loss of heat from the cylinder if the air velocity is (a) 10 m/sec, (b) 50 m/sec, and (c) 150 m/sec.

8-21. Assuming that a man can be approximated by a cylinder 30 mm in diameter and 1.8 m high with a surface temperature of 24°C, calculate the heat he would lose while standing in a 50 km/hr wind at −10°C.

8-22. Air at 25 m/sec flows across a tube. The tube could be square with a side of 5 cms or circular with a diameter of 5 cms. Compare the rate of heat transfer in each case. Consider both cases of the square tube as shown in Table 8-6. Assume $T_f = 80°C$.

8-23. Air at atmospheric pressure and 20°C flows across a bank of tubes 12 rows high and 6 rows deep at a velocity of 10 m/sec measured upstream of the tubes. The surfaces of the tubes are maintained at 100°C. The diameter of the tubes is 26 mm, and the tubes are arranged in a staggered fashion. The centers of the tubes form corners of an equilateral triangle of sides 45 mm. Calculate the total heat transfer rate if the tubes are 4 m long.

REFERENCES

[1] Kays, W. M., *Convective Heat and Mass Transfer*, McGraw-Hill Book Company, New York, (1966).

[2] Sellers, J. M., Tribus, M., and Klein, J. S., "Heat Transfer to Laminar Flow in a Round Tube or Flat Conduit," *Trans. ASME*, **78**, p. 441, (1956).

[3] Deissler, R. G., "Analysis of Turbulent Heat Transfer, Mass Transfer and Friction in Smooth Tubes at High Prandtl and Schmidt Numbers," NACA Report *1210*, (1955).

[4] Sparrow, E. M., Hallman, T. M., and Seigel, R., "Turbulent Heat Transfer in the Thermal Entrance Region of a Pipe With Uniform Heat Flux," *Appl. Sci. Res.*, **A7**, p. 37, (1957).

[5] Webb, R. L., "A Critical Evaluation of Analytical Solutions and Reynolds Analogy Equations for Turbulent Heat and Mass Transfer in Smooth Tubes," *Warme-und Stoffubertragung*, Bd. **4**, pp. 197-204, (1971).

[6] Petukhov, B. S., and Popov, V. N., "Theoretical Calculation of Heat Exchange and Frictional Resistance in Turbulent Flow in Tubes of an Incompressible Fluid with Variable Physical Properties," *Trans. in High Temperature*, **1**, No. *1*, (1963).

[7] Sleicher, C. A. and Tribus, M., "Heat Transfer in a Pipe With Turbulent Flow and Arbitrary Wall-Temperature Distribution," *Trans. ASME*, **79**, p. 789, (1957).

[8] Rohsenow, W. M., and Hartnett, J. P., *Handbook of Heat Transfer*, McGraw-Hill Book Company, (1973).

[9] Deissler, R. G., "Analysis of Turbulent Heat Transfer and Flow in the Entrance Regions of Smooth Passages," *NACA TN 3016*, (1953).

[10] Boelter, L. M. K., Young, G., and Iversen, H. W., "An Investigation of Aircraft Heaters XXVII–Distribution of Heat Transfer Rate in the Entrance Section of a Circular Tube," *NACA TN 1451*, (1948).

[11] Dittus, F. W., and Boelter, L. M. K., Univ. of California-Berkeley, *Pub. Eng.* **2**, p. 443, (1930).

[12] Colburn, A. P., "A Method of Correlating Forced Convection Heat Transfer Data and a Comparison with Fluid Friction," *Trans. A.I.Ch.E.*, **29**, p. 174, (1933).

[13] Seider, E. M., and Tate, C. E., "Heat Transfer and Pressure Drop of Liquids in Tubes," *Ind. Eng. Chem.*, **28**, p. 1429, (1936).

[14] Petukhov, B. S., "Heat Transfer and Friction in Turbulent Pipe Flow with Variable Physical Properties," *Advances in Heat Transfer*, **6**, p. 503, (1970).

[15] Dipprey, D. F. and Sabersky, R. H., "Heat and Momentum Transfer in Smooth and Rough Tubes at Various Prandtl Numbers," *Int. J. Heat-Mass Transfer*, **6**, p. 329, (1963).

[16] Nunner, W., VDI Forschungsheft 455, **B22:5** *(AERE Lib./Trans. 786)*, (1956).

[17] Hausen, H., "Darstellung des Warmeuber ganges in Rohren durch Verallgemeinerte Potanzbeziehungen," *VDIZ*, No. *4*, p. 91, (1943).

[18] Kestin, J., and Richardson, P. D., "Heat Transfer Across Turbulent Incompressible Boundary Layers," *Int. J. of Heat & Mass Transfer*, **6**, p. 147, (1963).

[19] Coles, D., "The Law of the Wake in Turbulent Boundary Layer," *J. of Fluid Mechanics*, **1**, p. 191, (1956).

[20] Spalding, D. B., "Heat Transfer to a Turbulent Stream from a Surface with a Stepwise Discontinuity in Wall Temperature," International Developments in Heat Transfer, Conf. Int. Dev. in Heat Transfer, pt. 2, p. 439, *ASME*, (1961).

[21] Van Driest, E. R., "The Turbulent Boundary Layer with Variable Prandtl Number," North American Aviation *Report A1-1914*, (1954).

[22] Schlichting, H. *Boundary Layer Theory*, 2nd ed., McGraw-Hill Book Company, p. 143, (1968).

[23] Zhukauskas, A. A., and Ambrazyavichyus, A. B., "Heat Transfer of a Plate in a Liquid Flow," *Int. J. Heat & Mass Transfer*, **3**, p. 305, (1961).

[24] Whitaker, S., "Forced Convection Heat Transfer Correlations for Flow in Pipes, Past Flat Plates, Single Cylinders, Single Spheres and for Flow in Packed Beds and Tube Bundles," *A.I.Ch.E.,* **18**, No. *2*, p. 361, (1972).

[25] Seban, R. A., and Shimazaki, T. T., "Heat Transfer to a Fluid Flowing Turbulently in a Smooth Pipe with Walls at Constant Temperature," *Trans. ASME*, **73**, p. 803, (1951).

[26] Skupinshi, E., Tortel, J., and Vantrey, L., "Determination des Coefficients de Convection d'un Alliage Sodium-potassium dans un Tube Circulaire," *Int. J. Heat & Mass Transfer,* **8**, p. 937, (1965).

[27] Stein, R. "Liquid Metal Heat Transfer," *Advanced Heat Transfer,* **3**, (1966).

[28] Giedt, W. H., "Investigation of Variation of Point Unit-heat-transfer Coefficient Around a Cylinder Normal to an Air Stream," *Trans. ASME,* **71**, p. 375, (1949).

[29] Hilpert, R., "Warmeabgabe von geheizen Drahten und Rohren," *Forsch. Gebiete Ingenieurwesen,* **4**, p. 220, (1933).

[30] Knudsen, J. D., and Katz, D. L., *Fluid Dynamics and Heat Transfer,* McGraw-Hill Book Company, New York, (1958).

[31] Fand, R. M., "Heat Transfer by Forced Convection from a Cylinder to Water in Crossflow," *Int. J. Heat & Mass Transfer*,**8**, p. 995, (1965).

[32] Jakob, M., *Heat Transfer, Vol. 1,* John Wiley & Sons, Inc., New York, (1949).

[33] McAdams, W. H., *Heat Transmission,* 3rd ed., McGraw-Hill Book Company, New York, (1954).

[34] Kramers, H., "Heat Transfer from Spheres to Flowing Media," *Physica*, **12**, p. 61, (1946).

[35] Vliet, G. C., and Leppert, G., "Forced Convection Heat Transfer from an Isothermal Sphere to Water," *J. Heat Transfer, Trans. ASME,* **83**, p. 163, (1961).

[36] Grimison, E. D., "Correlation and Utilization of New Data on Flow of Gases over Tube Banks," *Trans. ASME,* **59**, p. 538, (1937).

[37] Kays, W. M., and Lo, R. K., "Basic Heat Transfer and Flow Friction Data for Gas Flow Normal to Banks of Staggered Tubes: Use of a Transient Technique," Stanford Univ. Tech. *Report 15*, (1952).

9

Natural Convection

9-1 INTRODUCTION

The preceding chapter dealt with heat transfer between a fluid and a surface under conditions where the fluid flow was maintained by a pump, blower, or other pressure-producing device. There is another class of convective heat transfer problems when fluid motions occur without such overt pumping. For instance, a hot plate will eventually assume a temperature equal to that of the surrounding air even though there is no air blowing over it. The mode of heat transfer involved is termed as *natural* or *free convection.* Fluid particles in the immediate vicinity of the hot plate become warmer than the ambient fluid, thus resulting in localized changes in density. The change in density activates the body force due to gravity, which, in turn, gives rise to a motion of the fluid particles in the region close to the hot surface. Natural convection is the principle mode of heat transfer in baseboard heaters or steam "radiators" in houses.

The body forces need not be due to gravity alone. In the case of rotating turbine blades, the centrifugal force field serves as the body force that drives fluid motions in the presence of density differences that are caused by temperature differences.

9-2 AN EXPERIMENT FOR THE DETERMINATION OF THE COEFFICIENT OF HEAT TRANSFER FOR NATURAL CONVECTION

One can gain some insight into the factors influencing the natural convection heat transfer coefficient by considering a basic experiment. Attention may be directed to an experimental set-up shown in Figure 9-1. It consists of a large tank in which a small plate is suspended vertically. The surfaces of both the plate and the tank are of black matte finish resulting in very high surface emissivities close to 0.95. It is assumed that the tank is filled with a transparent gas like nitrogen, oxygen, or air. An electrical heater is embedded in the flat plate, which, when connected to a battery, heats the plate uniformly. Thermocouple wires are connected to the plate and to the interior of the tank, which is equipped with two valves so that it can be purged and charged with any gas. The electrical power supply can be varied through the use of a rheostat, and the power is measured using an ammeter and a voltmeter. Once the switch is closed, a power setting, P, is chosen, and sufficient time is allowed for the temperatures to stabilize.

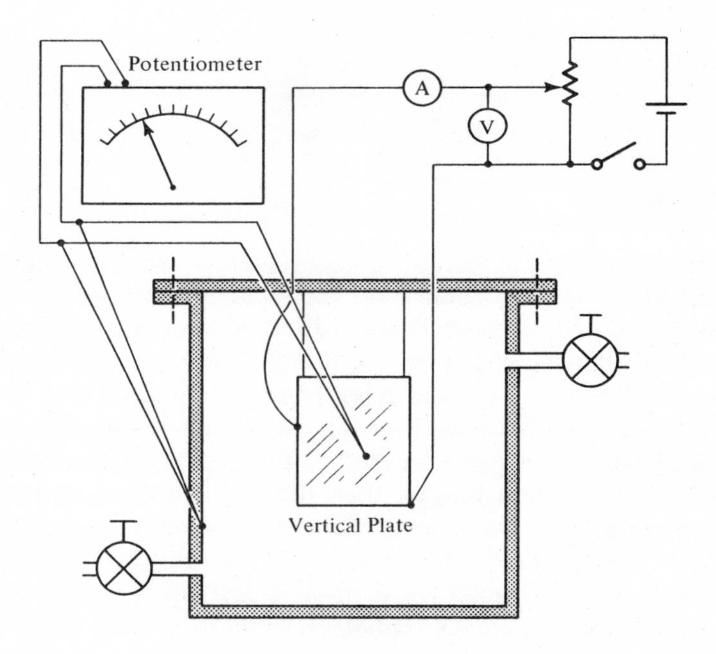

Figure 9-1 Natural convection experiment.

Next one can apply the principle of conservation of energy (first law of thermodynamics) and conclude that

$$\text{rate of energy supply to the plate} = \left\{ \begin{array}{l} \text{rate of energy carried away from the} \\ \text{plate by conduction, convection, and} \\ \text{radiation} \end{array} \right.$$

or

$$P = Q_{\text{cond}} + Q_{\text{conv}} + Q_{\text{rad}}$$

If the power leads, the thermocouple wires, and the tank cover suspension wires are of very small cross section, then the amount of heat conducted away through these members (Q_{cond}) will be very small and, therefore, as a first approximation, can be neglected. If the surface area of the interior of the tank is very large compared to the total surface area, A_p, of the plate, then the heat energy carried away from the plate due to radiation, Q_{rad}, is given by

$$Q_{\text{rad}} = \sigma A_p \epsilon_p (T_w^4 - T_t^4) \tag{9-1}$$

where

σ = Stefan-Boltzmann constant

T_w = temperature of the plate surface, °R or °K

T_t = temperature of the tank, °R or °K

and

ϵ_p = emissivity of the surface of the plate

The heat energy convected away from the plate to the air in the tank, Q_{conv}, can be expressed using Newton's law for convective cooling as

$$Q_{\text{conv}} = hA_p(T_w - T_\infty) \tag{9-1a}$$

where T_∞ is the temperature of the air.

T_∞ and T_t will be almost identical if the tank is very well insulated. Then, the coefficient of natural convection is determined from the equation

$$h = \frac{Q_{\text{conv}}}{A_p(T_w - T_\infty)} \quad \text{or} \quad h = \frac{P - Q_{\text{rad}}}{A_p(T_w - T_\infty)} \tag{9-2}$$

Although the forgoing experiment is a rather simple one, it can provide values of the heat transfer coefficient that are fairly accurate. If the power, P, supplied to the heating element in the plate is increased, it raises the temperature of the plate, and an increase in the value of the coefficient of natural convection, h, is observed. If the air in the tank is replaced by helium, all other things being equal, it is found that the value of h increases. Also, if the plate height is doubled, the coefficient of natural convection, h, decreases by approximately 19.0 percent.

An experiment, such as the one described above, can be used to obtain correlations for the average heat transfer coefficient for natural convection. Different parameters that affect the value of h for natural convection are the properties of the gas (specific heat c_p, viscosity μ, thermal conductivity k, coefficient of thermal expansion β, density ρ), gravitational acceleration g, temperature difference $(T_w - T_\infty)$ or ΔT, mean temperature of the gas T_∞, and characteristic height of the
$(T_w - T_\infty)$ or ΔT, mean temperature of the gas T_∞, and characteristic height of the
plate, L_c. Then, one may perform dimensional analysis by applying Buckingham's Pi Theorem (Section 8-6) and obtain the following dimensionless groups

$$\text{Nu} = \frac{hL_c}{k}; \quad \text{Pr} = \frac{\mu c_p}{k}; \quad \text{Gr} = \frac{g\beta\rho^2 L_c^3 \Delta T}{\mu^2} \tag{9-3}$$

The reader is already familiar with the Nusselt number, Nu, and the Prandtl number, Pr. The group referred to as Gr is the Grashof number and is of great importance in natural convection. It will be discussed in detail in the next section. One can determine the effect of changing these variables on the value of h by varying the temperature difference ΔT (through variation of power P), by introducing different gases into the tank, and by introducing plates of various heights. The results can be expressed in the form of an equation by correlating the three dimensionless groups, namely, Nu, Pr, and Gr. More sophisticated versions of this experiment are used to determine the values of convective heat transfer coefficients under a variety of conditions and provide data for verification of analytical solutions of different models.

As in forced convection, our objective in studying natural convection is to obtain information about the heat transfer coefficient, h, or the Nusselt number, Nu. From the standpoint of analysis, a mathematical model may be set up and solutions obtained by applying exact or approximate methods. To illustrate this approach, we will analyze the classical problem of natural convection from a vertical wall. Then, experimentally obtained results for various configurations will be presented.

9-3 NATURAL CONVECTION FROM A VERTICAL WALL

L. Lorenz (1881) was the first to solve the problem of heat transfer from a heated vertical wall due to natural convection. He postulated that the fluid close to the wall moves straight up vertically and that the horizontal components of velocity are negligible. Experimental work by Schmidt and Beckman (Reference 1) in 1930 demonstrated that the assumptions made by Lorenz needed refinement. This was further confirmed by Saunder's data (Reference 2) for pressures from 0.043 to 65 atmospheres. Pohlhausen (Reference 3) converted the governing partial differential equations into an ordinary differential equation and obtained a solution for air. The governing differential equations for the natural convection problem were solved by Ostrach (Reference 4) for a wide range of Prandtl numbers in 1952. Squire (Reference 3) presented an approximate solution to the problem using the integral technique.

Consider a vertical flat plate as shown in Figure 9-2. The coordinates are chosen so that x is in the streamwise direction and y is in the transverse direction. A thin layer exists adjacent to the hot surface of the vertical plate within which the variations in temperature and velocity are confined. This layer is called the *boundary layer*. Laminar flow exists within the boundary layer up to a certain height of the plate, beyond which turbulence gradually develops. An interesting feature of the velocity distribution is that the velocity is zero at the plate surface and at the outer edge of the boundary layer, while within the boundary layer it exhibits a maximum. The temperature, however, changes monotonically within the boundary layer.

It was stated earlier that the Grashof number, defined in equation (9-3), is the relevant dimensionless group for natural convection. It can be interpreted physically as the ratio of the buoyancy forces to the viscous forces. The Grashof number is to natural convection what the Reynolds number is to forced convection. In addition, the magnitude of the Grashof number serves to indicate what region the flow is in–laminar, transitional, or turbulent. The critical Grashof number for transition ranges between 10^8 and 10^9. A value of Grashof number higher than 10^9 would indicate a turbulent flow in natural convection, while a value of less than 10^8 would signify laminar flow. For air at a temperature of 80°F, the Grashof number would be about 2×10^7 for a plate of height one foot and at a temperature 10°F above or below the air temperature.

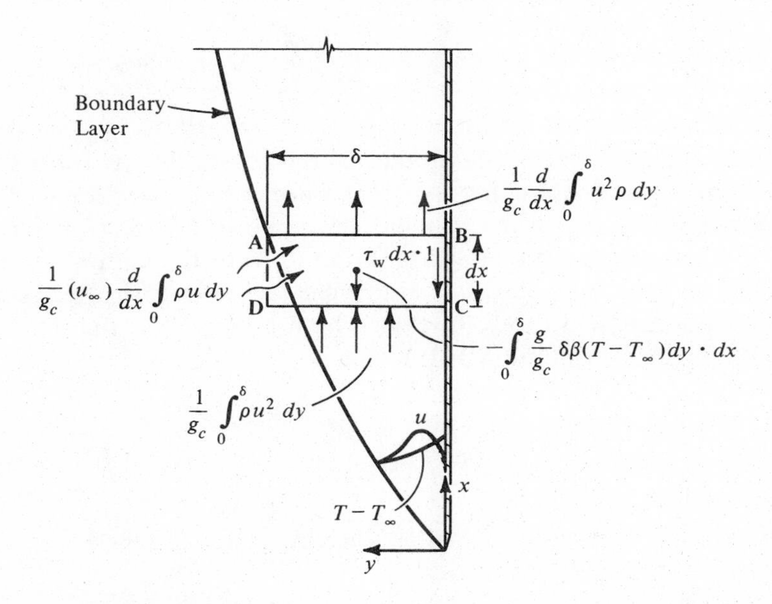

Figure 9-2 Natural convection from a vertical wall - integral analysis.

Another dimensionless quantity called the Rayleigh number is increasingly being used in the literature. It is defined as

$$\mathrm{Ra} = \mathrm{Gr}\,\mathrm{Pr} = \frac{g\beta L_c^3 \Delta T}{\nu\alpha} \tag{9-4}$$

For Prandtl numbers greater than 5, the Nusselt number is either independent of or slightly depends upon the Prandtl number if the Nusselt number is expressed as

$$\mathrm{Nu} = f(\mathrm{Re}, \mathrm{Pr}, \mathrm{Ra})$$

We will now present the integral method of solution following essentially Eckert's (Reference 5) procedure.

9-3.1 Integral Method for Natural Convection from a Vertical Wall

Consider a control volume **ABCD** that includes the entire thickness of the laminar boundary layer in natural convection, as shown in Figure 9-2. We seek the velocity and temperature distributions in the boundary layer and an expression for the Nusselt number.

The equation for conservation of momentum for flow over a flat plate with negligible pressure gradient is given by equation (7-19). It takes the following form for an elemental control volume **ABCD** of width equal to the boundary layer thickness, δ, and of height, dx

$$\tau_w\, dx = \frac{1}{g_c}\frac{d}{dx}\left[\int_0^\delta \rho u(u_\infty - u)\,dy\right] dx \tag{9-5}$$

The left-hand side of equation (9-5) represents the net force acting on the control volume shown in Figure 9-2 in the negative x direction and the right-hand side represents the net influx of x momentum into the control volume.

When we want to adapt equation (9-5) to the natural convection problem, we have to take into account additional forces due to a pressure gradient and the gravitational field (body forces). The net force, F_p, due to the pressure gradient on the control volume **ABCD** in the positive x direction is given by

$$F_p = \delta\left(\frac{dp}{dx}\right)dx = \left(\int_0^\delta \frac{dp}{dx}\,dy\right)dx$$

since (dp/dx) is assumed to be constant across the boundary layer. According to the principles of hydrostatics

$$\frac{dp}{dx} = -\rho_\infty\left(\frac{g}{g_c}\right)$$

Thus

$$F_p = \left[- \int_0^\delta \rho_\infty \left(\frac{g}{g_c} \right) dy \right] dx$$

The net body force, F_B, acting on the control volume **ABCD** in the positive x direction is given by

$$F_B = \left[\int_0^\delta \rho \left(\frac{g}{g_c} \right) dy \right] dx$$

Hence

$$F_p + F_B = \left[- \int_0^\delta \rho_\infty \left(\frac{g}{g_c} \right) dy \right] dx + \left[\int_0^\delta \rho \left(\frac{g}{g_c} \right) dy \right] dx$$

or

$$F_p + F_B = - \left[\int_0^\delta \frac{g}{g_c} (\rho_\infty - \rho) dy \right] dx \tag{9-6}$$

It is possible to eliminate the term, $(\rho_\infty - \rho)$, in equation (9-6) in favor of $(T - T_\infty)$ through the use of the coefficient of thermal expansion, β, which is defined as

$$\beta = - \frac{1}{\rho} \left(\frac{\partial \rho}{\partial T} \right)_p$$

As an approximation, the derivative $(\partial \rho / \partial T)$ is written as

$$\frac{\partial \rho}{\partial T} \simeq \frac{\rho_\infty - \rho}{T_\infty - T}$$

yielding

$$\beta \simeq - \frac{\rho_\infty - \rho}{\rho(T_\infty - T)} \tag{9-6a}$$

We may note that for an ideal gas

$$p = \rho RT \qquad \text{or} \qquad \rho = \frac{p}{RT}$$

Taking the partial derivative of ρ with respect to T, we obtain

$$\left(\frac{\partial \rho}{\partial T}\right)_p = -\frac{p}{RT^2} = -\frac{\rho}{T}$$

Hence

$$\beta = -\frac{1}{\rho}\left(\frac{\partial \rho}{\partial T}\right)_p = \frac{1}{\rho}\left(\frac{\rho}{T}\right) = \frac{1}{T} \tag{9-6b}$$

Note that T in equation (9-6b) is the absolute temperature in °R or °K. For fluids that do not obey the ideal gas law, values of β from the property tables have to be used.

Substituting for $(\rho_\infty - \rho)$ from the above into equation (9-6), there results

$$F_p + F_B = -\left[\int_0^\delta \frac{g}{g_c} \beta\rho (T_\infty - T)\, dy\right] dx \tag{9-6c}$$

Equation (9-5), where the left-hand side represents the net force acting in the negative x direction, can now be rewritten for the natural convection boundary layer to take into account the contribution of the pressure forces and body forces.

$$\tau_w\, dx - (F_p + F_B) = \frac{1}{g_c}\frac{d}{dx}\left[\int_0^\delta \rho u (u_\infty - u)\, dy\right] dx$$

Substituting for $(F_p + F_B)$ from equation (9-6c) and observing that $u_\infty = 0$ for natural convection, the above equation becomes

$$\tau_w\, dx - \frac{g}{g_c}\left[\int_0^\delta \rho\beta(T - T_\infty) dy\right] dx = -\frac{1}{g_c}\left[\frac{d}{dx}\left(\int_0^\delta \rho u^2\, dy\right)\right] dx$$

Substituting for τ_w from equation (7-2) and rearranging, we arrive at the following integral equation for the natural convection boundary layer.

$$\int_0^\delta \frac{g}{g_c} \rho\beta (T - T_\infty) dy - \frac{\mu}{g_c}\left(\frac{\partial u}{\partial y}\right)_{y=0} = \left(\frac{1}{g_c}\right)\frac{d}{dx}\left(\int_0^\delta \rho u^2 dy\right) \tag{9-7}$$

The integral energy equation, equation (8-29), is directly applicable to the present problem. The equation is

$$\frac{d}{dx}\int_0^\delta u(T_\infty - T)\,dy = \frac{k}{c_p\rho}\left(\frac{\partial T}{\partial y}\right)_{y=0} \tag{9-7a}$$

Equations (9-7) and (9-7a) are coupled equations since the unknowns, u and T, appear in both of them. As discussed earlier in this chapter, the velocity profile and temperature profile will be as depicted in Figure 9-2. The velocity profile is approximated by

$$u = U\left(\frac{y}{\delta}\right)\left[1 - \frac{y}{\delta}\right]^2 \tag{9-8}$$

where U and δ are functions of x to be determined later and have dimensions of velocity and distance, respectively. This approximation satisfies the requirement that u vanish at $y = 0$ and at $y = \delta$, and also that $(\partial u/\partial y) = 0$ at $y = \delta$. The temperature profile is approximated by

$$\frac{T - T_\infty}{T_w - T_\infty} = \Theta = \left[1 - \frac{y}{\delta}\right]^2 \tag{9-8a}$$

which satisfies the boundary conditions at $y = 0$, $T = T_w$, and at $y = \delta$, $T = T_\infty$ and $(\partial T/\partial y) = 0$.

Substitution of the velocity and temperature profiles in equations (9-7) and (9-7a) gives, respectively

$$\int_0^\delta \frac{g}{g_c}\rho\beta\left(1 - \frac{y}{\delta}\right)^2 (T_w - T_\infty)dy - \frac{\mu}{g_c}\frac{U}{\delta} = \frac{1}{g_c}\frac{d}{dx}\int_0^\delta \rho U^2\left(\frac{y^2}{\delta^2}\right)\left(1 - \frac{y}{\delta}\right)^4 dy$$

and

$$\frac{d}{dx}\int_0^\delta \left[-U\left(\frac{y}{\delta}\right)\left(1 - \frac{y}{\delta}\right)^4 (T_w - T_\infty)\right]dy = \alpha\left(-\frac{2}{\delta}\right)(T_w - T_\infty)$$

Observing that

$$\int_0^\delta \left(1 - \frac{y}{\delta}\right)^2 dy = \frac{\delta}{3}, \qquad \int_0^\delta \frac{y^2}{\delta^2}\left(1 - \frac{y}{\delta}\right)^4 dy = \frac{\delta}{105}$$

and

$$\int_0^\delta \frac{y}{\delta}\left(1 - \frac{y}{\delta}\right)^4 dy = \frac{\delta}{30}$$

we obtain

$$\frac{g}{g_c}\rho\beta(T_w - T_\infty)\frac{\delta}{3} - \frac{\mu U}{g_c\delta} = \frac{\rho}{g_c}\frac{d}{dx}\left[U^2\frac{\delta}{105}\right] \tag{9-9}$$

and

$$\frac{1}{30}(T_w - T_\infty)\frac{d}{dx}(U\delta) = 2\alpha(T_w - T_\infty)\frac{1}{\delta} \tag{9-9a}$$

A rearrangement of the above two equations gives

$$\frac{d}{dx}(U^2)\frac{\delta}{105} = g\beta(T_w - T_\infty)\frac{\delta}{3} - \frac{\nu U}{\delta} \tag{9-10}$$

and

$$\frac{1}{30}\frac{d}{dx}(U\delta) = \frac{2\alpha}{\delta} \tag{9-10a}$$

One way to solve equations (9-10) is to assume trial solutions for U and δ. Power functions of x are assumed for the unknowns, U and δ, as follows

$$U = Ax^m \qquad \delta = Bx^n \tag{9-11}$$

Substituting the above equations in equations (9-10), we get

$$\left(\frac{2m + n}{105}\right)A^2Bx^{(2m+n-1)} = g\beta(T_w - T_\infty)\frac{Bx^n}{3} - \frac{A}{B}\nu x^{(m-n)} \tag{9-12}$$

and

$$\left(\frac{m + n}{30}\right)ABx^{(m+n-1)} = \frac{2\alpha}{B}x^{-n} \tag{9-12a}$$

In equation (9-12), the power of x that appears in each terms must be identical; similarly for equation (9-12a). Therefore

$$2m + n - 1 = m - n$$

$$m + n - 1 = -n$$

The solution to the above set of equations is $m = 1/2, n = 1/4$. Thus

$$U = Ax^{1/2} \text{ and } \delta = Bx^{1/4}$$

Next, the constants A and B can be determined from equations (9-12) after substituting for m and n

$$\frac{A^2 B}{84} = g\beta(T_w - T_\infty)\frac{B}{3} - \left(\frac{A}{B}\right)\nu$$

and

$$\frac{AB}{40} = \frac{2\alpha}{B}$$

yielding

$$A = 5.17\,\nu\left(\frac{20}{21} + \frac{\nu}{\alpha}\right)^{-1/2}\left[\frac{g\beta(T_w - T_\infty)}{\nu^2}\right]^{1/2}$$

$$B = 3.93\left(\frac{20}{21} + \frac{\nu}{\alpha}\right)^{1/4}\left[\frac{g\beta(T_w - T_\infty)}{\nu^2}\right]^{-1/4}\left(\frac{\nu}{\alpha}\right)^{-1/2}$$

Hence, the boundary layer thickness is given by

$$\delta/x = 3.93\,\text{Pr}^{-1/2}\,(0.952 + \text{Pr})^{1/4}\,\text{Gr}_x^{-1/4} \tag{9-13}$$

where Pr is the Prandtl number, (ν/α), and

Gr_x is the Grashof number, $\dfrac{g\beta\,(T_w - T_\infty)x^3}{\nu^2}$

The local heat transfer coefficient, h, may be evaluated from

$$-k\left(\frac{\partial T}{\partial y}\right)_{y=0} = h(T_w - T_\infty)$$

Using the parabolic temperature profile of equation (9-8a), one obtains

$$h = \frac{2k}{\delta}$$

or

$$\frac{hx}{k} = \text{Nu} = 2\left(\frac{x}{\delta}\right) = 0.508\ \text{Pr}^{1/2}\ (0.952 + \text{Pr})^{-1/4}\ \text{Gr}_x^{1/4} \qquad \text{(9-14)}$$

The average heat transfer coefficient h_{av} is then obtained by integrating h over the height, L, of the wall since the temperature of the wall is uniform. It is

$$h_{av} = \frac{1}{L}\int_0^L h\,dy = \left(\frac{4}{3}\right)h_{x=L}$$

and

$$\frac{h_{av}L}{k} = \text{Nu}_{av} = 0.677\ \text{Pr}^{1/2}\ (0.952 + \text{Pr})^{-1/4}\ \text{Gr}_L^{1/4} \qquad \text{(9-14a)}$$

For air, with a Prandtl number Pr = 0.7, the above equations simplify to

$$\text{Nu} = 0.378\text{Gr}_x^{1/4} \qquad \text{(9-14b)}$$

$$\text{Nu}_{av} = 0.504\text{Gr}_L^{1/4} \qquad \text{(9-14c)}$$

Sample Problem 9-1 Determine the average coefficient for natural convection heat transfer for a vertical plate 1 foot high at 125°F. The surrounding air is at 75°F. Also, calculate the boundary layer thickness at $x = L$.

Solution: We need to evaluate the Grashof number and the Prandtl number and then determine the Nusselt number from equation (9-14a). The air properties to be used in the computation are those of the film temperature, that is, (1/2) (75 + 125)°F or 100°F. They are

$$\mathrm{Pr} = 0.706$$

$$\nu = 18.09 \times 10^{-5} \text{ ft}^2/\text{sec}$$

$$k = 0.0154 \text{ Btu/hr-ft °F}$$

Also

$$\Delta T = T_w - T_\infty = 125 - 75 = 50°\text{F} \quad \text{and} \quad L = 1 \text{ ft}$$

$$\beta = \frac{1}{T} = \frac{1}{460 + 100} = 1.785 \times 10^{-3} \text{ per °F}$$

$$\mathrm{Gr}_L = \frac{g\beta\Delta T L^3}{\nu^2}$$

$$= \frac{32.2 \times 1.785 \times 10^{-3} \times 50 \times 1^3}{(18.09) \times 10^{-5})^2}$$

$$= 0.88 \times 10^8$$

$$\mathrm{Nu}_{av} = 0.677\, \mathrm{Pr}^{1/2} (0.952 + \mathrm{Pr})^{-1/4}\, \mathrm{Gr}_L^{1/4}$$

$$= 0.677 \times (0.706)^{1/2} (0.952 + 0.706)^{-1/4} (0.88 \times 10^8)^{1/4}$$

$$= 48.5$$

$$h_{av} = \frac{\mathrm{Nu}_{av} k}{L} = \frac{48.5 \times 0.0154}{1} = 0.75 \text{ Btu/hr-ft}^2\text{°F}$$

The boundary layer thickness is given by equation (9-13)

$$\frac{\delta}{L} = 3.93\, \mathrm{Pr}^{-1/2} (0.952 + \mathrm{Pr})^{1/4}\, \mathrm{Gr}_L^{-1/4}$$

Substituting for L, Pr, and Gr_L gives

$$\delta = 1 \times 3.93 \times (0.706)^{-1/2} (0.952 + 0.706)^{1/4} (0.88 \times 10^8)^{-1/4}$$

or

$$\delta = 0.055 \text{ ft} \quad \text{or} \quad 0.66 \text{ inch}$$

The product of the Grashof number and the Prandtl number for this problem is 0.62×10^8, below the critical value of 10^9. Hence use of the equation for laminar flow is justified.

9-3.2 Differential Formulation for Natural Convection from a Vertical Wall

We consider an elemental control volume of dimensions dx by dy and of unit depth into the plane of the paper (Figure 9-3) and proceed to apply the principles of conservation of mass, momentum, and energy. The continuity equation is exactly the same as that obtained in Chapter 7, namely

$$\frac{\partial u}{\partial x} + \frac{\partial v}{\partial y} = 0 \tag{9-15}$$

The forces acting on the elemental volume include pressure forces, viscous forces, and body forces. In Chapter 8, the body forces were neglected, whereas in natural convection they are of prime importance. The body force acting on a unit volume is $\rho(g/g_c)$. The acceleration terms are the same as those in Section 7-5.2. Thus, the momentum equation, equation (7-37), takes the following form for the elemental control volume.

$$\left[u\left(\frac{\partial u}{\partial x}\right) + v\left(\frac{\partial u}{\partial y}\right)\right]\left(\frac{\rho}{g_c}\right) dxdy = \left(-\frac{\partial p}{\partial x}\right) dxdy - \rho\left(\frac{g}{g_c}\right) dxdy + \frac{\mu}{g_c}\left(\frac{\partial^2 u}{\partial y^2}\right) dxdy$$

or

$$\frac{\rho}{g_c}\left[u\left(\frac{\partial u}{\partial x}\right) + v\left(\frac{\partial u}{\partial y}\right)\right] = -\frac{\partial p}{\partial x} - \rho\left(\frac{g}{g_c}\right) + \frac{\mu}{g_c}\left(\frac{\partial^2 u}{\partial y^2}\right) \tag{9-16}$$

Since there are no externally imposed pressure gradients, the changes in the pressure are due only to the elevation or height of the column of fluid. According to the principles of hydrostatics

$$\frac{\partial p}{\partial x} = -\rho_\infty\left(\frac{g}{g_c}\right)$$

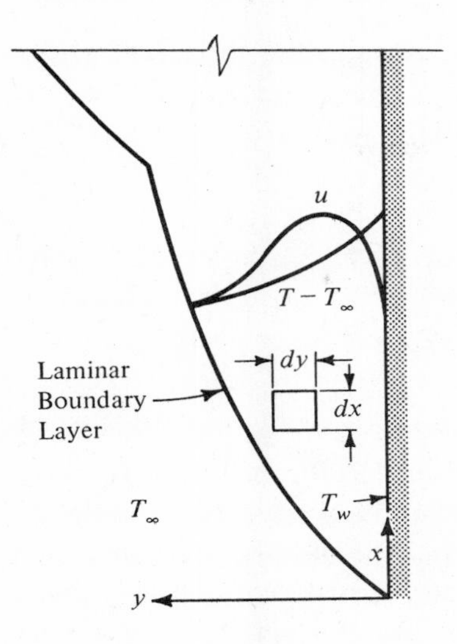

Figure 9-3 Natural convection from a vertical wall - differential analysis.

Substituting the above in equation (9-16), we obtain

$$\frac{\rho}{g_c}\left[u\left(\frac{\partial u}{\partial x}\right) + v\left(\frac{\partial u}{\partial y}\right)\right] = \frac{g}{g_c}(\rho_\infty - \rho) + \frac{\mu}{g_c}\left(\frac{\partial^2 u}{\partial y^2}\right) \tag{9-17}$$

The forgoing equation contains, in addition to the velocity variables, u and v, the density as yet another variable. It is possible to eliminate the term $(\rho_\infty - \rho)$ in favor of $(T - T_\infty)$ through the use of the coefficient of thermal expansion, β. Substituting for $(\rho_\infty - \rho)$ from equation (9-5a) into equation (9-17), we have

$$\rho\left[u\left(\frac{\partial u}{\partial x}\right) + v\left(\frac{\partial u}{\partial y}\right)\right] = g\rho\beta(T - T_\infty) + \mu\left(\frac{\partial^2 u}{\partial y^2}\right) \tag{9-18}$$

The energy equation for natural convection is the same as the simplified energy equation for forced convection, equation (8-40), thus the appropriate energy equation is

$$u\left(\frac{\partial T}{\partial x}\right) + v\left(\frac{\partial T}{\partial y}\right) = \alpha\left(\frac{\partial^2 T}{\partial y^2}\right) \tag{9-19}$$

In order to obtain a solution to the problem of natural convection from a vertical wall, one has to solve equations (9-15), (9-18), and (9-19) simultaneously with the following boundary conditions:

$$\text{at } y = 0, \quad T = T_w, \quad u = 0 \tag{9-20}$$

$$\text{as } y \to \infty, \quad T \to T_\infty, \quad u = 0 \tag{9-20a}$$

E. Pohlhausen (Reference 1) solved these equations for air by introducing a stream function and a similarity parameter. Figure 9-4 shows the numerical results of Pohlhausen and the experimental data due to Schmidt and Beckman (Reference 1) for the velocity profile while Figure 9-5 shows the comparison for the temperature profile. Figure 9-4 verifies that the velocity profile indeed exhibits a maximum value within the boundary layer.

As noted earlier, Ostrach solved the governing differential equations for a range of Prandtl numbers. His results are presented in Figures 9-6 and 9-7 for dimensionless velocity and temperature profiles.

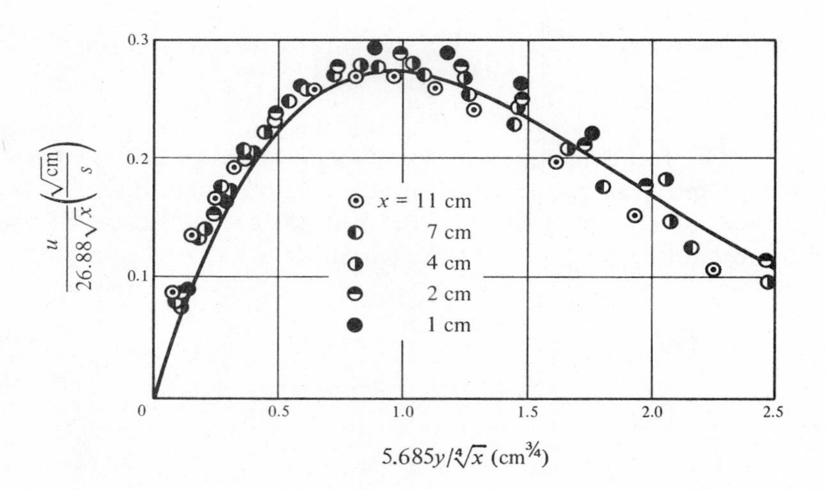

Figure 9-4 Velocity profiles in natural convection of air on a vertical plate - analytical results of Pohlhausen and experimental work of Schmidt and Beckman, from *Analysis of Heat & Mass Transfer* by Eckert, E.R.G. and R. Drake, Jr. Copyright 1972, by McGraw-Hill, Inc. Used with permission.

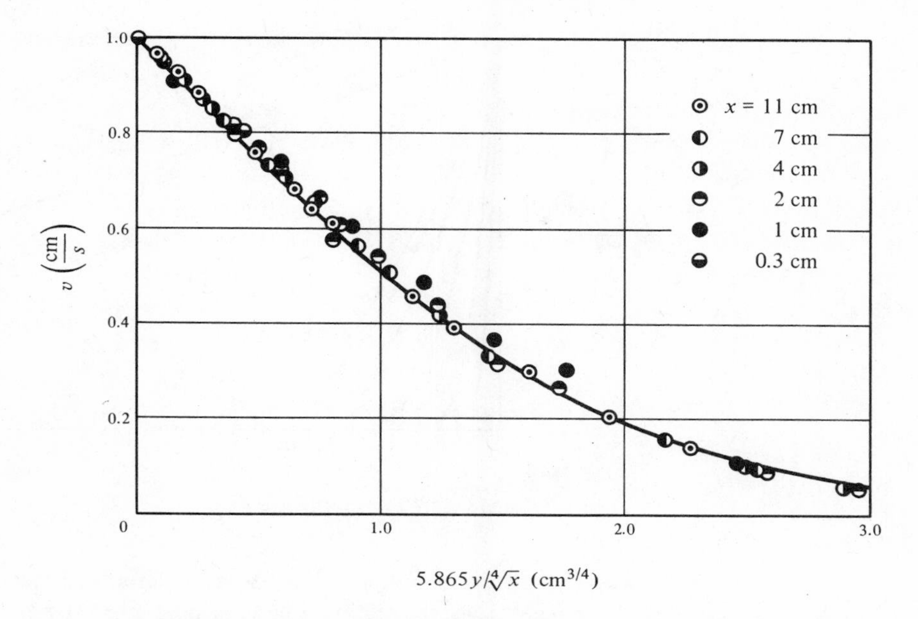

Figure 9-5 Temperature profiles in natural convection of air on a vertical plate - analytical results of Pohlhausen and experimental work of Schmidt and Beckman, from *Analysis of Heat & Mass Transfer* by Eckert, E.R.G. and R. Drake, Jr. Copyright 1972, by McGraw-Hill, Inc. Used with permission.

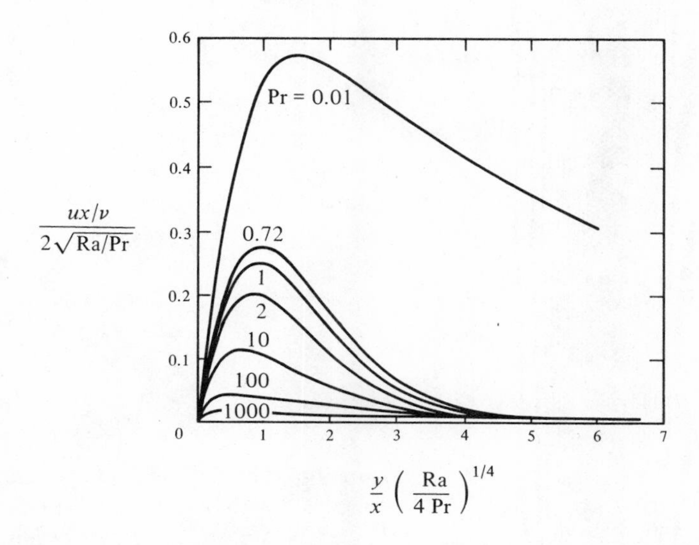

Figure 9-6 Analytical results for velocity profiles in laminar free convection on a vertical plate from S. Ostrach, *NACA* TN 2635, 1952.

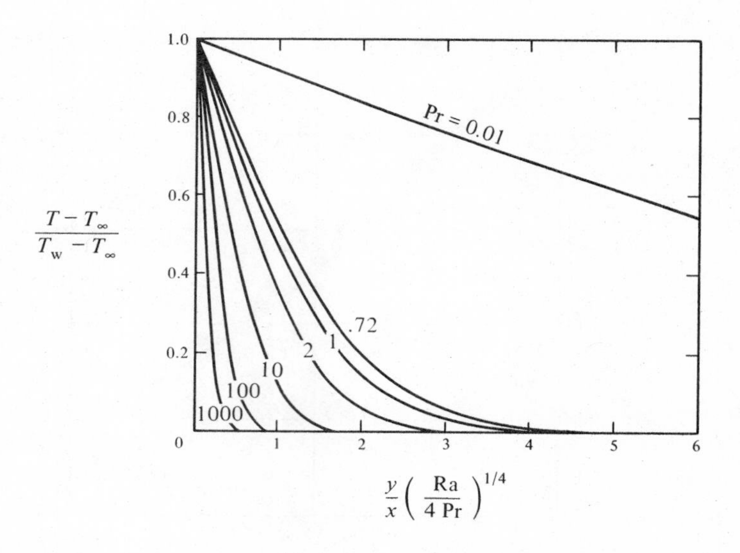

Figure 9-7 Analytical results for temperature profiles in laminar free convection on a vertical plate from S. Ostrach, *NACA* TN 2635, 1952.

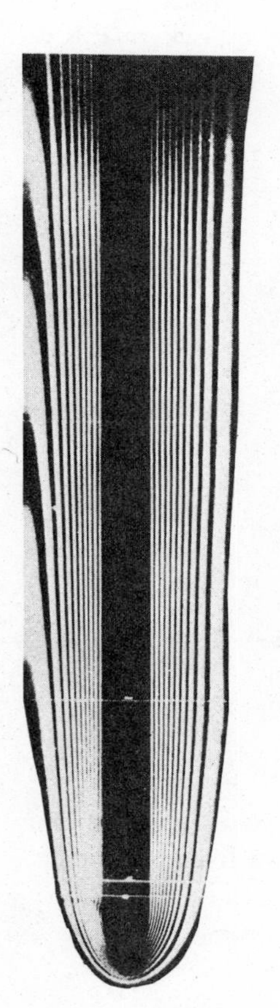

Figure 9-8 Lines of constant temperature around a heated vertical plate due to natural convection flow of air. Courtesy of Professor E. Soehngen.

9-4 INTERFEROMETER STUDIES OF NATURAL CONVECTION

Analytical solutions for many configurations are difficult and complex, and, therefore, much of the information used in design comes from experimental studies. The interferometer technique has been extensively used to determine temperature fields. A photograph taken using this technique is shown in Figure 9-8. The dark lines represent isotherms. Therefore, a region of tightly packed dark lines represents a region of rapidly changing temperature, and, conversely, a region of widely spaced dark lines represents a region of gradually changing temperature. In Figure 9-8, which depicts the temperature field in natural convection around a vertical plate, we observe rapidly changing temperatures near the bottom edge. We also observe that the boundary layer is thick in natural convection, and that the thickness of the boundary layer at $x = 0$ is not zero as implied by the analytical solution (Figure 9-6) to the natural convection problem.

In Figure 9-9, we see that the temperature field around a heated horizontal cylinder is not axisymmetric. The constant temperature lines tend to pull away from the top of the cylinder.

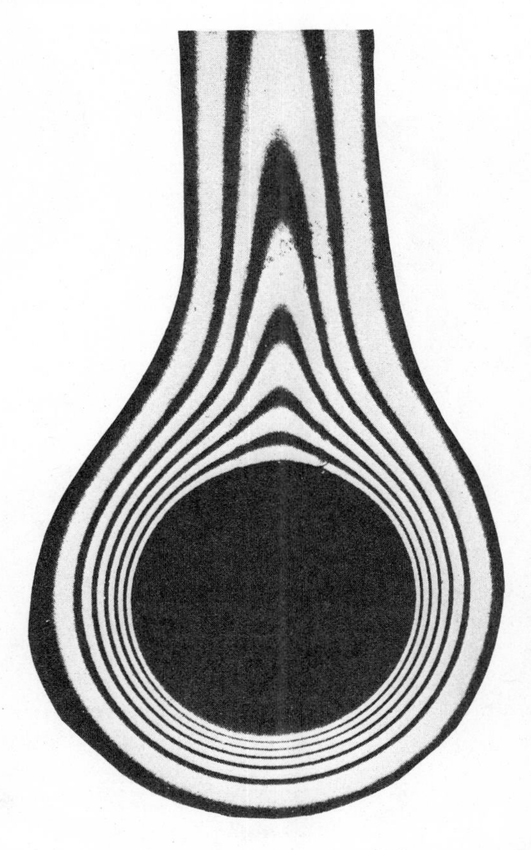

Figure 9-9 Lines of constant temperature around a heated horizontal cylinder due to natural convection flow of air. Courtesy of Professor E. Soehngen.

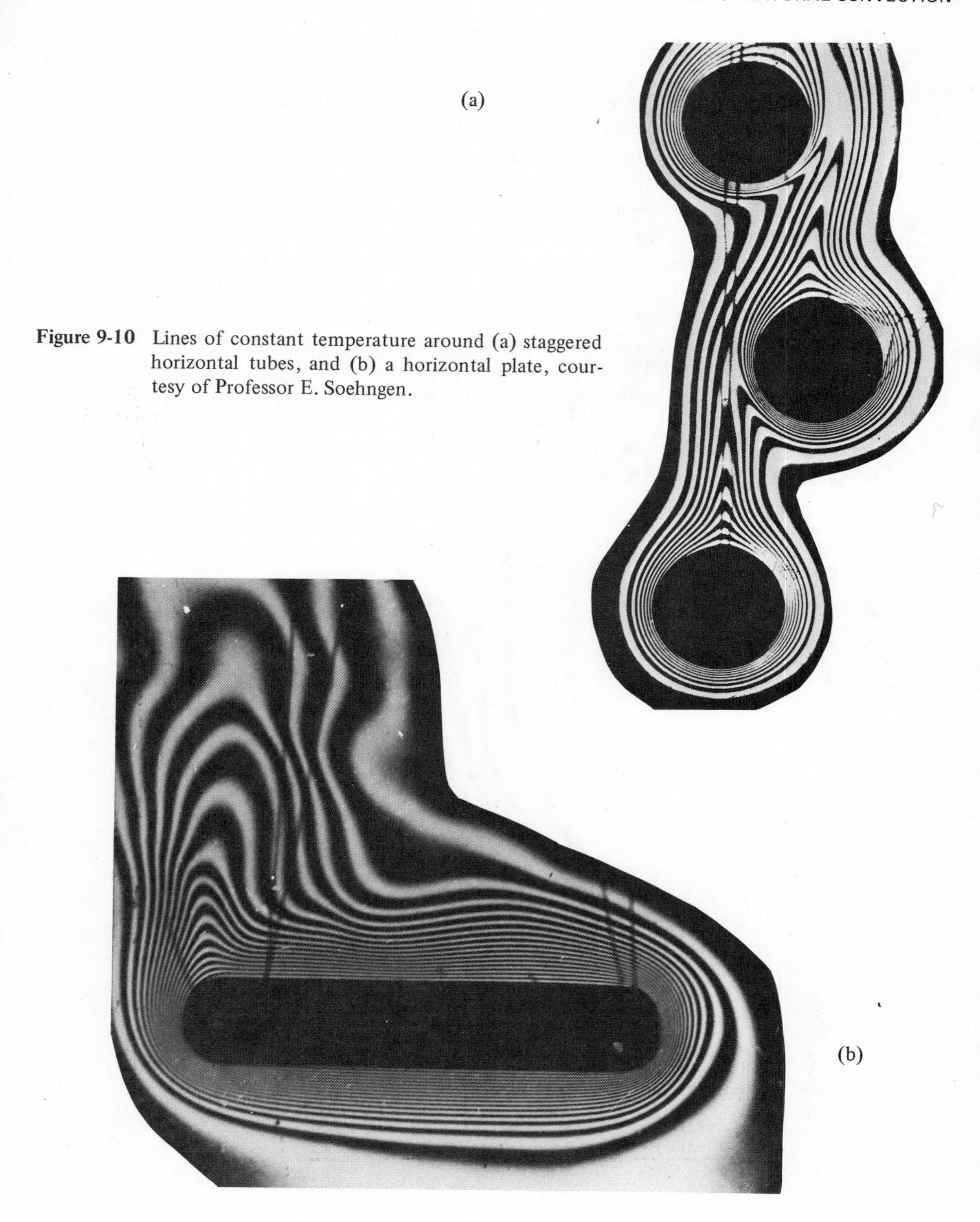

Figure 9-10 Lines of constant temperature around (a) staggered horizontal tubes, and (b) a horizontal plate, courtesy of Professor E. Soehngen.

Interferometer photographs of flow due to natural convection from staggered horizontal tubes and a horizontal flat plate are shown in Figure 9-10a and b, respectively. The interference in the flow field caused by the tubes is effectively demonstrated in Figure 9-10a.

9-5 DESIGN CORRELATIONS FOR NATURAL CONVECTION

Most of the correlations used for heat transfer design calculations involving natural convection are based on experimental results. The correlations consist of a functional relationship between the Nusselt number and the product of the Grashof number and the Prandtl number. As noted earlier, the product of the Grashof number and the Prandtl number is referred to as the Rayleigh number, Ra. We have already seen in the analysis of natural convection over a vertical wall that the aforementioned dimensionless groups are the most pertinent ones. The relationship is usually of the form:

$$\mathrm{Nu_{av}} = C\,(\mathrm{Gr\ Pr})^n = C\,\mathrm{Ra}^n$$

For isothermal walls the exponent n is typically 1/4 for laminar flow and is 1/3 for turbulent flow. In the latter case, it does not matter which dimension is used as the characteristic length in Nu and Gr, since the lengths in Nu and in $(\mathrm{Gr\ Pr})^{1/3}$ cancel out. The fluid properties like viscosity, thermal conductivity, etc., entering into these groups are evaluated at a temperature that is the arithmetic mean of the wall temperature, T_w, and the ambient fluid temperature, T_∞. This average temperature is called the *film temperature,* T_f.

Vertical plates, horizontal plates, cylinders, spheres, and enclosed spaces are some of the configurations that are of importance to a heat transfer engineer. Working formulas for these configurations are presented below for surfaces at a constant temperature.

9-5.1 Vertical Flat Plate

Correlations for heat transfer due to natural convection from isothermal vertical plates, recommended by McAdams (Reference 6), have been used extensively. [See Figure 9-11.] They are based on the height of the plate as the characteristic dimension, L_c. The relations for different ranges of the Rayleigh number are:

$$\mathrm{Nu_{av}}: \quad \text{Use Figure 9-11a} \quad \text{for } 10^{-1} < (\mathrm{Ra}_L) < 10^4 \qquad (9\text{-}21)$$

$$\mathrm{Nu_{av}} = 0.59(\mathrm{Ra}_L)^{1/4} \quad \text{for } 10^4 < (\mathrm{Ra}_L) < 10^9 \qquad (9\text{-}21a)$$

$$\mathrm{Nu_{av}} = 0.10(\mathrm{Ra}_L)^{1/3} \quad \text{for } 10^9 < (\mathrm{Ra}_L) < 10^{12} \qquad (9\text{-}21b)$$

For air at atmospheric pressure and at a film temperature, $T_f \simeq 300°F$ or $150°C$, the above become

$$\left.\begin{aligned} h_{av} &= 0.29\,(\Delta T/L)^{1/4} \text{ (English units)} \\ h_{av} &= 1.42\,(\Delta T/L)^{1/4} \text{ (SI units)} \end{aligned}\right\} \text{ for } 10^4 < (\text{Ra}_L) < 10^9 \quad (9\text{-}21c)$$

$$\left.\begin{aligned} h_{av} &= 0.19\,(\Delta T/L)^{1/3} \text{ (English units)} \\ h_{av} &= 1.32\,(\Delta T/L)^{1/3} \text{ (SI units)} \end{aligned}\right\} \text{ for } 10^9 < (\text{Ra}_L) < 10^{12} \quad (9\text{-}21d)$$

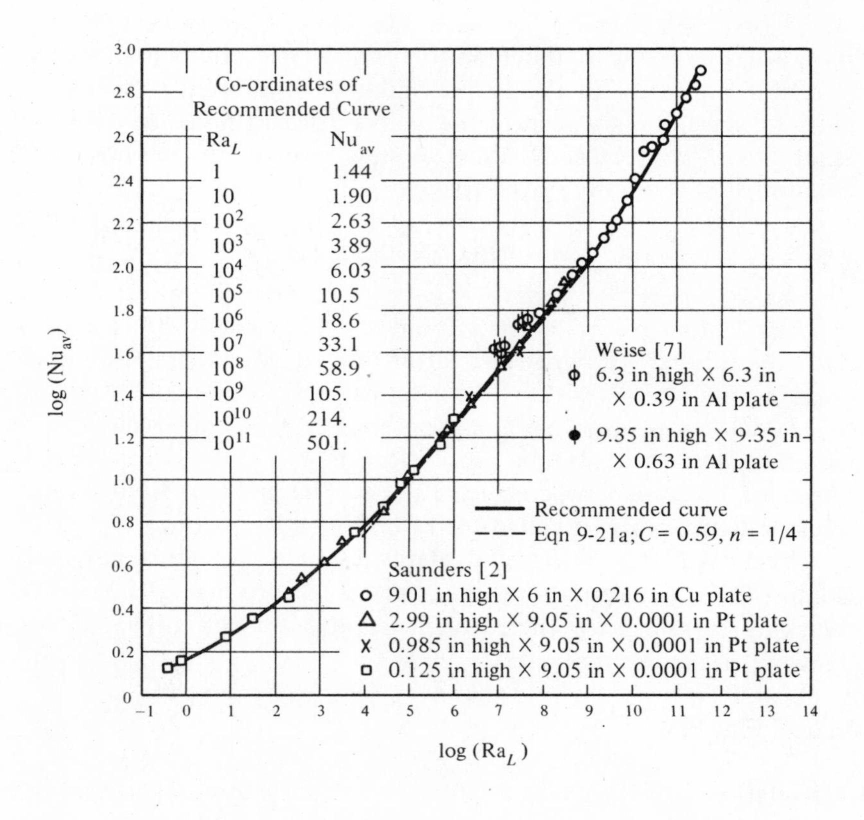

(a)

Figure 9-11 Correlation for natural convection heat transfer from vertical plates (a) due to McAdams [6] from *Heat Transmission* by W. H. McAdams. Copyright 1954 by W. H. McAdams. Used with permission. (b) and (c) due to Churchill and Chu [10].

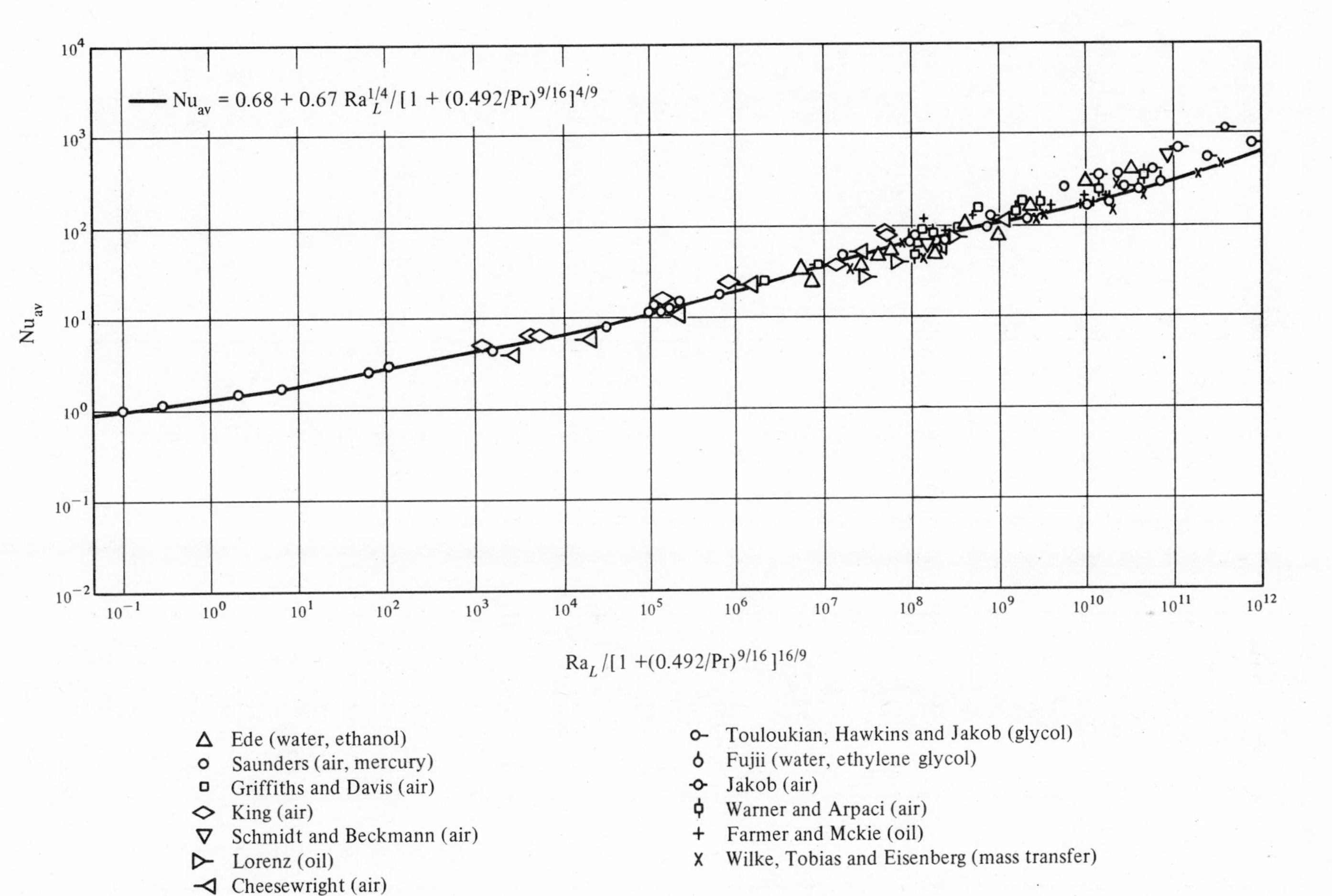

Figure 9-11 Continued

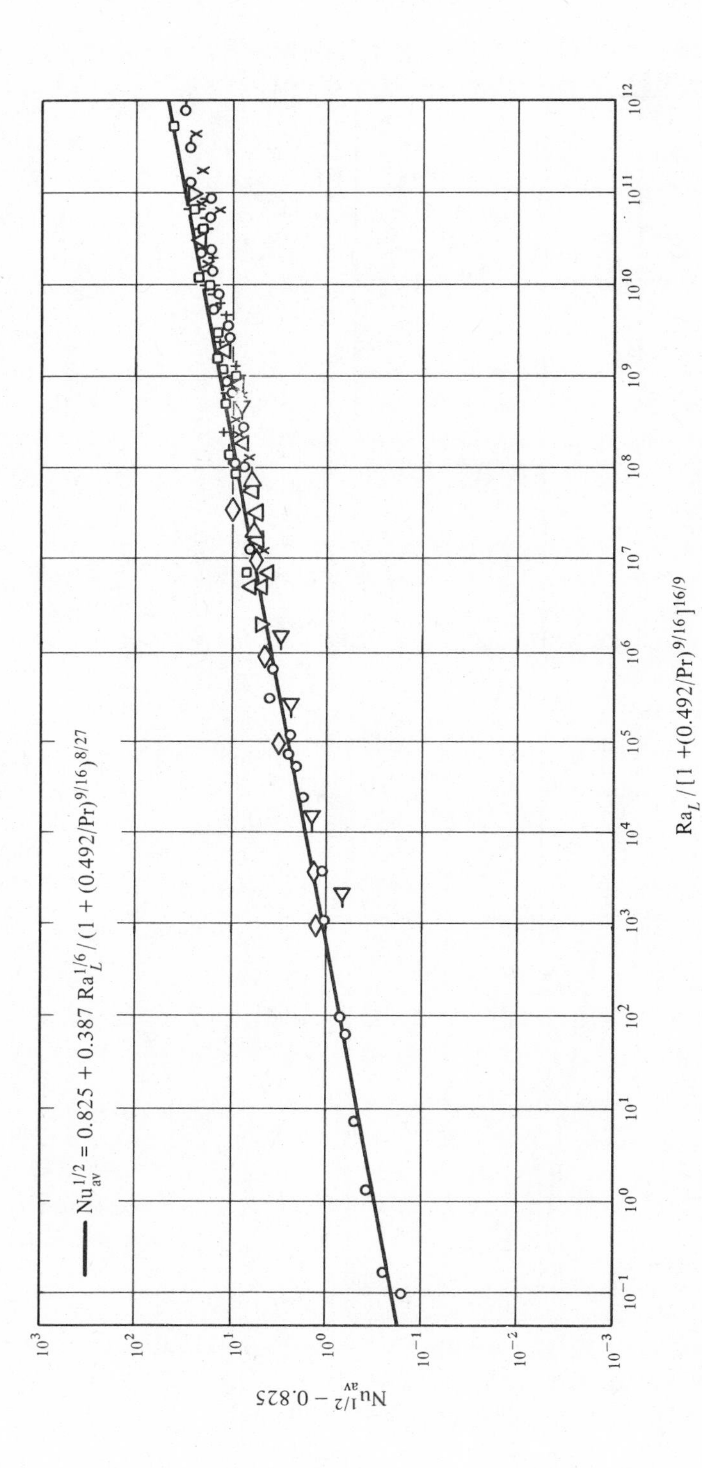

(c)

Figure 9-11 Continued

Recently Churchill and Chu (Reference 10) have demonstrated that the correlations, given below, fit experimental data very well.

$$\mathrm{Nu_{av}} = 0.68 + \frac{0.670\ \mathrm{Ra}_L^{1/4}}{[1 + (0.492/\mathrm{Pr})^{9/16}]^{4/9}} \qquad \text{for } \mathrm{Ra}_L < 10^9 \qquad (9\text{-}21e)$$

$$\mathrm{Nu_{av}} = \left[0.825 + \frac{0.387\ \mathrm{Ra}_L^{1/6}}{[1 + (0.492/\mathrm{Pr})^{9/16}]^{8/27}}\right]^2 \qquad \text{for } \mathrm{Ra}_L > 10^9 \qquad (9\text{-}21f)$$

***Sample Problem** 9-2* Determine the average heat transfer coefficient, h_{av} , for natural convection for the data of Sample Problem 9-1 but using equations (9-21).

Solution:

$$\mathrm{Gr}_L = 0.88 \times 10^8, \ \mathrm{Pr} = 0.706, \ (\mathrm{Gr}_L\,\mathrm{Pr}) = \mathrm{Ra}_L = 0.62 \times 10^8 < 10^9$$

Using equation (9-21a) gives

$$\mathrm{Nu_{av}} = 0.59(\mathrm{Ra}_L)^{1/4} = 88.1$$

and, hence

$$h_{av} = \frac{\mathrm{Nu_{av}}k}{L} = 0.81\ \mathrm{Btu/hr\text{-}ft^2\,°F}$$

If equation (9-21c) for air is used, we obtain

$$h_{av} = 0.29\,(\Delta T/L)^{1/4} = 0.29(50/1)^{1/4} = 0.77\ \mathrm{Btu/hr\text{-}ft^2\ °F.}$$

This value of h is not quite the same as that obtained from equation (9-21a), because the properties of air are evaluated at a somewhat different temperature. If equation (9-21e) is used, we obtain $\mathrm{Nu_{av}} = 46.3$ and $h_{av} = 0.7$ Btu/hr-ft^2 °F. Of the three values of h_{av} , we expect the last value to be more reliable.

***Sample Problem** 9-3* If ΔT is 200°F for natural convection from a vertical wall that is 2 feet in height and if the mean temperature of the air, T_f, is 100°F, determine the heat transfer coefficient, h.

Solution: Relative to Sample Problem 9-1, increasing ΔT from 50°F to 200°F and doubling the length, L, will result in

$$(Gr_L Pr) = (200/50) \times (2/1)^3 \times 0.62 \times 10^8 = 1.99 \times 10^9$$

Using equation (9-21d), we get

$$h_{av} = 0.19(\Delta T/L)^{1/3} = 0.88 \text{ Btu/hr-ft}^2 \text{ °F}$$

Use of equation (9-21f) gives

$$h_{av} = 1.17 \text{ Btu/hr-ft}^2 \text{ °F}$$

9-5.2 Horizontal Flat Plate

The value of the Nusselt number for heat transfer from horizontal surfaces depends on whether the plate is hot or cold in relation to the ambient fluid. When the direction of the heat flow is the same as the direction of the buoyancy force, which is vertically upward, the value of the Nusselt number is greater than that for the case where the directions of heat flow and of the bouyancy force are opposite.

Until recently the characteristic length for correlating the data was (a) the length of the side for a square, (b) the mean of the two sides for a rectangle, and (c) $0.9D$ for a circular disk of diameter D. Goldstein, Sparrow, and Jones (Reference 11) recently proposed the following definition of characteristic length.

$$L_c = \frac{\text{surface area of the flat plate}}{\text{perimeter of the flat plate}} \tag{9-22}$$

They showed that the use of the characteristic length as defined by the above equation makes it possible to correlate the data for a square plate, a circular plate, and a long rectangular plate of aspect ratio 7:1 by a single equation.
It was further demonstrated by Lloyd and Moran (Reference 12) that the above definition is applicable to symmetrical shapes such as a rectangle, a square, or a circle as well as to nonsymmetrical shapes such as a right triangle.

Hot Horizontal Plate Facing Upward A hot plate, with heating power turned on, belongs to this group. Correlations for this class of problems are due to Lloyd and Moran (Reference 12). The correlations are based on their recent experimental work employing electrochemical techniques, and they are given below.

$$Nu_{av} = 0.54 \, Ra^{1/4} \quad \text{for } 2.6 \times 10^4 < Ra < 10^7 \tag{9-23}$$

$$Nu_{av} = 0.15 \, Ra^{1/3} \quad \text{for } 10^7 < Ra < 3 \times 10^{10} \tag{9-23a}$$

The characteristic length to be used in Ra in the forgoing equations is defined by equation (9-22). For air at atmospheric pressure and at a film temperature of approximately 300°F (150°C), these become

$$\left.\begin{aligned} h_{av} &= 0.27(\Delta T/L_c)^{1/4} \text{ (English units)} \\ h_{av} &= 1.32(\Delta T/L_c)^{1/4} \text{ (SI units)} \end{aligned}\right\} \text{for } 2.6 \times 10^4 < Ra < 10^7 \qquad (9\text{-}23b)$$

$$\left.\begin{aligned} h_{av} &= 0.24(\Delta T)^{1/3} \quad \text{(English units)} \\ h_{av} &= 1.67\,(\Delta T)^{1/3} \quad \text{(SI units)} \end{aligned}\right\} \text{for} \quad 10^7 < Ra < 3 \times 10^{10} \qquad (9\text{-}23c)$$

Use of correlations given by McAdams (Reference 6) results in slightly smaller values of the Nusselt number than those obtained from equations (9-23).

Hot Horizontal Plate Facing Downward The correlations given below are recommended by McAdams (Reference 6). The characteristic length is that described in the paragraph preceding equation (9-22).

$$Nu_{av} = 0.27\,Ra^{1/4} \quad \text{for } 3 \times 10^5 < Ra < 3 \times 10^{10} \qquad (9\text{-}24)$$

For air under the same conditions as those for equations (9-23), the above becomes

$$\left.\begin{aligned} h_{av} &= 0.12(\Delta T/L_c)^{1/4} \text{ (English units)} \\ h_{av} &= 0.59\,(\Delta T/L_c)^{1/4} \text{ (SI units)} \end{aligned}\right\} \text{for } 3 \times 10^5 < Ra < 3 \times 10^{10} \qquad (9\text{-}24a)$$

Cold Horizontal Plate Facing Downward The ceiling of a room situated under the attic on a cold day and being warmed by air in the room will belong to this class of problems. Equations (9-23) and (9-23a) would be used for determining the Nusselt number.

Cold Horizontal Place Facing Upward Calculations for heat transfer from air to a slab of ice cream would require use of equation (9-24). Equation (9-24) should be used for determining the Nusselt number for cold horizontal plates facing upward.

9-5.3 Inclined Surfaces

Fuji and Imura (Reference 13) performed experiments on natural convection heat transfer from a plate of length, L, with an inclination angle, θ, ranging from 0° to 89° by employing deaerated pure water. The correlation based on their data for a downward-facing hot plate is

$$Nu_{av} = 0.56\,(Ra_L \cos\theta)^{1/4} \qquad 10^5 < Ra_L \cos\theta < 10^{11} \qquad (9\text{-}24b)$$

where θ is the angle of inclination of the plate with the vertical. Their data for a plate facing downward shows scatter, and a single correlation does not seem possible.

9-5.4 Rectangular Blocks

The characteristic dimension, L_c, is given by

$$L_c = \frac{L_h L_v}{L_h + L_v} \tag{9-25}$$

where L_h is the longer of the two horizontal dimensions and L_v is the vertical dimension. The correlation, due to King (Reference 14), is

$$\text{Nu}_{av} = 0.55\ \text{Ra}^{1/4} \quad \text{for } 10^4 < \text{Ra} < 10^9 \tag{9-25a}$$

The above correlation may also be employed for a short vertical solid cylinder. For air, the above becomes

$$\left.\begin{aligned} h_{av} &= 0.27(\Delta T/L_c)^{1/4} \text{ (English units)} \\ h_{av} &= 1.32(\Delta T/L_c)^{1/4} \text{ (SI units)} \end{aligned}\right\} \quad \text{for } 10^4 < \text{Ra} < 10^9 \tag{9-25b}$$

Sample Problem 9-4 A thin, 16-cm diameter horizontal plate is maintained at 130°C in a large body of water at 70°C. The plate convects heat from both its top and bottom surfaces. Determine the rate of heat input into the plate necessary to maintain the temperature of 130°C.

Solution: The rate of heat input into the plate, under steady-state conditons, equals the sum of heat lost by natural convection from the two surfaces of the plate. Thus, we need to determine the values of the heat transfer coefficients, $h_{av,\,t}$ and $h_{av,\,b}$ for the top and the bottom surfaces. The film temperature is (1/2) (130 + 70) or 100°C.

$$\rho = 960.63 \text{ kg/m}^3 \qquad \beta = 0.75 \times 10^{-3} \text{ per °K}$$

$$c_p = 4.216 \times 10^3 \text{ J/kg°K} \qquad \nu = 0.294 \times 10^{-6} \text{ m}^2\text{/sec}$$

$$k = 0.68 \text{ W/m°K} \qquad \alpha = 1.680 \times 10^{-7} \text{ m}^2\text{/sec}$$

The characteristic dimension, L_c, to be used is given by equation (9-22) as

$$L_c = \frac{(\pi/4)D^2}{\pi D} \quad \text{or} \quad L_c = D/4 = 16/4 = 4 \text{ cms}$$

$$\text{Pr} = \frac{\nu}{\alpha} = \frac{0.294 \times 10^{-6}}{1.68 \times 10^{-7}} = 1.75$$

$$\text{Gr} = \frac{g\beta\Delta T L_c^3}{\nu^2} = \frac{9.82 \times 0.75 \times 10^{-3} \times (130 - 70)(0.04)^3}{(0.294 \times 10^{-6})^2}$$

$$= 3.27 \times 10^8$$

$$\text{Ra} = \text{Gr Pr} = 3.27 \times 10^8 \times 1.75 = 5.72 \times 10^8 > 10^7$$

For the top surface, the use of equation (9-23a) gives

$$\text{Nu}_{av,t} = 0.15\,(\text{Ra})^{1/3} = 0.15(5.72 \times 10^8)^{1/3} = 124.5$$

resulting in

$$h_{av,t} = \frac{124.5 \times 0.68}{0.04} = 2117 \text{ W/m}^2\,°\text{K}$$

For the bottom surface, we use equation (9-24) to get

$$\text{Nu}_{av,b} = 0.27(5.72 \times 10^8)^{1/4} = 41.8$$

giving

$$h_{av,b} = \frac{41.8 \times 0.68}{0.04} = 710 \text{ W/m}^2\,°\text{K}$$

The rate of heat dissipation, Q, from the plate is obtained by summing the contributions from the top and the bottom surfaces.

$$Q = [h_{av,t}(T_w - T_\infty) + h_{av,b}(T_w - T_\infty)]\frac{\pi}{4}D^2$$

$$= [2117(130 - 70) + 710(130 - 70)]\frac{\pi}{4}(0.16)^2$$

$$= 3410 \text{ W or } 3.41 \text{ kW}$$

Consequently, the rate of heat input necessary to maintain the plate at 130°C is 3.41 kW.

9-5.5 Vertical Cylinders

The characteristic dimension for vertical cylinders is the height, L, of the cylinder. The heat transfer from a vertical cylinder can be calculated using the relations for a vertical plate provided the boundary layer thickness is small compared to the diameter, D, of the cylinder. The criterion for treating a vertical cylinder like a vertical plate as given by Gebhart (Reference 15) is

$$\frac{D}{L} \geqslant \frac{35}{(\mathrm{Gr}_L)^{1/4}} \tag{9-26}$$

For convenience, equations (9-21) are repeated here

$$\mathrm{Nu}_{av}: \quad \text{Use Figure 9-11a} \quad \text{for } 10^{-1} < (\mathrm{Ra}_L) < 10^4 \tag{9-21}$$

$$\mathrm{Nu}_{av} = 0.59(\mathrm{Ra}_L^{1/4}) \quad \text{for } 10^4 < (\mathrm{Ra}_L) < 10^9 \tag{9-21a}$$

$$\mathrm{Nu}_{av} = 0.10(\mathrm{Ra}_L^{1/3}) \quad \text{for } 10^9 < (\mathrm{Ra}_L) < 10^{12} \tag{9-21b}$$

For air:

$$\left.\begin{aligned} h_{av} &= 0.29\,(\Delta T/L)^{1/4} \text{ (English units)} \\ h_{av} &= 1.42\,(\Delta T/L)^{1/4} \text{ (SI units)} \end{aligned}\right\} \text{for } 10^4 < (\mathrm{Ra}_L) < 10^9 \tag{9-21c}$$

$$\left.\begin{aligned} h_{av} &= 0.19\,(\Delta T/L)^{1/3} \text{ (English units)} \\ h_{av} &= 1.32\,(\Delta T)^{1/3} \quad \text{(SI units)} \end{aligned}\right\} \text{for } 10^9 < (\mathrm{Ra}_L) < 10^{12} \tag{9-21d}$$

Minkowycz and Sparrow (Reference 16) obtained analytical results for isothermal vertical cylinders for which condition (9-26) is not fulfilled. They applied the local nonsimilarity method and obtained a numerical solution for $\mathrm{Pr} = 0.733$. They have shown that for $(D/L)(\mathrm{Gr}_L)^{1/4} = 0.6$, the overall heat transfer rate from an isothermal vertical cylinder is as much as four times that calculated for a vertical plate.

9-5.6 Horizontal Cylinders

The characteristic dimension for natural convective heat transfer from horizontal cylinders is the diameter, D, of the cylinder. McAdams (Reference 6) has compiled the data of a number of workers in the field. Correlations recommended by him are given below.

$$\mathrm{Nu}_{av} = 0.4 \quad \text{for } 0 < (\mathrm{Ra}_D) < 10^{-5} \tag{9-27}$$

$$\mathrm{Nu}_{av}: \quad \text{Use Figure 9-12} \quad \text{for } 10^{-5} < (\mathrm{Ra}_D) < 10^4 \tag{9-27a}$$

$$\text{Nu}_{\text{av}} = 0.53\,(\text{Ra}_D)^{1/4} \qquad \text{for } 10^4 < (\text{Ra}_D) < 10^9 \tag{9-27b}$$

$$\text{Nu}_{\text{av}} = 0.13\,(\text{Ra}_D)^{1/3} \qquad \text{for } 10^9 < (\text{Ra}_D) < 10^{12} \tag{9-27c}$$

For air, the forgoing equations become

$$\left.\begin{aligned} h_{\text{av}} &= 0.0063/D \quad \text{(English units)} \\ h_{\text{av}} &= 0.011/D \quad \text{(SI units)} \end{aligned}\right\} \text{for } 0 < (\text{Ra}_D) < 10^{-5} \tag{9-28}$$

$$h_{\text{av}}: \quad \text{Use Figure 9-12} \qquad \text{for } 10^{-5} < (\text{Ra}_D) < 10^4 \tag{9-28a}$$

$$\left.\begin{aligned} h_{\text{av}} &= 0.27\,(\Delta T/D)^{1/4} \quad \text{(English units)} \\ h_{\text{av}} &= 1.32\,(\Delta T/D)^{1/4} \quad \text{(SI units)} \end{aligned}\right\} \text{for } 10^4 < (\text{Ra}_D) < 10^9 \tag{9-28b}$$

$$\left.\begin{aligned} h_{\text{av}} &= 0.18\,(\Delta T)^{1/3} \quad \text{(English units)} \\ h_{\text{av}} &= 1.25\,(\Delta T)^{1/3} \quad \text{(SI units)} \end{aligned}\right\} \text{for } 10^9 < (\text{Ra}_D) < 10^{12} \tag{9-28c}$$

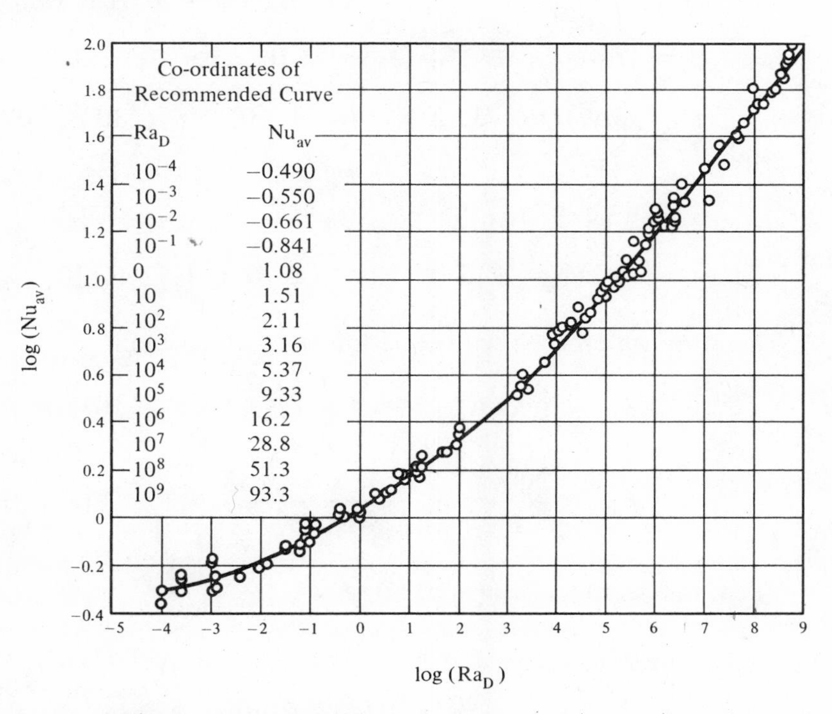

Figure 9-12 Correlations for natural convection heat transfer from horizontal cylinders, from *Heat Transmission* by W. H. McAdams. Copyright 1954, by W. H. McAdams. Used with permission.

The error involved in the use of equations (9-27) is ±15 percent. For low Prandtl numbers, a correlation due to Hyman, Bonilla, and Ehrlich (Reference 17) is believed to give a better result than equations (9-27). It is

$$\mathrm{Nu}_{av} = 0.53\left(\frac{\mathrm{Pr}}{\mathrm{Pr} + 0.952}\right)(\mathrm{Ra}_D)^{1/4} \tag{9-29}$$

Sprott, et al (Reference 18) demonstrated that sound pressure can produce an increase in the film coefficient by as much as 200 percent depending upon the decibel level and the frequency of the sound.

Sample Problem 9-5 A bare horizontal steam pipe with an outside diameter of 6 inches carries saturated steam at 230°F. The temperature of the air surrounding the pipe is 70°F. Calculate the quantity of condensate at the end of a 100-ft section of the pipe.

Solution: The film temperature of the air is (1/2) (230 + 70) or 150°F. Properties of air at 150°F are

$$\mathrm{Pr} = 0.70, \nu = 2.2 \times 10^{-4}\ \mathrm{ft^2/sec},\ k = 0.0168\ \mathrm{Btu/hr\text{-}ft\ °F},$$

$$\beta = 1/(460 + 150) = 1.64 \times 10^{-3}\ \mathrm{per\ °R}$$

$$\mathrm{Gr}_D = \frac{g\beta \Delta T D^3}{\nu^2} = \frac{32.2 \times 1.64 \times 10^{-3} \times (230 - 70) \times (6/12)^3}{(2.2 \times 10^{-4})^2}$$

$$= 2.18 \times 10^7$$

$$\mathrm{Ra}_D = \mathrm{Gr}_D \mathrm{Pr} = 2.18 \times 10^7 \times 0.70 = 1.52 \times 10^7 < 10^9$$

Therefore we proceed with equation (9-27b) to obtain

$$\mathrm{Nu}_{av} = 0.53\,(\mathrm{Gr}_D \mathrm{Pr})^{1/4} = 0.53(1.52 \times 10^7)^{1/4} = 33.1$$

$$h_{av} = \frac{\mathrm{Nu}_{av} k}{D} = \frac{33.1 \times 0.0168}{(6/12)} = 1.11\ \mathrm{Btu/hr\text{-}ft^2\,°F}$$

Then, the heat loss from a 100-ft length is

$$\pi D L h (T_w - T_\infty) = \pi(6/12) \times 100 \times 1.11(230 - 70)$$
$$= 27{,}900\ \mathrm{Btu/hr}$$

The latent heat of vaporization for steam at 230°F is 958.7 Btu/lb_m, the rate of condensate collection per hour equals

$$(27{,}900/958.7) = 29.1\ \mathrm{lb}_m/\mathrm{hour,\ per\ 100\text{-}ft\ length.}$$

9-5.7 Sphere

Based on the work by Yuge (Reference 19), the following relation is recommended for natural convection from a sphere

$$\mathrm{Nu}_{av} = 2 + 0.43\,(\mathrm{Ra}_D)^{1/4} \qquad \text{for } 1 < \mathrm{Ra}_D < 10^5 \qquad (9\text{-}30)$$

The properties to be used in the above equation are evaluated at the film temperature, and Ra_D signifies a Rayleigh number based on the diameter of the sphere. For air, the equation is simplified to read

$$h_{av} = \left[2 + 0.392\,\mathrm{Gr}_D^{1/4}\right]\frac{k}{D} \qquad \text{for } 1 < \mathrm{Gr}_D < 10^5 \qquad (9\text{-}30a)$$

9-5.8 Pressures Other Than Atmospheric Pressure

The simplified equations for air given amongst equations (9-21) through (9-30) are for one atmosphere pressure, p_a. When pressures, p, other than p_a are encountered, the heat transfer coefficient obtained by the use of the aforementioned equations should be multiplied by

$$\left(\frac{p}{p_a}\right)^{1/2} \qquad \text{for laminar boundary layers}$$

and by

$$\left(\frac{p}{p_a}\right)^{2/3} \qquad \text{for turbulent boundary layers}$$

∴ increased pressures increase h and Nu

9-6 CORRELATIONS FOR ENCLOSED SPACES

The characteristic dimension, L_c, for the Grashof number for natural convection in an enclosed space is the spacing, δ, between the two walls that bound the enclosed space. Since one is interested in the heat transfer across the gap from one wall to the other, the temperature difference ΔT is taken to be equal to $(T_1 - T_2)$, where T_1 and T_2 are the temperatures of the two walls bounding the enclosure.

9-6.1 Vertical Spaces

For small values of δ, leading to a Grashof number of less than 2000, the heat transfer takes place essentially by conduction, and the ratio of the effective thermal conductivity k_e to the actual thermal conductivity is essentially unity.

The effective thermal conductivity is defined by

$$\frac{Q}{A} = q = k_e\left(\frac{T_1 - T_2}{\delta}\right) \tag{9-31}$$

For higher Grashof numbers, empirical formulas due to Jakob (Reference 8) for the enclosed space between two vertical walls at T_1 and T_2 are

$$k_e = 0.18\, k\, \mathrm{Gr}_\delta^{1/4}\, (\delta/L)^{1/9} \qquad \text{for } 20{,}000 < \mathrm{Gr}_\delta < 200{,}000 \tag{9-32}$$

$$k_e = 0.065\, k\, \mathrm{Gr}_\delta^{1/3}\, (\delta/L)^{1/9} \qquad \text{for } 200{,}000 < \mathrm{Gr}_\delta < 10 \times 10^6 \tag{9-32a}$$

Properties entering into the Grashof number and the thermal conductivity are evaluated at a temperature that is the mean of T_1 and T_2 .

Neimann's results for the ratio (k_e/k) are presented by Grober (Reference 20) for a variety of orientations of enclosed spaces. Results of more recent investigations will now be presented.

9-6.2 Horizontal Spaces

Consider an enclosure whose surfaces are separated by a small gap. When the temperature of the upper surface of the enclosure is greater than that of the lower surface, convective currents are more or less absent, and the heat transfer process is due to conduction across the space. The Nusselt number in such a case is unity.

When the lower surface is warmer than the upper surface and the Grashof number, based on the spacing, δ, as the characteristic length, is less than 1700, pure conduction is the only mode of heat transfer. For $\mathrm{Gr}_\delta > 1700$, interesting convection patterns are set up. A pattern of hexagonal cells is created as shown in Figure 9-13. These were first observed by Benard (Reference 21) and are, therefore, referred to as Benard cells. For $\mathrm{Gr}_\delta > 50{,}000$, turbulence sets in and the cell pattern is destroyed.

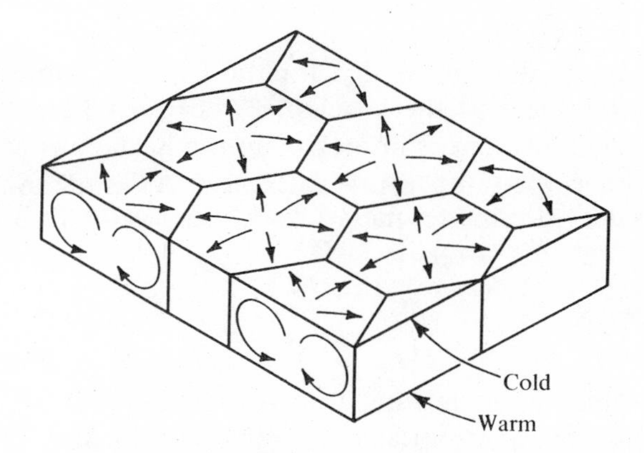

Figure 9-13 Benard cells for natural convection in horizontal spaces, $1700 < \mathrm{Gr}_\delta < 50{,}000$.

Goldstein and Chu (Reference 22) have discussed the work of several researchers. They found that the results of their experimental work agreed very well with the O'Toole and Silveston correlation (Reference 23), which is given below.

$$Nu_{av} = 0.104\, Ra_{\delta}^{0.305} Pr^{0.084} \qquad \text{for } 6 \times 10^6 < Ra_{\delta} < 10^8 \qquad (9\text{-}33)$$

It may be noted that when the Rayleigh number is high, the effect of the aspect ratio is negligibly small.

9-6.3 Inclined Spaces

Catton, et al, (Reference 24) obtained an analytical solution to the problem of natural convection in an inclined rectangular region for $0° < \theta < 120°$, where the angle θ is defined in Figure 9-14. They found that their results were in reasonable agreement with the data of the other researchers. Ayyaswamy and Catton (Reference 25) showed that the Nusselt number for an inclined space, Nu_{θ}, can be expressed as

$$Nu_{\theta} = Nu_{\theta=\pi/2} (\sin\theta)^{1/4} \qquad \text{for } 45 < \theta < 110° \qquad (9\text{-}33a)$$

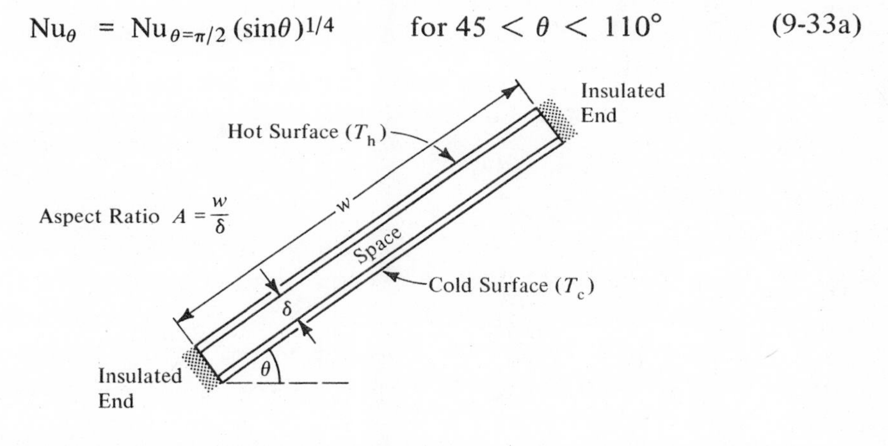

Figure 9-14 Nomenclature for natural convection in inclined space.

In 1974, Arnold, et al, (Reference 26) conducted experiments with silicone oil (Pr = 2000), and water (Pr = 4.5) for an aspect ratio of 6 for values of θ ranging from 15° to 180°. The investigators found a reasonable agreement with equation (9-33a) for $60° < \theta < 120°$. By assuming that the convective portion of the Nusselt number, that is (Nu − 1), is proportional to $g \sin \theta$, the investigators proposed the following correlation

$$Nu_{\theta} = 1 + (Nu_{\theta=\pi/2} - 1) \sin \theta \qquad (9\text{-}33b)$$

which is in excellent agreement with their data for

$$0 < \theta \leqslant \pi/2 \text{ and } 10^3 < Ra_{\delta} < 10^6$$

Free convection inside a horizontal cylindrical space containing air was investigated by Martini and Churchill (Reference 27). They found that the Nusselt number equaled 7 for the conditions they investigated.

9-6.4 Annular Space between Concentric Spheres

Scanlon, et al, (Reference 28) performed experiments on natural convective heat transfer through concentric spheres. They expressed their correlations in terms of the effective conductivity, k_e. It is defined by the equation

$$Q = \frac{4\pi k_e r_i r_o (T_o - T_i)}{r_o - r_i} \tag{9-34}$$

where Q is the actual heat transfer by convection, r_i and r_o are the inner and the outer radii of the enclosure, and T_i and T_o are the temperatures on the inside and the outside surfaces of the enclosure. We note that if k_e in equation (9-34) is replaced by the thermal conductivity, k, the quantity Q will represent the rate of heat conduction in a spherical shell.

The Rayleigh number is based on the gap width, δ, which equals $(r_o - r_i)$. The correlations are

$$\frac{k_e}{k} = 0.117\,\mathrm{Ra}_\delta^{0.276} \qquad \begin{cases} 1.4 \times 10^4 < \mathrm{Ra}_\delta < 2.5 \times 10^6 \\ \mathrm{Pr} = 0.7 \end{cases} \tag{9-35}$$

$$\frac{k_e}{k} = 0.033\,\mathrm{Ra}_\delta^{0.328} \qquad \begin{cases} 2.4 \times 10^4 < \mathrm{Ra}_\delta < 5.4 \times 10^8 \\ 4.7 < \mathrm{Pr} < 12.1 \end{cases} \tag{9-35a}$$

$$\frac{k_e}{k} = 0.031\,\mathrm{Ra}_\delta^{0.353} \qquad \begin{cases} 2.4 \times 10^4 < \mathrm{Ra}_\delta < 9.7 \times 10^7 \\ 148 < \mathrm{Pr} < 336 \end{cases} \tag{9-35b}$$

$$\frac{k_e}{k} = 0.056\,\mathrm{Ra}_\delta^{0.330} \qquad \begin{cases} 1.3 \times 10^3 < \mathrm{Ra}_\delta < 5.6 \times 10^6 \\ 1954 < \mathrm{Pr} < 4148 \end{cases} \tag{9-35c}$$

$$\frac{k_e}{k} = 0.228\,(\mathrm{Ra}^*)^{0.226} \qquad \begin{cases} 1.2 \times 10^2 < \mathrm{Ra}^* < 1.1 \times 10^9 \\ 0.7 < \mathrm{Pr} < 4148 \end{cases} \tag{9-35d}$$

where

$$\mathrm{Ra}^* = \frac{g\beta\delta^4 \Delta T}{\nu\alpha r_i} = \mathrm{Ra}_\delta \left(\frac{\delta}{r_i}\right)$$

Powe, et al, (Reference 29) investigated the flow patterns due to natural convection in cylindrical annuli. However, they did not present any correlation for the average Nusselt number.

9-7 NATURAL CONVECTION UNDER UNIFORM HEAT FLUX

Situations involving a uniform heat flux distribution along a surface are encountered when the surface is heated by an electric heater and when the conduction along the surface is small. Generally speaking, a surface subjected to such a uniform heat flux will not be at a uniform temperature, and therefore, the correlations presented earlier will not be applicable with high precision. Vliet (Reference 30) and Vliet and Lin (Reference 31) have reported extensive experimentation on natural convection with uniform surface heat flux, q.

In presenting the correlations for the Nusselt number, the Rayleigh number has to be redefined since ΔT is no longer known in advance. The modified Rayleigh number, Ra^*, is given by

$$\mathrm{Ra}^* = \frac{g\beta L_c^3}{\nu\alpha} \cdot \frac{qL_c}{k} = \frac{g\beta L_c^4 q}{k\nu\alpha} \tag{9-36}$$

where q is the uniform heat transfer per unit time and area released (or absorbed) by the surface. Note that the quantity (qL_c/k) has the dimensions of ΔT. The correlations, although initially determined for water, are believed valid for air as well. For vertical and inclined surfaces, *local* Nusselt numbers are given by:

$$\mathrm{Nu} = 0.60\,(\mathrm{Ra}_L^* \sin\theta)^{0.20} \quad \text{for } 10^5 < \mathrm{Ra}_L^* < 10^{11} \tag{9-37}$$

$$\mathrm{Nu} = 0.568\,(\mathrm{Ra}_L^* \sin\theta)^{0.22} \quad \text{for } 2 \times 10^{13} < \mathrm{Ra}_L^* < 10^{16} \tag{9-37a}$$

where θ is the angle made by the plate with the horizontal. Equations (9-37) and (9-37a) are, respectively, for the laminar and turbulent regions. One observes that for isothermal surfaces and $(\mathrm{Ra}_x^*) < 10^9$, laminar flow is encountered. On the other hand, laminar flow in the case of a prescribed heat flux prevails when $(\mathrm{Ra}_x^*) < 10^{12}$.

The correlations for natural convection in a vertical space of height, L, and width, δ, for uniform heat flux conditions are given by MacGregor and Emery (Reference 32). These are given below

$$\mathrm{Nu}_\delta = \frac{h\delta}{k} = 0.42\,(\mathrm{Ra}_\delta)^{1/4}\,\mathrm{Pr}^{0.012}\,(L/\delta)^{-0.30} \tag{9-38}$$

$$\text{for} \quad 10^4 < (\text{Ra}_\delta) < 3 \times 10^6$$
$$1 < \text{Pr} < 20{,}000$$
$$10 < (L/\delta) < 40$$

and

$$\text{Nu}_\delta = 0.046\,(\text{Ra}_\delta)^{1/3} \quad \text{for} \quad 10^6 < (\text{Ra}_\delta) < 10^9 \tag{9-38a}$$
$$1 < \text{Pr} < 20$$
$$1 < (L/\delta) < 40$$

9-8 COMBINED NATURAL AND FORCED CONVECTION

There are certain forced convection situations where the velocities due to the forced flow are comparable with the velocities due to natural convective currents. Such situations may be encountered, for instance, when the velocity of air is of the order 1 ft/sec. Under such conditions, it is necessary to deal with a superposition of forced and natural convection, and this is called *mixed* convection. As noted earlier, the pertinent dimensionless groups for forced convection and for natural convection are the Reynolds number and the Grashof number, respectively. One would, therefore, expect the criterion for the existence of mixed convection to involve these two groups. Indeed, an order of magnitude analysis of the boundary layer equations shows that $(\text{Gr}/\text{Re}^2) >> 1$ is the condition for natural convection to be of greater significance than forced convection.

Figure 9-15, due to Metais and Eckert (Reference 33) delineates the regions of free (natural), forced, and mixed convection for flow through horizontal tubes, while such delineation for flow through vertical tubes is shown in Figure 9-16. The correlation for mixed convection through horizontal tubes with laminar flow is given by Brown and Gauvin (Reference 34) as

$$\text{Nu}_{av} = 1.75\left(\frac{\mu_b}{\mu_w}\right)^{0.14} [\text{Gz} + 0.012\,(\text{Gz}\,\text{Gr}_D^{1/3})^{4/3}]^{1/3} \tag{9-39}$$

where

Gz = Graetz number = $\text{RePr}\,(D/L)$

μ_b = viscosity of the fluid at the bulk temperature

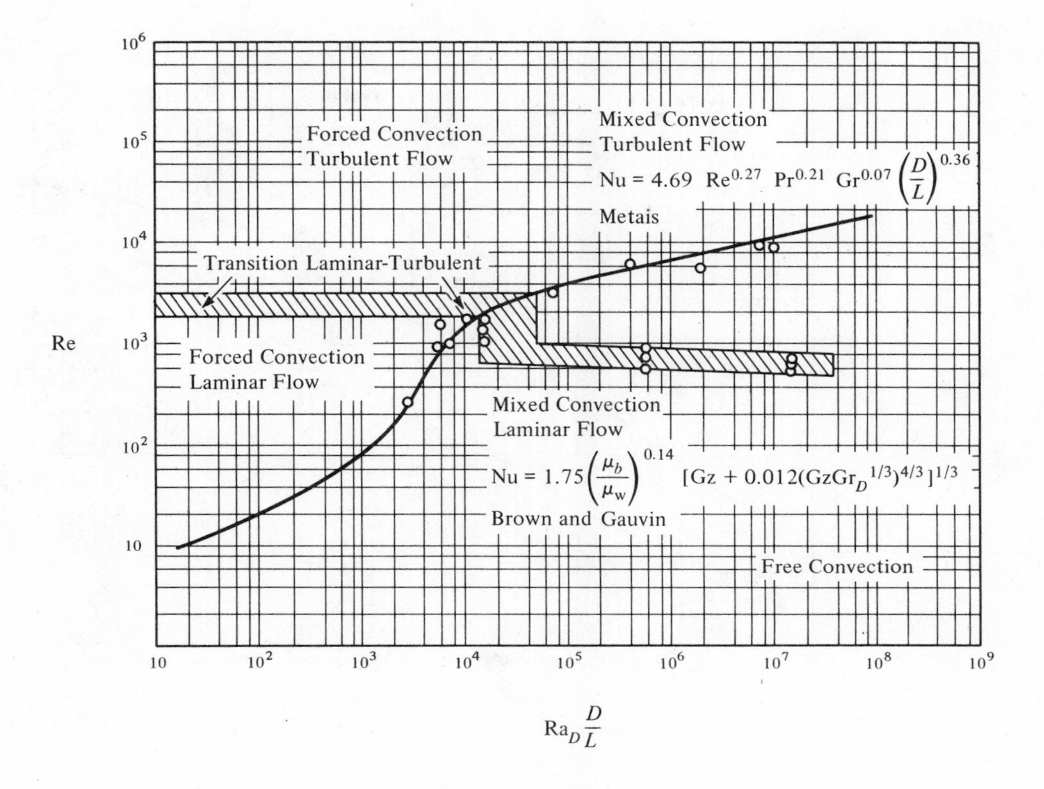

Figure 9-15 Regimes of natural, forced, and mixed convection for flow through *horizontal* tubes, according to Metais, B. and E. R. G. Eckert [33].

and

μ_w = viscosity of the fluid at the wall temperature.

For mixed convection with turbulent flow in horizontal tubes the correlation due to Metais (Reference 33) is

$$\mathrm{Nu}_{av} = 4.69\,\mathrm{Re}^{0.27}\mathrm{Pr}^{0.21}\mathrm{Gr}^{0.07}(D/L)^{0.36} \qquad (9\text{-}40)$$

The correlations for pure forced convection flow through tubes were discussed in Chapter 8.

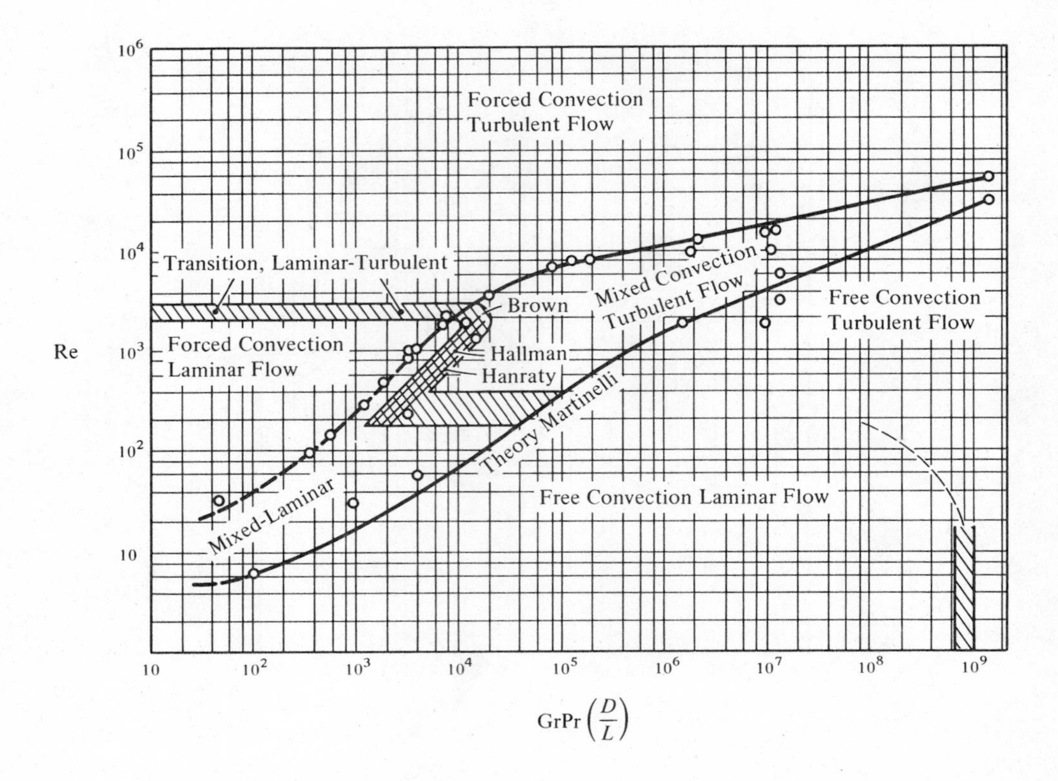

Figure 9-16 Regimes of natural, forced and mixed convection for flow through *vertical* tubes $[10^{-2} < \mathrm{Pr}(D/L) < 1]$ according to Metais, B. and E. R. G. Eckert [33].

PROBLEMS (English Engineering System of Units)

9-1. In an experiment, similar to the one described at the beginning of this chapter, the following data was obtained: A_p = 4 in^2, ϵ_p = 0.95, T_p = 340°F, T_t = 72°F, power input = 8.1 watts, and the pressure of the air in the tank = 29.9 inches of Hg.

Determine the coefficient of heat transfer for natural convection, and compare your value with the correlation for a vertical wall if the plate is 4 inches in height.

9-2. Estimate the thickness of the boundary layer at x = 2 inches for the vertical plate in problem 9-1. Also determine the maximum velocity of air within the boundary layer for x = 2 inches.

9-3. A large vertical steel tank 8 feet in height and 4 feet in diameter is used as a receiver for the output of an air compressor. The air in the tank is at 210°F and at 55 psia.

(a) If the inside walls of the tank are at 100°F, what is the rate of heat transfer by natural convection from the air to the inside vertical walls of the tank?

(b) If the wall thickness is 1/2 inch, what is the temperature drop across the wall? $k = 31$ Btu/hr-ft°F.

9-4. On a cold winter day, the environmental air is at 0°F. The interior of a house is at 70°F. The walls of the house have a conductance of 0.8 Btu/hr-ft^2°F. Using a trial-and-error procedure, estimate the h value for the interior and exterior of the walls of a 12-foot high house and, hence, determine the temperatures of the interior and exterior surfaces of the wall.

9-5. A thin, 12 inches by 20 inches, vertical steel plate is maintained at 300°F in air. The ambient air temperature is 65°F. Determine the rate of heat input into the plate, which loses heat by natural convection to the surrounding air.

9-6. Rework problem 9-5 for the surrounding fluid being water.

9-7. Determine the rate of heat gain by a thin horizontal plate, 8 inches in diameter, with a surface temperature of 100°F if it is immersed in engine oil at 300°F.

9-8. Solar energy at a rate of 105 Btu/hr-ft^2 is incident on a roof inclined at an angle of 30° with the horizontal. Assume that the back of the roof is perfectly insulated, that there is no wind, and that the surface of the roof behaves like a blackbody. Determine the equilibrium temperature of the roof if the ambient air is at 10°F. The length, L, is 9 ft.

9-9. Determine the coefficient of heat transfer due to natural convection in air for a short vertical cylinder of height 3 inches and diameter 2 inches. T_f = 170°F, T_∞ = 300°F.

9-10. If a vertical pipe of 4 inch O.D., at a surface temperature of 200°F, is in a room where the air is at 80°F, what is the rate of heat loss per foot length of the pipe? L_c = 12 ft.

9-11. Based on certain experimental results, the following equation is proposed for the coefficient of heat transfer due to natural convection from horizontal cylinders to air.

$$h = 1.016\, \Delta T^{0.27} D^{-0.19} T^{-0.27}$$

Convert the proposed equation into the form $\mathrm{Nu} = C\mathrm{Gr}^m \mathrm{Pr}^n$ by determining appropriate values of C, m, and n.

9-12. In the lumped parameter analysis of transient heat conduction problems (Chapter 4), it is customary to assume that the h value is constant. For natural convection cooling, the error introduced by such an assumption could be significant if the initial value of $(T_w - T_\infty)$ is large. Assuming that $h = C\,(\Delta T)^{1/3}$, where C is a constant, derive an expression for the time required for the ΔT to reduce to 50 percent of its initial value. Compare it with the time requirement if the analysis were made under an assumption of constant $h(=C)$. Assume $\Delta T = 100°F$ at $\tau = 0$.

9-13. A horizontal tube of 1/2-inch diameter is at 400°F. The surrounding air is at 50°F. Calculate the heat loss due to natural convection.

9-14. A 40-gallon tank, full of water at 40°F, is to be heated to 120°F by means of steam passing through a 3/8-inch O.D. copper coil having 20 turns of 18-inch diameter. The steam is at 220°F. Assume that there is perfect mixing of the water and that the coil can be approximated by a horizontal cylinder immersed in water. Estimate the heating time.

9-15. Estimate the time required to heat the water in problem 9-14 if the heating surface area was 1 foot by 4 feet at 900°F, with the 4 foot dimension vertical.

9-16. Determine the coefficient of heat transfer due to natural convection for a horizontal wire of 0.05-inch diameter immersed in water at 60°F if the wire surface is maintained at 500°F.

9-17. Saturated steam at 240°F enters a horizontal steel pipe of 3/4-inch I.D. and 1/8-inch wall thickness. The temperature of the air surrounding the pipe is 72°F. Determine the value of h for natural convection and the amount of heat lost to the room if the pipe is covered with 1/2-inch thick insulation with $k = 0.09$ Btu/hr-ft°F.

9-18. A 1-inch O.D. electrical transmission line carrying 100 amps and having a resistance of 1.25×10^{-3} ohms per foot is situated horizontally in the atmosphere. Neglecting the radiation losses, determine the temperature of the surface of the cable if the ambient temperature is (a) 75°F, and (b) −50°F.

9-19. A 1/2-inch diameter spherical steel ball at 300°F is immersed in engine oil at 100°F. Calculate the rate of convective heat loss.

9-20. Determine the rate of heat loss from a steel ball of 1/4-inch diameter immersed in water at 50°F with $\Delta T = 30°F$.

9-21. The vertical space between the aluminum foil of a foil-backed insulation and the surface of a plywood panel is 1/8-inch in width. What will the rate of heat transfer be across the gap if the temperatures of the surfaces separated by the gap are 50°F and 0°F. $L = 8$ ft.

9-22. Rework problem 9-21 if the spacing between the foil and the panel is doubled.

9-23. A vertical air space 3-inches thick, 15-inches deep, and 2-feet high is bounded by surfaces at 50°F and 30°F. Compare the convection heat transfer rate per unit area with that which would result if the space were subdivided by placing a thin metal foil vertically in the middle of the space.

9-24. Turbine blades can be cooled by free convection by machining slots in them. A certain slot is 0.5 inch by 0.15 inch in cross section and 3 inches in length. The circumferential rotational velocity at the mean radius of the slot (18 inches) is 800 fps. The air in the slot is at 100°F, while the blade temperature is 1600°F. Estimate the h value for free convection. (*Hint:* Replace g in the Grashof number by the centrifugal acceleration, u^2/r.)

9-25. The rate of movement of nitrogen gas in a hydraulic accumulator during charging is of the order of 1 ft/sec. If toward the end of the charging process the gas is at 280°F, and if the horizontal wall of the accumulator (6 inches in diameter and 60 inches in length) is at 70°F, determine the region to which the convective heat transfer process belongs. Then, calculate the heat transfer coefficient. Neglect the effect of the variation of pressure on the heat transfer coefficient during the charging process.

9-26. If at the end of the charging process the pressure in the accumulator in the preceding problem is 1500 psia and the temperatures of the gas and the wall are 310°F and 75°F, respectively, determine the heat transfer coefficient.

PROBLEMS (SI System of Units)

9-1. The following data were obtained in an experiment similar to the one described at the beginning of this chapter: A_p = 20 cm^2, ϵ_p = 0.90, T_p = 200°C, T_t = 18°C, power input = 8.5 W, and the pressure of the nitrogen gas in the tank = 758 mm of Hg. Determine the coefficient of heat transfer for natural convection, and compare your value with the correlation for a vertical wall if the plate is 10 cm in height.

9-2. What would be the thickness of the boundary layer at x = 4 cm for the vertical plate in problem 9-1? Also determine the maximum velocity of air within the boundary layer for x = 4 cm.

9-3. A large vertical steel tank 3 m in height and 1 m in diameter is used as a receiver for the output of an air compressor. The air in the tank is at 100°C and at 3 atmospheres.

(a) If the inside walls of the tank are at 30°C, what is the rate of heat transfer by natural convection from the air to the inside vertical walls of the tank?

(b) If the wall thickness is 1 cm, what is the temperature drop across the wall? k = 54 W/m°K.

9-4. The interior of a house is maintained at 25°C on a cold winter day when the environmental air is at −10°C. The walls of the house have a conductance of 4.0 W/m^2°C. Using a trial-and-error procedure, estimate the h value for the interior and exterior of the walls of a 4-m high house and, hence, determine the respective temperatures of the surfaces of the wall.

9-5. Determine the rate of heat input into a vertical steel plate that loses heat by natural convection to the surrounding air. The plate is 30 cm by 50 cm and is maintained at 150°C. The ambient air temperature is 18°C.

9-6. Rework problem 9-5 for the surrounding fluid being engine oil.

9-7. Determine the rate of heat gain from a thin horizontal plate, 20 cm in diameter, with a surface temperature of 100°C, if it is immersed in engine oil at 140°C.

9-8. Solar energy at a rate of 280 W/m^2 is incident on a roof inclined at an angle of 40° with the horizontal. Assume that the back of the roof is perfectly insulated, that there is no wind, and that the surface of the roof behaves like a blackbody. Determine the equilibrium temperature of the roof if the ambient air is at 0°C. The length, L, is 3 m.

9-9. Determine the coefficient of heat transfer due to natural convection in air for a solid whose dimensions are 5 cm by 8 cm by 12 cm. The vertical dimension is 12 cm. $T_f = 50°C$ and $T_\infty = 0°C$.

9-10. A vertical pipe of 10-cm O.D., at a surface temperature of 100°C, is in a room where the air is at 20°C. What is the rate of heat loss per meter length of the pipe? $L_c = 3$ m.

9-11. Based on certain experimental results, the following equation is proposed for the coefficient of heat transfer due to natural convection from horizontal cylinders to air.

$$h = 4.7\,\Delta T^{0.27} D^{-0.19} T^{-0.27}$$

Convert the proposed equation into the form $Nu = C Gr^m Pr^n$ by determining appropriate values of C, m, and n.

9-12. In the lumped parameter analysis of transient heat conduction problems (Chapter 4), it is customary to assume that the h value is constant. For natural convection cooling, the error introduced by such an assumption could be significant if the initial value of $(T_w - T_\infty)$ is large. Assuming that $h = C\,(\Delta T)^{1/3}$, where C is a constant, derive an expression for the time required for the ΔT to reduce to 50 percent of its initial value. Compare it with the time requirement if the analysis were made under an assumption of constant $h(=C)$. Assume $\Delta T = 55.5°$ at $\tau = 0$.

9-13. Calculate the heat loss due to natural convection from a horizontal tube of 15-mm O.D. at 400°C. The surrounding air is at 30°C.

9-14. A 0.2 m^3 tank, full of engine oil at 10°C, is to be heated to 60°C by means of steam passing through a 10-mm O.D. copper coil having 10 turns of 50-cm diameter. The steam is at 110°C. Assume that there is perfect mixing of the oil and that the coil can be approximated by a horizontal cylinder immersed in oil. Estimate the heating time.

9-15. Determine the time required to heat the engine oil in problem 9-14 if the heating surface area was 0.5 m by 1.5 m at 500°C, with the 1.5-m dimension vertical.

9-16. Estimate the coefficient of heat transfer due to natural convection for a horizontal wire of 2-mm diameter immersed in water at 20°C if the wire surface is maintained at 300°C.

9-17. Saturated steam at 120°C enters a horizontal steel pipe of 18-mm I.D. and 2-mm wall thickness. The temperature of the air surrounding the pipe is 18°C. Determine the value of h for natural convection and the amount of heat lost to the room if the pipe is covered with 12-mm thick insulation (k = 0.2 W/m°K).

9-18. A 25-mm O.D. electrical transmission line carrying 100 amps and having a resistance of 400×10^{-5} ohms per meter is situated horizontally in the atmosphere. Neglecting the radiation losses, determine the temperature of the surface of the cable if the ambient temperature is (a) 26°C, and (b) −26°C.

9-19. A 10-mm diameter spherical steel ball at 150°C is immersed in water at 30°C. Calculate the rate of convective heat loss.

9-20. Determine the rate of heat loss from a steel ball of 5-mm diameter placed in air at 10°C with $\Delta T = 40$°C.

9-21. The vertical space between the paper of a paper-backed insulation and the surface of a dry wall is 4 mm in width. What will the rate of heat transfer be across the gap if the temperatures of the surfaces separated by the gap are 10°C and −20°C. $L = 3$ m.

9-22. Rework problem 9-21 if the spacing between the insulation and the dry wall is doubled.

9-23. A vertical air space 8-cm thick, 50-cm deep, and 1-m high is bounded by surfaces at 20°C and 0°C. Compare the convection heat transfer rate per unit area with that which would result if the space were subdivided by placing a thin metal foil vertically in the middle of the space.

9-24. Turbine blades can be cooled by free convection by machining slots in them. A certain slot is 10 mm by 3 mm in cross section and 8 cm in length. The circumferential rotational velocity at the mean radius of the slot of 50 cm is 300 m/sec. The air in the slot is at 30°C, while the blade temperature is 1000°C. Estimate the h value for free convection. (*Hint:* Replace g in the Grashof number by the centrifugal acceleration, u^2/r.)

9-25. The rate of movement of nitrogen gas in a hydraulic accumulator during charging is of the order of 40 cm/sec. If toward the end of the charging process the gas is at 150°C and the horizontal wall of the accumulator (15 cm in diameter and 2 m in length) is at 18°C, determine the region to which the convective heat transfer process belongs. Then, calculate the heat transfer coefficient. Neglect the effect of the variation of pressure on the heat transfer coefficient during the charging process.

9-26. If at the end of the charging process the pressure in the accumulator in the preceding problem is 100 atmospheres and the temperatures of the gas and the wall are 175°C and 25°C, respectively, determine the heat transfer coefficient.

REFERENCES

[1] Schmidt, E., and Beckmann, W., "Das Temperatur-und Geschwindigkeitsfelt vor einer Warmer abgebenden senkrechten Platte bei naturlicher Konvection," *Tech. Mech. und Thermodynamik,* **1**, pp. 341, (1930).

[2] Saunder, O. A., *Proc. Royal Society* (London), **A157**, pp. 278, (1936).

[3] Squire, H. B., Unpublished work, see Goldstein, S., *Modern Developments in Fluid Dynamics,* **II**, Oxford at the Clarendon Press, pp. 641, (1938).

[4] Ostrach, S., "An Analysis of Laminar Free-Convection Flow and Heat Transfer about a Flat-Plate Parallel to the Direction of the Generating Body Force," *NACA Tech. Note No. 2635,* (1951).

[5] Eckert, E. R. G., and Drake, Jr., R. M. *Heat and Mass Transfer,* McGraw-Hill Book Company, 2nd ed., New York, (1949).

[6] McAdams, W. H., *Heat Transmission,* McGraw-Hill Book Company, 3rd ed., New York, (1954).

[7] Weise, R., *Forschungs Gebiete Ingenieurwesen*, **6**, pp. 281, (1935).

[8] Jakob, Max, *Heat Transfer*, John Wiley Sons, Inc., New York, (1958).

[9] Warner, C. Y., and Arpaci, V. S., "An Experimental Investigation of Turbulent Natural Convection in Air at Low Pressure along a Vertical Heated Flat Plate," *Int. Journal of Heat & Mass Transfer,* **2**, pp. 397, (1968).

[10] Churchill, S. W., and Chu, H. H. S., "Correlating Equations for Laminar and Turbulent Free Convection from a Vertical Plate," *Int. Journal of Heat & Mass Transfer,* **18**, pp. 1323, (1975).

[11] Goldstein, R. J., Sparrow, E. M., and Jones, D. C., "Natural Convection Mass Transfer Adjacent to Horizontal Plates," *Int. Journal of Heat & Mass Transfe,* **16**, No. *5* pp. 1025, (1973).

[12] Lloyd, J. R., and Moran, W. R., "Natural Convection Adjacent to Horizontal Surface of Various Planforms," *ASME* Paper No. *74-WA/HT-66*, (1974).

[13] Fuji, T., and Imura, H., "Natural-Convection Heat Transfer from a Plate with Arbitrary Inclination," *Int. Journal of Heat & Mass Transfer,* **15**, pp. 755, (1972).

[14] King, W. J., "The Basic Laws and Data of Heat Transmission," *Mechanical Engineering,* **54**, pp. 347, (1932).

[15] Gebhert, B., *Heat Transfer*, 2nd ed., McGraw-Hill Book Company, New York, (1969).

[16] Minkowycz, W. J., and Sparrow, E. M., "Local Nonsimilar Solutions for Natural Convection on a Vertical Cylinder," *ASME* Paper No. *74-HT-Y*, (1973).

[17] Hyman, S. C., Bonilla, C. F., and Ehrlich, S. W., "Natural Convection Transfer Process: I Heat Transfer to Liquid Metals and Nonmetals at Horizontal Cylinders," *Chemical Engineering Progress Symposium Series,* **49**, No. *5*, pp. 21, (1953).

[18] Sprott, A. L., Holman, J. P., and Durand, F. L., Reprint 60-HT-19, *ASME 4th Heat Transfer Conference,* (1960).

[19] Yuge, T., "Experiments on Heat Transfer from Spheres Including Combined Natural and Forced Convection," *ASME Journal of Heat Transfer,* **82** pp. 214, (1960).

[20] Grober, H., Erk, S., and Grigull, U., *Fundamentals of Heat Transfer,* 3rd ed., McGraw-Hill Book Company, New York, (1961).

[21] Benard, H., "Les Tourbillons Cellulaires dans une Liquide Transportant de la Chaleur par Convection en Rigime Permanent," *Annual Chim. Phys.* **23**, pp. 62, (1901).

[22] Goldstein, R. J., and Chu, T. Y., "Thermal Convection in a Horizontal Layer of Air," *Progress in Heat and Mass Transfer,* **2**, Pergamon Press, London, pp. 55, (1969).

[23] O'Toole, J. L., and Silveston, P. L., "Correlation of Convective Heat Transfer in Confined Horizontal Layers," *Chemical Engineering Progress Symposium,* **57**, No. *32*, pp. 81, (1961).

[24] Catton, I., Ayyaswamy, P. S., and Clever, R. M., "Natural Convection in a Finite, Rectangular Slot Arbitrarily Oriented with Respect to the Gravity Vector," *Int. Journal of Heat & Mass Transfer,* **17**, pp. 173, (1974).

[25] Ayyaswamy, P. S., and Catton, I., "The Boundary-Layer Regime for Natural Convection in a Differentially Heated, Tilted Rectangular Cavity," *ASME Journal of Heat Transfer,* **95**, pp. 543, (1973).

[26] Arnold, J. N., Bonaparte, P. N., Catton, I., and Edwards, D. K., "Experimental Investigation of Natural Convection in a Finite Rectangular Region Inclined at Various Angles from 0° to 180°," *Proceedings of Heat Transfer and Fluid Mechanics Institute*, Stanford Press, pp. 321, (1974).

[27] Martini, W. R., and Churchill, S. W., "Natural Convection Inside a Horizontal Cylinder," *AIChE Journal,* **6**, No. *2*, pp. 251, (1960).

[28] Scanlan, J. A., Bishop, E. H., and Powe, R. E., "Natural Convection Heat Transfer Between Concentric Spheres," *Int. Journal of Heat & Mass Transfer,* **13**, pp. 1857, (1970).

[29] Powe, R. E., Carley, C. T., and Bishop, E. H., "Free Convective Flow Patterns in Cylindrical Annulii," *ASME Journal of Heat Transfer,* pp. 310, (1969).

[30] Vliet, G. C., "Natural Convection Local Heat Transfer on Constant Heat Flux Inclined Surfaces," *ASME Journal of Heat Transfer,* **91**, pp. 511,(1969).

[31] Vliet, G. C., and Lin, C. K., "An Experimental Study of Turbulent Natural Convection Boundary Layers," *ASME Journal of Heat Transfer,* **91**, pp. 517, (1969).

[32] MacGregor, R. K., and Emery, A. F., "Free Convection Through Vertical Plane Layers–Moderate and High Prandtl Numbers," *ASME Journal of Heat Transfer,* **91**, pp. 391, (1969).

[33] Metais, B., and Eckert, E. R. G., "Forced, Mixed and Free Convection Regimes," *ASME Journal of Heat Transfer*, **86**, pp. 295, (1964).

[34] Brown, C. K. and Gauvin, W. H., "Combined Free and Forced Convection, Parts I and II," *Canadian Journal of Chemical Engineering,* **43**, pp. 306, 313, (1965).

10

Fins and Heat Exchangers

10-1 INTRODUCTION

Fins or extended surfaces are used to increase the rate of heat transfer from a surface. Fins attached to a surface, in effect, increase the total area available for heat transfer. Heat exchangers serve to transfer heat energy from a hot fluid to a cold fluid, the objective being either to heat a given fluid, like the saturated steam entering a superheater in a steam power plant, or to cool a given fluid, like the cooling water passing through a radiator of an automobile. Finned surfaces are devices that increase the efficiency of heat exchangers.

In the analysis and the design of a finned surface, the quantity of heat energy dissipated by a single fin of a given geometry is determined from the temperature gradient and the cross-sectional area available for heat flow at the base of the fin. The differential equation for the temperature distribution in a fin results from an energy balance on an elemental section of the fin that is conducting and convecting at the same time. The total number of fins necessary to dissipate a given quantity of heat is determined from the heat transfer load. In a heat exchanger, one is usually interested in determining the total surface area separating the hot and the cold fluid. If the surface happens to have fins on it, then the effective area of the heat exchanger is increased, resulting in a compact heat exchanger. A finned surface is usually used when the convective fluid available is a gas, since convective heat transfer coefficients for a gas are usually smaller than those for a liquid. Examples of a finned surface are the finned cylinders of a motorcycle engine and the baseboard heater in a house. When heat energy is to be dissipated in a space vehicle, where convection is absent, finned surfaces radiating heat energy are extensively used.

Fins and heat exchangers come in various designs. Fins can be of rectangular cross sections, like ribs attached along the length of a tube, called *longitudinal fins*, or concentric annular discs around a tube called *circumferential fins*. The thickness of fins may be uniform or variable. Heat exchangers can be considered basically of two types, the cross-flow exchanger, such as the automobile radiator where cool air flows across the coils containing the hot water, and the in-line exchanger where there exists a tube within a tube. In the latter, the hot and the cold fluids may flow in the same direction (parallel flow) or in the opposite direction (counter flow). In this chapter, we will first examine the heat flow from fins before investigating heat exchangers.

10-2 RECTANGULAR FIN OF CONSTANT CROSS SECTION

Consider a thin rectangular fin of rectangular cross section as shown in Figure 10-1. The problem of heat conduction in a fin losing heat to the surrounding ambient air by convection is, strictly speaking, not a one-dimensional problem. The principal heat conduction is along the x-axis, while convective loss from the top and bottom surfaces of the fin imply a nonzero temperature gradient in the y direction at $y = \pm(t/2)$.

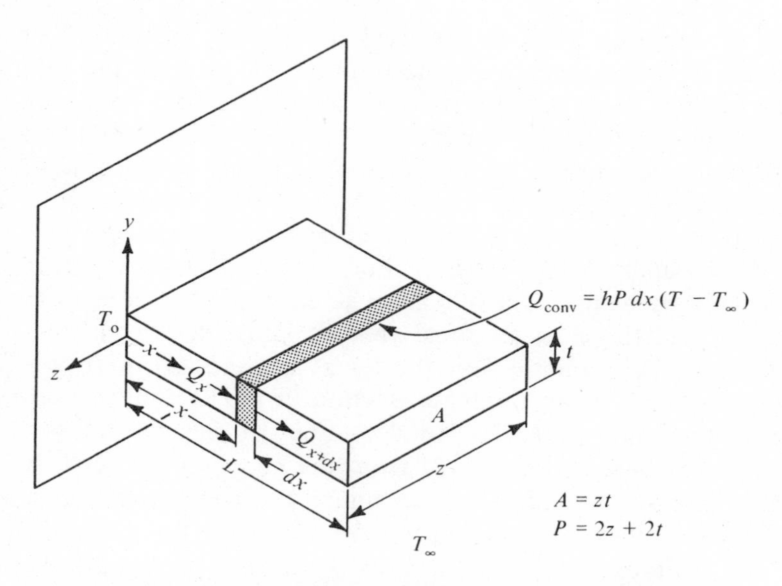

Figure 10-1 Rectangular fin of constant cross section.

However, if the fin is very thin, then at any value of x, the temperature will be essentially constant over the cross-sectional area, A. We assume that the temperature at the base of the fin is uniform and that the thermal conditions at the tip of the fin do not vary along the z direction. This means that the temperature will not be a function of y or z, and if there are no heat sources, steady-state conditions will prevail. Thus the temperature will depend on the x coordinate, the properties of the fin and the surrounding fluid, and the convective heat transfer coefficient.

10-2.1 The Governing Differential Equation and Its General Solution

To determine the differential equation that will yield the temperature as a function of x, an energy balance is made on a small differential element of width, dx, and perimeter, P, as follows:

$$\begin{pmatrix}\text{heat conducted in}\\ \text{at } x = x\end{pmatrix} = \begin{pmatrix}\text{heat conducted out}\\ \text{at } x = x + dx\end{pmatrix} + \begin{pmatrix}\text{heat lost by}\\ \text{convection over}\\ \text{the width, } dx\end{pmatrix}$$

$$(\text{heat conducted in at } x = x) = -kA\left(\frac{dT}{dx}\right)$$

$$(\text{heat conducted out at } x = x + dx) = -kA\left(\frac{dT}{dx}\right) + \frac{d}{dx}\left[-kA\left(\frac{dT}{dx}\right)\right]dx$$

$$= -kA\left(\frac{dT}{dx}\right) - kA\left(\frac{d^2T}{dx^2}\right)dx$$

since k and A are constants

$$(\text{heat lost by convection over the width, } dx) = h(Pdx)(T - T_\infty)$$

where P is the perimeter and Pdx is the area for convection. This energy balance gives

$$-kA\left(\frac{dT}{dx}\right) = -kA\left(\frac{dT}{dx}\right) - kA\left(\frac{d^2T}{dx^2}\right)dx + h(Pdx)(T - T_\infty)$$

or

$$\frac{d^2T}{dx^2} - m^2(T - T_\infty) = 0 \tag{10-1}$$

where

$$m^2 = \frac{hP}{kA}$$

Equation (10-1) is the differential equation that describes the temperature as a function of x and m. The quantity m is a function of the properties of the fin material and the surrounding fluid. To solve this ordinary differential equation, we note that it is a second-order, linear, nonhomogeneous differential equation with constant coefficients. Rewriting equation (10-1), we have

$$\frac{d^2 T}{dx^2} - m^2 T = - m^2 T_\infty$$

A general solution is given by $T = T_c + T_p$, where T_c is the complimentary solution and T_p is the particular solution. Recalling the operator technique for solving such an equation, we set the right-hand side equal to zero, examine the remaining left-hand portion of the equation (the homogeneous portion), and proceed to solve

$$\frac{d^2 T}{dx^2} - m^2 T = 0$$

Letting "p" be the differential operator, we write

$$p^2 - m^2 = 0$$

or

$$p = \pm m$$

and the complimentary portion, T_c, of the solution to equation (10-1) becomes

$$T_c = C_1 e^{-mx} + C_2 e^{mx}$$

Next we note that the nonhomogeneous portion, $m^2 T_\infty$, is a constant, so that in order to obtain the particular solution, T_p, to equation (10-1), we assume

$$T_p = A \quad \text{(constant)}$$

and substitute the above assumed solution for T_p into equation (10-1):

$$\frac{d^2 A}{dx^2} - m^2 A = -m^2 T_\infty$$

or

$$A = T_\infty \quad \text{and} \quad T_p = T_\infty$$

Now the general solution is $T = T_c + T_p$ or

$$T = C_1 e^{-mx} + C_2 e^{mx} + T_\infty$$

which is normally rewritten as

$$T - T_\infty = C_1 e^{-mx} + C_2 e^{mx} \tag{10-2}$$

10-2.2 Specific Solutions for Different Boundary Conditions

Having obtained the general solution, two boundary conditions are needed to solve for the unknown constants C_1 and C_2. There are actually three possible sets of boundary conditions that may be used. They are:

Case (I):
(a) at $x = 0$, $T = T_o$
(b) at $x = L \rightarrow \infty$, $T = T_\infty$

This is for an infinitely long fin, since the temperature at the end of the fin ($x = L$) equals that of the surrounding fluid. Actually a small diameter (1/8 inch) steel rod several inches long approximates this condition.

Case (II):
(a) at $x = 0$, $T = T_o$
(b) at $x = L$, $dT/dx = 0$

This is the case where the end of the fin is insulated or where, for all practical purposes, the loss of heat through the fin tip is negligible. The solution to this problem is important when performing chart solutions for real fins.

Case (III):
(a) at $x = 0$, $T = T_o$
(b) at $x = L$, $-kA(dT/dx) = hA(T - T_\infty)$

This is referred to as the condition for a finite fin where the heat conducted to the end is convected away to the surrounding fluid. Such a situation rigorously describes what actually happens in practice.

We now proceed to solve *Case I*.

$$T - T_\infty = C_1 e^{-mx} + C_2 e^{mx}$$

Applying boundary condition (b), we obtain

$$T_\infty - T_\infty = C_1 e^{-\infty} + C_2 e^{+\infty}$$

$$0 = C_1(0) + C_2(\infty)$$

This equality can only hold if C_2 is identically zero. Hence

$$C_2 = 0$$

and

$$T - T_\infty = C_1 e^{-mx}$$

Application of boundary condition (a) yields

$$T_o - T_\infty = C_1 e^0 = C_1$$

Hence

$$C_1 = T_o - T_\infty$$

and

$$\frac{T - T_\infty}{T_o - T_\infty} = e^{-mx} \tag{10-3}$$

To solve *Case II*, we proceed with the equation

$$T - T_\infty = C_1 e^{-mx} + C_2 e^{mx}$$

Applying boundary condition (a) for this case, we write

$$T_o - T_\infty = C_1 + C_2 \tag{10-4}$$

and applying boundary condition (b), we obtain

$$dT/dx = -mC_1 e^{-mx} + mC_2 e^{mx}$$

$$0 = -mC_1 e^{-mL} + mC_2 e^{mL}$$

or

$$C_1 = C_2 e^{2mL} \tag{10-4a}$$

Combining equations (10-4) and (10-4a) gives

$$T_o - T_\infty = C_2 (1 + e^{2mL})$$

or

$$C_2 = \frac{T_o - T_\infty}{1 + e^{2mL}} \quad \text{and} \quad C_1 = \frac{T_o - T_\infty}{e^{-2mL} + 1}$$

Hence

$$T - T_\infty = \frac{T_o - T_\infty}{e^{-2mL} + 1} e^{-mx} + \frac{T_o - T_\infty}{1 + e^{2mL}} e^{mx}$$

or

$$\frac{T - T_\infty}{T_o - T_\infty} = \frac{e^{-mx}}{e^{-2mL} + 1} + \frac{e^{mx}}{1 + e^{2mL}} \tag{10-4b}$$

Equation (10-4b) can be written in a more compact form if hyperbolic functions are used. Multiply the numerator and the denominator of the first term on the right-hand side of equation (10-4b) by e^{+mL}, and multiply the numerator and the denominator of the second term on the right-hand side of equation (10-4b) by e^{-mL}, giving

$$\frac{T - T_\infty}{T_o - T_\infty} = \frac{e^{-m(x-L)}}{e^{-mL} + e^{mL}} + \frac{e^{m(x-L)}}{e^{-mL} + e^{mL}}$$

or

$$\frac{T - T_\infty}{T_o - T_\infty} = \frac{e^{m(L-x)} + e^{-m(L-x)}}{e^{mL} + e^{-mL}}$$

Note that the hyperbolic cosine is defined as

$$\cosh(\theta) = \frac{e^{\theta} + e^{-\theta}}{2}$$

Therefore, we have

$$\frac{T - T_\infty}{T_0 - T_\infty} = \frac{\cosh[m(L - x)]}{\cosh(mL)} \tag{10-5}$$

The solution to *Case III* is algebraically very tedious. It is

$$\frac{T - T_\infty}{T_0 - T_\infty} = \frac{\cosh[m(L - x)] + (h/mk)\sinh[m(L - x)]}{\cosh(mL) + (h/mk)\sinh(mL)} \tag{10-6}$$

10-2.3 Heat Dissipated by a Fin

Once the temperature distribution is known, it is then possible to calculate the heat dissipated by a given fin. The heat dissipated must all come into the fin at its base where it is attached to the wall and be convected away to the surrounding fluid. This is very much analogous to water entering a hose and leaving through holes punched along its periphery in that the total water flow enters at the hose inlet and is dissipated through the holes along its length. To calculate the heat lost by a fin we may write

$$Q = -kA(dT/dx)_{x=0} \tag{10-7}$$

which is the heat conducted in at the base of the fin. Alternately we may write

$$Q = \int_{x=0}^{L} hP(T - T_\infty)dx \tag{10-8}$$

The above integral is the summation of all the heat lost by convection along the length of the fin.

Case (I):

To calculate the heat dissipated by an infinitely long fin we proceed as follows

$$Q = -kA(dT/dx)_{x=0}$$

Since

$$T = T_\infty + (T_o - T_\infty)e^{-mx}$$

$$dT/dx = -m\,(T_o - T_\infty)e^{-mx}$$

or

$$(dT/dx)_{x=0} = -m(T_o - T_\infty)$$

Hence, substituting the above result in equation (10-7), we have

$$Q = kAm(T_o - T_\infty)$$

Since

$$m = \sqrt{\frac{hP}{kA}}$$

the heat dissipated by the fin is given by

$$Q = \sqrt{hPkA}\,(T_o - T_\infty) \tag{10-9}$$

Equation (10-9) tells us how much heat will be dissipated by an infinitely long fin of constant cross-sectional area. An alternate approach would be to start with equations (10-8) and to proceed as follows

$$Q = \int_{x=0}^{L\to\infty} hP(T_o - T_\infty)e^{-mx}\,dx$$

Integration yields

$$Q = -\frac{1}{m}hP\,[(T_o - T_\infty)e^{-mx}]_0^\infty$$

$$Q = -\frac{1}{m}hP\,(T_o - T_\infty)(0 - 1)$$

or

$$Q = \sqrt{hPkA}\,(T_o - T_\infty)$$

which is the same as equation (10-9).

Case (II):

It may not seem practical to use a fin with an insulated tip since such insulation will decrease heat transfer from the fin. However, in many situations, because of rather large values of the ratio, (L/t), the flow of heat at the tip is negligibly small. Mathematically, this is the same as if $(dT/dx)_{x=L} \simeq 0$. Such a condition is realistic for large (L/t) ratios and also yields a manageable solution.

The analysis of this case is also important because of its application to the chart solutions associated with real fins. This point will be further explored in Section 10-3 where real fins are discussed.

To determine the heat lost by fins insulated at the tip, we proceed as follows. From equation (10-4b) we have

$$T - T_\infty = (T_o - T_\infty)\left[\frac{e^{-mx}}{e^{-2mL} + 1} + \frac{e^{mx}}{1 + e^{2mL}}\right]$$

$$\left(\frac{dT}{dx}\right)_{x=0} = (T_o - T_\infty)(m)\left[\frac{1}{1 + e^{2mL}} - \frac{1}{1 + e^{-2mL}}\right]$$

$$= (T_o - T_\infty)(m)\frac{e^{-mL} - e^{mL}}{e^{-mL} + e^{mL}}$$

Substituting the above result in equation (10-7), we obtain

$$Q = +kAm(T_o - T_\infty)\frac{e^{mL} - e^{-mL}}{e^{mL} + e^{-mL}}$$

Since

$$\frac{e^{mL} - e^{-mL}}{e^{mL} + e^{-mL}} = \tanh(mL)$$

we have

$$Q = \sqrt{hPkA}\,(T_o - T_\infty)\tanh(mL) \tag{10-10}$$

Case (III):

For the finite fin with convection off the end, the final expression for the heat loss from the fin is

$$Q = \sqrt{hPkA}\,(T_o - T_\infty)\frac{\sinh(mL) + (h/mk)\cosh(mL)}{\cosh(mL) + (h/mk)\sinh(mL)} \tag{10-11}$$

The preceding analysis was developed for a rectangular fin. It is also applicable to the case of a pin fin of diameter D. The area of the surface element for a pin fin is $(\pi D dx)$ as compared to $2(1 \cdot dx)$ for a rectangular fin of unit depth. The cross-sectional area for heat conduction is $(\pi/4)D^2$ as compared to $(t \cdot 1)$ for a rectangular fin. Thus, the quantity m^2 for a pin fin takes the form $(4h/kD)$ where D is the diameter of the pin fin.

Sample Problem 10-1 One end of a very long copper rod is connected to a wall at 400°F, while the other end protrudes into a room whose air temperature is 70°F. Estimate the heat lost by the rod if it is 1/4 inch in diameter and the heat transfer coefficient between its surface and the surrounding air is 5 Btu/hr-ft^2°F.

Solution: We will assume that the rod is long enough so that it approximates the condition of an infinite fin [*Case (I)*]. Therefore

$$Q = \sqrt{hPkA}(T_o - T_\infty),\ k = 225 \text{ Btu/hr-ft°F},\ P = \pi D = 0.065 \text{ ft}$$

$$A = \pi D^2/4 = 3.41 \times 10^{-4} \text{ ft}^2,\ h = 5 \text{ Btu/hr-ft°F}$$

$$T_o = 400°\text{F},\ T_\infty = 70°\text{F}$$

It follows from $Q = \sqrt{hPkA}\ (T_o - T_\infty)$ that

$$Q = \sqrt{(5)(0.065)(225)(3.41)(10^{-4})}\ (400 - 70)$$

$$= (0.158)\,(330)$$

$$Q = 52.1 \text{ Btu/hr}$$

Sample Problem 10-2 To determine the thermal conductivity of a long, solid 1-inch diameter rod, its base is placed in a furnace with a large portion of it projecting into room air at 70°F. After steady-state conditions prevail, the temperature at two points, 4 inches apart, are found to be 327°F and 187°F, respectively. The convective heat transfer coefficient between the rod surface and the surrounding air is 5 Btu/hr-ft^2°F. Determine the thermal conductivity of the rod material.

Solution: We can arbitrarily choose $x = 0$ at the point where the temperature is given to be 327°F. The point $x = 0$ does not have to be at the physical end of the bar. Also, since the bar is long, we can treat it as an infinitely long rod. For such a case, we can use equation (10-3).

$$\frac{T - T_\infty}{T_o - T_\infty} = e^{-mx} \qquad (10\text{-}3)$$

For the rod

$$\text{at } x = 0, T = T_o = 327°\text{F}$$

$$\text{at } x = 4 \text{ in.}, T = 187°\text{F}$$

Hence

$$\frac{187 - 70}{327 - 70} = e^{-(m4/12)}$$

giving

$$m = 2.36 = \sqrt{\frac{hP}{kA}} \quad \text{and} \quad m^2 = 5.57$$

Also

$$P = \pi D = \pi 1/12 = 0.262 \text{ ft}$$

$$A = (\pi/4)D^2 = \pi/4(1/12)^2 = 5.45 \times 10^{-3} \text{ ft}^2$$

and

$$h = 5 \text{ Btu/hr-ft}^2\text{°F}$$

Hence

$$k = \frac{hP}{5.57A} = \frac{5\,(0.262)}{(5.57)\,(5.45)\,(10^{-3})} = 43.1 \text{ Btu/hr-ft°F}$$

Sample Problem 10-3 Rework Sample Problem 10-1 for *Case (II)*, the fin with the insulated tip, if the fin length were 6 inches.

Solution:

$$Q = \sqrt{hPkA}\,(T_o - T_\infty) \tanh (mL)$$

$$m = \sqrt{\frac{hP}{kA}} = \sqrt{\frac{(5)\,(0.065)}{(225)\,(3.4)\,(10^{-4})}}$$

$$m = 2.06$$

From Sample Problem 10-1, $\sqrt{hPkA} = 0.158$

$$mL = (2.06)(1/2) = 1.03 \quad \text{and} \quad \tanh(1.03) = 0.77$$

Hence

$$Q = (0.158)(330)(0.77) = 40 \text{ Btu/hr}$$

Or, the heat loss from the insulated fin is 77% of that lost by the infinite fin.

The question is often asked if there are situations where adding fins may actually decrease the heat transfer from a given surface? The answer is yes. Consider the surface sketched in Figure 10-2.

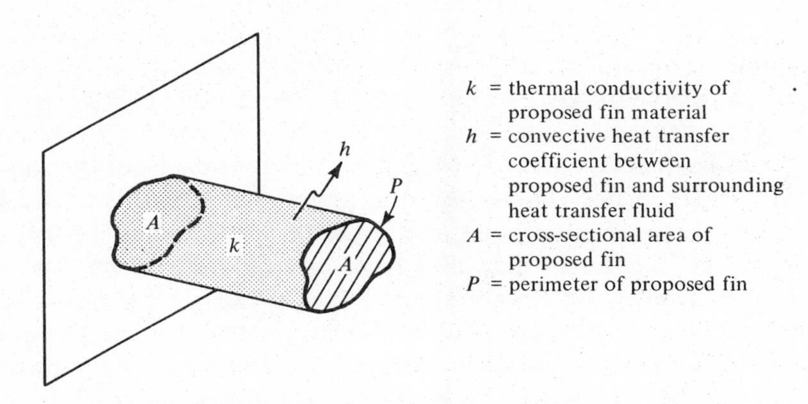

Figure 10-2 Surface to be finned.

If the value of the parameter (mL) is sufficiently large, $\tanh(mL) \simeq 1.0$, and the amount of heat transferred from the fin, from equation (10-10), is

$$Q_{\text{fin}} \simeq \sqrt{hPkA}\,(T_o - T_\infty)$$

If no fin were attached to the main body, the amount of heat transfer in such a situation would be

$$Q_{\text{no fin}} = hA(T_o - T_\infty)$$

Hence

$$\frac{Q_{\text{no fin}}}{Q_{\text{fin}}} \simeq \sqrt{\frac{hA}{kP}}$$

is a ratio of the heat flow without finning to the heat flow with finning and should have a value much less than unity to justify the addition of fins. This means that finning is only justified in systems where h is small. That is to say that in systems where boiling or condensation occur, finning will probably not be beneficial since h is very large, whereas in systems where gases are present, finning will help if h is sufficiently small to make the quantity $(hA/kP)^{1/2} \ll 1$. Also, it is apparent that the greater the value of the thermal conductivity of the fin material, the more likely $(hA/kP)^{1/2} \ll 1$ and the more likely the addition of fins will be advantageous.

10-3 HEAT TRANSFER FROM RECTANGULAR, TRIANGULAR, AND CIRCUMFERENTIAL FINS

Fins come in many shapes and cross sections (see Figure 10-3). The fin discussed earlier was a longitudinal rectangular fin. There are longitudinal trapezoidal or longitudinal triangular fins, whose thickness t is a linear function of x. If the functional form relating thickness, t, and the distance from the root of the fin, x, is nonlinear, then one can also have hyperbolic or circular profiles. When the fins are in the shape of an annular disc fitted on a tube, they are known as *circumferential fins.* Such fins may also have varying thicknesses. The procedure for analyzing triangular or circumferential fins is exactly the same as that used to analyze rectangular fins; an additional complexity, however, arises due to a continual change in the area available for heat conduction in the x direction. The area available for heat conduction decreases for a triangular (or trapezoidal) fin and increases for a circumferential fin of rectangular section, as one goes from the base to the tip of the fin. The solutions to these cases involve Bessel functions. Chart solutions for such cases are presented in Section 10-31.

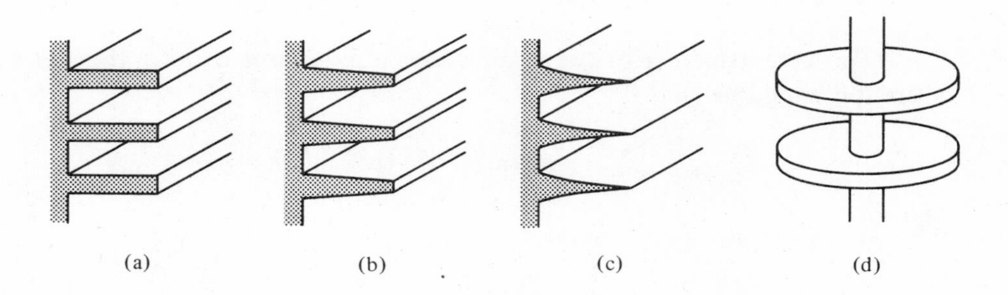

Figure 10-3 Fins of various shapes. (a) Rectangular. (b) Trapezoidal. (c) Arbitrary profile. (d) Circumferential.

It is necessary to know something about the heat transfer characteristics of various fins so that a proper selection may be made in designing a piece of heat transfer equipment. When selecting fins for a given application, the available space, weight, and cost must all be considered. In addition, the thermal properties of the fluid flowing over the fins must be considered along with the pump work necessary to pump the fluid across the fins if they are used in a forced convection system.

10-3.1 Fin Efficiency

In selecting fins, one of the parameters used is fin efficiency, η_f.

$$\eta_f = \frac{\text{actual heat transferred by fin}}{\left(\begin{array}{l}\text{the amount of heat transferred}\\ \text{if the entire fin were at the}\\ \text{base temperature}\end{array}\right)}$$

For an infinite rectangular fin [*Case (I)*]

$$\eta_f = \frac{\sqrt{hPkA}\,(T_o - T_\infty)}{hPL\,(T_o - T_\infty)}$$

$$\eta_f = \sqrt{\frac{kA}{hPL^2}} \qquad (10\text{-}12)$$

For a rectangular fin with an insulated tip [*Case (II)*]

$$\eta_f = \frac{\sqrt{hPkA}\,(T_o - T_\infty)\tanh(mL)}{hPL\,(T_o - T_\infty)}$$

$$\eta_f = \frac{\tanh(mL)}{mL} \qquad (10\text{-}12a)$$

For a real rectangular fin, Reference 1 indicates that the fin efficiency may be calculated from equation (10-12a) if

(1) the fin is long, wide, and thin,
(2) L is replaced by a corrected length, L_c, which is equal to $[L + (t/2)]$. This corrected length compensates for the fact that there is a convective loss of heat from the tip of a real fin.
(3) The quantity P/A is taken to be approximately equal to $(2/t)$, where t is the thickness of the fin.

Then we write

$$\eta_f = \frac{\tanh[\sqrt{2h/kt}\,(L + (t/2))]}{\sqrt{2h/kt}\,[L + (t/2)]} \tag{10-12b}$$

and the heat flow becomes

$$Q = \sqrt{hPkA}\,(T_o - T_\infty)\tanh[\sqrt{2h/kt}\,(L + (t/2)]$$

$$= \eta_f hPL_c(T_o - T_\infty)$$

In plotting various fin efficiencies, the profile area, A_m, is often used. The value of A_m and L_c are given in Figures 10-4 and 10-5, which give fin efficiencies for straight rectangular and triangular fins and for circumferential fins of rectangular profile.

The approach for calculating the heat flow rates from real fins as presented in Figures 10-4 and 10-5 is summarized below:

(1) Calculate the required parameters.
(a) For a straight rectangular fin

$$L_c = L + (t/2)$$

$$A_m = tL_c$$

(b) For a straight triangular fin

$$L_c = L$$

$$A_m = (t/2)L_c$$

(c) For a circumferential fin of rectangular cross section having inner radius, r_1, and outer radius, r_2

$$L = r_2 - r_1$$

$$L_c = L + (t/2)$$

$$r_{2c} = r_1 + L_c$$

$$A_m = t(r_{2c} - r_1)$$

Form the ratio (r_{2c}/r_1) in order to read Figure 10-5.

(2) Calculate the parameter

$$L_c^{3/2}\left(\frac{h}{kA_m}\right)^{1/2}$$

(3) Using the charts, determine η_f.

(4) Calculate the heat flow for the fin using its corrected fin length assuming its entire surface to be at the base temperature.
(a) For a straight rectangular fin of unit depth

$$Q_{max} = hPL_c(T_o - T_\infty)$$

(b) For a straight triangular fin of unit depth

$$Q_{max} = hPL_c(T_o - T_\infty)$$

(c) For circumferential fins of rectangular cross section

$$Q_{max} = 2\pi h(r_{2c}^2 - r_1^2)(T_o - T_\infty)$$

(5) Find the actual heat flow by multiplying Q_{max} by η_f.

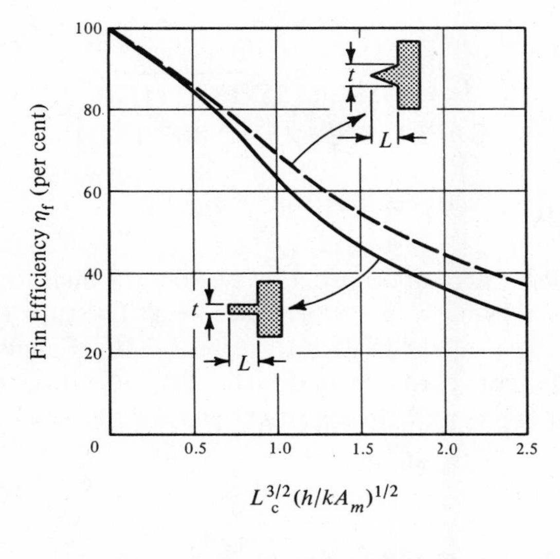

$$L_c = \begin{cases} L + t/2 & \text{rectangular fin} \\ L & \text{triangular fin} \end{cases}$$

$$A_m = \begin{cases} tL_c & \text{rectangular fin} \\ (t/2)L_c & \text{triangular fin} \end{cases}$$

Figure 10-4 Efficiencies of rectangular and triangular fins.

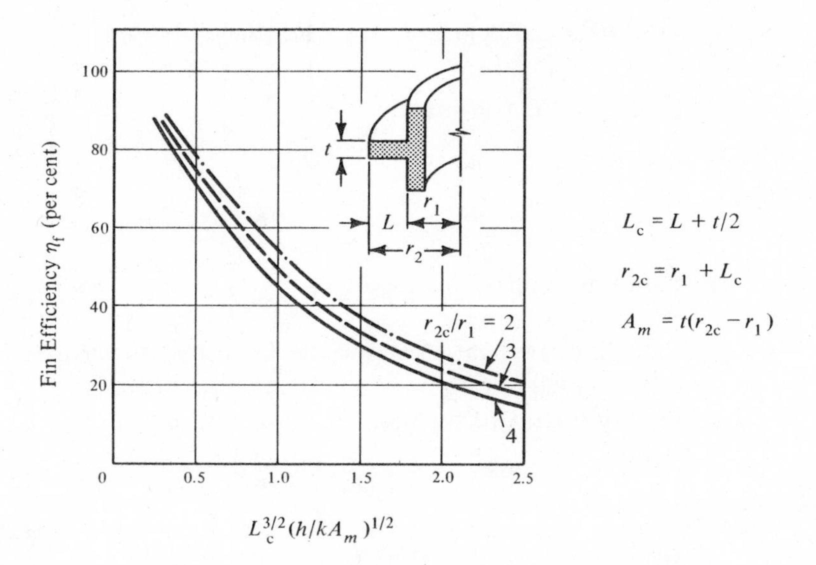

Figure 10-5 Efficiencies of circumferential fins of rectangular profile.

Kern and Kraus (Reference 2) give an excellent discussion of heat transfer from extended surfaces.

Sample Problem 10-4 Determine the fin efficiency for the fin described in Sample Problem 10-1 if its length were 6 inches.

Solution:

$$\eta_f = \sqrt{\frac{kA}{hPL^2}} = \sqrt{\frac{(225)\,(3.41)\,(10^{-8})}{(5)\,(0.065)\,(1/2)^2}} \tag{10-12}$$

$$\eta_f = 0.97 \equiv 97\%$$

Sample Problem 10-5 Circumferential steel fins 3/8-inch long and 1/4-inch thick are placed on a 1-inch tube to dissipate heat. The tube surface temperature is 400°F, and the ambient air temperature is 70°F. Calculate the heat dissipated per fin if the convective heat transfer coefficient between the fin and the air is 20 Btu/hr-ft^2°F and the thermal conductivity of steel is taken to be 30 Btu/hr-ft°F.

Solution:

$$L_c = L + (t/2) = \frac{3}{8} + \frac{1}{8} = \frac{1}{2} \text{ inch} = 0.0417 \text{ ft}$$

$$r_{2c} = r_1 + L_c = \frac{1}{2} + \frac{1}{2} = 1 \text{ inch}$$

$$\frac{r_{2c}}{r_1} = \frac{1}{1/2} = 2$$

$$A_m = t(r_{2c} - r_1) = \frac{1}{4}\left(\frac{1}{2}\right)\frac{1}{144} = 8.68 \times 10^{-4} \text{ ft}^2$$

$$L_c^{3/2}\left(\frac{h}{kA_m}\right)^{1/2} = (0.0417)^{3/2}\left[\frac{20}{(30)(8.68)(10^{-4})}\right]^{1/2} = 0.24$$

From Figure 10-5 with $r_{2c}/r_1 = 2$, we read the value of η_f to be 0.92.

The heat that would have been transferred from the fin if the entire fin were all at the base temperature is

$$Q_{max} = 2h\pi(r_{2c}^2 - r_1^2)(400 - 70)$$

$$Q_{max} = (2)(20)(\pi)\left[\left(\frac{1}{12}\right)^2 - \left(\frac{1}{24}\right)^2\right](330)$$

Hence

$$Q_{max} = 216 \text{ Btu/hr}$$

and the actual heat flow is

$$Q = \eta_f Q_{max} = 199 \text{ Btu/hr}$$

10-4 HEAT EXCHANGERS

The term *heat exchanger* implies a device that will serve to transfer heat energy from a hot fluid to a cold fluid. Heat exchangers can be classified in many different ways. One way is to base the classification on the relative directions of the flow of the hot and cold fluids, giving rise to terms like *parallel flow*, when both the fluids move in the same direction; *counter flow*, when the fluids move in parallel but in opposite directions; and *cross flow*, when the directions of flow are mutually perpendicular. These configurations are shown in Figure 10-6.

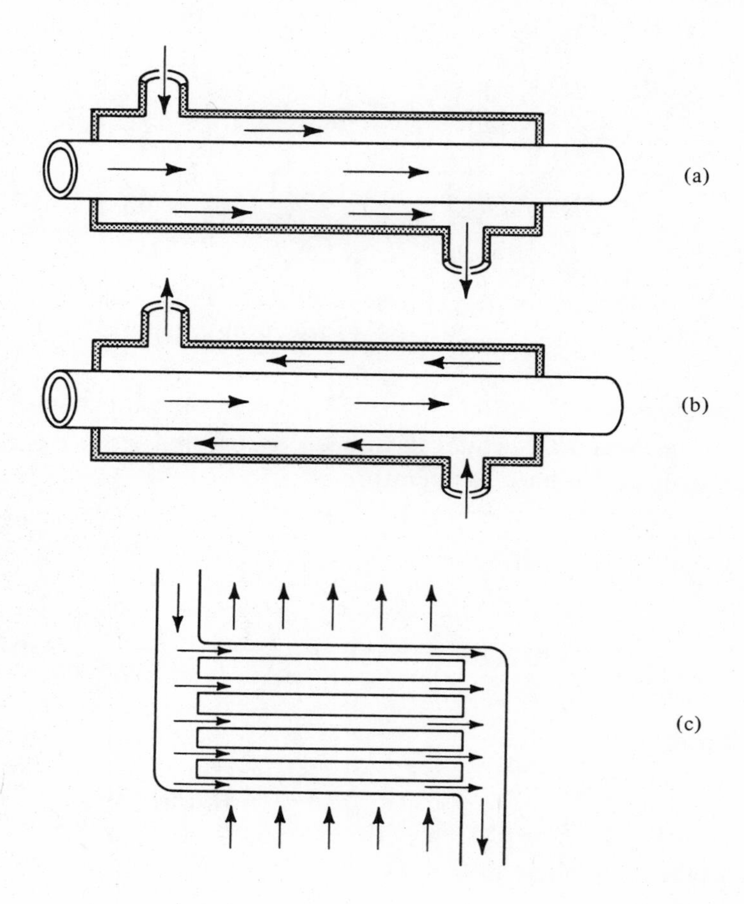

Figure 10-6 Heat exchangers. (a) Parallel flow. (b) Counter flow. (c) Cross flow.

The parallel and the counter flow modes usually involve two concentric tubes with one fluid flowing in the central tube and the other flowing in the annulus. In practical applications, one seldom finds a truly parallel- or counter-flow type exchanger.

One of the important parameters controlling the net heat transfer from the hot fluid to the cold fluid is the surface area separating the two fluids across which heat transfer takes place. When large quantities of heat are to be transferred, computations usually indicate a requirement of large heat transfer areas. Increasing the area necessarily means either increasing the total path length travelled by the fluids in the heat exchanger or decreasing the diameter of tubes and increasing the number of tubes at the same time. The second alternative can lead to large pressure drops. From a practical point of view, a long path for the first alternative cannot be one dimensional. One therefore finds situations as depicted in Figure 10-7, where there are more than one pass for the two fluids. The fluid flowing in the tubes is called the *tube fluid*, while the fluid flowing outside the tubes is referred to as the *shell fluid*.

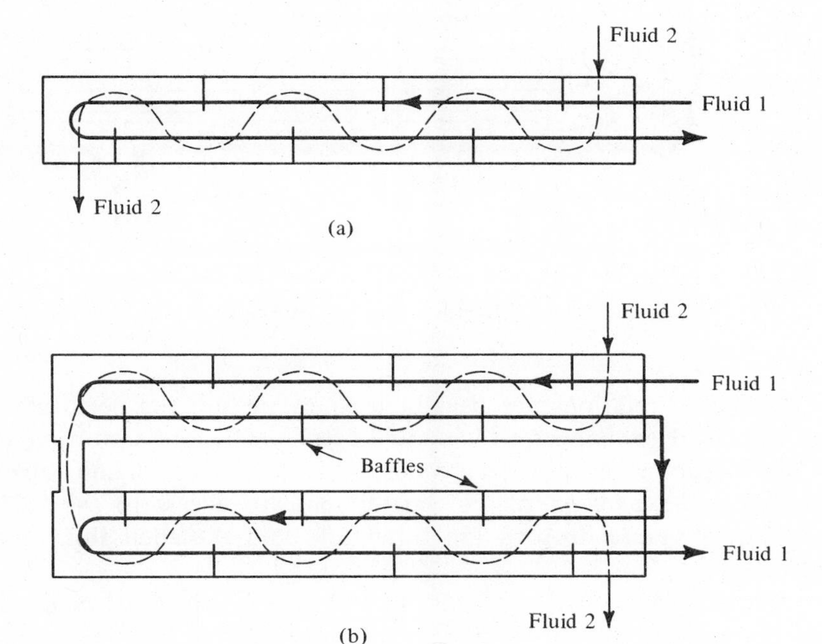

Figure 10-7 Heat exchangers. (a) Two-tube pass, one-shell pass heat exchanger. (b) Four-tube pass, two-shell pass heat exchanger.

Fluids may be turned around once, twice, or many more times to achieve a compact design resulting in exchangers with two tube passes, four tube passes, etc. Such configurations result in heat exchangers that are not purely parallel flow nor counter flow arrangements. The baffles shown in Figure 10-7 serve to create turbulence in the shell fluid, thereby enhancing the heat transfer rate. Over a period of time, deposits or scales build up on the inside surface of the tubes, thus requiring periodic cleaning.

For heavy duty installations, shell and tube type heat exchangers, (Figure 10-8), are used. The shell usually contains baffles or vertical plates with very little end clearances, the purpose of which is to accelerate the shell fluid to high velocities and thus bring about an improvement in heat transfer coefficients. They also serve to increase the path length of the shell fluid. A tube bundle consisting of a large number of tubes through which the tube fluid moves is housed within the shell. The tube-ends may have floating or fixed heads. In the latter case, only moderate temperature changes can be used so as to limit thermal stresses.

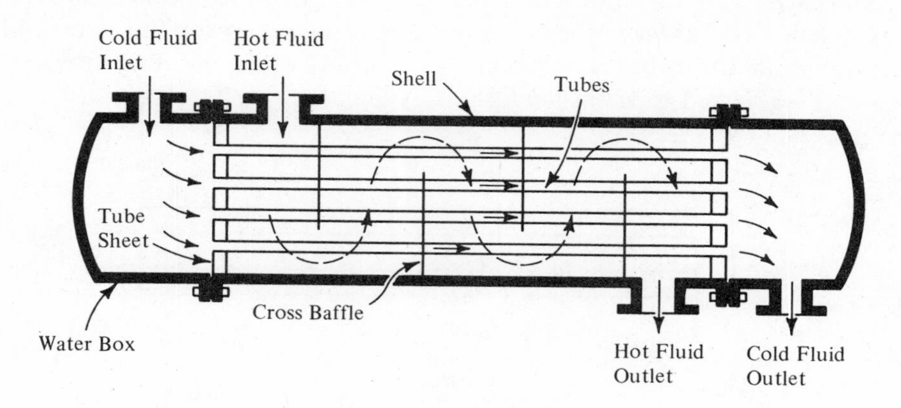

Figure 10-8 Schematic drawing of one-shell pass, one-tube pass heat exchanger.

Empirical relations are usually used to predict the performance of the shell and tube heat exchanger, which could be of a single- or a multiple-pass type. Some of the factors governing the mechanical design of a shell and tube heat exchanger are thermal expansion of the heater, ease of access to the tubes for cleaning purposes, pressure drop in the tube fluid and the shell fluid, clearances at the baffles, etc.

We have seen earlier in this chapter that longitudinal or circumferential fins can be used effectively to increase the rate of heat transfer from surfaces. When the heat exchanger tubes have such fins, pins, or spiral grooves, they lead to a different class of heat exchangers that are usually evaluated and rated by the manufacturer. Such heat exchangers are used when a liquid flows inside the tubes and a gas flows outside the tubes over the extended surfaces. An automobile radiator belongs to this category. Usually the gas on the outside will not cause any significant scaling, and therefore there will not be any need to clean the fins. The extended surface becomes a necessity in view of the rather low values of convective heat transfer coefficients for gases. Some of the extended surfaces used are shown in Figure 10-9a. Still other types of heat exchangers result from forming passages for flow by inserting a number of tubes of circular or elliptic sections through a number of parallel plates. They may also be manufactured by forming, bending, and welding or brazing a number of metal sheets resulting in compact heat exchangers (Figure 10-9b). They usually are of the cross-flow type and are used when the pressure drops available for moving the fluids are limited and when the heat exchanger load is moderate.

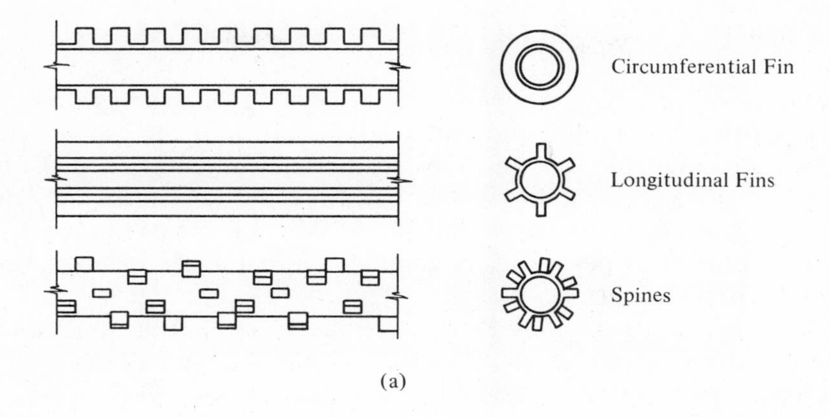

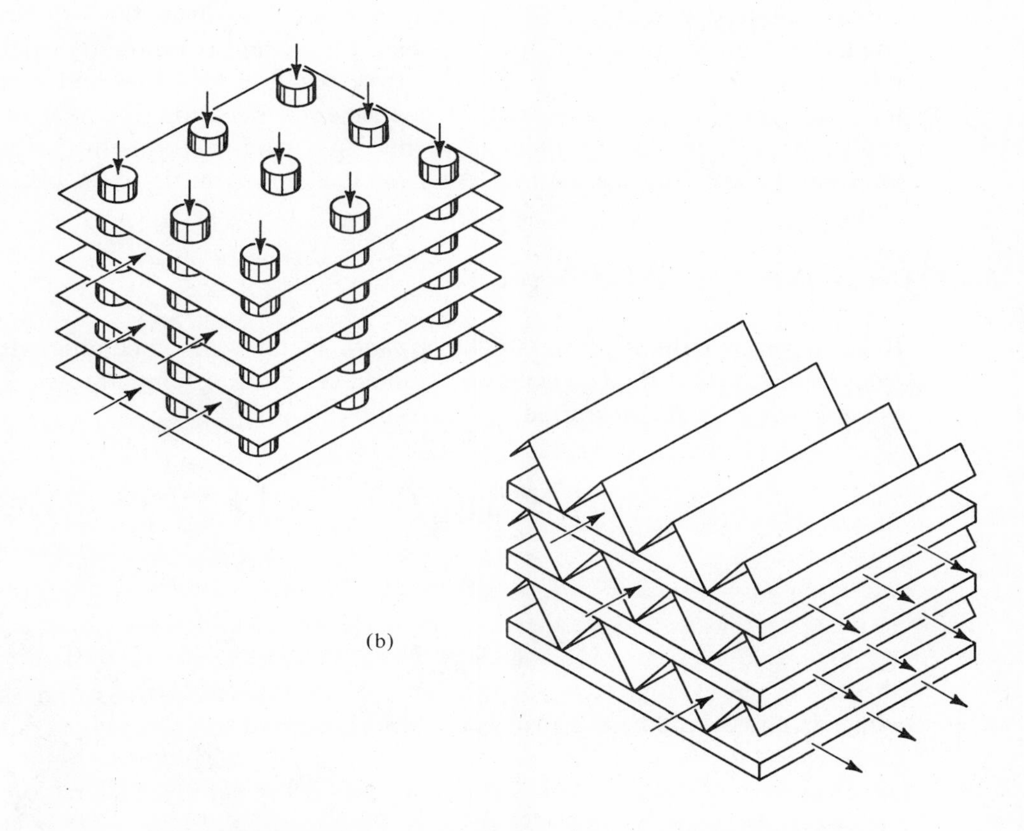

Figure 10-9 (a) Extended surfaces used to promote heat transfer. (b) Cross-flow compact exchanger.

10-5 PERFORMANCE VARIABLES FOR HEAT EXCHANGERS

The variables affecting the performance of a heat exchanger are the mass flow rates, specific heats, inlet and outlet temperatures of the hot and the cold fluids, surface area available for heat transfer, thermal conductivity of the tube material, degree of deposits or scaling on the inside of the tubes, and the convective heat transfer coefficients on the inside and outside surfaces of the tubes. The effect of the last four quantities is usually combined into a single quantity, the overall heat transfer coefficient, U.

Once the heat capacities, the inlet temperatures and the quantity of heat to be transferred are specified, the principles of thermodynamics will immediately fix the outlet temperatures. Heat transfer computations will, however, be necessary in order to determine how much surface area is needed provided the U value is known. Such calculations will also be needed to determine if a given heat exchanger will do a required job or not. The pressure drop through the heat exchanger is directly related to the pumping power and is indirectly related to the rate of heat transfer since it does control the flow velocities and therefore the mass flow rate and the heat transfer coefficients. Selection of a heat exchanger is really an optimization problem involving such parameters as the pressure drop, pumping power, surface area of the heat exchanger, initial cost, cleaning costs, etc.

10-5.1 Overall Heat Transfer Coefficient, U

If we have two fluids separated by a *plane* metal wall of conductivity, k, and thickness, t, and if the fluids are maintained at constant temperatures, T_h and T_c, the amount of heat transferred, Q, is given by

$$Q/A = h_h(T_h - T_1) = k\left(\frac{T_1 - T_2}{t}\right) = h_c(T_2 - T_c) \tag{10-13}$$

or

$$Q/A = U(T_h - T_c) \tag{10-13a}$$

where U, the overall heat transfer coefficient, is given by

$$\frac{1}{U} = \frac{1}{h_h} + \frac{t}{k} + \frac{1}{h_c} \tag{10-13b}$$

In the above, h_h and h_c are the convective heat transfer coefficients on the hot side and on the cold side of the metal wall, and T_1 and T_2 represent the temperatures of the two faces of the wall. In this case, we see that the overall heat transfer coefficient, U, is dependent upon h_h and h_c for a given wall. The overall heat transfer coefficient for heat exchangers depends on the convective heat transfer coefficients, h_i and h_o, on the inside and the outside surfaces of the tube and also on the surface areas, A_i and A_o, of the inside and the outside of the tubes. Before developing an expression for the overall heat transfer coefficient for a heat exchanger, we give expressions for the Nusselt numbers under various conditions to facilitate a computation of h_i and h_o:

(1) Laminar flow inside tubes at constant wall temperature, equation (8-26):

$$\mathrm{Nu}_{av} = 1.86(\mathrm{Re\ Pr})^{1/3} (D/L)^{1/3} (\mu/\mu_w)^{0.14} \quad (10\text{-}14)$$

(2) Turbulent flow inside tubes, equation (8-23):

For $2 \leqslant \mathrm{Pr} \leqslant 140 \quad 10^4 \leqslant \mathrm{Re} \leqslant 1.25 \times 10^5$

$0.08 \leqslant (\mu/\mu_w) \leqslant 40$

$$\underset{\text{(heating)}}{\mathrm{Nu}} = \frac{(f/8)\ \mathrm{Re\ Pr}}{1.07 + 12.7\sqrt{f/8}\ (\mathrm{Pr}^{2/3} - 1)} \left(\frac{\mu}{\mu_w}\right)^{0.11} \quad (10\text{-}14a)$$

$$\underset{\text{(cooling)}}{\mathrm{Nu}} = \frac{(f/8)\ \mathrm{Re\ Pr}}{1.07 + 12.7\sqrt{f/8}\ (\mathrm{Pr}^{2/3} - 1)} \left(\frac{\mu}{\mu_w}\right)^{0.25} \quad (10\text{-}14b)$$

where

$$f = [1.82 \log_{10}(\mathrm{Re}) - 1.64]^{-2}$$

(3) Flow of gases across a bundle of tubes: Use Tables 8-7 and 8-8 and equations (8-59).

(4) Filmwise condensation on horizontal tubes, equation (11-14):

$$\mathrm{Nu}_{av} = 0.73\left[\frac{g\rho_{liq}^2 h_{fg} d^3}{\mu_{liq}\Delta T k_{liq}}\right]^{1/4} \quad \text{for} \quad \rho_{liq} >> \rho_{vap} \quad (10\text{-}14c)$$

All the properties appearing in equations (10-14, a, and b), except μ_w, are evaluated at the bulk temperature, and the quantity, μ_w, is evaluated at the wall temperature. Properties with the subscript, liq, appearing in equation (10-14c) represent properties of the liquid film, evaluated at the film temperature; the quantity, h_{fg}, is the latent heat of vaporization of the liquid at the saturation temperature.

When we consider heat exchanger fluids flowing outside and inside a tube, we can write the following equation for the rate of heat transfer between the two fluids.

$$Q = 2\pi r_o L h_o (T_1 - T_o) = \frac{2\pi k L (T_o - T_i)}{\ln (r_o/r_i)} = 2\pi r_i L h_i (T_i - T_2)$$

where

T_1 = the temperature of the outside fluid

T_2 = the temperature of the inside fluid

L = the length of the tube

r_o = the outer radius of the tube

r_i = the inner radius of the tube

k = the thermal conductivity of the tube material

Now, if we let

$$\Delta T = T_1 - T_2, A_o = 2\pi r_o L, \text{ and } A_i = 2\pi r_i L$$

the above equation can be recast in the following form.

$$Q = A_o U_o \Delta T = A_i U_i \Delta T \tag{10-15}$$

where

$$\frac{1}{U_o} = \frac{1}{h_o} + \frac{r_o}{k} \ln\left(\frac{r_o}{r_i}\right) + \left(\frac{r_o}{r_i}\right)\frac{1}{h_i} \tag{10-15a}$$

and

$$\frac{1}{U_i} = \left(\frac{r_i}{r_o}\right)\frac{1}{h_o} + \frac{r_i}{k} \ln\left(\frac{r_o}{r_i}\right) + \frac{1}{h_i} \tag{10-15b}$$

The forgoing expressions for the overall heat transfer coefficients are for clean tubes. The overall heat transfer coefficient, U_o, when multiplied by the product of the surface area of the outside of the tube and ΔT will give the total amount of heat transferred. The quantity, U_o, therefore, is called the overall heat transfer coefficient based on the *outside* area, A_o. Likewise, the quantity, U_i, given by equation (10-15b), is termed the overall heat transfer coefficient based on the *inside* area of the tube. Such a distinction is necessary, because the area available for heat transfer is not constant but increases as one goes radially from the inside of the tube to the outside of the tube. Some representative values of U are given in Table 10-1.

TABLE 10-1 Approximate Values of Overall Heat Transfer Coefficients

Physical Situation	Btu/hr-ft^2°F	W/m^2°K
Brick exterior wall, plaster interior:		
uninsulated	0.45	2.50
Frame exterior wall, plaster interior:		
uninsulated	0.25	1.40
with rock-wool insulation	0.07	0.40
Plate-glass window	1.10	6.20
Double plate-glass window	0.40	2.30
Steam condenser	200-1000	1000-5000
Feedwater heater	200-1500	1000-8000
Freon 12 condenser with water coolant	50-150	300-850
Water-to-water heat exchanger	150-300	850-1700
Finned-tube heat exchanger, water in tubes, air across tubes	5-10	30-55
Water-to-oil heat exchanger	20-60	110-340

Representative values of fouling factors are listed in Table 10-2.

TABLE 10-2 Fouling Factors*

Type of fluid	hr-ft^2°F/Btu	m^2°K/W
Sea water below 125°F (50°C)	0.0005	0.0001
Sea water above 125°F (50°C)	0.001	0.0002
Treated boiler feedwater above 125°F (50°C)	0.001	0.0002
Fuel oil	0.005	0.0010
Quenching oil	0.004	0.0008
Alcohol vapors	0.0005	0.0001
Steam, non-oil-bearing	0.0005	0.0001
Industrial air	0.002	0.0004
Refrigerating liquid	0.001	0.0002

*From Reference 3.

It is a well-known fact that the inside surfaces of the tubes of a heat exchanger do not remain clean after several months of operation. Scaling or deposits form over the interior surface. Scaling or deposits on the inside of the tubes is really a gradual build-up of layers of dirt due to impurities in the fluid, chemical reaction between the fluid and the metal, rust, etc. The deposits can severely affect the value of U. The effect of the deposits is represented quantitatively by the *fouling factor*, R_f, which must be determined experimentally. Its net effect is to increase the resistance to heat flow. It is related to the overall heat transfer coefficient under clean conditions and under fouled conditions by the equation

$$\frac{1}{U_{\text{foul}}} = R_f + \frac{1}{U_{\text{clean}}}$$

10-6 ANALYSIS OF A HEAT EXCHANGER

Our objective in analyzing a heat exchanger is to be able to express the total amount of heat transferred, Q, from the hot fluid to the cold fluid in terms of the overall heat transfer coefficient, U; the surface area of the heat exchanger, A; and the inlet and the exit temperatures of the hot and the cold fluids. An energy balance on the two fluids gives

energy lost by the hot fluid = energy gained by the cold fluid

or

$$\dot{m}_h c_h (T_{h,i} - T_{h,o}) = Q = \dot{m}_c c_c (T_{c,o} - T_{c,i}) \tag{10-16}$$

where

$\dot{m}_h$ = mass flow rate of the hot fluid

c_h = constant pressure specific heat of the hot fluid

$T_{h,i}$ = inlet temperature of the hot fluid

$T_{h,o}$ = exit temperature of the hot fluid

$\dot{m}_c$ = mass flow rate of the cold fluid

c_c = constant pressure specific heat of the cold fluid

$T_{c,o}$ = exit temperature of the cold fluid

$T_{c,i}$ = inlet temperature of the cold fluid

The product ($\dot{m}c$) appears frequently in the analysis of heat exchangers and is often called the *heat capacity rate* C. It should be noted that equation (10-16) is valid for all types of exchangers.

Analysis of a truly parallel-flow or counter-flow heat exchanger is relatively straightforward and will be presented in the following section. The cross-flow heat exchanger with both fluids moving in mutually perpendicular directions presents complexities of an integro-differential equation, which are beyond the scope of this text. The multiple-pass heat exchangers present formidable difficulties for analysis, and one relies heavily on charts and empirical formulas that will be presented later in the chapter.

10-6.1 Parallel Flow Heat Exchanger

In analyzing the parallel flow exchanger, we note that the properties are no longer constant along the length of the tube, since both the fluids experience either a loss or a gain of energy, and the fluid temperatures change. The quantity, U, as a first approximation may be treated as a constant quantity. Let us consider a small length, dx, of the tube shown in Figure 10-10. The amount of heat transferred across an elemental area $2\pi r_o dx$ can be written as

$$dQ = 2\pi r_o dx \cdot U_o \cdot \theta \tag{10-17}$$

where

$$\theta = T_h - T_c$$

$$r_o = \text{outside radius of the tube}$$

and

$$U_o = \text{overall heat transfer coefficient based on the outside area}$$

Mathematically speaking, equation (10-17) contains dQ and dx as differentials of the variables Q and x; r_o and U as constants; and θ as a variable. We would like to have one more relationship among the variables Q, x, and θ or among their differentials. As the hot fluid moves a distance, dx, its temperature changes by dT_h. The change is negative, since there is a drop in the energy of the hot fluid, which must be equal to the amount of energy transported across the wall, dQ. The quantity, dQ, in turn is responsible for changing the energy of the cold fluid by bringing about a change in its temperature, which equals dT_c. Thus an energy balance gives

$$-\dot{m}_h c_h dT_h = dQ = \dot{m}_c c_c dT_c \tag{10-17a}$$

where $\dot{m}$ and c have been defined earlier in Section 10-5. Rearranging, we obtain

$$\frac{dT_h}{dQ} = -\frac{1}{\dot{m}_h c_h} \quad \text{and} \quad \frac{dT_c}{dQ} = \frac{1}{\dot{m}_c c_c} \tag{10-17b}$$

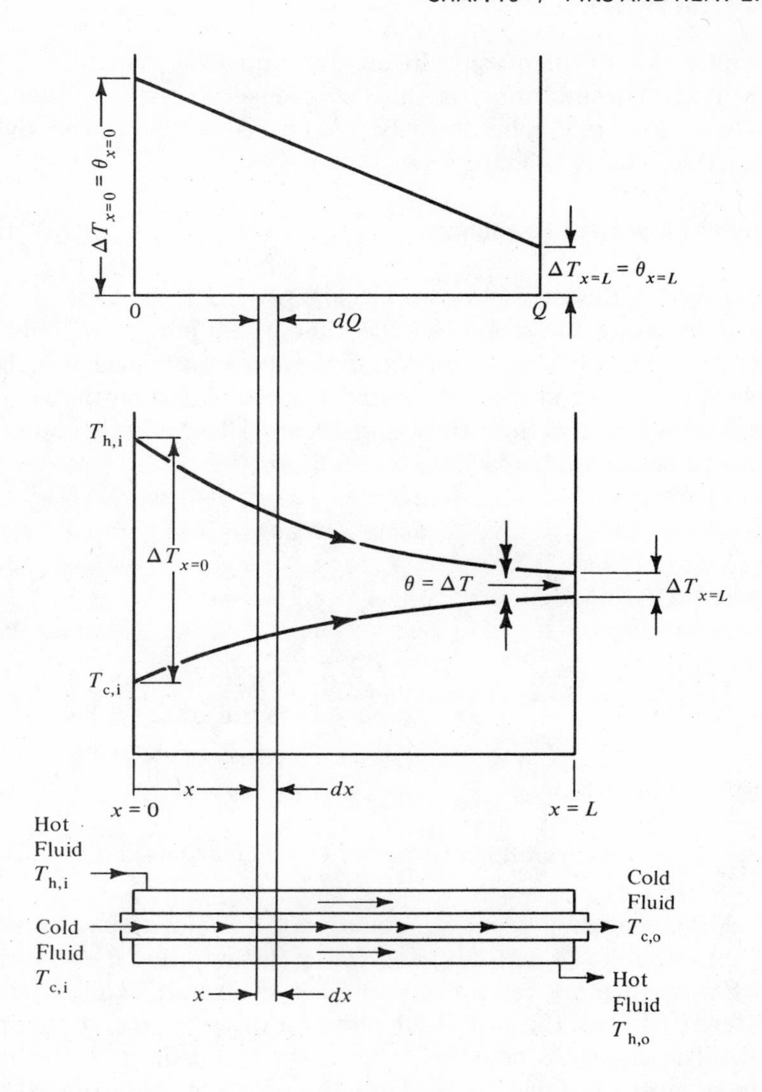

Figure 10-10 Temperature distribution in a parallel-flow heat exchanger.

If a graph of T_h versus Q and T_c versus Q is plotted (Figure 10-10), one obtains straight lines in view of equations (10-17b), since the mass flow rates and specific heats are considered as constants. The last two equations can be combined to become

$$\frac{dT_h}{dQ} - \frac{dT_c}{dQ} = -\frac{1}{\dot{m}_h c_h} - \frac{1}{\dot{m}_c c_c} = \lambda$$

where λ is a constant. Hence

$$\frac{d}{dQ}(T_h - T_c) = \lambda$$

or

$$\frac{d\theta}{dQ} = \lambda = \frac{\theta_{x=L} - \theta_{x=0}}{Q} \quad \text{[See Figure 10-10]} \tag{10-17c}$$

where

$$\theta_{x=0} = T_{h,i} - T_{c,i} \quad \text{and} \quad \theta_{x=L} = T_{h,o} - T_{c,o}$$

Equation (10-17c) offers another relationship between θ, the temperature difference across the tube wall, and Q implying that the graph of θ versus Q is a straight line with a slope equal to λ. The slope is expressed as a ratio of the change in θ as one goes from $x = 0$ to $x = L$ to the quantity of heat transferred, Q, over the entire length, L.

Now, eliminating dQ from equations (10-17) and (10-17c), we obtain

$$\frac{d\theta}{\lambda} = 2\pi r_o U_o \theta dx$$

or

$$\frac{d\theta}{\lambda\theta} = 2\pi r_o U_o dx$$

Integration gives

$$\frac{1}{\lambda}\Big[\ln\theta\Big]_{\theta_{x=0}}^{\theta_{x=L}} = \Big[2\pi r_o U_o x\Big]_{x=0}^{x=L}$$

or

$$\frac{1}{\lambda}\left[\ln\left(\frac{\theta_{x=L}}{\theta_{x=0}}\right)\right] = 2\pi r_o U_o L = U_o A_o$$

where A_o, the surface area of the heat exchanger, equals $2\pi r_o L$. Substituting for λ from equation (10-17c), the above becomes

$$\frac{Q}{\theta_{x=L} - \theta_{x=0}} \ln\left(\frac{\theta_{x=L}}{\theta_{x=0}}\right) = U_o A_o$$

or

$$Q = U_o A_o \frac{\theta_{x=L} - \theta_{x=0}}{\ln (\theta_{x=L}/\theta_{x=0})} = U_o A_o \frac{\theta_{x=0} - \theta_{x=L}}{\ln [\theta_{x=0}/\theta_{x=L}]} \quad (10\text{-}18)$$

Since, from equation (10-15), $U_o A_o = U_i A_i = UA$, we drop the subscript "o" in the U_o and A_o in the above equation and write

$$Q = UA \frac{\theta_{x=0} - \theta_{x=L}}{\ln [\theta_{x=0}/\theta_{x=L}]}$$

Substituting for $\theta_{x=0}$ and $\theta_{x=L}$, we obtain

$$Q = UA \left[\frac{(T_{h,i} - T_{c,i}) - (T_{h,o} - T_{c,o})}{\ln [(T_{h,i} - T_{c,i})/(T_{h,o} - T_{c,o})]}\right] \quad (10\text{-}18a)$$

For a parallel-flow heat exchanger, the Logarithmic Mean Temperature Difference, LMTD, is defined as the quantity in the brackets in equation (10-18a). It can be written in a more general form, applicable to any situation, in the following manner.

$$\text{LMTD} = \frac{(\Delta T)_{x=0} - (\Delta T)_{x=L}}{\ln[(\Delta T)_{x=0}/(\Delta T)_{x=L}]} \quad (10\text{-}19)$$

Thus

$$Q = UA(\text{LMTD}) \quad (10\text{-}20)$$

The quantity, ΔT, in equation (10-19) denotes the temperature difference between the hot and the cold fluids at a given location in the heat exchanger. The above definition of LMTD is used in multiple pass exchangers also. Equations (10-16) and (10-20) allow one to solve for any two unknown quantities, typically one of the four temperatures and the heat transfer area, A.

10-6.2 Counter-flow Heat Exchanger

Although equation (10-20) was derived for a parallel flow heat exchanger, is is equally valid for a counter-flow heat exchanger (Figure 10-11) where the fluids move in parallel but in opposite directions.

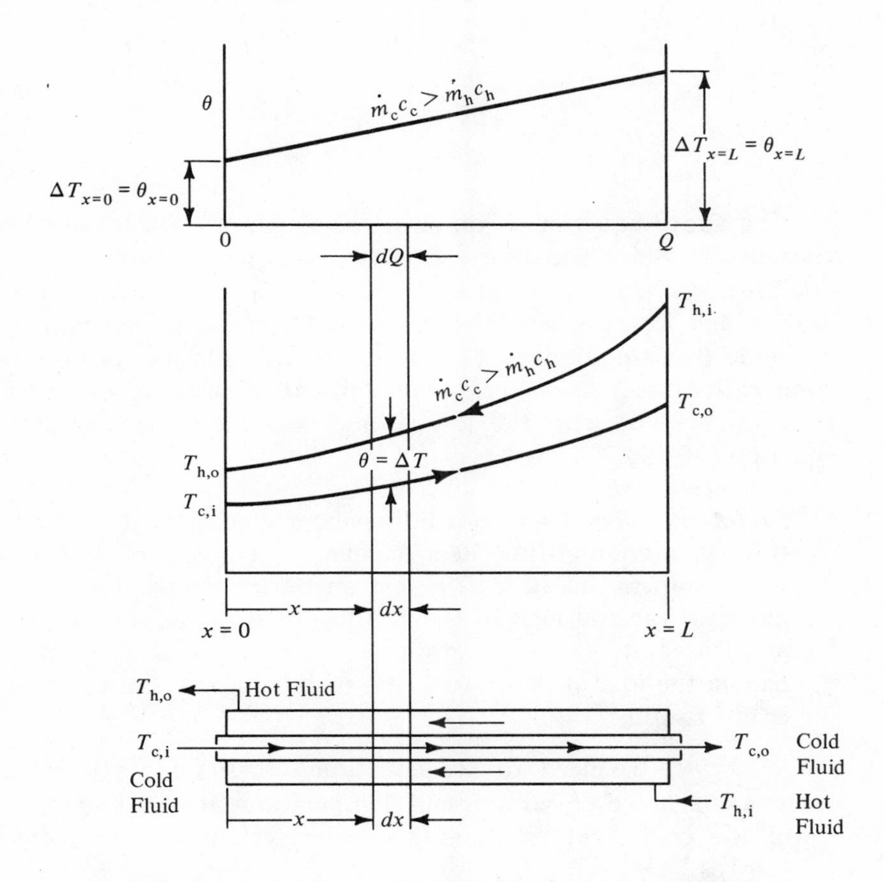

Figure 10-11 Temperature distribution in a counter-flow heat exchanger.

Equation (10-16) remains the same as before. The hot fluid is entering at $x = L$, while the cold fluid is entering at $x = 0$. Consequently, temperature variations are as depicted in Figure 10-11. If an observer moves along the x coordinate a distance dx, he will find that both fluids register a temperature rise. It does not mean that both fluids are gaining energy! Performing an energy balance, we may write

$$+ \dot{m}_h c_h dT_h = dQ = + \dot{m}_c c_c dT_c$$

We observe that the above equation is the same as equation (10-17a) except for the sign of the left- most quantity. Rearranging gives

$$\frac{d}{dQ}(T_h - T_c) = \left[\frac{1}{\dot{m}_h c_h} - \frac{1}{\dot{m}_c c_c}\right] = \lambda'$$

or

$$\frac{d\theta}{dQ} = \lambda' = \frac{\theta_{x=L} - \theta_{x=0}}{Q}$$

The above equation is the same as equation (10-17c) except for a different constant λ'. Since equation (10-18a) was obtained from equations (10-17) and (10-17c), which we have shown to be valid for counter-flow heat exchangers and since λ and λ' are given by the same expression, i.e., $[(\theta_{x=L} - \theta_{x=0})/Q]$, we conclude that equations (10-17c) and (10-18) are valid for both parallel-flow and counter-flow heat exchangers. The amount of heat transferred is then given by equation (10-20) with the logarithmic mean temperature difference given by equation (10-19).

Sample Problem 10-6 Hot oil having a specific heat of 0.5 Btu/lb$_m$ °F flows through a counter-flow heat exchanger at a rate of 50,000 lb$_m$/hr with an inlet temperature of 380°F and an outlet temperature of 150°F. Cold oil having a specific heat of 0.4 Btu/lb$_m$ °F flows in at a rate of 80,000 lb$_m$/hr and leaves at 300°F. Determine the area of the heat exchanger necessary to handle the load if the overall heat transfer coefficient based on the inside area is 135 Btu/hr-ft^2 °F.

Solution: We need to use equations (10-19) and (10-20) to calculate the area, A_i; however, all four end- temperatures are not known. The temperature of the cold fluid entering the exchanger can be computed from equation (10-16).

$$\dot{m}_c c_c (T_{c,o} - T_{c,i}) = \dot{m}_h c_h (T_{h,i} - T_{h,o})$$

$$\dot{m}_c = 80{,}000 \text{ lb}_m/\text{hr},\ T_{c,o} = 300°\text{F},\ c_c = 0.4 \text{ Btu/lb}_m\,°\text{F}$$

$$\dot{m}_h = 50{,}000 \text{ lb}_m/\text{hr},\ c_h = 0.5 \text{ Btu/lb}_m\,°\text{F}$$

$$T_{h,i} = 380°\text{F},\ T_{h,o} = 150°\text{F}$$

Substitution gives

$$80{,}000 \times 0.4(300 - T_{c,i}) = 50{,}000 \times 0.5(380 - 150)$$

or

$$T_{c,i} = 120°F$$

Hence

$$(\Delta T)_{x=0} = T_{h,o} - T_{c,i} = 150 - 120 = 30°F$$

$$(\Delta T)_{x=L} = T_{h,i} - T_{c,o} = 380 - 300 = 80°F$$

$$\text{LMTD} = \frac{(\Delta T)_{x=0} - (\Delta T)_{x=L}}{\ln[(\Delta T)_{x=0}/(\Delta T)_{x=L}]} = \frac{30 - 80}{\ln(30/80)} = 51°F$$

$$Q = \dot{m}_h c_h (T_{h,i} - T_{h,o}) = U_i A_i \text{LMTD}$$

or

$$50{,}000 \times 0.5 \times (380 - 150) = 135 \times A_i \times 51$$

or

$$A_i = 835 \text{ ft}^2$$

Sample Problem 10-7 A certain heat exchanger has a total outside surface area of 17.5 m^2. It is to be used for cooling oil at 200°C with a mass flow rate of 10,000 kg/hr having a specific heat of 1900 J/ (kg°K). Water at a flow rate of 3000 kg/hr is available at 20°C as a cooling agent. If the overall heat transfer coefficient is 300 W/m^2°K based on the outside area, estimate the temperature of the oil as it exits from the heat exchanger if the heat exchanger is operated (1) in a parallel-flow mode, and (2) in a counter-flow mode.

Solution: We observe that the exit temperatures of both the oil and the water are not given, and therefore neither LMTD nor Q can be directly calculated. However, we do have two expressions for Q, one based on an energy balance and the other based on the LMTD, which can be equated. It may seem that a trial-and-error solution is at hand, but such is not the case, as shown below.

$$\dot{m}_h = 10{,}000 \text{ kg/hr}, \quad T_{h,i} = 200°C, \quad U_o = 300 \text{ W/m}^2{}°K$$

$$\dot{m}_c = 3000 \text{ kg/hr}, \quad T_{c,i} = 20°C, \quad A_o = 17.5 \text{ m}^2$$

$$c_h = 1900 \text{ J/kg}°K, \quad c_c = 4186 \text{ J/kg}°K \text{ (for water)}$$

$$\dot{m}_h c_h (T_{h,i} - T_{h,o}) = Q = \dot{m}_c c_c (T_{c,o} - T_{c,i})$$

$$10{,}000 \times 1900 (200 - T_{h,o}) = 3000 \times 4186 (T_{c,o} - 20)$$

or

$$T_{c,o} = 322.6 - 1.513T_{h,o}$$

The above equation is valid for parallel-flow as well as for counter-flow operation.

(1) *Parallel Flow Operation:*

$$\dot{m}_h c_h (T_{h,i} - T_{h,o}) = Q = U_o A_o \left\{ \frac{(T_{h,i} - T_{c,i}) - (T_{h,o} - T_{c,o})}{\ln [(T_{h,i} - T_{c,i})/(T_{h,o} - T_{c,o})]} \right\}$$

Substituting, we obtain

$$10{,}000 \times 1900 (200 - T_{h,o}) = 300 \times 17.5 \times \frac{(200 - 20) - (T_{h,o} - T_{c,o})}{\ln[(200 - 20)/(T_{h,o} - T_{c,o})]} \times 3600$$

Substituting for $T_{c,o}$, and simplifying, we have

$$\ln \frac{180}{2.513T_{h,o} - 322.6} = (1.01) \frac{2.513(200 - T_{h,o})}{(200 - T_{h,o})} = 2.54$$

or

$$\frac{180}{2.513T_{h,o} - 322.6} = e^{2.54} = 12.68$$

or

$$180 = 12.68 (2.513T_{h,o} - 322.6)$$

from which $T_{h,o} = 134°C$. Hence

$$T_{c,o} = 322.6 - (1.513 \times 134) = 120°C$$

(2) *Counter-flow Operation:*

$$\dot{m}_h c_h (T_{h,i} - T_{h,o}) = Q = U_o A_o \left\{ \frac{(T_{h,i} - T_{c,o}) - (T_{h,o} - T_{c,i})}{\ln [(T_{h,i} - T_{c,o})/(T_{h,o} - T_{c,i})]} \right\}$$

Substituting values of various parameters, we get

$$10{,}000 \times 1900\,(200 - T_{h,o}) = 300 \times 17.5 \times \frac{(200 - T_{c,o}) - (T_{h,o} - 20)}{\ln[(200 - T_{c,o})/(T_{h,o} - 20)]} \times 3600$$

Substituting for $T_{c,o}$ and rearranging, we have

$$\ln\left[\frac{1.513T_{h,o} - 122.6}{T_{h,o} - 20}\right] = \frac{0.513\,(T_{h,o} - 200)}{0.986(200 - T_{h,o})} = -0.518$$

Further simplification gives

$$T_{h,o} = 120.6°\text{C}$$

Finally, the energy balance equation yields

$$T_{c,o} = 140.1°\text{C}$$

Note that in the parallel-flow heat exchanger, the temperature $T_{h,o}$ was greater than the temperature $T_{c,o}$; these two will be equal only when the area of the parallel-flow heat exchanger is infinitely large. This is due to a progressively decreasing driving potential $(T_h - T_c)$ that is available for heat transfer as one moves from the inlet to the outlet. In the case of the counter-flow exchanger, the temperature of the exiting cold fluid was found to be greater than the temperature of the exiting hot fluid. It can be seen from this example that a counter-flow heat exchanger will transfer a greater quantity of heat than a parallel-flow heat exchanger, operating under otherwise similar conditions.

10-6.3 Multiple-Pass Heat Exchangers

In a double-pass heat exchanger, Figure 10-7a, the fluid in the tube experiences both parallel flow and counter flow. McAdams (Reference 4) gives an expression for the rate of heat transfer in this case. For multiple-tube and shell passes, charts are used. The heat transfer is calculated from the equation

$$Q = UAF\,(\text{LMTD}) \tag{10-20a}$$

where F is a correction factor to be read from an appropriate chart. And

$$\text{LMTD} = \frac{(T_{h,i} - T_{c,o}) - (T_{h,o} - T_{c,i})}{\ln[(T_{h,i} - T_{c,o})/(T_{h,o} - T_{c,i})]} \tag{10-20b}$$

The factor, F, takes into account the effects of the multiple-shell or multiple-tube passes. One should try to select parameters so that the value of F is greater than 0.75. If a smaller value is determined for a given configuration, another configuration should be analyzed. When the cold fluid flows through the tubes the parameters used in the charts for the correction factor are

$$P = \frac{T_{c,o} - T_{c,i}}{T_{h,i} - T_{c,i}}$$

$$R = \frac{T_{h,i} - T_{h,o}}{T_{c,o} - T_{c,i}}$$

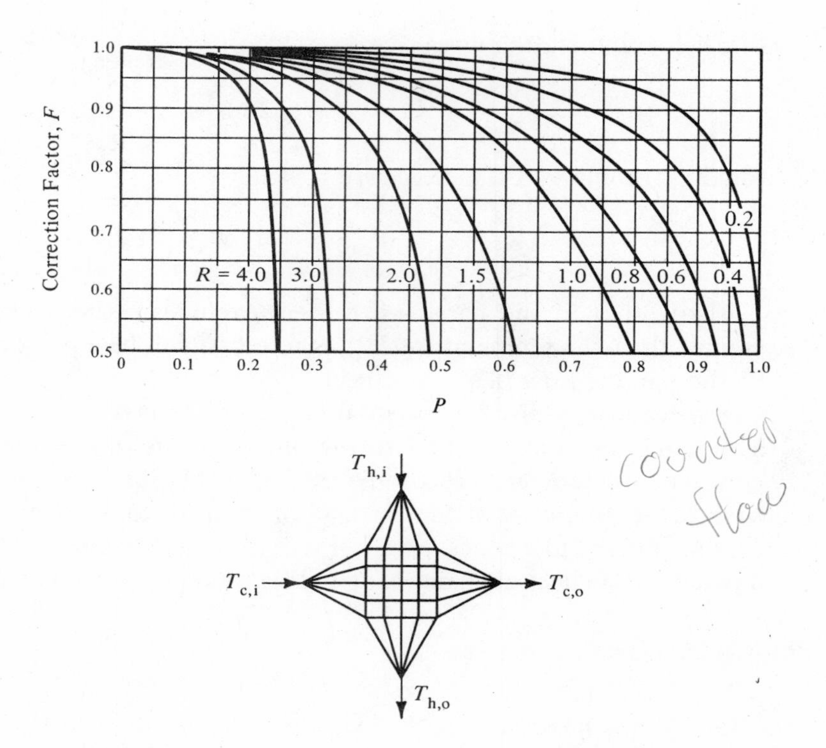

$$P = \frac{T_{c,o} - T_{c,i}}{T_{h,i} - T_{c,i}}$$

$$R = \frac{T_{h,i} - T_{h,o}}{T_{c,o} - T_{c,i}}$$

When hot fluid flows in tubes, interchange subscripts c and h.

Figure 10-12 Correction factor, F, for a cross-flow heat exchanger with both fluids unmixed, from *Mean Temperature Difference in Design* by Bowman, R. A., A. C. Mueller, and W. M. Nagel, *ASME Trans.*, Vol. 62, 1940.

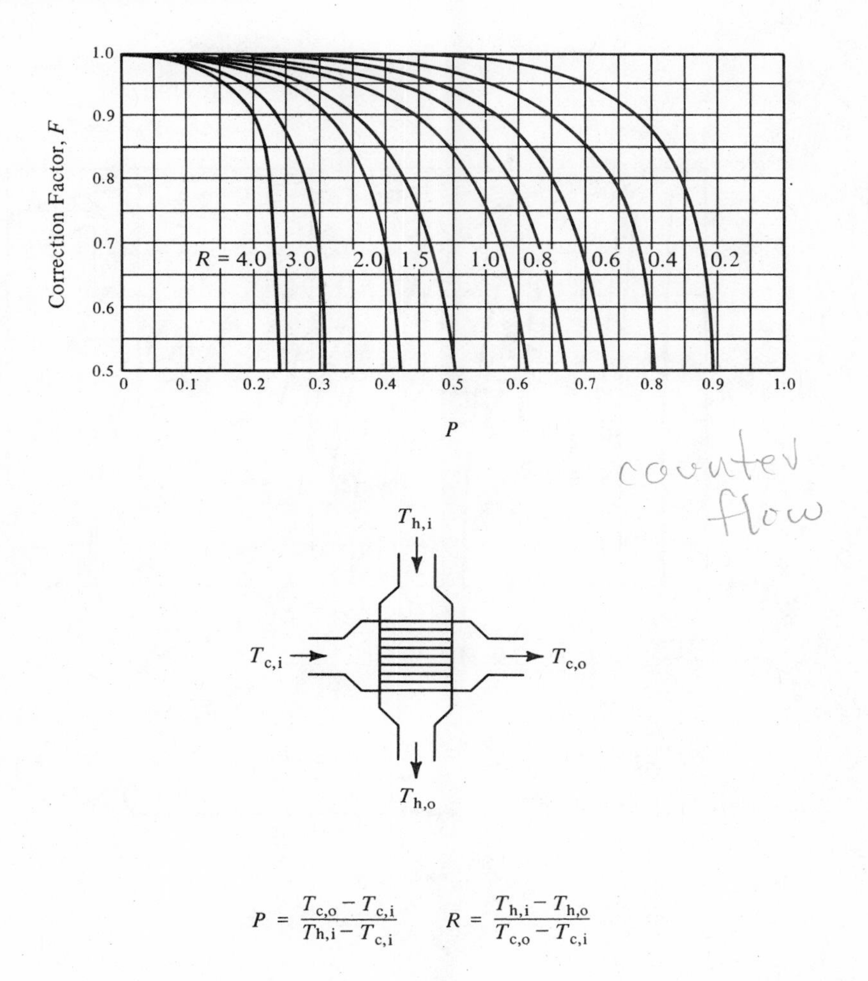

When hot fluid flows in tubes, interchange subscripts c and h.

Figure 10-13 Correction factor, F, for single pass cross-flow heat exchanger with one fluid mixed and the other unmixed, from *Mean Temperature Difference in Design* by Bowman, R. A., A. C. Mueller, and W. M. Nagel, *ASME Trans.*, Vol. 62, 1940.

When the hot fluid flows through the tubes the subscripts h and c in the above definitions should be interchanged. Correction factor charts for different configura- are presented in Figures 10-12 through 10-15.

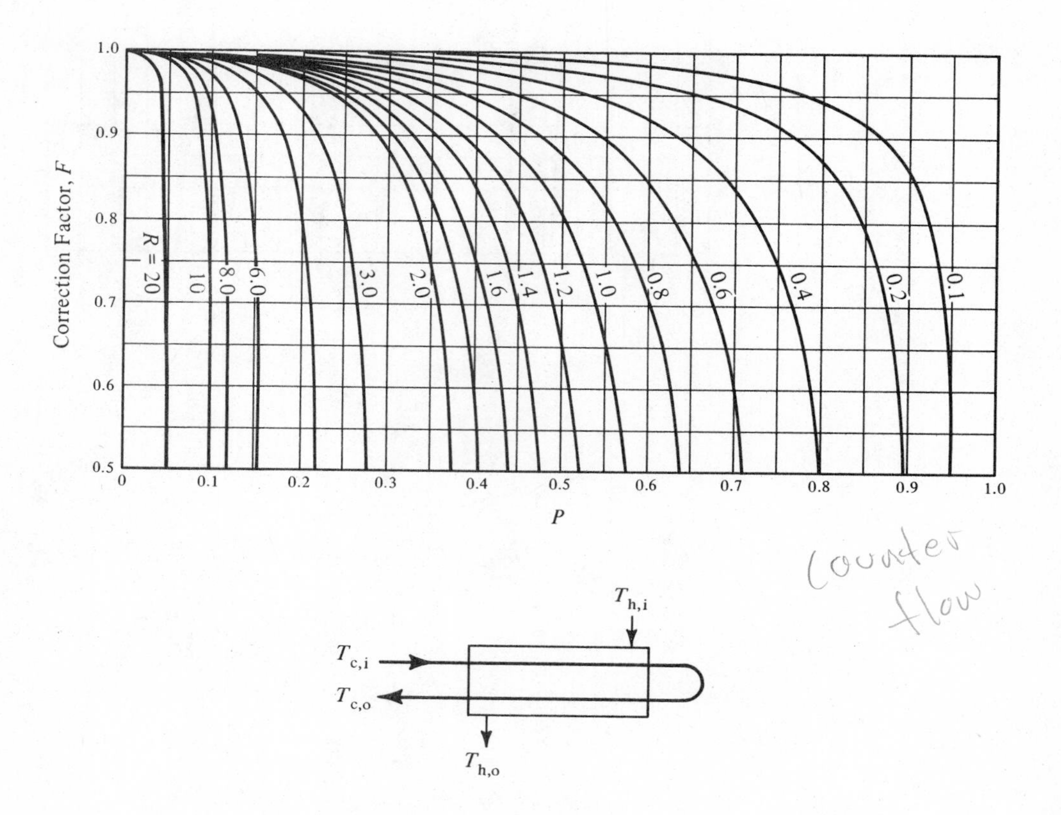

$$P = \frac{T_{c,o} - T_{c,i}}{T_{h,i} - T_{c,i}}$$

$$R = \frac{T_{h,i} - T_{h,o}}{T_{c,o} - T_{c,i}}$$

When hot fluid flows in tubes, interchange subscripts c and h.

Figure 10-14 Correction factor, F, for a heat exchanger with one shell pass and two (or a multiple of two) tube passes. Courtesy of the Tubular Exchanger Manufacturer's Association.

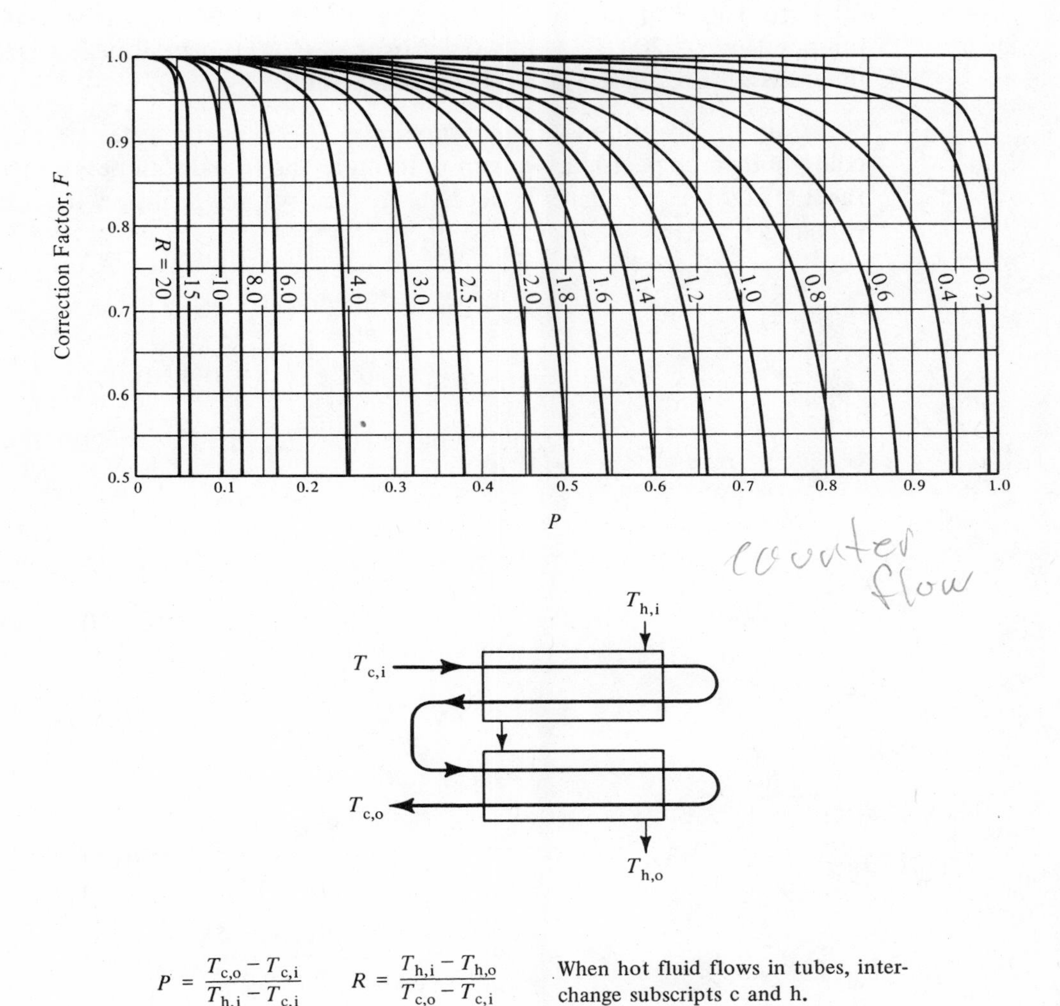

$$P = \frac{T_{c,o} - T_{c,i}}{T_{h,i} - T_{c,i}} \qquad R = \frac{T_{h,i} - T_{h,o}}{T_{c,o} - T_{c,i}}$$

When hot fluid flows in tubes, interchange subscripts c and h.

Figure 10-15 Correction factor, F, for a heat exchanger with two shell passes and a multiple of two tube passes. Courtesy of the Tubular Exchanger Manufacturer's Association.

Sample Problem 10-8 Determine the area of a two-shell pass and four-tube pass heat exchanger to heat oil with a mass flow rate of 10,000 lb_m/hr from 60°F to 150°F if hot water at a flow rate of 8,000 lb_m/hr is available at a temperature of 200°F. c_{oil} = 0.4 Btu/lb_m°F. The overall heat transfer coefficient is 95 Btu/hr-ft^2°F. Oil is the tube fluid.

Solution: Since three temperatures and the capacity rates of the hot and cold fluids are specified, we can determine the fourth temperature and hence the LMTD. Use of charts is necessary, since we are dealing with a multi-pass exchanger.

$$\dot{m}_c = 10{,}000 \text{ lb}_m/\text{hr}, \quad T_{c,o} = 150°\text{F},$$

$$T_{c,i} = 60°\text{F}, \quad c_c = 0.4 \text{ Btu/lb}_m °\text{F}$$

$$\dot{m}_h = 8000 \text{ lb}_m/\text{hr}, \quad T_{h,i} = 200°\text{F},$$

$$c_h = 1 \text{ Btu/lb}_m °\text{F}$$

$$\dot{m}_h c_h (T_{h,i} - T_{h,o}) = \dot{m}_c c_c (T_{c,o} - T_{c,i})$$

$$8000 \times 1 \times (200 - T_{h,o}) = 10{,}000 \times 0.4 \times (150 - 60)$$

or

$$T_{h,o} = 155°\text{F}$$

$$P = \frac{T_{c,o} - T_{c,i}}{T_{h,i} - T_{c,i}} = \frac{150 - 60}{200 - 60} = 0.64$$

$$R = \frac{T_{h,i} - T_{h,o}}{T_{c,o} - T_{c,i}} = \frac{200 - 155}{150 - 60} = 0.5$$

From Figure 10-15, for the above values of P and R, one reads $F = 0.96$. Also, from equation (10-20b)

$$(\text{LMTD}) = \frac{(200 - 150) - (155 - 60)}{\ln[(200 - 150)/(155 - 60)]} = 70.1°\text{F}$$

The amount of heat transferred is given by equation (10-20a).

$$Q = \dot{m}_h c_h (T_{h,i} - T_{h,o}) = UA \cdot F \cdot \text{LMTD}$$

$$10{,}000 \times 1 \times (200 - 155) = 95 \times A \times 0.96 \times 70.1$$

$$A = 70.4 \text{ ft}^2$$

10-7 HEAT EXCHANGER EFFECTIVENESS

The function of a heat exchanger may be to heat the cold fluid coming in at $T_{c,i}$ to as high a temperature as possible, which is $T_{h,i}$. Alternately, the purpose may be the removal of energy from the hot fluid, the limiting case being when $T_{h,o}$ equals $T_{c,i}$. With a counter-flow heat exchanger having an infinitely large surface area, either one of the forgoing two objectives can be met depending on which one of the two heat capacity rates is greater, $(\dot{m}_c c_c)$ or $(\dot{m}_h c_h)$. In any case, equation (10-16), which reflects the energy balance, must be satisfied.

$$\dot{m}_h c_h (T_{h,i} - T_{h,o}) = Q = \dot{m}_c c_c (T_{c,o} - T_{c,i}) \tag{10-16}$$

We consider three distinct cases.

(1) $(\dot{m}_h c_h) > (\dot{m}_c c_c)$ If the heat capacity rate of the hot fluid is greater than that of the cold fluid, the temperature gain of the cold fluid will be greater than the temperature drop of the hot fluid. If we also have an infinitely large area available for heat transfer, making it possible to have identical temperatures of the two fluids at either the inlet or the outlet of the exchanger, one can see that the only possible temperature distribution is that shown in Figure 10-16a. This is because the highest temperature in the system cannot exceed $T_{h,i}$.

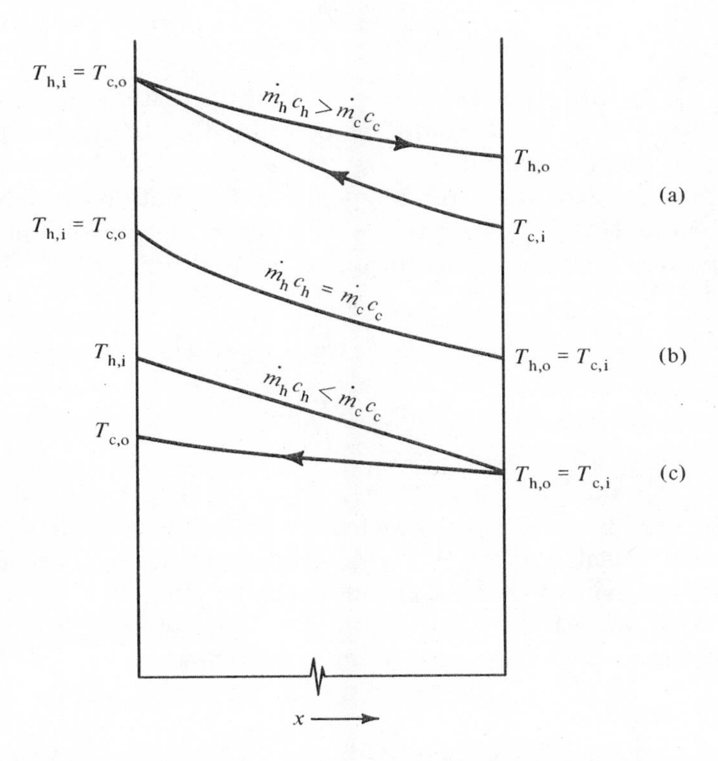

Figure 10-16 Possible temperature distribution in a counter-flow exchanger with infinitely large area.

The maximum amount of heat transfer, Q_{max}, is then given by

$$Q_{max} = \dot{m}_c c_c (T_{c,o} - T_{c,i})$$

But from Figure 10-16a,

$$T_{c,o} = T_{h,i}$$

$$Q_{max} = (\dot{m}_c c_c)(T_{h,i} - T_{c,i}) \tag{10-21}$$

(2) $(\dot{m}_h c_h) = (\dot{m}_c c_c)$ It is clear from equation (10-16) that the temperature gain of the cold fluid and temperature drop of the hot fluid will be identical in this situation. Also, because of the assumption of an infinitely large surface area of the heat exchanger, the temperatures of the two fluids will be the same at any cross section in the heat exchanger (Figure 10-16b). The ideal or maximum possible amount of heat transfer, Q_{max}, is given by

$$Q_{max} = \dot{m}_c c_c (T_{c,o} - T_{c,i}) = \dot{m}_h c_h (T_{h,i} - T_{h,o})$$

$$= \dot{m}_c c_c (T_{h,i} - T_{c,i}) = \dot{m}_h c_h (T_{h,i} - T_{c,i}) \tag{10-21a}$$

(3) $(\dot{m}_h c_h) < (\dot{m}_c c_c)$ In this situation the temperature gain of the cold fluid will be less than the temperature drop of the hot fluid in view of the principle of conservation of energy [equation (10-16)]. With an infinitely large area of the heat exchanger, the temperature distribution for this case will be as shown in Figure 10-16c, since the lowest possible temperature in the system is $T_{c,i}$. The ideal or maximum amount of heat transfer is given by

$$Q_{max} = \dot{m}_c c_c (T_{c,o} - T_{c,i}) = \dot{m}_h c_h (T_{h,i} - T_{h,o})$$

$$= \dot{m}_h c_h (T_{h,i} - T_{c,i}) \tag{10-21b}$$

Another way of looking at Q_{max}, is that it represents the maximum amount of heat transfer possible, which is the product of the largest temperature difference available, $(T_{h,i} - T_{c,i})$, and the smaller heat capacity rate, smaller because the first law of thermodynamics must be obeyed.

If we let $(\dot{m}c)_{min}$ denote the lesser of the two values, $\dot{m}_h c_h$ and $\dot{m}_c c_c$, equations (10-21, a, and b) can be written as

$$Q_{max} = (\dot{m}c)_{min}(T_{h,i} - T_{c,i}) \tag{10-22}$$

One can use equation (10-22) to represent the maximum possible amount of heat energy that can be transferred by any heat exchanger. We define heat exchanger effectiveness, ϵ:

$$\epsilon = \frac{\text{actual heat transferred}}{\text{maximum possible heat transfer}}$$

Since either side of equation (10-16) represents the actual heat transfer, we can write:

For $(\dot{m}_h c_h) > (\dot{m}_c c_c)$

$$\epsilon = \frac{\dot{m}_c c_c (T_{c,o} - T_{c,i})}{(\dot{m}c)_{\min}(T_{h,i} - T_{c,i})} = \frac{T_{c,o} - T_{c,i}}{T_{h,i} - T_{c,i}} \tag{10-23}$$

For $(\dot{m}_h c_h) < (\dot{m}_c c_c)$

$$\epsilon = \frac{\dot{m}_h c_h (T_{h,i} - T_{h,o})}{(\dot{m}c)_{\min}(T_{h,i} - T_{c,i})} = \frac{T_{h,i} - T_{h,o}}{T_{h,i} - T_{c,i}} \tag{10-23a}$$

In view of the above equations, effectivenss, ϵ, can also be stated as

$$\epsilon = \frac{\left[\begin{array}{c}\text{temperature change of the fluid}\\ \text{with minimum capacity rate } (\dot{m}c)\end{array}\right]}{\left[\begin{array}{c}\text{largest temperature difference in the}\\ \text{heat exchanger } (T_{h,i} - T_{c,i})\end{array}\right]} \tag{10-23b}$$

Observe that only three temperatures appear in the expressions for the effectiveness, the inlet temperatures of the two fluids, and the exit temperature of either the cold fluid or the hot fluid. Recall that all four terminal temperatures appear in the expression for the log mean temperature difference. Therefore, we find the concept of heat exchanger effectiveness useful when all four temperatures are not prescribed. We next seek expressions for the effectiveness in terms of the overall heat transfer coefficient, U, surface area, A, and the heat capacity rates, $\dot{m}_h c_h$ and $\dot{m}_c c_c$.

10-7.1 Effectiveness for a Parallel-Flow Heat Exchanger

We begin by equating expressions for the rate of heat transfer obtained on the basis of the energy balance and the LMTD [equations (10-16) and (10-18a)]

$$Q = (UA)\frac{(T_{h,i} - T_{c,i}) - (T_{h,o} - T_{c,o})}{\ln[(T_{h,i} - T_{c,i})/(T_{h,o} - T_{c,o})]} = \dot{m}_h c_h (T_{h,i} - T_{h,o})$$

Rearrangement gives

$$\ln\left(\frac{T_{h,i} - T_{c,i}}{T_{h,o} - T_{c,o}}\right) = \frac{UA}{\dot{m}_h c_h}\left(1 + \frac{T_{c,o} - T_{c,i}}{T_{h,i} - T_{h,o}}\right) \tag{10-24}$$

The above equation contains all four terminal temperatures. We can eliminate the ratio of temperature differences on the right side by rewriting equation (10-16), $\dot{m}_h c_h (T_{h,i} - T_{h,o}) = \dot{m}_c c_c (T_{c,o} - T_{c,i})$

as

$$\frac{T_{c,o} - T_{c,i}}{T_{h,i} - T_{h,o}} = \frac{\dot{m}_h c_h}{\dot{m}_c c_c} = R \tag{10-24a}$$

Substituting into equation (10-24) and multiplying both sides by (−1), we obtain

$$-\ln \frac{T_{h,i} - T_{c,i}}{T_{h,o} - T_{c,o}} = -\frac{UA}{\dot{m}_h c_h}(1 + R)$$

or

$$\frac{T_{h,o} - T_{c,o}}{T_{h,i} - T_{c,i}} = \exp\left[-\frac{UA}{\dot{m}_h c_h}(1 + R)\right] \tag{10-24b}$$

We observe that the left side of the above equation, except for the term $T_{c,o}$ is similar to the expression for the effectiveness, ϵ, equation (10-23a), for the case where the hot fluid is the one with the minimum heat capacity rate. From equation (10-24a)

$$T_{c,o} = T_{c,i} + R(T_{h,i} - T_{h,o}) \tag{10-24c}$$

Also, the quantity $(T_{h,o} - T_{c,o})$ in the numerator of the left side of equation (10-24b) can be rewritten as

$$T_{h,o} - T_{c,o} = T_{h,o} - T_{h,i} + T_{h,i} - T_{c,o}$$

Thus

$$\frac{T_{h,o} - T_{c,o}}{T_{h,i} - T_{c,i}} = \frac{(T_{h,o} - T_{h,i}) + (T_{h,i} - T_{c,i}) - R(T_{h,i} - T_{h,o})}{T_{h,i} - T_{c,i}}$$

where equation (10-24c) has been used to eliminate $T_{c,o}$. By carrying out the division and employing equation (10-23a), we obtain

$$(T_{h,o} - T_{h,i})/(T_{h,i} - T_{c,i}) = -\epsilon + 1 - R\epsilon = 1 - (R + 1)\epsilon$$

Combining the above equation and equation (10-24b), we write

$$1 - (R + 1)\epsilon = \exp\left[-\frac{UA}{\dot{m}_h c_h}(1 + R)\right]$$

or

$$\epsilon = \frac{1 - \exp[-(UA/\dot{m}_h c_h)(1 + R)]}{(1 + R)}$$

where

$$R = (\dot{m}_h c_h/\dot{m}_c c_c) < 1$$

The above expression for ϵ can also be obtained more directly by working on equations (10-17b) and (10-17c). Exactly the same form as the above, with R defined as $(\dot{m}_c c_c/\dot{m}_h c_h)$, is obtained when the cold fluid happens to have the minimum heat capacity rate. Therefore, the effectiveness is often expressed as

$$\epsilon = \frac{1 - \exp\{-(UA/C_{min})\,[1 + (C_{min}/C_{max})]\}}{1 + (C_{min}/C_{max})} \tag{10-25}$$

where

$$C = \dot{m}c$$

Notice that the expression for ϵ contains U, A, and the heat capacities only. The above form is more convenient to use when all four temperatures are not prescribed. In essence, Sample Problem 10-7 has been solved by the above procedure.

10-7.2 Effectiveness for a Counter-flow Heat Exchanger

The approach for obtaining an expression for the effectiveness, ϵ, for a counter-flow heat exchanger is similar to the one used for the parallel-flow heat exchanger. Equating rates of heat flow as given by the LMTD formulation and the energy balance, one has

$$Q = UA \frac{(T_{h,i} - T_{c,o}) - (T_{h,o} - T_{c,i})}{\ln[(T_{h,i} - T_{c,o})/(T_{h,o} - T_{c,i})]} = \dot{m}_h c_h (T_{h,i} - T_{h,o})$$

Simplifying and rearranging, one has

$$\ln\left(\frac{T_{h,i} - T_{c,o}}{T_{h,o} - T_{c,i}}\right) = \frac{UA}{\dot{m}_h c_h}\left(1 - \frac{T_{c,o} - T_{c,i}}{T_{h,i} - T_{h,o}}\right) = \frac{UA}{\dot{m}_h c_h}(1 - R)$$

or

$$\frac{T_{h,o} - T_{c,i}}{T_{h,i} - T_{c,o}} = \exp\left[-\frac{UA}{\dot{m}_h c_h}(1 - R)\right] = \beta$$

or

$$(T_{h,o} - T_{c,i}) = \beta (T_{h,i} - T_{c,o})$$

We now substitute the expression for $T_{c,o}$, as obtained in the preceding derivation of ϵ for parallel flow, and assume that the hot fluid is the one with the minimum heat capacity rate. Thus, we have

$$T_{h,o} - T_{c,i} = \beta [T_{h,i} - T_{c,i} - R(T_{h,i} - T_{h,o})]$$

Rearrangement gives

$$T_{h,o}(1 - R\beta) + R\beta T_{h,i} = T_{c,i}(1 - \beta) + \beta T_{h,i}$$

Subtracting $T_{h,i}$ from both sides of the above equation, multiplying by (-1), and regrouping terms, one obtains

$$(1 - R\beta)(-T_{h,o} + T_{h,i}) = (1 - \beta)(-T_{c,i} + T_{h,i})$$

or

$$\frac{T_{h,i} - T_{h,o}}{T_{h,i} - T_{c,i}} = \frac{1 - \beta}{1 - R\beta}$$

Recognizing the left side as ϵ[equation 10-23a)] and substituting for β, we finally obtain

$$\epsilon = \frac{1 - \exp\left[-(UA/\dot{m}_h c_h)(1 - R)\right]}{1 - R\exp\left[-(UA/\dot{m}_h c_h)(1 - R)\right]}$$

where

$$R = (\dot{m}_h c_h)/(\dot{m}_c c_c) < 1$$

If the cold fluid were the one with the minimum value of heat capacity rate, we would obtain exactly the same expression as above with R defined as $(\dot{m}_c c_c/\dot{m}_h c_h)$. Hence the effectiveness for a counter-flow exchanger can be written as

$$\epsilon = \frac{1 - \exp\{-(UA/C_{min})[1 - (C_{min}/C_{max})]\}}{1 - (C_{min}/C_{max})\exp\{-(UA/C_{min})[1 - (C_{min}/C_{max})]\}} \quad (10\text{-}26)$$

where

$$C = \dot{m}c$$

A closer look at equations (10-25) and (10-26) reveals that they contain three dimensionless quantities, namely ϵ, (C_{min}/C_{max}), and (UA/C_{min}). The last group is called the Number of Transfer Units or NTU and is indicative of the size of the heat exchanger. The LMTD and the NTU approaches are really two faces of the same coin; depending on the nature of the data given, one chooses either the LMTD approach or the NTU approach.

10-7.3 Effectiveness for Cross-Flow Heat Exchangers

In order to discuss the effectiveness of a cross-flow heat exchanger, it is necessary to introduce the concept of mixed and unmixed fluids. If a fluid entering the exchanger is divided into separate paths by tubes or channels, it is called an unmixed fluid. If such tubes or channels do not exist, the fluid is said to be a mixed fluid.

Expressions for the heat exchanger effectiveness for various situations in cross flow are given below as presented in Reference 6

(1) Both fluids unmixed (approximate expression):

$$\epsilon = 1 - \exp\left[\frac{R}{\eta}\{\exp[-R\,\eta\mathrm{NTU}] - 1\}\right];\ \eta = \mathrm{NTU}^{-0.22} \tag{10-27}$$

(2) Both fluids mixed:

$$\epsilon = \mathrm{NTU}\left[\frac{\mathrm{NTU}}{1 - \exp(-\mathrm{NTU})} + \frac{R\,\mathrm{NTU}}{1 - \exp[-R\,\mathrm{NTU}]} - 1\right]^{-1} \tag{10-27a}$$

(3) Fluid with minimum capacity rate unmixed:

$$\epsilon = (1/R)(1 - \exp\{-R[1 - \exp(-\mathrm{NTU})]\}) \tag{10-27b}$$

(4) Fluid with maximum capacity rate unmixed:

$$\epsilon = 1 - \exp\{-1/R\,[1 - \exp(-R\,\mathrm{NTU})]\} \tag{10-27c}$$

Graphs of the variation in effectiveness, ϵ, with respect to NTU having (C_{min}/C_{max}) as a parameter are presented in Figures 10-17 through 10-22, for parallel-flow, counter-flow, and cross-flow heat exchangers.

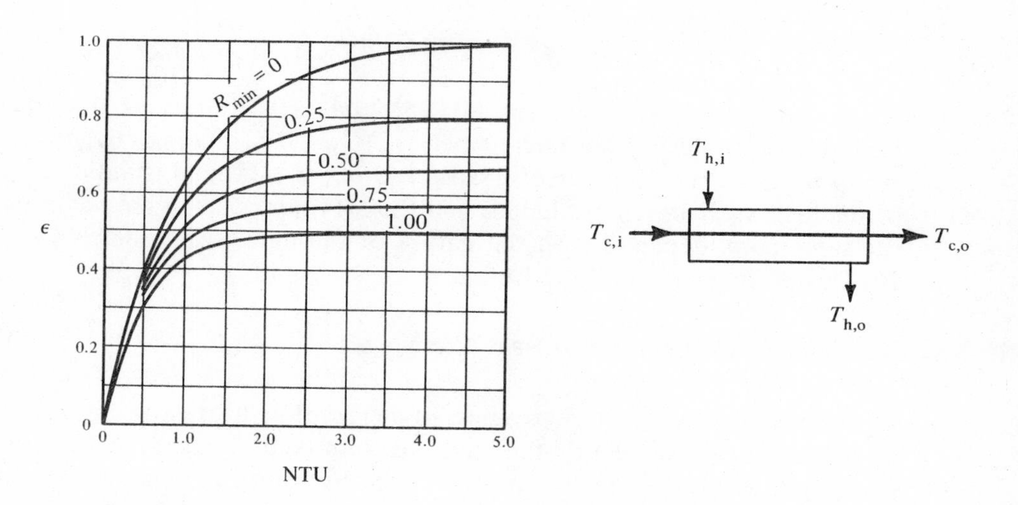

Figure 10-17 Effectiveness for parallel-flow heat exchanger, from *Compact Heat Exchangers* by Kays, W. M. and A. L. London. Copyright 1964, by McGraw-Hill, Inc. Used with permission.

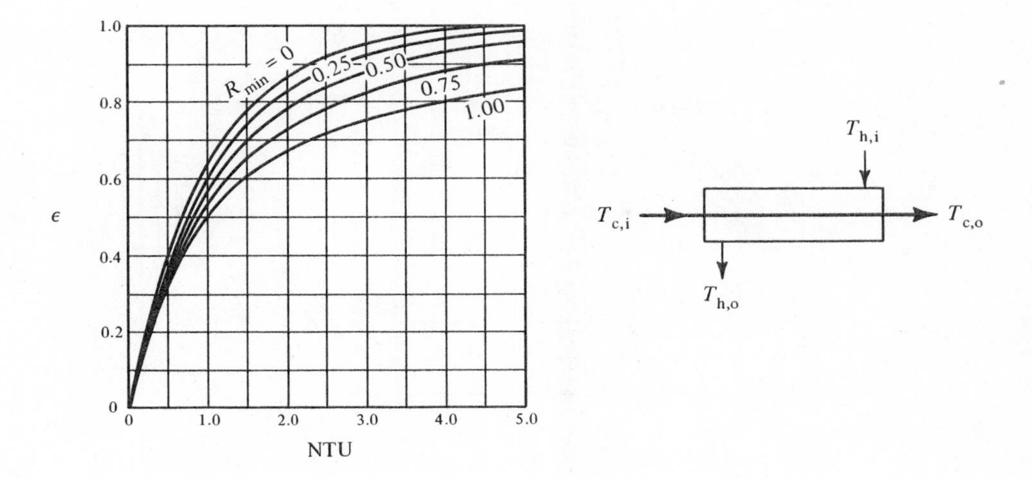

Figure 10-18 Effectiveness for a counter-flow heat exchanger, from *Compact Heat Exchangers* by Kays, W. M. and A. L. London. Copyright 1964, by McGraw-Hill, Inc. Used with permission.

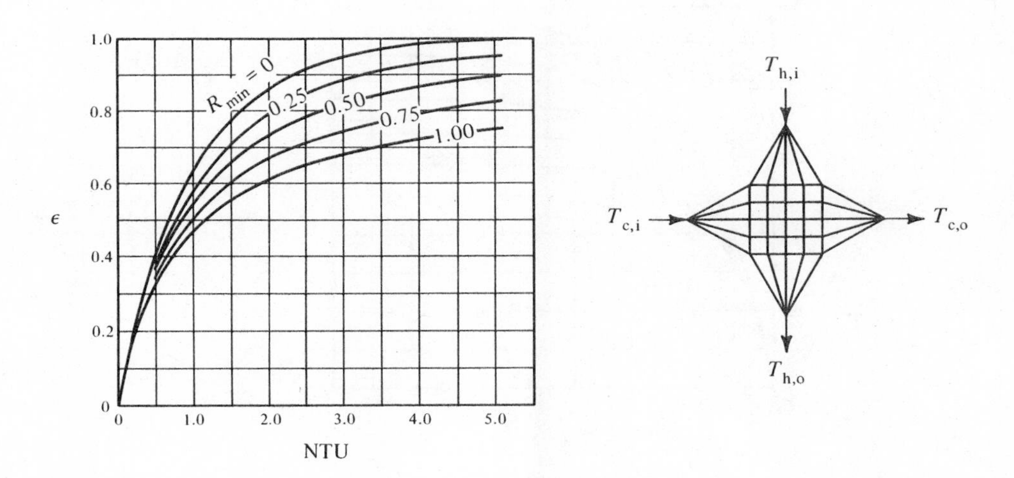

Figure 10-19 Effectiveness for cross-flow heat exchanger with both fluids unmixed, from *Compact Heat Exchangers* by Kays, W. M. and A. L. London. Copyright 1964, by McGraw- Hill, Inc. Used with permission.

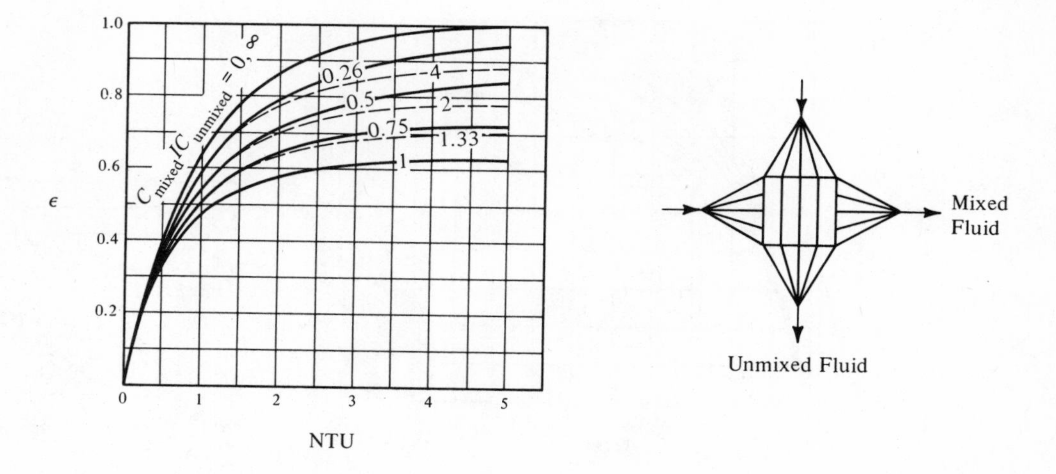

Figure 10-20 Effectiveness for cross-flow heat exchanger with one fluid mixed, from *Compact Heat Exchangers* by Kays, W. M. and A. L. London. Copyright 1964 by McGraw-Hill, Inc. Used with permission.

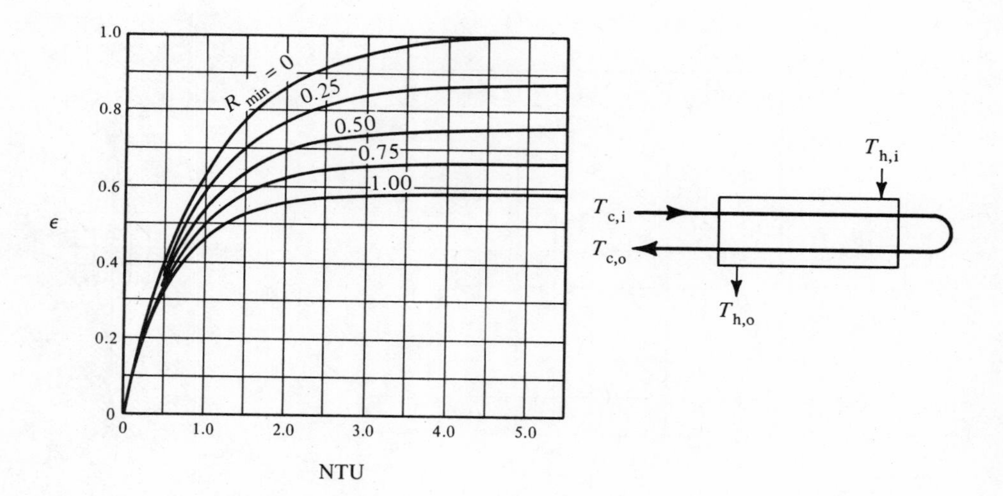

Figure 10-21 Effectiveness for heat exchanger with one shell pass and two (or multiples of two) tube passes, from *Compact Heat Exchangers* by Kays, W. M. and A. L. London. Copyright 1964, by McGraw-Hill, Inc. Used with permission.

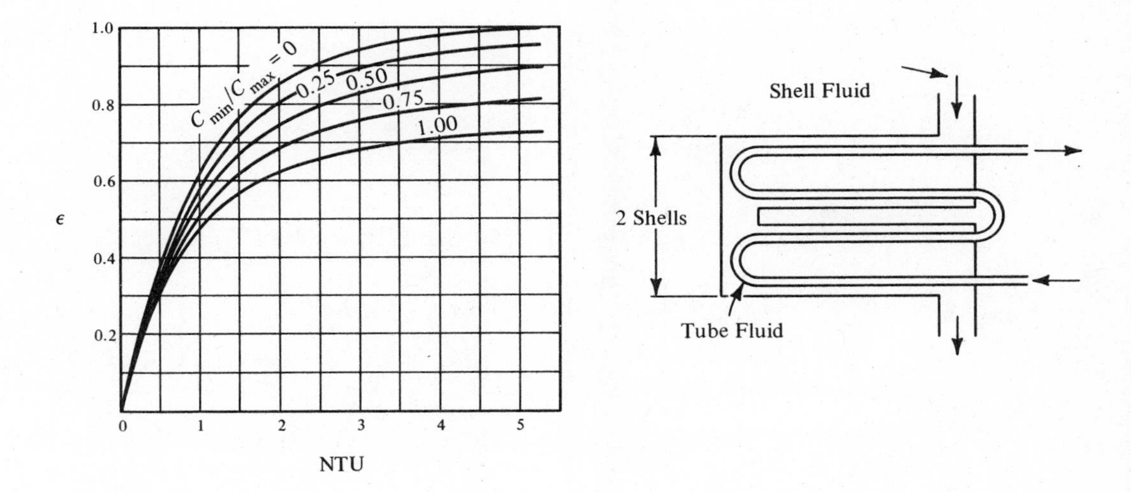

Figure 10-22 Effectiveness for heat exchanger with two shell passes and four (or multiples of four) tube passes, from *Compact Heat Exchanger* by Kays, W. M. and A. L. London, Copyright 1964, by McGraw-Hill, Inc. Used with permission.

Sample Problem 10-9 Calculate the effectivenss of the heat exchanger of Sample Problem 10-7 when used (1) in parallel mode, (2) in counter-flow mode.

Solution:

$$\dot{m}_c c_c = 3000 \times 4186 = 12.6 \times 10^6 \text{ J/°C hr}$$

$$\dot{m}_h c_h = 10{,}000 \times 1900 = 19 \times 10^6 \text{ J/°C hr}$$

$$C_{\min} = \dot{m}_c c_c = 12.6 \times 10^6$$

and

$$\frac{C_{\min}}{C_{\max}} = \frac{12.6 \times 10^6}{19 \times 10^6} = 0.66$$

$$\text{NTU} = \frac{UA}{C_{\min}} = \frac{300 \times 3600 \times 17.5}{12.6 \times 10^6} = 1.5$$

(1) *Parallel Flow:* Substituting the forgoing values into equation (10-25), we obtain

$$\epsilon = \frac{1 - \exp\,[-1.5\,(1 + 0.66)]}{1 + 0.66} = 0.55$$

This value checks with the one read from Figure 10-17. Hence

$$T_{c,o} - T_{c,i} = 0.55\,(T_{h,i} - T_{c,i})$$

$$T_{c,o} = 20 + 0.55\,(200 - 20) = 120°C$$

This answer checks with the one obtained earlier in Sample Problem 10-7.

(2) *Counter-Flow:* Employing equation (10-26), we have

$$\epsilon = \frac{1 - \exp\,[-1.5\,(1 - 0.6609)]}{1 - (0.66)\exp\,[-1.5\,(1 - 0.66)]} = 0.67$$

This value checks with the one read from Figure 10-18. Hence

$$T_{c,o} - T_{c,i} = 0.67\,(T_{h,i} - T_{c,i})$$

$$T_{c,o} = 20 + 0.67\,(200 - 20) = 140°C$$

This value of $T_{c,o}$ again checks with the one obtained in Sample Problem 10-7.

10-8 VARIABLE OVERALL HEAT TRANSFER COEFFICIENT

So far we have discussed the parallel and the counter-flow exchangers with a constant U value. In reality, the U value is strongly dependent on the temperature and the physical properties (thermal conductivity, viscosity, etc.) of the fluids used. Analytical results for the case where U varies linearly with the temperature difference $(T_h - T_c)$ according to the relation

$$U = a + b\Delta T, \qquad \Delta T = T_h - T_c$$

is given by Reference 4 as

$$\frac{Q}{A} = \frac{U_{x=L}(\Delta T)_{x=o} - U_{x=o}(\Delta T)_{x=L}}{\ln\{[U_{x=L}(\Delta T)_{x=o}]/[U_{x=o}(\Delta T)_{x=L}]\}} \tag{10-28}$$

Sample Problem 10-10 A certain heat exchanger has one shell pass and four tube passes, with a total surface area of 100 ft^2. Oil with a specific heat of 0.6 Btu/lb$_m$ °F comes in at a temperature of 350°F at a rate of 16 gallons per minute. Cooling water at a rate of 6000 lb$_m$/hr is available at 60°F. Determine the exit temperatures of the oil and the water if the U value is 180 Btu/hr-ft^2 °F.

Solution:

$$T_{h,i} = 350°F, \dot{m}_h = 16 \times 0.1337 \times 62.4 \times 60 = 8009 \text{ lb}_m/\text{hr}$$

$$c_h = 0.6 \text{ Btu/lb}_m °F, \text{ giving } C_h = 0.6 \times 8009 = 4805 \text{ Btu/hr°F}$$

$$C_c = 1 \times 6000 = 6000 \text{ Btu/hr°F}$$

$$R = \frac{C_{min}}{C_{max}} = \frac{4805}{6000} = 0.8$$

$$\text{NTU} = UA/C_{min} = 180 \times 100/4805 = 3.74$$

From Figure 10-21, for the above values of R and NTU, one obtains

$$\epsilon = \frac{T_{h,i} - T_{h,o}}{T_{h,i} - T_{c,i}} = 0.64$$

or

$$\frac{350 - T_{h,o}}{350 - 60} = 0.64 \quad \text{or} \quad T_{h,o} = 164.4°F$$

The exit temperature of the water may now be determined from the relation

$$C_c(T_{c,o} - T_{c,i}) = C_h(T_{h,i} - T_{h,o})$$

$$6000(T_{c,o} - 60) = 4805(350 - 164.4)$$

or

$$T_{c,o} = 208.6°F$$

In the preceding sample problems, we assumed that the U value was known. In reality, one has to calculate the convective heat transfer coefficients on the inside and the outside of the tubes and then compute the U value from equation

(10-15a) or (b). The computation of the U value requires one to assume a mean film temperature or a mean surface temperature that can be verified only after all the temperatures at the inlet and the outlet of the exchanger are calculated. The following sample problem illustrates the procedure to be used to solve a heat exchanger problem when the U value is not known.

Sample Problem 10-11 A steam condensor is made of 5/8-inch diameter, 18 gage, brass tubes. There are two tube passes of 125 tubes per pass, each pass being 6 ft long. The cooling water enters at 75°F and has an average velocity of 4.5 ft/sec. Determine the number of pounds of saturated steam at 2 psia that will be condensed. The outside film coefficient is known to be 2000 Btu/hr-ft^2°F.

Solution: A 5/8 inch, 18 gage, tube has an inner diameter, D_i = 0.527 in = 0.0439 ft and an outer diameter, D_o = 0.625 in = 0.052 ft. The total surface area of the exchanger is given by

$$A_o = \pi D_o L \text{ (number of tubes per pass)} \cdot \text{(number of passes)}$$

$$= \pi 0.052 \times 6 \times 125 \times 2$$

$$= 245 \text{ ft}^2$$

The total cross-sectional area for water flow is

$$a_i = (\pi/4) D_i^2 \text{ (number of tubes per pass)}$$

$$= (\pi/4)(.0439)^2(125)$$

$$= 0.1892 \text{ ft}^2$$

The saturation temperature of the steam at 2 psia is 126°F, which will be the temperature on the outside surface of the tubes throughout the exchanger, $T_{h,i} = T_{h,o} = 126°\text{F}$. The latent heat of vaporization at 2 psia is 1022.1 Btu/lb$_m$. At this point, we do not know $T_{c,o}$, $\dot{m}_h$, h_i, and U. We can calculate h_i and hence U if we assume $T_{c,o}$. Let us assume $T_{c,o} = 100°\text{F}$.

$$\text{LMTD} = \frac{(126 - 75) - (126 - 100)}{\ln[(126 - 75)/(126 - 100)]} = 37.11°\text{F}$$

Since the hot fluid is the condensing steam, $\dot{m}_h c_h \to \infty$, and $R \to 0$. Hence the F factor is unity.

The mean temperature of the water is (75 + 100)/2 or 87.5°F, for which c_c = 0.997 Btu/lb$_m$ °R, ρ = 62.13 lb$_m$/ft^3, ν = 0.0307 ft^2/hr, and k = 0.359 Btu/hr-ft°F. Thus

$$\dot{m}_c = \rho a_i u_{av} = 62.13 \times 0.1892 \times 4.5 = 52.9 \text{ lb}_m/\text{sec}$$

$$= 190{,}430 \text{ lb}_m/\text{hr}$$

$$\text{Re} = \frac{u_{av} D_i}{\nu} = \frac{4.5 \times 3600 \times 0.0439}{0.0307} = 23{,}165$$

$$\text{Pr} = \frac{\mu c_p}{k} = 5.33$$

From equation (10-14a)

$$\text{Nu}_{av} = \frac{(f/8)\,\text{Re Pr}}{1.07 + 12.7\sqrt{f/8}\,(\text{Pr}^{2/3} - 1)}\left(\frac{\mu}{\mu_w}\right)^{0.11}$$

where

$$f = (1.82 \log_{10}(\text{Re}) - 1.64)^{-2}$$

Substitution for Re in the preceding equation gives

$$f = 0.025$$

Also

$$\mu = 1.91 \text{ lb}_m/\text{ft-hr} \quad \text{and} \quad \mu_w = 1.30 \text{ lb}_m/\text{ft-hr}$$

and we obtain

$$\text{Nu}_{av} = \frac{h_i D_i}{k} = 159 \quad \text{and} \quad h_i = 1300 \text{ Btu/hr-ft}^2\,^\circ\text{F}$$

$$\frac{1}{U_o} = \frac{r_o}{r_i h_i} + \frac{r_o \ln(r_o/r_i)}{k} + \frac{1}{h_o}$$

$$U_o = [0.00091 + 0.01235 + 0.0005]^{-1} = 73 \text{ Btu/hr-ft}^2\,^\circ\text{F}$$

The above value is based on the outside area of the tubes.

At this point, we can calculate the mean temperature of the outside surface $T_{o,s}$ of the tube from the equation

$$h_o(T_{steam} - T_{o,s}) = U(T_{steam} - T_{c,mean})$$

$$2000\,(126 - T_{o,s}) = 73\,(126 - 87.5)$$

$$T_{o,s} = 124.6°F$$

If the heat transfer coefficient on the outside surface had not been specified and if we were to calculate it, we would have assumed a value of $T_{o,s}$ for the computation of the Nusselt number for condensation.

Now we may check the assumed value of $T_{c,o}$ as follows:

$$\dot{m}_c c_c (T_{c,o} - T_{c,i}) = UA\,(\text{LMTD})$$

$$190{,}430 \times 0.997 \times (T_{c,o} - 75) = 73 \times 245 \times 37.11$$

or

$$T_{c,o} = 78.5°F$$

Taking the actual value as the mean of the assumed and the calculated values, we have

$$T_{c,o} = (1/2)\,(100 + 78.5) = 89°F$$

The amount of steam condensed per hour, $\dot{m}_{steam}$, is given by

$$(\dot{m}_{steam})(\text{latent heat}) = \dot{m}_c c_c (T_{c,o} - T_{c,i})$$

or

$$\dot{m}_{steam} = \frac{190{,}430 \times 0.997 \times (89 - 75)}{1022.1}$$

$$= 2600\ \text{lb}_m/\text{hr}$$

The student should rework the problem until the value of $T_{c,o}$ converges.

PROBLEMS (English Engineering System of Units)

10-1. A long steel rod has a cross section of a 1/4 inch × 1/2 inch and has one end maintained at 400°F, while the remainder of the rod is exposed to a convective environment. The thermal conductivity of steel is taken to be 30

Btu/hr-ft°F and the convective heat transfer coefficient is 5 Btu/hr-ft2°F. Determine the heat lost by the rod if the ambient temperature is 80°F.

10-2. A copper rod 1/8 inch in diameter and 6 inches long protrudes from a wall whose temperature is 400°F. The rod is exposed to an environment at 75°F, and the convective heat transfer coefficient between the rod surface and the environment is 25 Btu/hr-ft2°F. Estimate the total heat dissipated by the rod; list your basic assumptions.

10-3. Derive the equation for heat dissipated by a fin with an insulated tip, $Q = \sqrt{hPkA}\ (T_o - T_\infty) \tanh (mL)$, by integrating the convective losses along its surface.

10-4. A thin rod is attached to two walls whose temperatures are T_1 and T_2, respectively, as shown in EFigure 10-4.

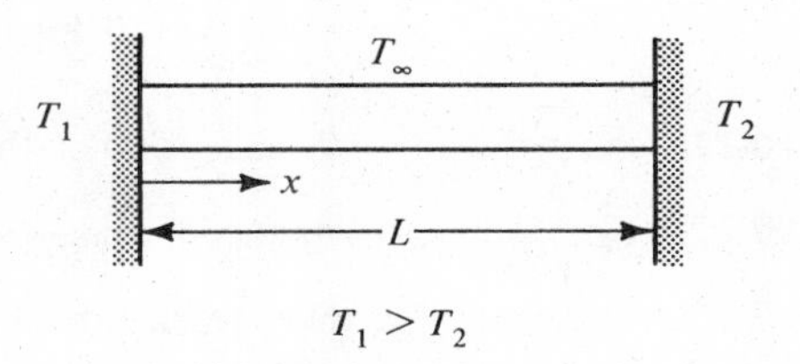

E Figure 10-4

It loses heat by convection to the ambient air at T_∞. Determine and solve the differential equation for temperature as a function of x.

10-5. Do problem 10-4 with $T_1 = T_2 = T_o$, and let the rod be $2L$ units long. (*Hint:* You should get the same solution for this problem as we did for the insulated fin problem of length L.)

10-6. Consider the insulated fin problem only this time; let there be a uniform heat generation per unit time and volume, $\dot{q}$, distributed throughout the fin. Determine the temperature as a function of x for this modification to the standard problem.

10-7. A rectangular fin 1 inch long and 1/4 inch thick made of carbon steel ($k = 24$ Btu/hr-ft°F) is attached to a plane wall, which is maintained at a temperature of 700°F. The surrounding environment is at 150°F, and the convective heat transfer coefficient is 15 Btu/hr-ft2°F. Calculate the heat dissipated by the fin per unit depth.

10-8. A straight rectangular fin 3/4 inch thick and 6 inches long protrudes from a wall whose temperature is maintained at 500°F. If the fin is made of aluminum, the ambient temperature is 70°F, and the convective heat transfer coefficient is 4 Btu/hr-ft2°F, determine the fin efficiency and the heat lost from the fin per unit depth.

10-9. A circumferential aluminum fin, 1/8 inch thick and 1/2 inch long, surrounds a 1-inch O.D. tube whose temperature is maintained at 500°F. If the ambient temperature is 70°F and the convective heat transfer coefficient is 8 Btu/hr-ft^2°F, calculate the heat dissipated by the fin.

10-10. Derive the differential equation for the temperature distribution in a straight triangular fin. For convenience, use the coordinate system shown in EFigure 10-10.

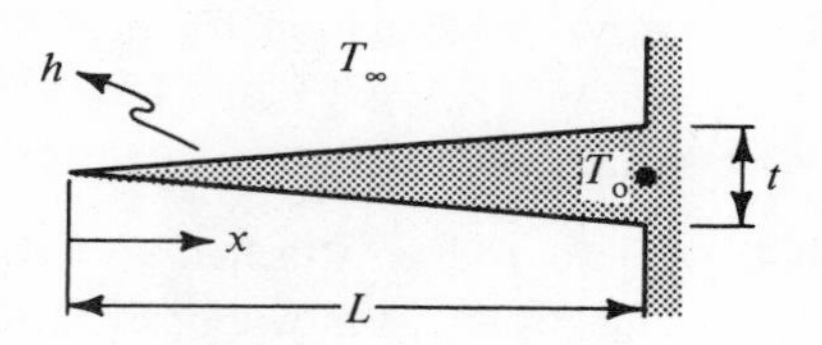

E Figure 10-10

10-11. Do problem 10-7, except make the fin triangular.

10-12. Do problem 10-8, except make the fin triangular.

10-13. The total efficiency of a finned surface may be defined as the ratio of the actual heat transferred by the combined finned and the unfinned surfaces to the heat that would be transferred if this entire surface area were maintained at the base temperature, T_o. If

η_t = total efficiency

A_f = surface area of fins alone

A = total heat transfer area of the finned and the unfinned surfaces

η_f = fin efficiency

Show that $\eta_t = 1 - [(A_f/A)(1 - \eta_f)]$.

10-14. What is the expression for LMTD in a case where the capacity rates ($\dot{m}c$) for the hot and the cold fluids are the same? Assume a counter flow operation.

10-15. In a heat exchanger, cold fluid enters at 90°F and leaves at 400°F, whereas the hot fluid enters at 700°F and leaves at 500°F. Find the LMTD values for (a) parallel flow, and (b) counter flow.

10-16. If the hot fluid problem 10-15 has a capacity rate of 8000 Btu/hr°F, and if the overall heat transfer coefficient, U_i, based on the inside area, is 75 Btu/hr-ft^2°F, determine the surface area of the heat exchanger required for (a) parallel flow, and (b) counter flow.

10-17. After the exchanger in problem 10-16 has been in operation in the parallel-flow mode for over a year, it is discovered that the cold fluid gets heated to only 300°F, and the hot fluid emerges from the exchanger at a temperature greater than 500°F. The capacity rates are the same as before. What is the cause of deterioration in the performance of the heat exchanger? What is the value of R_f?

10-18. Determine the surface area of a single-pass heat exchanger capable of heating 5000 lb_m/hr of water from 70°F to 190°F. Saturated steam at 40 psia is condensing on the outer tube surface. The U_o value is 300 Btu/hr-ft²°F. Also determine the rate of condensation of steam. h_{fg} = 933.7 Btu/lb_m.

10-19. An oil having a specific heat of 0.5 Btu/lb_m°F enters an oil cooler at 180°F at the rate of 24,000 lb_m/hr. The cooler is a counter-flow unit with water as the coolant, the heat transfer surface being 350 ft². The water enters the exchanger at 80°F. The U_i value is 110 Btu/hr-ft²°F. Determine the mass flow rate of water if the oil leaves the cooler at 100°F.

10-20. An old heat exchanger with a surface area of 200 ft² is available for cooling oil with a capacity rate of 7000 Btu/hr°F. The inlet temperature of the oil is 210°F while that of the water is 60°F. Assume c_{oil}= 0.45 Btu/lb_m°F. Water is available at a maximum flow rate of 10 gpm. Determine the exit temperatures in (a) parallel-flow, and (b) counter-flow arrangements if the U_o value is 90 Btu/hr-ft²°F. Plot a graph of the temperature of the oil leaving the counter-flow arrangement as a function of the mass flow rate of the cooling water. (*Hint:* Use NTU approach.)

10-21. A shell-and-tube exchanger consists of 80 twelve-foot lengths of standard 1 in 18 BWG copper condenser tubing (I.D. = 0.902 in; O.D. = 1 in). Steam at 15 psia is to be condensed using water at a rate of 500,000 lb_m/hr flowing inside the tubes. The water inlet temperature is 55°F. The unit surface conductances in Btu/hr-ft²°F, based on the actual area, are (a) water side, 500; (b) steam side, 2000; (c) scale on steam side, 1900; and (d) scale on water side: 2200. Estimate the rate of condensation of steam. h_{fg} = 970 Btu/lb_m.

10-22. Show that ϵ = NTU/(1 + NTU) for counter flow and $\epsilon = (1/2)(1 - e^{-2\,NTU})$ for parallel flow if the capacity rates of the hot and cold fluids are equal.

10-23. A heat exchanger has one shell pass, 4 tube passes, and 80 tubes per pass with each pass 8 feet long. The tubes are 1 inch O.D. and 0.834 inch I.D. Cooling water enters the tubes at 50°F at a rate of 1 ft/sec. The U_o value is 100 Btu/hr-ft²F. Oil (c_p = 0.45 Btu/lb_m°F) at 180°F enters the shell at a flow rate of 70,000 lb_m/hr. What is the outlet temperature of the cooling water?

10-24. A cross-flow exchanger with both fluids unmixed is used to cool air that is initially at 140°F and flowing at a rate of 10,000 lb_m/hr. Water is used as the coolant at 60°F with a flow rate of 9600 lb_m/hr. Assume the U_i value to be 24 Btu/hr-ft^2°F. The surface area of the exchanger is 200 ft^2. Calculate the exit temperatures of the air and the water.

10-25. In a one-shell pass, two-tube pass heat exchanger, water at 70°F enters at a rate of 10,000 lb_m/hr in the shell. Engine oil flows through the tubes at a rate of 7,000 lb_m/hr. Assume c_h = 0.4 Btu/lb_m °F and U_o = 40 Btu/hr-ft^2°F. The inlet and the outlet temperatures of the oil are 300°F and 200°F respectively. Determine the surface area of the heat exchanger.

10-26. Work problem 10-25 by the LMTD method.

10-27. A cross-flow exchanger with both fluids unmixed has a surface area of 400 ft^2. It is used to cool air from 280°F by using air at 70°F. The hot air flows at a rate of 8000 lb_m/hr, while the cold air rate is 13,000 lb_m/hr. Assuming the U value to be 12 Btu/hr-ft^2°F, calculate the exit temperatures of both the streams.

10-28. Derive an expression for the U factor if the inside and outside surfaces of the tubes in a heat exchanger are finned and have fin efficiencies of η_i and η_o on the inside and on the outside, and A_i and A_o as the total area (inclusive of the respective fin surfaces) on the inside and the outside, respectively.

10-29. In a shell-and-tube exchanger consisting of 64 eight-foot lengths of tubes (I.D. = 1 inch, O.D. = 1.2 inch), steam at 5 psia is to be condensed using water at a rate of 400,000 lb_m/hr flowing inside the tubes. Water inlet temperature is 60°F. Estimate the inside and the outside convective heat transfer coefficients, and hence determine the rate at which steam can be condensed. Also determine the outlet temperature of the water. h_{fg} = 1000 Btu/lb_m. Assume film condensation.

10-30. If circumferential fins of 1/8-inch thickness and 1/4-inch length are attached to the outside surface of the tubes of the heat exchanger in problem 10-29, determine the rate of condensation of steam. The fins are spaced 3/8 inch apart and are made of brass.

PROBLEMS (SI System of Units)

10-1. An aluminum rod is exposed to an environment at 15°C and the convective heat transfer coefficient between the rod surface and the environment is 300 W/m^2 °K. The rod, 3 mm in diameter and 8 cm long, protrudes from a wall whose temperature is 140°C. Estimate the total heat dissipated by the rod; list your basic assumptions.

10-2. A rectangular copper fin has one end maintained at 200°C, while the remainder of the fin is exposed to a convective environment at 25°C. If the thermal conductivity of copper is taken to be 386 W/m2°K and the convective heat transfer coefficient is 35 W/m2°K, determine the heat lost by the fin per unit depth. The length of the fin is 5 cm and the thickness is 4 mm. Assume the fin tip to be insulated.

10-3. Derive the equation for heat dissipated by a fin with an insulated tip, $Q = \sqrt{hPkA}\,(T_o - T_\infty)\tanh(mL)$, by integrating the convective losses along its surface.

10-4. A thin rod is attached to two walls whose temperatures are T_1 and T_2, respectively, as shown in SIFigure 10-4.

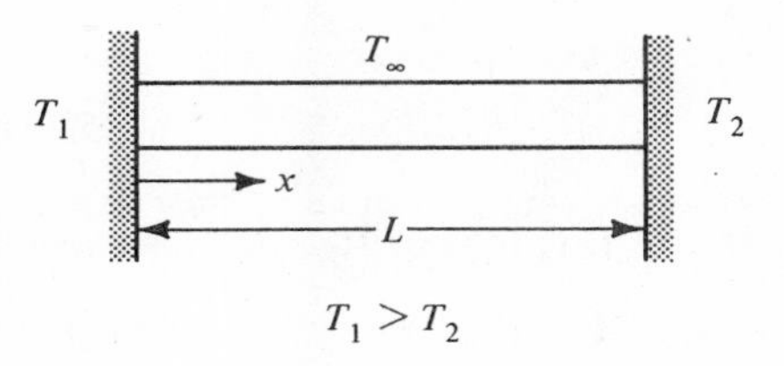

SI Figure 10-4

It loses heat by convection to the ambient air at T_∞. Determine and solve the differential equation for temperature as a function of x.

10-5. Do problem 10-4 with $T_1 = T_2 = T_o$, and let the rod be $2L$ units long. (*Hint:* You should get the same solution for this problem as we did for the insulated fin problem of length, L.)

10-6. Consider the insulated fin problem only this time; let there be a uniform heat generation per unit time and volume, $\dot{q}$, distributed throughout the fin. Determine the temperature as a function of x for this modification to the standard problem.

10-7. A carbon steel (k = 54 W/m°K) rod with a cross section of an equilateral triangle (each side - 5 mm) is 8 cm long. It is attached to a plane wall, which is maintained at a temperature of 400°C. The surrounding environment is at 50°C, and the convective heat coefficient is 90 W/m2°K. Calculate the heat dissipated by the rod.

10-8. An aluminum fin 18 mm thick and 16 cm long has a root temperature of 300°C. The ambient temperature is 20°C, and the convective heat transfer coefficient is 30 W/m2°K. Determine the fin efficiency and the heat lost from the fin per unit depth.

10-9. An aluminum fin 3 mm thick and 12 mm long surrounds a 25 mm O.D. tube whose temperature is maintained at 250°C. If the ambient temperature is 10°C and the convective heat transfer coefficient is 48 W/m²°K, calculate the heat dissipated by the fin.

10-10. Derive the differential equation for the temperature distribution in a straight triangular fin. For convenience, use the coordinate system shown in SIFigure 10-10.

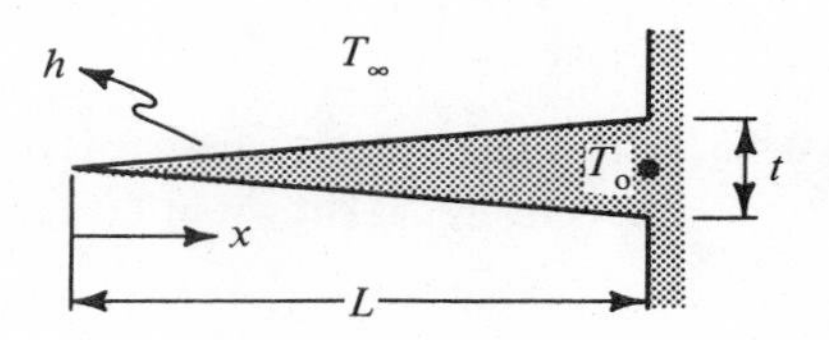

SI Figure 10-10

10-11. Do problem 10-8, except make the fin triangular.

10-12. The total efficiency of a finned surface may be defined as the ratio of the actual heat transferred by the combined finned and the unfinned surfaces to the heat that would be transferred if this entire surface area were maintained at the base temperature, T_o. If

η_t = total efficiency

A_f = surface area of fins alone

A = total heat transfer area of the finned and the unfinned surfaces

η_f = fin efficiency

show that $\eta_t = 1 - [(A_f/A)(1 - \eta_f)]$.

10-13. What is the expression for LMTD in a case where the capacity rates ($\dot{m}c$) for the hot and the cold fluids are the same? Assume a counter flow operation.

10-14. In a heat exchanger, cold fluid enters at 30°C and leaves at 200°C, whereas the hot fluid enters at 360°C and leaves at 300°C. Find the LMTD values for (a) parallel flow, and (b) counter flow.

10-15. If the hot fluid in problem 10-14 has a capacity rate of 2500 W/°K and if the overall heat transfer coefficient U_o is 800 W/m²°K, determine the surface area of the heat exchanger required for (a) parallel flow, and (b) counter flow.

10-16. After the exchanger in problem 10-15 has been in operation in the parallel-flow mode for over a year, it is discovered that the cold fluid gets heated to only 120°C, and the hot fluid emerges from the exchanger at a temperature greater than 300°C. The capacity rates are the same as before. What is the cause of deterioration in the performance of the heat exchanger? What is the value of R_f?

10-17. Saturated steam at 120°C is condensing on the outer tube surface of a single-pass heat exchanger. The U_o value is 1800 W/m²°K. Determine the surface area of the heat exchanger capable of heating 2000 kg/hr of water from 20°C to 90°C. Also determine the rate of condensation of steam. h_{fg} = 2202 kJ/kg.

10-18. An oil having specific heat of 2000 J/kg°K enters an oil cooler at 80°C at the rate of 8000 kg/hr. The cooler is a counter-flow unit with water as the coolant, the heat transfer surface being 7.5 m². The water enters the exchanger at 10°C. The U_i value is 600 W/m²°K. Determine the mass flow rate of water if the oil leaves the cooler at 42°C.

10-19. An old heat exchanger with a surface area of 1 m² is available for cooling oil with a capacity rate of 300 kg/hr. The specific heat of the oil is 2200 J/kg°K. The inlet temperature of the oil is 110°C while that of the water is 15°C. Water is available at a maximum flow rate of 0.4 liters/sec. Determine the lowest possible temperature in (a) parallel-flow, and (b) counter-flow arrangements if the U value is 600 W/m²°K. Plot a graph of the temperature of the oil leaving the counter-flow arrangement as a function of mass flow rate of the cooling water. (*Hint:* Use the NTU approach.)

10-20. A shell-and-tube exchanger consists of 60 four-meter lengths of standard copper condenser tubing (I.D. = 23 mm; O.D. = 25mm). Steam at 100°C is to be condensed using water at a rate of 200,000 kg/hr flowing inside the tubes. The water inlet temperature is 8°C. The unit surface conductances in W/m²°K, based on actual area, are (a) water side, 35,000; (b) steam side, 15,000; (c) scale on steam side, 14,000; and (d) scale on water side, 12,200. Estimate the rate of condensation of steam.

$$h_{fg} = 2257 \text{ kJ/kg}$$

10-21. Show that $\epsilon = \text{NTU}/(1 + \text{NTU})$ for counter flow and $\epsilon = (1/2)(1 - e^{-2\text{NTU}})$ for parallel flow if the capacity rates of the hot and cold fluids are equal.

10-22. A cross-flow exchanger with both fluids unmixed is used to cool air that is initially at 40°C and flowing at a rate of 4000 kg/hr. Water is used as the coolant at 5°C with a flow rate of 4600 kg/hr. Assume the U value to be 150 W/m²°C. The surface area of the exchanger is 25 m². Calculate the exit temperatures of the air and water.

10-23. In a one-shell pass, two-tube pass heat exchanger, water at 18°C enters at a rate of 4000 kg/hr in the shell. Engine oil with c_p = 2600 J/kg°K flows at a rate of 2000 kg/hr through the tubes. The surface area of the exchanger is 14

m^2. The U_o value may be assumed to be 200 W/m^2°K. Determine the exit temperatures of the two fluids if the oil enters at 150°C.

10-24. Work problem 10-23 by the LMTD method.

10-25. A cross-flow exchanger with cold fluid unmixed is used to cool air from 100°C by using air at 20°C. The hot air flows at a rate of 3000 kg/hr, while the cold air rate is 7000 kg/hr. The hot air leaves the exchanger at 55°C. Assuming the U value to be 70 W/m^2°K, calculate the exit temperature of the cold fluid and the area of the heat exchanger.

10-26. Derive an expression for the U factor if the inside and outside surfaces of the tubes in a heat exchanger are finned and have fin efficiencies of η_i and η_o on the inside and on the outside, and A_i and A_o as the total area (inclusive of fin surfaces) on the inside and the outside, respectively.

10-27. Circumferential fins of 3 mm thickness and 6 mm length are attached to the outside of the tubes of the heat exchanger in problem 10-20. Determine the rate of condensation of steam by neglecting the effect of scales if the fins are spaced 8 mm apart.

REFERENCES

[1] Harper, W. P., and Brown, D. R., "Mathematical Equations for Heat Conduction in the Fins of Air-Cooled Engines," *NACA Report 158*, (1922).

[2] Kern, Donald Q., and Kraus, A. D., *Extended Surface Heat Transfer*, McGraw-Hill Book Company, New York, (1972).

[3] *Standards of Tubular Exchanger Manufacturers Association,* 4th ed., Tubular Exchanger Manufacturers Association, Inc., New York, (1959).

[4] McAdams, W. H., *Heat Transmission*, McGraw-Hill Book Company, New York, (1954).

[5] Bowman, R. A., Mueller, A. C., and Nagle, W. M., "Mean Temperature Difference in Design," *Trans. ASME.*, **62**, p. 288, (1940).

[6] Kays, W. M., and London, A. L., *Compact Heat Exchangers,* McGraw-Hill Book Company, New York, (1968).

11

Heat Transfer with Change of Phase

11-1 INTRODUCTION

In the heat transfer processes that have been studied in the forgoing chapters, we have considered only homogeneous single-phase systems. However, for many engineering problems, heat transfer occurs during changes of phase. In the generation of electrical power, water is turned to steam in a boiler, and the boiling process must be well understood for effective design. Furthermore, after the steam has been used to drive a turbine, it is condensed from a vapor to a liquid so that it may be pumped back to the boiler to repeat the process. This necessitates an understanding of condensation heat transfer. Both of these processes, boiling and condensation, involve a change of phase. Also, the processes of evaporation and condensation are primary mechanisms in the heat pipe, a device that allows a large heat transfer to occur through a small area. Another example is the heat that must be dissipated by a space vehicle when it reenters the earth's atmosphere. In this case, the frictional heating of the air molecules causes the solid nose cone of the space capsule to undergo a change of phase.

In the following sections, we will discuss condensation and boiling and briefly describe the operation of a heat pipe. The first topic to be presented will be that of condensation. Condensation is important in many heat exchanger applications in addition to its role in power plant steam condensers.

The material presented in this chapter is mainly descriptive in nature, and the reader is referred to the various references for more detailed information.

11-2 CONDENSATION HEAT TRANSFER

In heating or in power generating systems using steam, the spent steam must be condensed for reuse. The efficiency of the condensing unit is determined by the mode of condensation that prevails: *dropwise*, in which the vapor condenses into small liquid droplets of various sizes; and *filmwise*, in which the vapor condenses into a continuous film covering the entire surface. In practice, either may be found in a condenser. However, it is more desirable to maintain dropwise condensation, because its heat transfer coefficient is five to ten times that of filmwise condensation. This increase is due to greater heat flow through the curved surface of the drops than through a film of the same mass, greater heat transfer between the vapor and the bare surface, and more rapid removal of the condensate due to coalescence and roll-off of drops than would occur in a film flow.

Since the pioneering work of Nusselt (Reference 1) in 1916 on filmwise condensation and of Jakob (Reference 2) in 1936 on dropwise condensation, much effort has been expended in describing, analyzing, predicting, and improving heat transfer rates. Their successors have investigated the effects on condensation of such variables as heat flux level, surface temperature, noncondensable gases, superheating, surface orientation, wettability, thermal conductivity of the surface, and the presence of dropwise condensation promoters. Mechanical methods of mimicking dropwise condensation have also been devised.

11-2.1 Filmwise Condensation

The basic theory of filmwise condensation was proposed by Nusselt (Reference 1) in 1916. The Nusselt model dealt with a pure vapor at its saturation temperature. The vapor condenses on a vertical surface whose temperature is less than the saturation temperature of the vapor. It was assumed that the condensate forms a film on the surface. The thickness of the liquid film was determined from a balance between the gravity and friction forces. Nusselt's analysis led to the following expressions for the local and average heat transfer coefficients.

$$h = \left[\frac{k_{\text{liq}}^3 \rho_{\text{liq}} (\rho_{\text{liq}} - \rho_{\text{vap}}) g h_{fg}}{4\mu_{\text{liq}} (T_{\text{sat}} - T_{\text{w}}) x}\right]^{1/4} \qquad (11\text{-}1)$$

and

$$h_{\text{av}} = \frac{4}{3} \text{h}_{x=L} \qquad (11\text{-}1a)$$

where

g = acceleration due to gravity

ρ_{liq} = density of the liquid phase

ρ_{vap} = density of the vapor phase

h_{fg} = latent heat of condensation

k_{liq} = thermal conductivity of the liquid

μ_{liq} = dynamic viscosity of the liquid

$T_{sat} - T_w = T_{saturated\ vapor} - T_{surface}$

x = distance from the top of the condensing surface

To obtain this relationship, Nusselt assumed that the flow in the liquid was laminar and that the velocities were so small that the fluid inertia could be neglected. Furthermore, he postulated that heat conduction was the dominant heat transfer mechanism, so that the temperature profile in the film was linear.

Nusselt's model oversimplifies the actual phenomenon of filmwise condensation. More recent analyses have removed many of his simplifying assumptions. Rohsenow (Reference 3), in 1956, analyzed the linear temperature profile assumption and took account of the convective transport in the condensate film. In 1959, Sparrow and Gregg (Reference 4) further refined the analysis by accounting for the effects of fluid inertia. This is an important contribution in dropwise condensation as well.

Madejski (Reference 5), in 1966, analyzed the usually neglected part of the temperature differential, which is due to molecular-kinetic mass transfer in change of phase. He concluded that this normally small contribution to the temperature differential is significant at low pressures and in vacuum condensers. Bayley et al. (Reference 6) noted that if turbulent flow prevails in the film ($Re_{film} \geqslant 1800$), the average heat transfer coefficient is larger than that predicted by Nusselt's laminar model.

The Reynolds number is given by

$$Re = \frac{\rho_{liq} u_{av} L_c}{\mu_{liq}}$$

The characteristic length, L_c, is

$$L_c = \frac{4 \text{ cross-sectional area}}{\text{wetted perimeter}} = \frac{4A}{P}$$

Also, the mass flow rate is

$$\dot{m} = \rho A u_{av}$$

Combining the forgoing equations, we obtain the following expressions for the Reynolds number for the condensate flow.

$$\text{Re} = \frac{4\dot{m}}{P\mu_{\text{liq}}} \tag{11-2}$$

We note that the energy convected to the cool surface as a result of the condensation process is equal to $h_{av}(T_{\text{sat}} - T_{\text{w}}) = \dot{m}h_{\text{fg}}$. Thus

$$\text{Re} = \frac{4h_{av}L(T_{\text{sat}} - T_{\text{w}})}{h_{fg}\mu_{\text{liq}}} \tag{11-3}$$

For Re > 1800, the average convective heat transfer coefficient is given by Reference 7 as

$$h_{av} = 0.0077\,\text{Re}^{0.4}\left[\frac{\mu_{\text{liq}}^2}{k_{\text{liq}}^3\,\rho_{\text{liq}}(\rho_{\text{liq}} - \rho_{\text{vap}})g}\right]^{-1/3} \tag{11-4}$$

Henderson and Marchello's (Reference 8) correction of the Nusselt equation for condensation on a horizontal tube was based on liquid drop breakoff at the bottom of the tube and circumferentially variable film thickness due to the weight of the droplet. They used an empirical correction factor based on the supportable droplet weight, i.e., the surface tension multiplied by the circumference of the supporting area.

There are various factors influencing filmwise condensation that were not considered by Nusselt. Minkowycz and Sparrow (Reference 9) concluded that a concentration of one-half percent of a noncondensable gas (e.g., air) in steam could reduce the condensation rate by fifty percent or more. This is an important contribution to the condensation resistance in dropwise as well as filmwise condensation. Felicion and Seban (Reference 10) studied the effect of a noncondensable gas heavier than the condensing vapor and found a slight effect due to solubility of the gas in the condensate. However, their conclusions were limited to values of the film Reynolds number, Re, less than twenty-five.

Whereas there have been many refinements and extensions of Nusselt's analysis, it is evident that his work was basic to all that followed. His equations are sufficiently accurate to be used, in many cases, in their original forms.

We shall conclude our discussion of filmwise condensation by presenting the derivation of Nusselt's equation, equation (11-1) for filmwise condensation on a vertical flat plate. Figure 11-1 is a schematic of the physical situation.

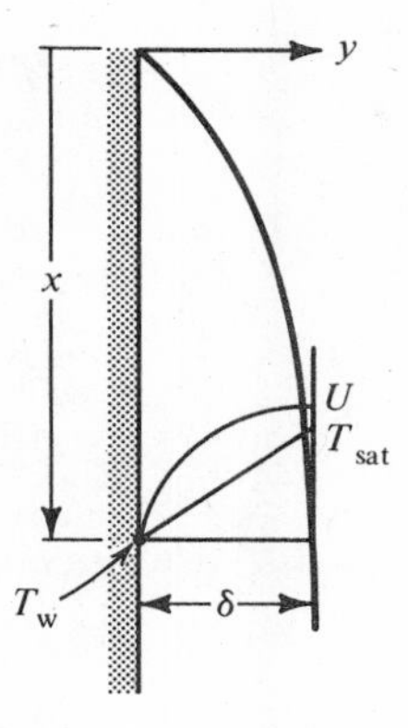

Figure 11-1 Condensation on a vertical flat plate.

As noted earlier in this section, Nusselt assumed laminar flow in the liquid film, neglected fluid inertia, and postulated heat conduction across the film as the dominant mechanism for energy transfer. These assumptions lead to

$$u = U\left[2\frac{y}{\delta} - \frac{y^2}{\delta^2}\right] \tag{11-5}$$

$$dQ = \left(\frac{k_{liq}}{\delta}\right) dx\,(T_{sat} - T_w) \tag{11-6}$$

where U is the velocity at the edge of the film, δ is the film thickness, and dQ is the heat transfer rate in the length, dx. Equation (11-5) indicates a parabolic velocity profile within the film.

Note that the convective heat transfer coefficient between the wall surface and the vapor is given by

$$dQ = h\,dx\,(T_{sat} - T_w) \tag{11-7}$$

Since dQ from equations (11-6) and (11-7) may be set equal to each other

$$h = \frac{k_{liq}}{\delta} \tag{11-8}$$

To complete the derivation of equation (11-1), it is necessary to determine δ. The first step is to employ a force balance in which the wall shear force is set equal to the downward pull of gravity.

$$\frac{1}{g_c}\rho_{liq} g \delta dx = \tau_w dx \tag{11-9}$$

Since $\tau_w = (1/g_c)\,\mu_{liq}\,(\partial u/\partial y)_{x=0}$ equation (11-5) gives

$$\tau_w = \frac{\mu_{liq}}{g_c}\left(\frac{2U}{\delta}\right)$$

and substituting this result into equation (11-9), we obtain

$$\rho_{liq} g \delta dx = \mu_{liq}\left(\frac{2U}{\delta}\right) dx$$

or

$$U = \frac{\rho_{liq} g}{2\mu_{liq}}(\delta^2)$$

Now, equation (11-5) can be rewritten

$$u = \frac{\rho_{liq} g \delta^2}{2\mu_{liq}}\left(\frac{2y}{\delta} - \frac{y^2}{\delta^2}\right) \tag{11-10}$$

Since the velocity distribution is parabolic, its average value, u_{av}, at a given value of x will be two-thirds of its maximum value, U, at that location.

$$u_{av} = \frac{2}{3}(U) = \frac{\rho_{liq} g \delta^2}{3\mu_{liq}}$$

Consequently, the mass flow rate of the fluid flowing through the cross-sectional area at x is

$$\dot{m} = \rho_{liq} u_{av} \delta = \frac{\rho_{liq}^2 g \delta^3}{3\mu_{liq}}$$

If we move down the plate a distance, dx, we find an increase in the mass flow by the amount

$$d\dot{m} = \frac{\rho_{\text{liq}}^2 g}{\mu_{\text{liq}}}(\delta^2 d\delta) \tag{11-11}$$

This increase is caused by the condensation of vapor. The amount of vapor condensed is equal to the heat transfer rate divided by the latent heat of condensation; that is

$$d\dot{m} = dQ/h_{\text{fg}} \tag{11-12}$$

Combining equations (11-6), (11-11), and (11-12), we have

$$\frac{k_{\text{liq}}dx}{\delta}(T_{\text{sat}} - T_{\text{w}}) = (h_{fg})\frac{\rho_{\text{liq}}^2 g\delta^2 d\delta}{\mu_{\text{liq}}}$$

so that

$$\delta^3\left(\frac{d\delta}{dx}\right) = \frac{\mu_{\text{liq}}k_{\text{liq}}}{\rho_{\text{liq}}^2 g h_{fg}}(T_{\text{sat}} - T_{\text{w}})$$

Separating variables and integrating yields

$$\frac{\delta^4}{4} = \frac{\mu_{\text{liq}}k_{\text{liq}}}{\rho_{\text{liq}}^2 g h_{\text{fg}}}(T_{\text{sat}} - T_{\text{w}})x + C$$

where C is the constant of integration and is equal to zero, since at $x = 0$, $\delta = 0$. Solving for δ gives

$$\delta = \left[\frac{4\mu_{\text{liq}}k_{\text{liq}}}{\rho_{\text{liq}}^2 g h_{\text{fg}}}(T_{\text{sat}} - T_{\text{w}})x\right]^{1/4}$$

Next, equation (11-8) is

$$h = \frac{k_{\text{liq}}}{\delta}$$

which results in Nusselt's equation, equation (11-1)

$$h = \left[\frac{k_{liq}^3 \rho_{liq}(\rho_{liq} - \rho_{vap})g h_{fg}}{4\mu_{liq}(T_{sat} - T_w)x}\right]^{1/4} \tag{11-1}$$

Equation (11-1) gives the local heat transfer coefficient at a distance, x, from the top of the plate. The average value over the entire distance, x, for the case of uniform wall temperature is given by

$$h_{av} = \frac{1}{L}\int_o^L h dx = \left(\frac{4}{3}\right) h_{x=L}$$

If the vertical surface were replaced by a horizontal tube, then

$$h_{av} = 0.73 \left[\frac{k_{liq}^3 \rho_{liq}(\rho_{liq} - \rho_{vap})g h_{fg}}{\mu_{liq}(T_{sat} - T_w)d}\right]^{1/4} \tag{11-14}$$

where d is the diameter of the tube. When $L/d = 2.76$, the value of h_{av} for the two surfaces is the same.

Sample Problem 11-1 Assuming laminar flow, find the surface temperature of a 4-foot vertical surface upon which steam is condensing into a film covering the entire surface. The partial pressure of the water vapor is 14.7 psia, and the average heat transfer coefficient is 800 Btu/hr-ft^2 °F.

Solution:

$$h_{av} = \left(\frac{4}{3}\right) h_{x=L}.$$

$$h_{av} = \frac{4}{3}\left[\frac{k_{liq}^3 \rho_{liq}(\rho_{liq} - \rho_{vap})g h_{fg}}{4\mu_{liq}(T_{sat} - T_w)L}\right]^{1/4}$$

$$h_{av} = 800 \text{ Btu/hr-ft}^2\,°\text{F}$$

$$g = 32 \text{ ft/sec}^2 = 4.147 \times 10^8 \text{ ft/hr}^2$$

$$\rho_{liq} = 59.97 \text{ lb}_m/\text{ft}^3$$

$$\rho_{vap} = 0.0372 \text{ lb}_m/\text{ft}^3$$

$$h_{fg} = 970.3 \text{ Btu/lb}_m$$

$$k_{\text{liq}} = 0.393 \text{ Btu/hr-ft°F}$$

$$\mu_{\text{liq}} = 0.688 \text{ lb}_m\text{/hr-ft}$$

$$T_{\text{sat}} = 212°\text{F @ } 14.7 \text{ psia}$$

$$L = 4 \text{ feet}$$

$$T_{\text{w}} = ?$$

$$h_{\text{av}}^4 = \left(\frac{256}{81}\right)\frac{k_{\text{liq}}^3 \rho_{\text{liq}} (\rho_{\text{liq}} - \rho_{\text{vap}}) g h_{fg}}{4\mu_{\text{liq}} (T_{\text{sat}} - T_{\text{w}}) L}$$

Solving the above equation for T_{w}, we obtain

$$T_{\text{w}} = T_{\text{sat}} - \left(\frac{64}{81}\right)\frac{k_{\text{liq}}^3 \rho_{\text{liq}} (\rho_{\text{liq}} - \rho_{\text{vap}}) g h_{fg}}{\mu_{\text{liq}} L h_{\text{av}}^4}$$

$$= 212 - \left(\frac{64}{81}\right) \times \frac{(0.393)^3 (59.97)(59.97 - 0.0372)(4.147 \times 10^8)(970.3)}{(0.688)(4)(800)^4}$$

$$T_{\text{w}} = 212 - 61.5 = 150.5°\text{F}$$

11-2.2 Dropwise Condensation

Basic heat transfer texts do not treat dropwise condensation as fully as filmwise condensation, preferring to say that the process is still not fully understood. This statement is the result of the existence of several theories of the process, none of which has been fully proven or disproven.

The first theory, by Jakob (Reference 2), of dropwise condensation proposed the existence of a condensate film that broke into drops at nucleation sites. These drops grew *in situ* until they rolled away or broke off the surface. This process was, apparently, confirmed experimentally by Welch and Westwater (Reference 11). Later interpretations by Umur and Griffith (Reference 12) of the same experiment, however, indicated that no film was formed and that droplets were formed only at discrete nucleation sites.

Many factors influencing the rate of dropwise condensation have been analyzed, and several methods of inducing and enhancing dropwise condensation have been devised. There are four main methods of promoting dropwise condensation as given by Davies et al. (Reference 13):

(1) Injection of the promoter into the steam, e.g., straight chain primary amines;
(2) Use of chemically adhering promoters, e.g., fatty acids;
(3) Use of physically adhering promoters, e.g., Teflon; and
(4) Plating noble metals onto the surface.

11-3 BOILING HEAT TRANSFER

As in the case of condensation, boiling is of importance in the generation of power. In a typical electric power plant, water is turned into steam in a boiler so that it may be used to drive a turbine.

What follows is a short, qualitative description of the boiling heat transfer phenomena. The reader is referred to Reference 14 for an excellent summary of the general topic.

If the heat of evaporation is supplied to a saturated liquid at its free surface, such as radiant energy incident on a pail of water, vapor may be produced, and such a process is called *evaporation*. On the other hand, if heat is added to a liquid from a submerged solid surface, vapor may be produced that can form bubbles, which can grow, eventually be detached from the surface, and rise to the free surface due to buoyancy effects. Such a process is called *pool boiling*. We will now consider different types of pool boiling and present a correlation for pool boiling data.

Pool boiling may occur when a heating surface is submerged below the free surface of a liquid. A necessary condition for boiling to occur is that the temperature of the heating surface exceeds the saturation temperature of the liquid. The temperature of the liquid determines the type of boiling. If the temperature of the liquid is below its saturation temperature, the process is called *subcooled* or *local boiling*. However, if the temperature of the liquid is at the saturation temperature, the process is called *saturated* or *bulk boiling*.

The different regimes of pool boiling may be best understood by examining a general plot that shows how the surface heat flux varies with the difference between the surface temperature and the saturation temperature of the liquid (often referred to as the excess temperature). Figure 11-2 shows such a plot.

Referring to Figure 11-2, below point **A**, nonboiling convective heat transfer occurs with superheated liquid rising by natural convection to the free surface of the liquid where evaporation takes place. As $(T_w - T_{sat})$ is increased, bubbles begin to form and rise through the liquid. Near point **A**, these bubbles tend to collapse as they move toward the free surface. However, as the temperature excess increases, and point **B** is approached, the bubbles rise all the way to the free surface. Nucleate boiling is said to exist between points **A** and **B**.

Thus far, as the temperature difference is increased, the heat flux also increases. This trend is reversed between points **B** and **C**. As the temperature difference is increased beyond point **B**, bubbles are formed so rapidly that they tend to coalesce, forming a vapor film over the heating surface, which inhibits interaction between the liquid and the surface. Since the vapor film possesses a relatively low thermal condutivity, the heat flux drops off. Near point **B**, there is a mixture of nucleate and film boiling, while at point **C**, there is only film boiling. Boiling is unstable near point **B**, but at point **C**, stable film boiling prevails. Since the excess temperature is high during film boiling, a significant portion of the heat lost by the surface may be due to radiation, as is the case as we move up the curve toward point **D**.

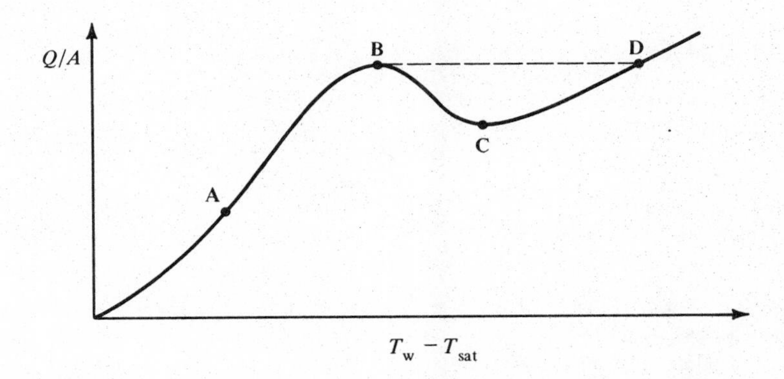

Figure 11-2 Heat flux from submerged surface as a function of $(T_w - T_{sat})$ for pool boiling.

Various attempts have been made to correlate pool boiling data. Equation (11-15), presented below, is recommended by Rohsenow in Reference 14.

$$\frac{c_{liq}(T_w - T_{sat})}{h_{fg}} = C_{sf}\left[\left(\frac{Q/A}{\mu_{liq}h_{fg}}\right)\sqrt{\frac{g_c\sigma}{g(\rho_{liq} - \rho_{vap})}}\right]^{1/3} \mathrm{Pr}_{liq}^{s} \qquad (11\text{-}15)$$

where

c_{liq} = specific heat of the liquid at constant pressure

μ_{liq} = dynamic viscosity of the liquid

g_c = gravitational constant

g = local acceleration due to gravity

$\mathrm{Pr}_{\mathrm{liq}}$ = Prandtl number of the liquid

C_{sf} = experimental constant

h_{fg} = heat of evaporation

ρ_{vap} = density of vapor

ρ_{liq} = density of liquid

σ = vapor-liquid surface tension

s = 1.0 for water, 1.7 for all other liquids

Table 11-1 gives values for C_{sf} for several surface-liquid combinations. Table 11-2 gives values of the surface tension for water as a function of temperature.

TABLE 11-1 Experimental Constant C_{sf} as a Function of Surface-Liquid Combination*

Surface Combination	C_{sf}	s
Water-nickel [15]	0.006	1.0
Water-platinum [16]	0.013	1.0
Water-copper [16]	0.013	1.0
Water-brass [17]	0.006	1.0
CCl_4-copper [18]	0.013	1.7
Benzene-chromium [19]	0.010	1.7
n-pentane-chromium [19]	0.015	1.7
Ethyl alcohol-chromium [19]	0.0027	1.7
Isopropyl alcohol-copper [18]	0.0025	1.7
35% K_2CO_3-copper [18]	0.0054	1.7
50% K_2CO_3-copper [18]	0.0027	1.7
n-butyl alchohol-copper [18]	0.0030	1.7

*Numbers in brackets refer to source of data listed at end of chapter.

TABLE 11-2 Vapor-Liquid Surface Tension for Water

Saturation Temperature, ° F	*Surface tension* σ *lb*$_f$*/ft*	*Saturation Temperature ° C*	*Surface tension* σ *N/m*^{2}
32	51.8×10^{-4}	0.0	7.6×10^{-2}
60	50.2×10^{-4}	15.5	7.3×10^{-2}
100	47.8×10^{-4}	37.8	7.0×10^{-2}
140	45.2×10^{-4}	60.0	6.6×10^{-2}
200	41.2×10^{-4}	93.3	6.0×10^{-2}
212	40.3×10^{-4}	100.0	5.9×10^{-2}
320	31.6×10^{-4}	160.0	4.6×10^{-2}
440	21.9×10^{-4}	226.7	3.2×10^{-2}
560	11.1×10^{-4}	293.3	1.6×10^{-2}
680	1.0×10^{-4}	360.0	0.15×10^{-2}
705.4	0	374.1	0

In many engineering applications, heat fluxes rather than surface temperatures are prescribed during boiling processes. Let us reexamine Figure 11-2 with this in mind. As long as the prescribed heat flux is less than that at point **B**, we find that as Q/A is increased, the excess temperature also increases in a continuous manner. However, if the prescribed heat flux is increased by only a small amount above that at point **B**, there is a large step increase in the required excess temperature (point **D**) to support the new heat flux. Frequently, the surface temperature at **D** is above that of the melting point of the submerged surface, resulting in a phenomena referred to as *burnout*.

A knowledge of the burnout heat flux is important to prevent failure of heating elements. Specifically, in nuclear power plants, the heat flux from the radioactive fuel is usually prescribed in the boiler, and engineers must be careful to keep this heat flux below the burnout level to avoid a severe accident. Reference 20 gives the following equation for heat flux, q_B, at burnout:

$$q_B = \frac{Q_B}{A} = \left(\frac{\pi}{24}\right) h_{fg}\,\rho_{vap}\left[\frac{\sigma g_c g(\rho_{liq} - \rho_{vap})}{\rho_{vap}^2}\right]^{1/4} \times \left[\frac{\rho_{liq} + \rho_{vap}}{\rho_{liq}}\right]^{-1/2} \qquad (11\text{-}16)$$

where all the items have been previously defined.

In closing this section, we note that if boiling occurs under forced convection conditions, such as in flow through a tube, the total heat flux can be approximated as the sum of the heat flux due to boiling and the heat flux due to forced convection. That is

$$\left(\frac{Q}{A}\right)_{total} = \left(\frac{Q}{A}\right)_{boiling} + \left(\frac{Q}{A}\right)_{convection} \qquad (11\text{-}17)$$

Sample Problem 11-2 Water is boiled at a rate of 40 lb_m/hr in a kettle at atmospheric pressure. The bottom of the kettle is flat, 9 inches in diameter, and copper. What is the temperature of the bottom surface of the kettle?

Solution: Assume the temperature of the water is at 212°F. From equation (11-15)

$$\frac{c_{liq}(T_w - T_{sat})}{h_{fg}} = C_{sf}\left\{\frac{Q/A}{\mu_{liq} h_{fg}}\left[\frac{g_c\sigma}{g(\rho_{liq} - \rho_{vap})}\right]^{1/2}\right\}^{0.33} Pr_{liq}^{s}$$

$c_{liq} = 1.0058$ Btu/lb_m °F

$T_w = ?$

$$T_{sat} = 212°F$$

$$h_{fg} = 970.3 \text{ Btu/lb}_m$$

$$C_{sf} = 0.013$$

$$Q = 970.3 \text{ Btu/lb}_m \times 40 \text{ lb}_m/\text{hr} = 38812 \text{ Btu/hr}$$

$$A = \frac{\pi d^2}{4} = 0.442 \text{ ft}^2$$

$$\mu_{liq} = 0.688 \text{ lb}_m/\text{hr-ft}$$

$$g_c = 32.2 \text{ lb}_m \text{ ft/lb}_f\text{sec}^2$$

$$\sigma = 40.3 \times 10^{-4} \text{ lb}_f/\text{ft}$$

$$g = 32.2 \text{ ft/sec}^2$$

$$\rho_{liq} = 59.97 \text{ lb}_m/\text{ft}^3$$

$$\rho_{vap} = 0.0372 \text{ lb}_m/\text{ft}^3$$

$$\text{Pr}_{liq} = 1.756$$

$$s = 1$$

$$T_w = T_{sat} + \frac{C_{sf} h_{fg}}{c_{liq}} \left\{ \frac{Q/A}{\mu_{liq} h_{fg}} \left[\frac{g_c \sigma}{g(\rho_{liq} - \rho_{vap})} \right]^{1/2} \right\}^{0.33} \text{Pr}_{liq}^{s}$$

$$= 212°F + \frac{(0.013)(970.3)}{1.0058} \times$$

$$\left\{ \frac{38812/0.442}{(0.688)(970.3)} \left[\frac{32.2(40.3 \times 10^{-4}}{32.2(59.97 - 0.0372)} \right]^{1/2} \right\}^{0.33} 1.756$$

$$= 212°F + (12.54)[(131.5)(8.2 \times 10^{-3})]^{0.33}(1.756)$$

$$= 212°F + 22.6°F$$

$$T_w = 234.6°F$$

Sample Problem 11-3 Determine the burnout heat flux for nucleate boiling for Sample Problem 11-2.

Solution:
From equation (11-16)

$$q_B = \frac{Q_B}{A} = \frac{\pi}{24}(h_{fg}\rho_{vap})\left[\frac{\sigma g\, g_c(\rho_{liq} - \rho_{vap})}{\rho_{vap}^2}\right]^{1/4}\left[\frac{\rho_{liq} + \rho_{vap}}{\rho_{liq}}\right]^{-1/2}$$

$Q_B = ?$

$A = 0.442\ ft^2$

$h_{fg} = 970.3\ Btu/lb_m$

$\rho_{vap} = 0.0372\ lb_m/ft^3$

$\sigma = 40.3 \times 10^{-4}\ lb_f/ft$

$g = 32.2\ ft/sec^2$

$\rho_{liq} = 59.97\ lb_m/ft^3$

$g_c = 32.2\ lb_m\ ft/lb_f sec^2$

$$Q_B = \frac{\pi}{24}(Ah_{fg}\rho_{vap})\left[\frac{\sigma g g_c(\rho_{liq} - \rho_{vap})}{\rho_{vap}^2}\right]^{1/4}\left[\frac{\rho_{liq} + \rho_{vap}}{\rho_{liq}}\right]^{-1/2}$$

$$Q_B = \frac{(3.14)(0.442)(970.3)(0.0372)}{24}$$

$$\times \left[\frac{(40.3 \times 10^{-4})(32.2)(32.2)(59.97 - 0.0372)}{(0.0372)^2}\right]^{1/4}$$

$$\times \left[\frac{59.97 + 0.0372}{59.97}\right]^{-1/2}$$

$$Q_B = (2.087)(20.6)(1.0)$$

$$Q_B = 43.0\ (Btu/sec)(3600\ sec/hr)$$

$$Q_B = 1.55 \times 10^5\ Btu/hr$$

11-4 HEAT PIPE

In Chapter 10, we discussed fins and heat exchangers. The main objective in employing such devices was the efficient transfer of heat. It was stated that the smaller the heat exchanger used to handle a given heat load, the better its design. The heat pipe is an innovative device capable of transferring large quantities of heat through relatively small cross-sectional areas and with very small temperature differences. In fact, a typical heat pipe will transfer about ten times the heat that an equivilent volume of copper is capable of handling. Figure 11-3 shows a typical heat pipe.

In essence, a heat pipe consists of a closed pipe lined with a wicking material and containing a condensible gas. The center portion of the pipe is insulated, and its two noninsulated ends respectively serve as evaporators and condensers. At the hot end, fluid is evaporated and thereby absorbs a quantity of heat equal to its latent heat of evaporation. As fluid is evaporated, the vapor is forced through the central core of the heat pipe to the cool end, where it is condensed and gives up its latent heat of condensation, provided that the low temperature end is at a temperature below the saturation temperature of the vapor. Then, the condensed liquid is absorbed by the wicking material and travels back to the hot end by capillary action. The working of a heat pipe is independent of its spatial orientation, since gravitational forces do not play a significant role.

Modifications of the heat pipe allow it to be applied in different ways. For example, the insulated portion may be made of flexible tubing to permit accommodation of different physical constraints. Also, it can be applied to cool micro-electronic circuits where maintaining constant temperatures for different components is important. Under steady-state operating conditions, a power transistor transferring heat to the evaporator section of a heat pipe would have its temperature held constant at the saturation temperature of the vapor in the heat pipe. As heating and cooling needs become increasingly sophisticated, we can expect to see increased use of heat pipes.

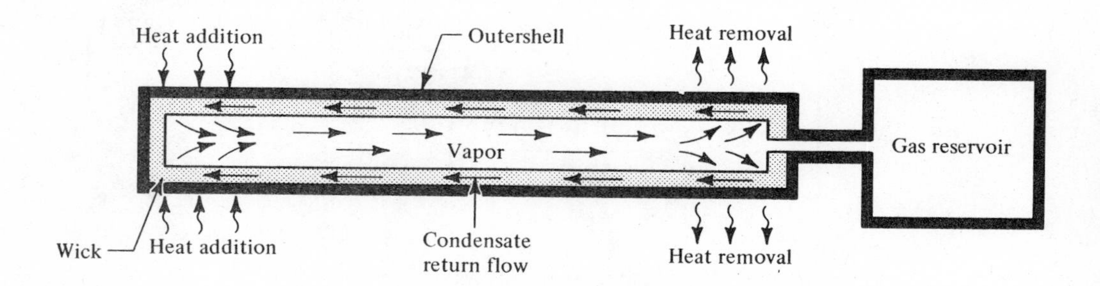

Figure 11-3 Typical heat pipe.

PROBLEMS

11-1. A 2 ft by 2 ft vertical plate is exposed to steam at atmospheric pressure ($T_{sat} = 212°F$, $h_{fg} = 970.3$ Btu/lb$_m$). The plate temperature is 210°F. Calculate the heat transfer rate and the mass condensed per hour.

11-2. Steam at 1 atmosphere (T_{sat} = 100°C., h_{fg} = 2.257 × 10^6 J/kg) condenses on a vertical plate 40 cm high, which is at 97°C. Determine the local heat transfer coefficient at the midpoint, at the top of the plate, and the average coefficient over the entire plate. Show that the flow is laminar everywhere on the plate.

11-3. Calculate the condensate thickness on a horizontal tube experiencing filmwise condensation by applying the basic assumptions utilized in deriving Nusselt's equation [equation (11-1)] and assuming the film thickness to be small as compared to the tube diameter.

11-4. Use equation (11-15) to predict the convective heat transfer coefficient in pool boiling when ($T_w - T_{sat}$) is 20°C, and the pressure is 1 atmosphere. Use a value of 0.01 for C_{sf} and take h_{fg} = 2.257 × 10^6 J/kg and σ = 0.057 N/m. Is the value of h so determined very sensitive to the value of C_{sf} used?

11-5. The condenser of a large steam power plant is made up of a horizontal bank of brass tubes having a 1.2-in. I.D. and a wall thickness of 0.05 inches. Each tube is 10 feet long. If the condenser is to condense 25,000 lb_m of water vapor per hour at 2 psia, and cooling water circulates through the tubes with a velocity of 6 ft/sec while entering the tubes at 60°F, estimate the number of tubes needed in the condenser. Assume ΔT_{H_2O} is small.

11-6. The condenser of a large steam power plant is made up of a horizontal bank of brass tubes having a 1.2-in. I.D. and a wall thickness of 0.05 inches. The cooling water inside the tubes enters at 60°F and at such a velocity as to give an average forced convection heat transfer coefficient of 350 Btu/hr-ft^2°F. Steam at 1.27 psia (T_{sat} = 110°F., h_{fg} = 1031.6 Btu/lb_m) experiences filmwise condensation on the outer surface of the tubes. If dropwise condensation could be induced, the outside convective heat transfer coefficient would increase by a factor of 5 to 10. Under such conditions, by what percentage could the heat transfer surface be reduced?

REFERENCES

[1] Nusselt, W., "The Surface Condensation of Water Vapor," (in German), *Zeitschrift des Vereines Deuschen Ingenieure*, **60**, p. 541, (1916).

[2] Jakob, M., "Heat Transfer in Evaporation and Condensation–II," *Mechanical Engineering*, **58**, p. 729, (1936).

[3] Rohsenow, W. M., "Heat Transfer and Temperature Distribution in Laminar Film Condensation," *Trans ASME*, **78**, p. 1654, (1956).

[4] Sparrow, E. M., and Gregg, J. L., "A Boundary-layer Treatment of Laminar-film Condensation," *J. of Heat Transfer, Trans ASME, Series C*, **81**, p. 13, (1959).

[5] Madejski, J., "Effect of Molecular Kinetic Resistances on Heat Transfer in Condensation," *Int. J. Heat and Mass Transfer*, **9**, p. 35, (1966).

[6] Bayley, F. J., Owen, J. M., and Turner, A. B., *Heat Transfer*, Barnes and Noble, Publishers, New York, (1972).

[7] Kirkbride, C. C., "Heat Transfer by Condensing Vapors on Vertical Tubes," *Trans. AIChE,* **30**, p. 170, (1934).

[8] Henderson, C. L., and Marchello, J. M., "Role of Surface Tension and Tube Diameter in Film Condensation on Horizontal Tubes," *AIChE Journal,* **13**, p. 613, (1967).

[9] Minkowycz, W. J., and Sparrow, E. M., "Condensation Heat Transfer in the Presence of Noncondensables, Interfacial Resistance, Superheating, Variable Properties, and Diffusion," *Int. J. Heat and Mass Transfer,* **9**, p. 1125, (1973).

[10] Felicion, F. S., and Seban, R. A., "Laminar Film Condensation of a Vapor Containing a Soluble, Non-condensing Gas," *Int. J. Heat and Mass Transfer,* **16**, p. 1601, (1973).

[11] Welch, J. F., and Westwater, J. W., "Microscopic Study of Dropwise Condensation," *Proceedings*, Second International Heat Transfer Conference, **II**, p. 302, (May 1961).

[12] Umur, A., and Griffith, P., "Mechanism of Dropwise Condensation," *Trans ASME J. of Heat Transfer*, p. 275, (May 1965).

[13] Davies, G. A., Mojyehedi, W., and Ponter, A. B., "Measurement of Contact Angles under Condensation Condition; The Prediction of Dropwise-Filmwise Transition," *Int. J. Heat and Mass Transfer*, **14**, p. 709, (1971).

[14] Rohsenow, W. M., and Hartnett, J. P., *Handbook of Heat Transfer*, Chapter 13, McGraw-Hill Book Company, (1973).

[15] Rohsenow, W. M., and Clark, J. A., "Heat Transfer and Pressure Drop Data, for High Heat Flux Densities at Sub-critical Pressures," *Heat Transfer Fluid Mech. Inst.*, Stanford, Calif., (1951).

[16] Addoms, J. N., "Heat Transfer at High Rates to Water Boiling Outside Cylinders," Doctoral Dissertation, MIT, (1948).

[17] Cryder, D. S., and Finalborqv, A. C., "Heat Transmission from Metal to Boiling Liquids: Effect of Temperature of the Liquid on Film Coefficient," *Trans. Am. Inst. Chem. Eng.,* **33**, (1937).

[18] Piret, E. L., and Isbin, H. S., "Two-phase Heat Transfer in Natural Circulation Evaporators," *AIChE Heat Transfer Symp.*, St. Louis, Missouri, (Dec. 1953).

[19] Cichelli, M. T., and Bonilla, C. F., "Heat Transfer to Liquids Boiling under Pressure," *Trans. AIChE,* **41**, (1945).

[20] Zuber, N., "On the Stability of Boiling Heat Transfer," *Trans ASME,* **80**, p. 711, (1958).

Appendixes

APPENDIX A Units and Dimensions

Quantity	*Symbol*	*English Units*	*SI Units*	*Dimensions*
Mass	m	lb_m	kg (kilogram)	m or M
Length	L	ft	m (meter)	L
Time	τ	sec or hr	sec (second)	τ
Temperature	T	°F	°C (degree Celsius)	T
Acceleration	a	ft/sec^2	m/s^2	$L\tau^{-1}$
Area	A	ft^2	m^2	L^2
Coefficient of thermal expansion	β	$°R^{-1}$	$°K^{-1}$	T^{-1}
Dynamic viscosity	μ	lb_m/ft-hr	kg/m hr	$mL^{-1}\tau^{-1}$
Emissive power	e	$Btu/hr\text{-}ft^2$	W/m^2	$m\tau^{-3}$
Energy	E	Btu	J (Joule)	$mL^2\tau^{-2}$
Force	F	lb_f	N (Newton)	$mL\tau^{-2}$
Heat flux	q	$Btu/hr\text{-}ft^2$	W/m^2	$m\tau^{-3}$
Heat transfer coefficient	h	$Btu/hr\text{-}ft^2°F$	$W/m^2°C$	$m\tau^{-3}T^{-1}$
Irradiation	G	$Btu/hr\text{-}ft^2$	W/m^2	$m\tau^{-3}$
Kinematic viscosity	ν	ft^2/hr	m^2/hr	$L^2\tau^{-1}$
Mass density	ρ	lb_m/ft^3	kg/m^3	mL^{-3}
Pressure	p	lb_f/ft^2	N/m^2	$mL^{-1}\tau^{-2}$
Quantity of heat per unit time	Q	Btu/hr	W (watt)	$mL^2\tau^{-3}$
Radiosity	J	$Btu/hr\text{-}ft^2$	W/m^2	$m\tau^{-3}$
Specific heat	c	$Btu/lb_m°F$	J/kg °C	$L^2\tau^{-2}T^{-1}$
Stefan-Boltzmann Constant	σ	$Btu/hr\text{-}ft^2°F^4$	$W/m^2°K^4$	$m\tau^{-3}T^{-4}$
Thermal capacity	C	Btu/°F	J/°C	$mL^2\tau^{-2}T^{-1}$
Thermal conductivity	k	Btu/hr-ft°F	W/m °C	$mL\tau^{-3}T^{-1}$
Thermal diffusivity	α	ft^2/hr	m^2/hr	$L^2\tau^{-1}$
Thermal resistance	R	hr °F/Btu	°C/W	$m^{-1}L^{-2}\tau^3T$
Velocity	u_∞	ft/sec	m/s	$L\tau^{-1}$
Volume	V	ft^3	m^3	L^3
Work	W	ft lb_f	Nm	$mL^2\tau^{-2}$

APPENDIX B Conversion Factors

Length	1 in = 0.08333 ft 1 mile = 1609.26 m 1 cm = 0.03281 ft 1 m = 3.281 ft
Mass	1 lb_m = 0.4535 kg 1 kg = 2.205 lb_m 1 slug = 32.1736 lb_m
Force	1 lb_f = 4.448 N 1 N = 0.2248 lb_f
Energy	1 Btu = 778.16 ft lb_f 1 Btu = 1055.07 J 1 Btu = 0.2520 Kcal 1 J = 0.9478×10^{-3} Btu 1 kw hr (kilowatt hour) = 3413 Btu
Power	1 Btu/hr = 0.293 W 1 hp (horsepower) = 2545 Btu/hr 1 hp = 745.7 W 1 Kcal = 1.1626 Watt hr 1 W = 3.413 Btu/hr
Pressure	1 atm = 2116 psf 1 ft water = 62.43 psf 1 in Hg = 70.77 psf 1 atm = 1.01325×10^5 N/m^2 1 lb_f/in^2 = 0.06895×10^5 N/m^2
Density	1 lb_m/in^3 = 1728 lb_m/ft^3 1 lb_m/in^3 = 2.77×10^4 kg/m^3 1 kg/m^3 = 0.06243 lb_m/ft^3
Temperature	T(°R) = T(°F) + 460 T(°K) = T(°C) + 273 T(°F) = (9/5) T(°C) + 32 1°K = 1.8°R
Viscosity	1 lb_f sec/ft^2 = 32.174 lb_m/sec ft 1 cp (centipoise) = 0.000672 lb_m/sec ft 1 cp = 2.42 lb_m/hr-ft
Volume	1 gal (U.S.) = 0.1337 ft^3 1 ft^3 = 28.32 liters = 0.02832 m^3
Heat flux	1 Btu/hr-ft^2 = 3.1537 W/m^2 1 W/m^2 = 0.317 Btu/hr-ft^2
Specific heat	1 Btu/lb_m°F = 4180 J/kg°C 1 J/kg°C = 2.392×10^{-4} Btu/lb_m°F
Thermal conductivity	1 Btu/hr-ft°F = 1.7303 W/m°C 1 W/m°C = 0.578 Btu/hr-ft°C
Heat transfer coefficient	1 Btu/hr-ft^2 °F = 5.6783 W/m^2 °C 1 W/m^2°C = 0.1761 Btu/hr-ft^2°F

APPENDIX C-1 Property Values for Metals (English System of Units)†

Metals	Properties at 68 F				k, thermal conductivity, Btu/hr ft F									
	ρ lb/ft³	c_p Btu/lb F	k Btu/hr ft F	α ft²/hr	−148 F −100 C	32 F 0 C	212 F 100 C	392 F 200 C	572 F 300 C	752 F 400 C	1112 F 600 C	1472 F 800 C	1832 F 1000 C	2192 F 1200 C
Aluminum:														
Pure	169	0.214	118	3.262	124	117	119	124	132	144				
Al-Cu (Duralumin) 94–96 Al, 3–5 Cu, trace Mg	174	0.211	95	2.587	73	92	105	112						
Al-Mg (Hydronalium) 91–95 Al, 5–9 Mg	163	0.216	65	1.846	54	63	73	82						
Al-Si (Silumin) 87 Al, 13 Si	166	0.208	95	2.751	86	94	101	107						
Al-Si (Silumin, copper bearing) 86.5 Al, 12.5 Si, 1 Cu.	166	0.207	79	2.299	69	79	83	88	93					
Al-Si (Alusil) 78–80 Al, 20–22 Si	164	0.204	93	2.779	83	91	97	101	103					
Al-Mg-Si 97 Al, 1 Mg, 1 Si, 1 Mn	169	0.213	102	2.833	.....	101	109	118						
Lead	710	0.031	20	0.908	21.3	20.3	19.3	18.2	17.2					
Iron:														
Pure	493	0.108	42	0.788	50	42	39	36	32	28	23	21	20	21
Wrought iron (C < 0.5 %)	490	0.11	34	0.630	.....	34	33	30	28	26	21	19	19	19
Cast iron (C ≈ 4 %)	454	0.10	30	0.660										
Steel (C max ≈ 1.5 %)														
Carbon steel C ≈ 0.5 %	489	0.111	31	0.571	.....	32	30	28	26	24	20	18	17	18
1.0 %	487	0.113	25	0.454	.....	25	25	24	23	21	19	17	16	17
1.5 %	484	0.116	21	0.376	.....	21	21	21	20	19	18	16	16	17
Nickel steel Ni ≈ 0 %	493	0.108	42	0.785										
10 %	496	0.11	15	0.279										
20 %	499	0.11	11	0.204										
30 %	504	0.11	7	0.126										
40 %	510	0.11	6	0.108										
50 %	516	0.11	8	0.140										
60 %	523	0.11	11	0.191										
70 %	531	0.11	15	0.258										
80 %	538	0.11	20	0.338										
90 %	547	0.11	27	0.448										
100 %	556	0.106	52	0.882										

APPENDIX C-1 Property Values for Metals (English System of Units) (continued)

Metals	Properties at 68 F				k, thermal conductivity, Btu/hr ft F									
	ρ lb/ft^3	c_p Btu/lb F	k Btu/hr ft F	α ft^2/hr	−148 F −100 C	32 F 0 C	212 F 100 C	392 F 200 C	572 F 300 C	752 F 400 C	1112 F 600 C	1472 F 800 C	1832 F 1000 C	2192 F 1200 C
Invar Ni = 36 %	508	0.11	6.2	0.111										
Chrome steel Cr = 0 %	493	0.108	42	0.785	50	42	39	36	32	28	23	21	20	21
1 %	491	0.11	35	0.645	.....	36	32	30	27	24	21	19	19	
2 %	491	0.11	30	0.559	.....	31	28	26	24	22	19	18	18	
5 %	489	0.11	23	0.430	.....	23	22	21	21	19	17	17	17	17
10 %	486	0.11	18	0.336	.....	18	18	18	17	17	16	16	17	
20 %	480	0.11	13	0.246	.....	13	13	13	13	14	14	15	17	
30 %	476	0.11	11	0.210										
Cr-Ni (chrome-nickel): 15 Cr, 10 Ni	491	0.11	11	0.204										
18 Cr, 8 Ni (V2A)	488	0.11	9.4	0.172	.....	9.4	10	10	11	11	13	15	18	
20 Cr, 15 Ni	489	0.11	8.7	0.161										
25 Cr, 20 Ni	491	0.11	7.4	0.140										
Ni-Cr (nickel-chrome): 80 Ni, 15 Cr	532	0.11	10	0.172										
60 Ni, 15 Cr	516	0.11	7.4	0.129										
40 Ni, 15 Cr	504	0.11	6.7	0.118										
20 Ni, 15 Cr	491	0.11	8 1	0.151	.....	8.1	8.7	8.7	9.4	10	11	13		
Cr-Ni-Al: 6 Cr, 1.5 Al, 0.5 Si (Sicromal 8)	482	0.117	13	0.230										
24 Cr, 2.5 Al, 0.5 Si (Sicromal 12)	479	0.118	11	0.194										
Manganese steel Ma = 0 %	493	0.118	42	0.722										
1 %	491	0.11	29	0.538										
2 %	491	0.11	22	0.407	.....	22	21	21	21	20	19			
5 %	490	0.11	13	0.247										
10 %	487	0.11	10	0.187										
Tungsten steel W = 0 %	493	0.108	42	0.785										
1 %	494	0.107	38	0.720										
2 %	497	0.106	36	0.683	.....	36	34	31	28	26	21			
5 %	504	0.104	31	0.591										
10 %	519	0.100	28	0.539										
20 %	551	0.093	25	0.484										
Silicon steel Si = 0 %	493	0.108	42	0.785										
1 %	485	0.11	24	0.451										
2 %	479	0.11	18	0.344										
5 %	463	0.11	11	0.215										

Copper:														
Pure	559	0.0915	223	4.353	235	223	219	216	213	210	204			
Aluminum bronze 95 Cu, 5 Al	541	0.098	48	0.903										
Bronze 75 Cu, 25 Sn	541	0.082	15	0.333										
Red brass 85 Cu, 9 Sn, 6 Zn	544	0.092	35	0.699	.....	34	41							
Brass 70 Cu, 30 Zn	532	0.092	64	1.322	51	.....	74	83	85	85				
German silver 62 Cu, 15 Ni, 22 Zn	538	0.094	14.4	0.284	11.1	.....	18	23	26	28				
Constantan 60 Cu, 40 Ni	557	0.098	13.1	0.237	12	.....	12.8	15						
Magnesium:														
Pure	109	0.242	99	3.762	103	99	97	94	91					
Mg-Al (electrolytic) 6–8 % Al, 1–2 % Zn	113	0.24	38	1.397	.....	30	36	43	48					
Mg-Mn 2 % Mn	111	0.24	66	2.473	54	64	72	75						
Molybdenum	638	0.060	71	1.856	80	72	68	66	64	63	61	59	57	53
Nickel:														
Pure (99.9 %)	556	0.1065	52	0.878	60	54	48	42	37	34				
Impure (99.2 %)	556	0.106	40	0.677	.....	40	37	34	32	30	32	36	39	40
Ni-Cr 90 Ni, 10 Cr	541	0.106	10	0.172	.....	9.9	10.9	12.1	13.2	14.2				
80 Ni, 20 Cr	519	0.106	7.3	0.133	.....	7.1	8.0	9.0	9.9	10.9	13.0			
Silver:														
Purest	657	0.0559	242	6.589	242	241	240	238						
Pure (99.9 %)	657	0.0559	235	6.418	242	237	240	216	209	208				
Tungsten	1208	0.0321	94	2.430	.....	96	87	82	77	73	65	44		
Zinc, pure	446	0.0918	64.8	1.591	66	65	63	61	58	54				
Tin, pure	456	0.0541	37	1.505	43	38.1	34	33						

† From Heat & Mass Transfer by Eckert & Drake.

APPENDIX C-2 Property Values for Metals (SI System of Units)†

Metal	Properties at 20 C				Thermal conductivity k, W/m K									
	ρ, kg/m³	c_p J/kg K	k, W/m K	α, m²/s	−100 C −148 F	0 C 32 F	100 C 212 F	200 C 392 F	300 C 572 F	400 C 752 F	600 C 1112 F	800 C 1472 F	1000 C 1832 F	1200 C 2192 F
Aluminum:														
Pure	2,707	0.896×10^3	204	8.418×10^{-5}	215	202	206	215	228	249				
Al-Cu (Duralumin) 94–96 Al, 3–5 Cu, trace Mg	2,787	0.883	164	6.676	126	159	182	194						
Al-Mg (Hydronalium) 91–95 Al, 5–9 Mg	2,611	0.904	112	4.764	93	109	125	142						
Al-Si (Silumin) 87 Al, 13 Si	2,659	0.871	164	7.099	149	163	175	185						
Al-Si (Silumin, copper bearing) 86.5 Al, 1 Cu	2,659	0.867	137	5.933	119	137	144	152	161					
Al-Si (Alusil) 78–80 Al, 20–22 Si	2,627	0.854	161	7.172	144	157	168	175	178					
Al-Mg-Si 97 Al, 1 Mg, 1 Si, 1 Mn	2,707	0.892	177	7.311		175	189	204						
Lead	11,373	0.130	35	2.343	36.9	35.1	33.4	31.5	29.8					
Iron:														
Pure	7,897	0.452	73	2.034	87	73	67	62	55	48	40	36	35	36
Wrought iron (C H 0.5 %)	7,849	0.46	59	1,626		59	57	52	48	45	36	33	33	33
Cast iron (C ≈ 4 %)	7,272	0.42	52	1.703										
Steel (C max ≈ 1.5 %)														
Carbon steel C ≈ 0.5 %	7,833	0.465	54	1.474		55	52	48	45	42	35	31	29	31
1.0 %	7,801	0.473	43	1.172		43	43	42	40	36	33	29	28	29
1.5 %	7,753	0.486	36	0.970		36	36	36	35	33	31	28	28	29
Nickel steel Ni ≈ 0 %	7,897	0.452	73	2.026										
10 %	7,945	0.46	26	0.720										
20 %	7,993	0.46	19	0.526										
30 %	8,073	0.46	12	0.325										
40 %	8,169	0.46	10	0.279										
50 %	8,266	0.46	14	0.361										
60 %	8,378	0.46	19	0.493										
70 %	8,506	0.46	26	0.666										
80 %	8,618	0.46	35	0.872										
90 %	8,762	0.46	47	1.156										
100 %	8,906	0.448	90	2.276										

Invar Ni = 36 %	8,137	0.46	10.7	0.286										
Chrome Steel Cr = 0 %	7,897	0.452	73	2.026	87	73	67	62	55	48	40	36	35	36
1 %	7,865	0.46	61	1.665		62	55	52	47	42	36	33	33	
2 %	7,865	0.46	52	1.443		54	48	45	42	38	33	31	31	
5 %	7,833	0.46	40	1.110		40	38	36	36	33	29	29	29	
10 %	7,785	0.46	31	0.867		31	31	31	29	29	28	28	29	
20 %	7,689	0.46	22	0.635		22	22	22	22	24	24	26	29	
30 %	7,625	0.46	19	0.542										
Cr-Ni (chrome-nickel): 15 Cr, 10 Ni	7,865	0.46	19	0.526										
18 Cr, 8 Ni (V2A)	7,817	0.46	16.3	0.444		16.3	17	17	19	19	22	26	31	
20 Cr, 15 Ni	7,833	0.46	15.1	0.415										
25 Cr, 20 Ni	7,865	0.46	12.8	0.361										
Ni-Cr (nickel-chrome): 80 Ni, 15 Cr	8,522	0.46	17	0.444										
60 Ni, 15 Cr	8,266	0.46	12.8	0.333										
40 Ni, 15 Cr	8,073	0.46	11.6	0.305										
20 Ni, 15 Cr	7,865	0.46	14.0	0.390		14.0	15.1	15.1	16.3	17	19	22		
Cr-Ni-Al: 6 Cr, 1.5 Al, 0.55 Si (Sicromal 8(	7,721	0.490	22	0.594										
24 Cr, 2.5 Al, 0.55 Si (Sicromal 12)	7,673	0.494	19	0.501										
Manganese steel Mn = 0 %	7,897	0.494	73	1.863										
1 %	7,865	0.46	50	1.388										
2 %	7,865	0.46	38	1.050		38	36	36	36	35	33			
5 %	7,849	0.46	22	0.637										
10 %	7,801	0.46	17	0.483										
Tungsten steel W = 0 %	7,897	0.452	73	2.026										
1 %	7,913	0.448	66	1.858										
2 %	7,961	0.444	62	1.763		62	59	54	48	45	36			
5 %	8,073	0.435	54	1.525										
10 %	8,314	0.419	48	1.391										
20 %	8,826	0.389	43	1.249										
Silicon steel Si = 0 %	7,897	0.452	73	2.026										
1 %	7,769	0.46	42	1.164										
2 %	7,673	0.46	31	0.888										
5 %	7,417	0.46	19	0.555										

APPENDIX C-2 Property Values for Metals (SI System of Units) (continued)

Metal	*Properties at 20 C*				*Thermal conductivity k*, W/m K									
	ρ, kg/m³	c_p J/kg K	k, W/m K	α, m²/s	−100 C −148 F	0 C 32 F	100 C 212 F	200 C 392 F	300 C 572 F	400 C 752 F	600 C 1112 F	800 C 1472 F	1000 C 1832 F	1200 C 2192 F
Copper:														
Pure	8,954	0.3831×10^3	386	11.234×10^{-5}	407	386	379	374	369	363	353			
Aluminum bronze 95 cu, 5 Al	8,666	0.410	83	2.330										
Bronze 75 Cu, 25 Sn	8,666	0.343	26	0.859										
Red Brass 85 Cu, 9 Sn, 6 Zn	8,714	0.385	61	1.804		59	71							
Brass 70 Cu, 30 Zn	8,522	0.385	111	3.412	88		128	144	147	147				
German silver 62 Cu, 15 Ni, 22 Zn	8,618	0.394	24.9	0.733	19.2		31	40	45	48				
Constantan 60 Cu, 40 Ni	8,922	0.410	22.7	0.612	21		22.2	26						
Magnesium:														
Pure	1,746	1.013	171	9.708	178	171	168	163	157					
Mg-Al (electrolytic) 6–8% Al, 1–2% Zn	1,810	1.00	66	3.605		52	62	74	83					
Mg-Mn 2% Mn	1,778	1.00	114	6.382	93	111	125	130						
Mg-Mn 2% Mn	1,778	1.00	114	6.382	93	111	125	130						
Molybdenum	10,220	0.251	123	4.790	138	125	118	114	111	109	106	102	99	92
Nickel:														
Pure (99.9%)	8,906	0.4459	90	2.266	104	93	83	73	64	59				
Impure (99.2%)	8,906	0.444	69	1.747		69	64	59	55	52	55	62	67	69
Ni-Cr 90 Ni, 10 Cr	8,666	0.444	17	0.444		17.1	18.9	20.9	22.8	24.6				
80 Ni, 20 Cr	8,314	0.444	12.6	0.343		12.3	13.8	15.6	17.1	18.9	22.5			
Silver:														
Purest	10,524	0.2340	419	17.004	419	417	415	412						
Pure (99.9%)	10,524	0.2340	407	16.563	419	410	415	374	362	360				
Tungsten	19,350	0.1344	163	6.271		166	151	142	133	126	112	76		
Zinc, pure	7.144	0.3843	112.2	4.106	114	112	109	106	100	93				
Tin, pure	7,304	0.2265	64	3.884	74	65.9	59	57						

† From Analysis of Heat & Mass Transfer by Eckert & Drake. Copyright 1972 by McGraw-Hill, Inc. Used with permission of McGraw-Hill Book Co.

APPENDIX D-1 Property Values for Nonmetals (English System of Units)

Material	t, F	ρ, lb/ft^3	c_p, $\frac{\text{Btu}}{\text{lb F}}$	k, $\frac{\text{Btu}}{\text{hr ft F}}$	α, $\frac{\text{ft}^2}{\text{hr}}$
Aerogel, silica	248	8.5	.....	0.013	
Asbestos	−328	29.3	.....	0.043	
Asbestos	32	29.3	.....	0.090	
Asbestos	32	36.0	0.195	0.087	
Asbestos	212	36.0	0.195	0.111	
Asbestos	392	36.0	.....	0.120	
Asbestos	752	36.0	.....	0.129	
Asbestos	−328	43.5	.....	0.090	
Asbestos	32	43.5	.....	0.135	
Brick, dry	68	110–113	0.20	0.22–0.30	0.011–0.013
Bakelite	68	79.5	0.38	0.134	0.0044
Cardboard, corrugated	.....	....	.....	0.037	
Clay	68	91.0	0.21	0.739	0.039
Concrete	68	119–144	0.21	0.47–0.81	0.019–0.027
Coal, anthracite	68	75–94	0.30	0.15	0.005–0.006
Coal, powdered	86	46	0.31	0.067	0.005
Cotton	68	5	0.31	0.034	0.075
Cork, board	86	10	.....	0.025	
Cork, expanded scrap	68	2.8–7.4	0.45	0.021	0.006–0.017
Cork, ground	86	9.4	.....	0.025	
Diatomaceous earth	100	20.0	.....	0.036	
Diatomaceous earth	1600	20.0	.....	0.082	
Earth, coarse gravelly	68	128	0.44	0.30	0.0054
Felt, wool	86	20.6	.....	0.03	
Fiber, insulating board	70	14.8	.....	0.028	
Fiber, red	68	80.5	.....	0.27	
Glass plate	68	169	0.2	0.44	0.013
Glass, borosilicate	86	139	.....	0.63	
Glass, wool	68	12.5	0.16	0.023	0.011
Granite	.....	....	.....	1.0–2.3	
Ice	32	57	0.46	1.28	0.048
Marble	68	156–169	0.193	1.6	0.054
Rubber, hard	32	74.8	.....	0.087	
Sandstone	68	135–144	0.17	0.94–1.2	0.041–0.049
Silk	68	3.6	0.33	0.021	0.017
Wood, oak radial	68	38–50	0.57	0.10–0.12	0.0043–0.0047
Wood, fir (20% moisture) radial	68	26.0–26.3	0.65	0.08	0.0048

APPENDIX D-2 Property Values for Nonmetals (SI System of Units)†

Material	t, C	ρ, kg/m³	c_p, J/kg K	k, W/m K	α, m²/s
Aerogel, silica	120	136.2		0.022	
Asbestos	−200	469.3		0.074	
	0	469.3		0.156	
	0	576.7	0.816×10^3	0.151	
	100	576.7	0.816	0.192	
	200	576.7		0.208	
	400	576.7		0.223	
	−200	696.8		0.156	
	0	696.8		0.234	
Brick, dry	20	1,762–1,810	0.84	0.38–0.52	$0.028\text{–}0.034\times10^{-5}$
Bakelite	20	1,273.5	1.59	0.232	0.0114
Cardboard, corrugated				0.064	
Clay	20	1,457.7	0.88	1.279	0.101
Concrete	20	1,906–2,307	0.88	0.81–1.40	0.049–0.070
Coal, anthracite	20	1,201–1,506	1.26	0.26	0.013–0.015
Powdered	30	737	1.30	0.116	0.013
Cotton	20	80	1.30	0.059	0.194
Cork, board	30	160		0.043	
Expanded scrap	20	44.9–118.5	1.88	0.036	0.015–0.044
Ground	30	150.6		0.043	
Diatomaceous earth	38	320.4		0.062	
	871	320.4		0.142	
Earth, coarse gravelly	20	2,050	1.84	0.52	0.0139
Felt, wood	30	330.0		0.05	
Fiber, insulating board	21	237.1		0.048	
Red	20	1,289.5		0.47	
Glass plate	20	2,707	0.8	0.76	0.034
Glass, borosilicate	30	2,227		1.09	
Wool	20	200.2	0.67	0.040	0.028
Granite				1.7–4.0	
Ice	0	913	1.93	2.22	0.124
Marble	20	2,499–2,707	0.808	2.8	0.139
Rubber, hard	0	1,198.2		0.151	
Sandstone	20	2,162–2,307	0.71	1.63–2.1	0.106–0.126
Silk	20	57.7	1.38	0.036	0.044
Wood, oak radial	20	609–801	2.39	0.17–0.21	0.0111–0.0121
Fir (20% moisture) radial	20	416.5–421.3	2.72	0.14	0.0124

APPENDIX E-1 Property Values for Fluids in a Saturated State (English System of Units)†

t, F	ρ, lb/ft^3	c_p, $\frac{\text{Btu}}{\text{lb F}}$	ν, ft^2/sec	k, $\frac{\text{Btu}}{\text{hr ft F}}$	α, $\frac{\text{ft}^2}{\text{hr}}$	Pr	β, $\frac{1}{R}$
				Water (H_2O)			
32	62.57	1.0074	1.925×10^{-5}	0.319	5.07×10^{-3}	13.6	
68	62.46	0.9988	1.083	0.345	5.54	7.02	0.10×10^{-3}
104	62.09	0.9980	0.708	0.363	5.86	4.34	
140	61.52	0.9994	0.514	0.376	6.02	3.02	
176	60.81	1.0023	0.392	0.386	6.34	2.22	
212	59.97	1.0070	0.316	0.393	6.51	1.74	
248	59.01	1.015	0.266	0.396	6.62	1.446	
284	57.95	1.023	0.230	0.395	6.68	1.241	
320	56.79	1.037	0.204	0.393	6.70	1.099	
356	55.50	1.055	0.186	0.390	6.68	1.004	
392	54.11	1.076	0.172	0.384	6.61	0.937	
428	52.59	1.101	0.161	0.377	6.51	0.891	
464	50.92	1.136	0.154	0.367	6.35	0.871	
500	49.06	1.182	0.148	0.353	6.11	0.874	
537	46.98	1.244	0.145	0.335	5.74	0.910	
572	44.59	1.368	0.145	0.312	5.13	1.019	
				Ammonia (NH_3)			
−58	43.93	1.066	0.468×10^{-5}	0.316	6.75×10^{-3}	2.60	
−40	43.18	1.067	0.437	0.316	6.88	2.28	
−22	42.41	1.069	0.417	0.317	6.98	2.15	
−4	41.62	1.077	0.410	0.316	7.05	2.09	
14	40.80	1.090	0.407	0.314	7.07	2.07	
32	39.96	1.107	0.402	0.312	7.05	2.05	
50	39.09	1.126	0.396	0.307	6.98	2.04	
68	38.19	1.146	0.386	0.301	6.88	2.02	1.36×10^{-3}
86	37.23	1.168	0.376	0.293	6.75	2.01	
104	36.27	1.194	0.366	0.285	6.59	2.00	
122	35.23	1.222	0.355	0.275	6.41	1.99	
				Carbon dioxide (CO_2)			
−58	72.19	0.44	0.128×10^{-5}	0.0494	1.558×10^{-3}	2.96	
−40	69.78	0.45	0.127	0.0584	1.864	2.46	
−22	67.22	0.47	0.126	0.0645	2.043	2.22	
−4	64.45	0.49	0.124	0.0665	2.110	2.12	
14	61.39	0.52	0.122	0.0635	1.989	2.20	

APPENDIX E-1 Property Values for Fluids in a Saturated State (continued)

t, F	ρ, lb/ft^3	c_p, $\frac{\text{Btu}}{\text{lb F}}$	ν, ft^2/sec	k, $\frac{\text{Btu}}{\text{hr ft F}}$	α, $\frac{\text{ft}^2}{\text{hr}}$	Pr	β, $\frac{1}{R}$
			Carbon dioxide (CO_2) (*Continued*)				
32	57.87	0.59	0.117	0.0604	1.774	2.38	
50	53.69	0.75	0.109	0.0561	1.398	2.80	
68	48.23	1.2	0.098	0.0504	0.860	4.10	3.67×10^{-3}
86	37.32	8.7	0.086	0.0406	0.108	28.7	
			Sulfur dioxide (SO_2)				
−58	97.44	0.3247	0.521×10^{-5}	0.140	4.42×10^{-3}	4.24	
−40	95.94	0.3250	0.456	0.136	4.38	3.74	
−22	94.43	0.3252	0.399	0.133	4.33	3.31	
−4	92.93	0.3254	0.349	0.130	4.29	2.93	
14	91.37	0.3255	0.310	0.126	4.25	2.62	
32	89.80	0.3257	0.277	0.122	4.19	2.38	
50	88.18	0.3259	0.250	0.118	4.13	2.18	
68	86.55	0.3261	0.226	0.115	4.07	2.00	1.08×10^{-3}
86	84.86	0.3263	0.204	0.111	4.01	1.83	
104	82.98	0.3266	0.186	0.107	3.95	1.70	
122	81.10	0.3268	0.174	0.102	3.87	1.61	
			Methylchloride (CH_3Cl)				
−58	65.71	0.3525	0.344×10^{-5}	0.124	5.38×10^{-3}	2.31	
−40	64.51	0.3541	0.342	0.121	5.30	2.32	
−22	63.46	0.3564	0.338	0.117	5.18	2.35	
−4	62.39	0.3593	0.333	0.113	5.04	2.38	
14	61.27	0.3629	0.329	0.108	4.87	2.43	
32	60.08	0.3673	0.325	0.103	4.70	2.49	
50	58.83	0.3726	0.320	0.099	4.52	2.55	
68	57.64	0.3788	0.315	0.094	4.31	2.63	
86	56.38	0.3860	0.310	0.089	4.10	2.72	
104	55.13	0.3942	0.303	0.083	3.86	2.83	
122	53.76	0.4034	0.295	0.077	3.57	2.97	
			Dichlorodifluoromethane (Freon) (CCl_2F_2)				
−58	96.56	0.2090	0.334×10^{-5}	0.039	1.94×10^{-3}	6.2	1.4×10^{-4}
−40	94.81	0.2113	0.300	0.040	1.99	5.4	
−22	92.99	0.2139	0.272	0.040	2.04	4.8	
−4	91.18	0.2167	0.253	0.041	2.09	4.4	
14	89.24	0.2198	0.238	0.042	2.13	4.0	

APPENDIX E-1 Property Values for Fluids in a Saturated State (continued)

t, F	ρ, lb/ft^3	c_p, $\frac{\text{Btu}}{\text{lb F}}$	ν, ft^2/sec	k, $\frac{\text{Btu}}{\text{hr ft F}}$	α, $\frac{\text{ft}^2}{\text{hr}}$	Pr	β, $\frac{1}{R}$
			Dichlorodifluoromethane (Freon) (CCl_2F_2) (*Continued*)				
32	87.24	0.2232	0.230	0.042	2.16	3.8	
50	85.17	0.2268	0.219	0.042	2.17	3.6	
68	83.04	0.2307	0.213	0.042	2.17	3.5	
86	80.85	0.2349	0.209	0.041	2.17	3.5	
104	78.48	0.2393	0.206	0.040	2.15	3.5	
122	75.91	0.2440	0.204	0.039	2.11	3.5	
			Eutectic calcium chloride solution (29.9% $CaCl_2$)				
−58	82.39	0.623	39.13×10^{-5}	0.232	4.52×10^{-3}	312	
−40	82.09	0.6295	26.88	0.240	4.65	208	
−22	81.79	0.6356	18.49	0.248	4.78	139	
−4	81.50	0.642	11.88	0.257	4.91	87.1	
14	81.20	0.648	7.49	0.265	5.04	53.6	
32	80.91	0.654	4.73	0.273	5.16	33.0	
50	80.62	0.660	3.61	0.280	5.28	24.6	
68	80.32	0.666	2.93	0.288	5.40	19.6	
86	80.03	0.672	2.44	0.295	5.50	16.0	
104	79.73	0.678	2.07	0.302	5.60	13.3	
122	79.44	0.685	1.78	0.309	5.69	11.3	
			Glycerin [$C_3H_5(OH)_3$]				
32	79.66	0.540	0.0895	0.163	3.81×10^{-3}	84.7×10^3	
50	79.29	0.554	0.0323	0.164	3.74	31.0	
68	78.91	0.570	0.0127	0.165	3.67	12.5	0.28×10^{-3}
86	78.54	0.584	0.0054	0.165	3.60	5.38	
104	78.16	0.600	0.0024	0.165	3.54	2.45	
122	77.72	0.617	0.0016	0.166	3.46	1.63	
			Ethylene glycol [$C_2H_4(OH_2)$]				
32	70.59	0.548	61.92×10^{-5}	0.140	3.62×10^{-3}	615	
68	69.71	0.569	20.64	0.144	3.64	204	0.36×10^{-3}
104	68.76	0.591	9.35	0.148	3.64	93	
140	67.90	0.612	5.11	0.150	3.61	51	
176	67.27	0.633	3.21	0.151	3.57	32.4	
212	66.08	0.655	2.18	0.152	3.52	22.4	

APPENDIX E-1 Property Values for Fluids in a Saturated State (continued)

t, F	ρ, lb/ft^3	c_p, $\frac{\text{Btu}}{\text{lb F}}$	ν, ft^2/sec	k, $\frac{\text{Btu}}{\text{hr ft F}}$	α, $\frac{\text{ft}^2}{\text{hr}}$	Pr	β, $\frac{1}{R}$
				Engine oil (unused)			
32	56.13	0.429	0.0461	0.085	3.53×10^{-3}	47100	
68	55.45	0.449	0.0097	0.084	3.38	10400	0.39×10^{-3}
104	54.69	0.469	0.0026	0.083	3.23	2870	
140	53.94	0.489	0.903×10^{-3}	0.081	3.10	1050	
176	53.19	0.509	0.404	0.080	2.98	490	
212	52.44	0.530	0.219	0.079	2.86	276	
248	51.75	0.551	0.133	0.078	2.75	175	
284	51.00	0.572	0.086	0.077	2.66	116	
320	50.31	0.593	0.060	0.076	2.57	84	
				Mercury (Hg)			
32	850.78	0.0335	0.133×10^{-5}	4.74	166.6×10^{-3}	0.0288	
68	847.71	0.0333	0.123	5.02	178.5	0.0249	1.01×10^{-4}
122	843.14	0.0331	0.112	5.43	194.6	0.0207	
212	835.57	0.0328	0.0999	6.07	221.5	0.0162	
302	828.06	0.0326	0.0918	6.64	246.2	0.0134	
392	820.61	0.0375	0.0863	7.13	267.7	0.0116	
482	813.16	0.0324	0.0823	7.55	287.0	0.0103	
600	802	0.032	0.0724	8.10	316	0.0083	

† From Heat & Mass Transfer by Eckert and Drake. Copyright 1959 by McGraw-Hill Book Company, Inc. Used with permission of McGraw-Hill Book Co.

APPENDIX E-2 Property Values for Fluids in a Saturated State (SI System of Units)†

t, C	ρ, kg/m^3	c_p, J/kg K	ν, m^2/s	k, W/m K	α, m^2/s	Pr	β, K^{-1}
Water, H_2O							
0	1,002.28	4.2178×10^3	1.788×10^{-6}	0.552	1.308×10^{-7}	13.6	
20	1,000.52	4.1818	1.006	0.597	1.430	7.02	0.18×10^{-3}
40	994.59	4.1784	0.658	0.628	1.512	4.34	
60	985.46	4.1843	0.478	0.651	1.554	3.02	
80	974.08	4.1964	0.364	0.668	1.636	2.22	
100	960.63	4.2161	0.294	0.680	1.680	1.74	
120	945.25	4.250	0.247	0.685	1.708	1.446	
140	928.27	4.283	0.214	0.684	1.724	1.241	
160	909.69	4.342	0.190	0.680	1.729	1.099	
180	889.03	4.417	0.173	0.675	1.724	1.004	
200	866.76	4.505	0.160	0.665	1.706	0.937	
220	842.41	4.610	0.150	0.652	1.680	0.891	
240	815.66	4.756	0.143	0.635	1.639	0.871	
260	785.87	4.949	0.137	0.611	1.577	0.874	
280.6	752.55	5.208	0.135	0.580	1.481	0.910	
300	714.26	5.728	0.135	0.540	1.324	1.019	
Ammonia, NH_3							
−50	703.69	4.463×10^3	0.435×10^{-6}	0.547	1.742×10^{-7}	2.60	
−40	691.68	4.467	0.406	0.547	1.775	2.28	
−30	679.34	4.476	0.387	0.549	1.801	2.15	
−20	666.69	4.509	0.381	0.547	1.819	2.09	
−10	653.55	4.564	0.378	0.543	1.825	2.07	
0	640.10	4.635	0.373	0.540	1.819	2.05	
10	626.16	4.714	0.368	0.531	1.801	2.04	
20	611.75	4.798	0.359	0.521	1.775	2.02	2.45×10^{-3}
30	596.37	4.890	0.349	0.507	1.742	2.01	
40	580.99	4.999	0.340	0.493	1.701	2.00	
50	564.33	5.116	0.330	0.476	1.654	1.99	
Carbon dioxide, CO_2							
−50	1,156.34	1.84×10^3	0.119×10^{-6}	0.0855	0.4021×10^{-7}	2.96	
−40	1,117.77	1.88	0.118	0.1011	0.4810	2.46	
−30	1,076.76	1.97	0.117	0.1116	0.5272	2.22	
−20	1,032.39	2.05	0.115	0.1151	0.5445	2.12	
−10	983.38	2.18	0.113	0.1099	0.5133	2.20	
0	926.99	2.47	0.108	0.1045	0.4578	2.38	
10	860.03	3.14	0.101	0.0971	0.3608	2.80	
20	772.57	5.0	0.091	0.0872	0.2219	4.10	14.00×10^{-3}
30	597.81	36.4	0.080	0.0703	0.0279	28.7	

APPENDIX E-2 Property Values for Fluids in a Saturated State (continued)

t, C	ρ, kg/m^3	c_p, J/kg K	ν, m^2/s	k, W/m K	α, m^2/s	Pr	β, K^{-1}
Sulfur dioxide, SO_2							
−50	1,560.84	1.3595×10^3	0.484×10^{-6}	0.242	1.141×10^{-7}	4.24	
−40	1,536.81	1.3607	0.424	0.235	1.130	3.74	
−30	1,520.64	1.3616	0.371	0.230	1.117	3.31	
−20	1,488.60	1.3624	0.324	0.225	1.107	2.93	
−10	1,463.61	1.3628	0.288	0.218	1.097	2.62	
0	1,438.46	1.3636	0.257	0.211	1.081	2.38	
10	1,412.51	1.3645	0.232	0.204	1.066	2.18	
20	1,386.40	1.3653	0.210	0.199	1.050	2.00	1.94×10^{-3}
30	1,359.33	1.3662	0.190	0.192	1.035	1.83	
40	1,329.22	1.3674	0.173	0.185	1.019	1.70	
50	1,299.10	1.3683	0.162	0.177	0.999	1.61	
Methyl chloride, CH_3Cl							
−50	1,052.58	1.4759×10^3	0.320×10^{-6}	0.215	1.388×10^{-7}	2.31	
−40	1,033.35	1.4826	0.318	0.209	1.368	2.32	
−30	1,016.53	1.4922	0.314	0.202	1.337	2.35	
−20	999.39	1.5043	0.309	0.196	1.301	2.38	
−10	981.45	1.5194	0.306	0.187	1.257	2.43	
0	962.39	1.5378	0.302	0.178	1.213	2.49	
10	942.36	1.5600	0.297	0.171	1.166	2.55	
20	923.31	1.5860	0.293	0.163	1.112	2.63	
30	903.12	1.6161	0.288	0.154	1.058	2.72	
40	883.10	1.6504	0.281	0.144	0.996	2.83	
50	861.15	1.6890	0.274	0.133	0.921	2.97	
Dichlorodifluoromethane (Freon), CCl_2F_2							
−50	1,546.75	0.8750×10^3	0.310×10^{-6}	0.067	0.501×10^{-7}	6.2	2.63×10^{-3}
−40	1,518.71	0.8847	0.279	0.069	0.514	5.4	
−30	1,489.56	0.8956	0.253	0.069	0.526	4.8	
−20	1,460.57	0.9073	0.235	0.071	0.539	4.4	
−10	1,429.49	0.9203	0.221	0.073	0.550	4.0	
0	1,397.45	0.9345	0.214	0.073	0.557	3.8	
10	1,364.30	0.9496	0.203	0.073	0.560	3.6	
20	1,330.18	0.9659	0.198	0.073	0.560	3.5	
30	1,295.10	0.9835	0.194	0.071	0.560	3.5	
40	1,257.13	1.0019	0.191	0.069	0.555	3.5	
50	1,215.96	1.0216	0.190	0.067	0.545	3.5	
Eutectic calcium chloride solution, 29.9% $CaCl_2$							
−50	1,319.76	2.608×10^3	36.35×10^{-6}	0.402	1.166×10^{-7}	312	
−40	1,314.96	2.6356	24.97	0.415	1.200	208	
−30	1,310.15	2.6611	17.18	0.429	1.234	139	
−20	1,305.51	2.688	11.04	0.445	1.267	87.1	
−10	1,300.70	2.713	6.96	0.459	1.300	53.6	

APPENDIX E-2 Property Values for Fluids in a Saturated State (continued)

t, C	ρ, kg/m³	c_p, J/kg K	ν, m²/s	k, W/m K	α, m²/s	Pr	β, K⁻¹
Eutectic calcium chloride solution, 29.9% $CaCl_2$ (continued)							
0	1,296.06	2.738×10^3	4.39×10^{-6}	0.472	1.332×10^{-7}	33.0	
10	1,291.41	2.763	3.35	0.485	1.363	24.6	
20	1,286.61	2.788	2.72	0.498	1.394	19.6	
30	1,281.96	2.814	2.27	0.511	1.419	16.0	
40	1,277.16	2.839	1.92	0.523	1.445	13.3	
50	1,272.51	2.868	1.65	0.535	1.468	11.3	
Glycerin, $C_3H_5(OH)_3$							
0	1,276.03	2.261×10^3	0.00831	0.282	0.983×10^{-7}	84.7×10^3	
10	1,270.11	2.319	0.00300	0.284	0.965	31.0	
20	1,264.02	2.386	0.00118	0.286	0.947	12.5	0.50×10^{-3}
30	1,258.09	2.445	0.00050	0.286	0.929	5.38	
40	1,252.01	2.512	0.00022	0.286	0.914	2.45	
50	1,244.96	2.583	0.00015	0.287	0.893	1.63	
Ethylene glycol, $C_2H_4(OH_2)$							
0	1,130.75	2.294×10^3	57.53×10^{-6}	0.242	0.934×10^{-7}	615	
20	1,116.65	2.382	19.18	0.249	0.939	204	0.65×10^{-3}
40	1,101.43	2.474	8.69	0.256	0.939	93	
60	1,087.66	2.562	4.75	0.260	0.932	51	
80	1,077.56	2.650	2.98	0.261	0.921	32.4	
100	1,058.50	2.742	2.03	0.263	0.908	22.4	
Engine oil (unused)							
0	899.12	1.796×10^3	0.00428	0.147	0.911×10^{-7}	47,100	
20	888.23	1.880	0.00090	0.145	0.872	10,400	0.70×10^{-3}
40	876.05	1.964	0.00024	0.144	0.834	2,870	
60	864.04	2.047	0.839×10^{-4}	0.140	0.800	1,050	
80	852.02	2.131	0.375	0.138	0.769	490	
100	840.01	2.219	0.203	0.137	0.738	276	
120	828.96	2.307	0.124	0.135	0.710	175	
140	816.94	2.395	0.080	0.133	0.686	116	
160	805.89	2.483	0.056	0.132	0.663	84	
Mercury, Hg							
0	13,628.22	0.1403×10^3	0.124×10^{-6}	8.20	42.99×10^{-7}	0.0288	
20	13,579.04	0.1394	0.114	8.69	46.06	0.0249	1.82×10^{-4}
50	13,505.84	0.1386	0.104	9.40	50.22	0.0207	
100	13,384.58	0.1373	0.0928	10.51	57.16	0.0162	
150	13,264.28	0.1365	0.0853	11.49	63.54	0.0134	
200	13,144.94	0.1570	0.0802	12.34	69.08	0.0116	
250	13,025.60	0.1357	0.0765	13.07	74.06	0.0103	
315.5	12,847	0.134	0.0673	14.02	81.5	0.0083	

APPENDIX F-1 Property Values for Liquid Metals (English System of Units)†

Metal	Melting point, °F	Normal boiling point, °F	Temperature, °F	Density, lb_m/ft^3	Viscosity, $lb_m/(ft)(sec)$	Heat capacity, $Btu/(lb_m)(°F)$	Thermal conductivity, $Btu/(hr)(ft^2)(°F)/ft$	Prandtl number
Bismuth	520	2691	600	625	1.09×10^{-3}	0.0345	9.5	0.014
			1400	591	0.53	0.0393	9.0	0.0084
Lead	621	3159	700	658	1.61	0.038	9.3	0.024
			1300	633	0.92	0.037	8.6	0.016
Lithium	354	2403	400	31.6	0.40	1.0	22.0	0.065
			1800	27.6	0.28	1.0		
Mercury	−38.0	675	50	847	1.07	0.033	4.7	0.027
			600	802	0.58	0.032	8.1	0.0084
Potassium	147	1400	300	50.4	0.25	0.19	26.0	0.0066
			1300	42.1	0.09	0.18	19.1	0.0031
Sodium	208	1621	400	56.3	0.29	0.32	46.4	0.0072
			1300	48.6	0.12	0.30	34.5	0.0038
Sodium-Potassium, 22% Na	66.2	1518	200	53.0	0.330	0.226	14.1	0.019
			1400	43.1	0.0981	0.211		
56% Na	12	1443	200	55.4	0.390	0.270	14.8	0.026
			1400	46.2	0.108	0.249	16.7	0.058
Lead-Bismuth, 44.5% Pb	257	3038	550	646	1.18	0.035	6.20	0.024
			1200	614	0.772			

†From "Fluid Dynamics and Heat Transfer," by Knudsen and Katz, Copyright 1958 by McGraw-Hill Inc. Used with permission of McGraw-Hill Book Co.

APPENDIX F-2 Property Values for Liquid Metals (SI System of Units)†

Metal, composition, and melting point	*t*, C	*ρ*, kg/m³	*c_p*, J/kg K	*ν*, m²/s	*k*, W/m K	*α* m²/s	Pr
Bismuth (271.1 C)	316	10,011	0.1444×10^{3}	1.617×10^{-7}	16.4	1.138×10^{-5}	0.0142
	538	9,739	0.1545	1.133	15.6	1.035	0.0110
	760	9,467	0.1645	0.8343	15.6	1.001	0.0083
Lead (327.2 C)	371	10,540	0.159	2.276	16.1	1.084	0.024
	482	10,412	0.155	1.849	15.6	1.223	0.017
	704	10,140		1.347	14.9		
Mercury (−38.9) see Table B-3 Table B-3							
Potassium (63.9 C)	149	807.3	0.80	4.608	45.0	6.99	0.0066
	427	741.7	0.75	2.397	39.5	7.07	0.0034
	704	674.4	0.75	1.905	33.1	6.55	0.0029
Sodium (97.8 C)	93	929.1	1.38	7.516	86.2	6.71	0.011
	371	860.2	1.30	3.270	72.3	6.48	0.0051
	704	778.5	1.26	2.285	59.7	6.12	0.0037
NaK(56/44, 19 C)	93	887.4	1.130	6.522	25.6	2.552	0.026
	371	821.7	1.055	2.871	27.5	3.17	0.0091
	704	740.1	1.043	2.174	28.9	3.74	0.0058
NaK(22/78, −11.1 C)	93	849.0	0.946	5.797	24.4	3.05	0.019
	399	775.3	0.879	2.666	26.7	3.92	0.0068
	760	690.4	0.883	2.118			
PbBi(44.5/55.5, 125 C)	149	10,524	0.147		9.05	0.586	
	371	10,236	0.147	1.496	11.86	0.790	0.189
	649	9,835		1.171			

† Atomic Energy Commission, "Liquid Metals Handbook," 2nd ed., Department of the Navy, Washington, D.C., 1952.

APPENDIX G-1 Property Values of Gases at Atmospheric Pressure (English System of Units)†

T, F	ρ, lb/ft^3	c_p, Btu/lb F	μ, lb/sec ft	ν, ft^2/sec	k, Btu/hr ft F	α, ft^2/hr	Pr
			Air				
−280	0.2248	0.2452	0.4653×10^{-5}	2.070×10^{-5}	0.005342	0.09691	0.770
−190	0.1478	0.2412	0.6910	4.675	0.007936	0.2226	0.753
−100	0.1104	0.2403	0.8930	8.062	0.01045	0.3939	0.739
−10	0.0882	0.2401	1.074	10.22	0.01287	0.5100	0.722
80	0.0735	0.2402	1.241	16.88	0.01516	0.8587	0.708
170	0 0623	0.2410	1.394	22.38	0.01735	1.156	0.697
260	0.0551	0.2422	1.536	27.88	0.01944	1.457	0.689
350	0.0489	0.2438	1.669	31.06	0.02142	1.636	0.683
440	0.0440	0.2459	1.795	40.80	0.02333	2.156	0.680
530	0.0401	0.2482	1.914	47.73	0.02519	2.531	0.680
620	0.0367	0.2520	2.028	55.26	0.02692	2.911	0.680
710	0.0339	0.2540	2.135	62.98	0.02862	3.324	0.682
800	0.0314	0.2568	2.239	71.31	0.03022	3.748	0.684
890	0.0294	0.2593	2.339	79.56	0.03183	4.175	0.686
980	0.0275	0.2622	2.436	88.58	0.03339	4.631	0.689
1070	0.0259	0.2650	2.530	97.68	0.03483	5.075	0.692
1160	0.0245	0.2678	2.620	106.9	0.03628	5.530	0.696
1250	0.0232	0.2704	2.703	116.5	0.03770	6.010	0.699
1340	0.0220	0.2727	2.790	126.8	0.03901	6.502	0.702
1520	0.0200	0.2772	2.955	147.8	0.04178	7.536	0.706
1700	0.0184	0.2815	3.109	169.0	0.04410	8.514	0.714
1880	0.0169	0.2860	3.258	192.8	0.04641	9.602	0.722
2060	0.0157	0.2900	3.398	216.4	0.04880	10.72	0.726
2240	0.0147	0.2939	3.533	240.3	0.05098	11.80	0.734
2420	0.0138	0.2982	3.668	265.8	0.05348	12.88	0.741
2600	0.0130	0.3028	3.792	291.7	0.05550	14.00	0.749
2780	0.0123	0.3075	3.915	318.3	0.05750	15.09	0.759
2960	0.0116	0.3128	4.029	347.1	0.0591	16.40	0.767
3140	0.0110	0.3196	4.168	378.8	0.0612	17.41	0.783
3320	0.0105	0.3278	4.301	409.9	0.0632	18.36	0.803
3500	0.0100	0.3390	4.398	439.8	0.0646	19.05	0.831
3680	0.0096	0.3541	4.513	470.1	0.0663	19.61	0.863
3860	0.0091	0.3759	4.611	506.9	0.0681	19.92	0.916
4160	0.0087	0.4031	4.750	546.0	0.0709	20.21	0.972
			Helium				
−456		1.242	5.66×10^{-7}		0.0061		
−400	0.0915	1.242	33.7	3.68×10^{-5}	0.0204	0.1792	0.74
−200	0.211	1.242	84.3	39.95	0.0536	2.044	0.70
−100	0.0152	1.242	105.2	69.30	0.0680	3.599	0.694
0	0.0119	1.242	122.1	102.8	0.0784	5.299	0.70
200	0.00829	1.242	154.9	186.9	0.0977	9.490	0.71

APPENDIX G-1 Property Values of Gases at Atmospheric Pressure (continued)

T, F	ρ, lb/ft³	c_p, Btu/lb F	μ, lb/sec ft	ν, ft²/sec	k, Btu/hr ft F	α, ft²/hr	Pr
			Helium (*Continued*)				
400	0.00637	1.242	184.8	289.9	0.114	14.40	0.72
600	0.00517	1.242	209.2	404.5	0.130	20.21	0.72
800	0.00439	1.242	233.5	531.9	0.145	25.81	0.72
1000	0.00376	1.242	256.5	682.5	0.159	34.00	0.72
1200	0.00330	1.242	277.9	841.0	0.172	41.98	0.72
			Hydrogen				
−406	0.05289	2.589	1.079×10^{-6}	2.040×10^{-5}	0.0132	0.0966	0.759
−370	0.03181	2.508	1.691	5.253	0.0209	0.262	0.721
−280	0.01534	2.682	2.830	18.45	0.0384	0.933	0.712
−190	0.01022	3.010	3.760	36.79	0.0567	1.84	0.718
−100	0.00766	3.234	4.578	59.77	0.0741	2.99	0.719
−10	0.00613	3.358	5.321	86.80	0.0902	4.38	0.713
80	0.00511	3.419	6.023	117.9	0.105	6.02	0.706
170	0.00438	3.448	6.689	152.7	0.119	7.87	0.697
260	0.00383	3.461	7.300	190.6	0.132	9.95	0 690
350	0.00341	3.463	7.915	232.1	0.145	12.26	0.682
440	0.00307	3.465	8.491	276.6	0.157	14.79	0.675
530	0.00279	3.471	9.055	324.6	0.169	17.50	0.668
620	0.00255	3.472	9.599	376.4	0.182	20.56	0.664
800	0.00218	3.481	10.68	489.9	0.203	26.75	0.659
980	0.00191	3 505	11.69	612	0.222	33.18	0.664
1160	0.00170	3.540	12.62	743	0.238	39.59	0.676
1340	0.00153	3.575	13.55	885	0.254	46.49	0.686
1520	0.00139	3.622	14.42	1039	0.268	53.19	0.703
1700	0.00128	3.670	15.29	1192	0.282	60.00	0.715
1880	0.00118	3.720	16.18	1370	0.296	67.40	0.733
1940	0.00115	3.735	16.42	1429	0.300	69.80	0.736
			Oxygen				
−280	0.2492	0.2264	5.220×10^{-6}	2.095×10^{-5}	0.00522	0.09252	0.815
−190	0.1635	0.2192	7.721	4.722	0.00790	0.2204	0.773
−100	0.1221	0.2181	9.979	8.173	0.01054	0.3958	0.745
−10	0.0975	0.2187	12.01	12.32	0.01305	0.6120	0.725
80	0.0812	0.2198	13.86	17.07	0.01546	0.8662	0.709
170	0.0695	0.2219	15.56	22.39	0.01774	1.150	0.702
260	0.0609	0.2250	17.16	28.18	0.02000	1.460	0.695
350	0.0542	0.2285	18.66	34.43	0.02212	1.786	0.694
440	0.0487	0.2322	20.10	41.27	0.02411	2.132	0.697
530	0.0443	0.2360	21.48	48.49	0.02610	2.496	0.700
620	0.0406	0.2399	22.79	56.13	0.02792	2.867	0.704

APPENDIX G-1 Property Values of Gases at Atmospheric Pressure (continued)

T, F	ρ, lb/ft^3	c_p, Btu/lb F	μ, lb/sec ft	ν, ft^2/sec	k, Btu/hr ft F	α, ft^2/hr	Pr
Nitrogen							
−280	0.2173	0.2561	4.611×10^{-6}	2.122×10^{-5}	0.005460	0.09811	0.786
−100	0.1068	0.2491	8.700	8.146	0.01054	0.3962	0.747
80	0.0713	0.2486	11.99	16.82	0.01514	0.8542	0.713
260	0.0533	0.2498	14.77	27.71	0.01927	1.447	0.691
440	0.0426	0.2521	17.27	40.54	0.02302	2.143	0.684
620	0.0355	0.2569	19.56	55.10	0.02646	2.901	0.686
800	0.0308	0.2620	21.59	70.10	0.02960	3.668	0.691
980	0.0267	0.2681	23.41	87.68	0.03241	4.528	0.700
1160	0.0237	0.2738	25.19	98.02	0.03507	5.404	0.711
1340	0.0213	0.2789	26.88	126.2	0.03741	6.297	0.724
1520	0.0194	0.2832	28.41	146.4	0.03958	7.204	0.736
1700	0.0178	0.2875	29.90	168.0	0.04151	8.111	0.748
Carbon dioxide							
−64	0.1544	0.187	7.462×10^{-6}	4.833×10^{-5}	0.006243	0.2294	0.818
−10	0.1352	0.192	8.460	6.257	0.007444	0.2868	0.793
80	0.1122	0.208	10.051	8.957	0.009575	0.4103	0.770
170	0.0959	0.215	11.561	12.05	0.01183	0.5738	0.755
260	0.0838	0.225	12.98	15.49	0.01422	0.7542	0.738
350	0.0744	0.234	14.34	19.27	0.01674	0.9615	0.721
440	0.0670	0.242	15.63	23.33	0.01937	1.195	0.702
530	0.0608	0.250	16.85	27.71	0.02208	1.453	0.685
620	0.0558	0.257	18.03	32.31	0.02491	1.737	0.668
Carbon monoxide							
−64	0.09699	0.2491	9.295×10^{-6}	9.583×10^{-5}	0.01101	0.4557	0.758
−10	0.0525	0.2490	10.35	12.14	0.01239	0.5837	0.750
80	0.07109	0.2489	11.990	16.87	0.01459	0.8246	0.737
170	0.06082	0.2492	13.50	22.20	0.01666	1.099	0.728
260	0.05329	0.2504	14.91	27.98	0.01864	1.397	0.722
350	0.04735	0.2520	16.25	34.32	0.0252	1.720	0.718
440	0.04259	0.2540	17.51	41.11	0.02232	2.063	0.718
530	0.03872	0.2569	18.74	48.40	0.02405	2.418	0.721
620	0.03549	0.2598	19.89	56.04	0.02569	2.786	0.724
Ammonia (NH_3)							
−58	0.0239	0.525	4.875×10^{-6}	2.04×10^{-4}	0.0099	0.796	0.93
32	0.0495	0.520	6.285	1.27	0.0127	0.507	0.90
122	0.0405	0.520	7.415	1.83	0.0156	0.744	0.88
212	0.0349	0.534	8.659	2.48	0.0189	1.015	0.87
302	0.0308	0.553	9.859	3.20	0.0226	1.330	0.87
392	0.0275	0.572	11.08	4.03	0.0270	1.713	0.84

APPENDIX G-1 Property Values of Gases at Atmospheric Pressure (continued)

T, F	ρ, lb/ft^3	c_p, Btu/lb F	μ, lb/sec ft	ν, ft^2/sec	k, Btu/hr ft F	α, ft^2/hr	Pr
			Steam (H_2O vapor)				
224	0.0366	0.492	8.54×10^{-6}	2.33×10^{-4}	0.0142	0.789	1.060
260	0.0346	0.481	9.03	2.61	0.0151	0.906	1.040
350	0.0306	0.473	10.25	3.35	0.0173	1.19	1.010
440	0.0275	0.474	11.45	4.16	0.0196	1.50	0.996
530	0.0250	0.477	12.66	5.06	0.0219	1.84	0.991
620	0.0228	0.484	13.89	6.09	0.0244	2.22	0.986
710	0.0211	0.491	15.10	7.15	0.0268	2.58	0.995
800	0.0196	0.498	16.30	8.31	0.0292	2.99	1.000
890	0.0183	0.506	17.50	9.56	0.0317	3.42	1.005
980	0.0171	0.514	18.72	10.98	0.0342	3.88	1.010
1070	0.0161	0.522	19.95	12.40	0.0368	4.38	1.019

†From Heat & Mass Transfer by Eckert and Drake. Copyright 1959 by McGraw-Hill Book Company, Inc. Used with permission of McGraw-Hill Book Co.

APPENDIX G-2 Property Values of Gases at Atmospheric Pressure (SI System of Units)†

Air

T, K	ρ kg/m^3	c_p, J/kg K	μ, kg/ms	ν, m^2/s	k, W/m K	α, m^2/s	Pr
100	3.6010	1.0266 × 10^3	0.6924 × 10^{-5}	1.923 × 10^{-6}	0.009246	0.02501 × 10^{-4}	0.770
150	2.3675	1.0099	1.0283	4.343	0.013735	0.05745	0.753
200	1.7684	1.0061	1.3289	7.490	0.01809	0.10165	0.739
250	1.4128	1.0053	1.488	9.49	0.02227	0.13161	0.722
300	1.1774	1.0057	1.983	15.68	0.02624	0.22160	0.708
350	0.9980	1.0090	2.075	20.76	0.03003	0.2983	0.697
400	0.8826	1.0140	2.286	25.90	0.03365	0.3760	0.689
450	0.7833	1.0207	2.484	28.86	0.03707	0.4222	0.683
500	0.7048	1.0295	2.671	37.90	0.04038	0.5564	0.680
550	0.6423	1.0392	2.848	44.34	0.04360	0.6532	0.680
600	0.5879	1.0551	3.018	51.34	0.04659	0.7512	0.680
650	0.5430	1.0635	3.177	58.51	0.04953	0.8578	0.682
700	0.5030	1.0752	3.332	66.25	0.05230	0.9672	0.684
750	0.4709	1.0856	3.481	73.91	0.05509	1.0774	0.686
800	0.4405	1.0978	3.625	82.29	0.05779	1.1951	0.689
850	0.4149	1.1095	3.765	90.75	0.06028	1.3097	0.692
900	0.3925	1.1212	3.899	99.3	0.06279	1.4271	0.696
950	0.3716	1.1321	4.023	108.2	0.06525	1.5510	0.699
1000	0.3524	1.1417	4.152	117.8	0.06752	1.6779	0.702
1100	0.3204	1.160	4.44	138.6	0.0732	1.969	0.704
1200	0.2947	1.179	4.69	159.1	0.0782	2.251	0.707
1300	0.2707	1.197	4.93	182.1	0.0837	2.583	0.705
1400	0.2515	1.214	5.17	205.5	0.0891	2.920	0.705
1500	0.2355	1.230	5.40	229.1	0.0946	3.262	0.705
1600	0.2211	1.248	5.63	254.5	0.100	3.609	0.705
1700	0.2082	1.267	5.85	280.5	0.105	3.977	0.705
1800	0.1970	1.287	6.07	308.1	0.111	4.379	0.704
1900	0.1858	1.309	6.29	338.5	0.117	4.811	0.704
2000	0.1762	1.338	6.50	369.0	0.124	5.260	0.702
2100	0.1682	1.372	6.72	399.6	0.131	5.715	0.700
2200	0.1602	1.419	6.93	432.6	0.139	6.120	0.707
2300	0.1538	1.482	7.14	464.0	0.149	6.540	0.710
2400	0.1458	1.574	7.35	504.0	0.161	7.020	0.718
2500	0.1394	1.688	7.57	543.5	0.175	7.441	0.730

Helium

T, K	ρ kg/m^3	c_p, J/kg K	μ, kg/ms	ν, m^2/s	k, W/m K	α, m^2/s	Pr
3		5.200 × 10^3	8.42 × 10^{-7}		0.0106		
33	1.4657	5.200	50.2	3.42 × 10^{-6}	0.0353	0.04625 × 10^{-4}	0.74
144	3.3799	5.200	125.5	37.11	0.0928	0.5275	0.70
200	0.2435	5.200	156.6	64.38	0.1177	0.9288	0.694
255	0.1906	5.200	181.7	95.50	0.1357	1.3675	0.70
366	0.13280	5.200	230.5	173.6	0.1691	2.449	0.71
477	0.10204	5.200	275.0	269.3	0.197	3.716	0.72
589	0.08282	5.200	311.3	375.8	0.225	5.215	0.72
700	0.07032	5.200	347.5	494.2	0.251	6.661	0.72
800	0.06023	5.200	381.7	634.1	0.275	8.774	0.72
900	0.05286	5.200	413.6	781.3	0.298	10.834	0.72

APPENDIX G-2 Property Values of Gases at Atmospheric Pressure (continued)

Hydrogen

T, K	ρ, kg/m³	c_p, J/kg K	μ, kg/ms	ν, m²/s	k, W/m K	α, m²/s	Pr
30	0.84722	10.840 × 10^3	1.606 × 10^{-6}	1.895 × 10^{-6}	0.0228	0.02493 × 10^{-4}	0.759
50	0.50955	10.501	2.516	4.880	0.0362	0.0676	0.721
100	0.24572	11.229	4.212	17.14	0.0665	0.2408	0.712
150	0.16371	12.602	5.595	34.18	0.0981	0.475	0.718
200	0.12270	13.540	6.813	55.53	0.1282	0.772	0.719
250	0.09819	14.059	7.919	80.64	0.1561	1.130	0.713
300	0.08185	14.314	8.963	109.5	0.182	1.554	0.706
350	0.07016	14.436	9.954	141.9	0.206	2.031	0.697
400	0.06135	14.491	10.864	177.1	0.228	2.568	0.690
450	0.05462	14.499	11.779	215.6	0.251	3.164	0.682
500	0.04918	14.507	12.636	257.0	0.272	3.817	0.675
550	0.04469	14.532	13.475	301.6	0.292	4.516	0.668
600	0.04085	14.537	14.285	349.7	0.315	5.306	0.664
700	0.03492	14.574	15.89	455.1	0.351	6.903	0.659
800	0.03060	14.675	17.40	569	0.384	8.563	0.664
900	0.02723	14.821	18.78	690	0.412	10.217	0.676
1000	0.02451	14.968	20.16	822	0.440	11.997	0.686
1100	0.02227	15.165	21.46	965	0.464	13.726	0.703
1200	0.02050	15.366	22.75	1107	0.488	15.484	0.715
1300	0.01890	15.575	24.08	1273	0.512	17.394	0.733
1333	0.01842	15.638	24.44	1328	0.519	18.013	0.736

Oxygen

T, K	ρ, kg/m³	c_p, J/kg K	μ, kg/ms	ν, m²/s	k, W/m K	α, m²/s	Pr
100	3.9918	0.9479 × 10^3	7.768 × 10^{-6}	1.946 × 10^{-6}	0.00903	0.023876 × 10^{-4}	0.815
150	2.6190	0.9178	11.490	4.387	0.01367	0.05688	0.773
200	1.9559	0.9131	14.850	7.593	0.01824	0.10214	0.745
250	1.5618	0.9157	17.87	11.45	0.02259	0.15794	0.725
300	1.3007	0.9203	20.63	15.86	0.02676	0.22353	0.709
350	1.1133	0.9291	23.16	20.80	0.03070	0.2968	0.702
400	0.9755	0.9420	25.54	26.18	0.03461	0.3768	0.695
450	0.8682	0.9567	27.77	31.99	0.03828	0.4609	0.694
500	0.7801	0.9722	29.91	38.34	0.04173	0.5502	0.697
550	0.7096	0.9881	31.97	45.05	0.04517	0.6441	0.700
600	0.6504	1.0044	33.92	52.15	0.04832	0.7399	0.704

Nitrogen

T, K	ρ, kg/m³	c_p, J/kg K	μ, kg/ms	ν, m²/s	k, W/m K	α, m²/s	Pr
100	3.4808	1.0722 × 10^3	6.862 × 10^{-6}	1.971 × 10^{-6}	0.009450	0.025319 × 10^{-4}	0.786
200	1.7108	1.0429	12.947	7.568	0.01824	0.10224	0.747
300	1.1421	1.0408	17.84	15.63	0.02620	0.22044	0.713
400	0.8538	1.0459	21.98	25.74	0.03335	0.3734	0.691
500	0.6824	1.0555	25.70	37.66	0.03984	0.5530	0.684
600	0.5687	1.0756	29.11	51.19	0.04580	0.7486	0.686
700	0.4934	1.0969	32.13	65.13	0.05123	0.9466	0.691
800	0.4277	1.1225	34.84	81.46	0.05609	1.1685	0.700
900	0.3796	1.1464	37.49	91.06	0.06070	1.3946	0.711
1000	0.3412	1.1677	40.00	117.2	0.06475	1.6250	0.724
1100	0.3108	1.1857	42.28	136.0	0.06850	1.8591	0.736
1200	0.2851	1.2037	44.50	156.1	0.07184	2.0932	0.748

APPENDIX G-2 Property Values of Gases at Atmospheric Pressure (continued)

T, K	ρ, kg/m³	c_p, J/kg K	μ, kg/ms	ν, m²/s	k, W/m K	α, m²/s	Pr
Carbon dioxide							
220	2.4733	0.783×10^3	11.105×10^{-6}	4.490×10^{-6}	0.010805	0.05920×10^{-4}	0.818
250	2.1657	0.804	12.590	5.813	0.012884	0.07401	0.793
300	1.7973	0.871	14.958	8.321	0.016572	0.10588	0.770
350	1.5362	0.900	17.205	11.19	0.02047	0.14808	0.755
400	1.3424	0.942	19.32	14.39	0.02461	0.19463	0.738
450	1.1918	0.980	21.34	17.90	0.02897	0.24813	0.721
500	1.0732	1.013	23.26	21.67	0.03352	0.3084	0.702
550	0.9739	1.047	25.08	25.74	0.03821	0.3750	0.685
600	0.8938	1.076	26.83	30.02	0.04311	0.4483	0.668
Carbon monoxide							
220	1.55363	1.0429×10^3	13.832×10^{-6}	8.903×10^{-6}	0.01906	0.11760×10^{-4}	0.758
250	0.8410	1.0425	15.40	11.28	0.02144	0.15063	0.750
300	1.13876	1.0421	17.843	15.67	0.02525	0.21280	0.737
350	0.97425	1.0434	20.09	20.62	0.02883	0.2836	0.728
400	0.85363	1.0484	22.19	25.99	0.03226	0.3605	0.722
450	0.75848	1.0551	24.18	31.88	0.0436	0.4439	0.718
500	0.68223	1.0635	26.06	38.19	0.03863	0.5324	0.718
550	0.62024	1.0756	27.89	44.97	0.04162	0.6240	0.721
600	0.56850	1.0877	29.60	52.06	0.04446	0.7190	0.724
Ammonia, NH_3							
220	0.3828	2.198×10^3	7.255×10^{-6}	1.90×10^{-5}	0.0171	0.2054×10^{-4}	0.93
273	0.7929	2.177	9.353	1.18	0.0220	0.1308	0.90
323	0.6487	2.177	11.035	1.70	0.0270	0.1920	0.88
373	0.5590	2.236	12.886	2.30	0.0327	0.2619	0.87
423	0.4934	2.315	14.672	2.97	0.0391	0.3432	0.87
473	0.4405	2.395	16.49	3.74	0.0467	0.4421	0.84
Steam (H_2O vapor)							
380	0.5863	2.060×10^3	12.71×10^{-6}	2.16×10^{-5}	0.0246	0.2036×10^{-4}	1.060
400	0.5542	2.014	13.44	2.42	0.0261	0.2338	1.040
450	0.4902	1.980	15.25	3.11	0.0299	0.307	1.010
500	0.4405	1.985	17.04	3.86	0.0339	0.387	0.996
550	0.4005	1.997	18.84	4.70	0.0379	0.475	0.991
600	0.3652	2.026	20.67	5.66	0.0422	0.573	0.986
650	0.3380	2.056	22.47	6.64	0.0464	0.666	0.995
700	0.3140	2.085	24.26	7.72	0.0505	0.772	1.000
750	0.2931	2.119	26.04	8.88	0.0549	0.883	1.005
800	0.2739	2.152	27.86	10.20	0.0592	1.001	1.010
850	0.2579	2.186	29.69	11.52	0.0637	1.130	1.019

APPENDIX H Total Emissivity Data† I. Metals

Surface	°C	°F	ϵ
Aluminum			
polished, 98 percent pure	200–600	400–1100	0.04–0.06
commercial sheet	100	200	0.09
rough plate	40	100	0.07
heavily oxidized	100–550	200–1000	0.20–0.33
Antimony			
polished	40–250	100–500	0.28–0.31
Bismuth			
bright	100	200	0.34
Brass			
highly polished	250	500	0.03
polished	40	100	0.07
dull plate	40–250	100–500	0.22
oxidized	40–250	100–500	0.46–0.56
Chromium			
polished sheet	40–550	100–1000	0.08–0.27
Cobalt			
unoxidized	250–550	500–1000	0.13–0.23
Copper			
highly polished electrolytic	100	200	0.02
polished	40	100	0.04
slightly polished	40	100	0.12
polished, lightly tarnished	40	100	0.05
dull	40	100	0.15
black oxidized	40	100	0.76
Gold			
pure, highly polished	100–600	200–1100	0.02–0.035
Inconel			
X, stably oxidized	230–900	450–1600	0.55–0.78
B, stably oxidized	230–1000	450–1750	0.32–0.55
X and B, polished	150–300	300–600	0.20
Iron and Steel			
mild steel, polished	150–500	300–900	0.14–0.32
steel, polished	40–250	100–500	0.07–0.10
sheet steel, ground	1000	1700	0.55
sheet steel, rolled	40	100	0.66
sheet steel, strong rough oxide	40	100	0.80
steel, oxidized at 1100 °F	250	500	0.79
cast iron, with skin	40	100	0.70–0.80
cast iron, newly turned	40	100	0.44
cast iron, polished	200	400	0.21
cast iron, oxidized	40–250	100–500	0.57–0.66
iron, red rusted	40	100	0.61
iron, heavily rusted	40	100	0.85
wrought iron, smooth	40	100	0.35
wrought iron, dull oxidized	20–360	70–680	0.94
stainless, polished	40	100	0.07–0.17
stainless, after repeated heating and cooling	230–930	450–1650	0.50–0.70

APPENDIX H Total Emissivity Data I. Metals (continued)

Surface	°C	°F	ϵ
Lead			
polished	40–250	100–500	0.05–0.08
gray, oxidized	40	100	0.28
oxidized at 390 °F	200	400	0.63
oxidized at 1100 °F	40	100	0.63
Magnesium			
polished	40–250	100–500	0.07–0.13
Manganin			
bright rolled	100	200	0.05
Mercury			
pure, clean	40–100	100–200	0.10–0.12
Molybdenum			
polished	40–250	100–500	0.06–0.08
polished	550–1100	1000–2000	0.11–0.18
filament	550–2800	1000–5000	0.08–0.29
Monel			
after repeated heating and cooling	230–930	450–1650	0.45–0.70
oxidized at 1100 °F	200–600	400–1100	0.41–0.46
polished	40	100	0.17
Nickel			
polished	40–250	100–500	0.05–0.07
oxidized	40–250	100–500	0.35–0.49
wire	250–1100	500–2000	0.10–0.19
Platinum			
pure, polished plate	200–600	400–1100	0.05–0.10
oxidized at 1100 °F	250–550	500–1000	0.07–0.11
electrolytic	250–550	500–1000	0.06–0.10
strip	550–1100	1000–2000	0.12–0.14
filament	40–1100	100–2000	0.04–0.19
wire	200–1370	400–2500	0.07–0.18
Silver			
polished or deposited	40–550	100–1000	0.01–0.03
oxidized	40–550	100–1000	0.02–0.04
German silver,* polished	250–550	500–1000	0.07–0.09
Tin			
bright tinned iron	40	100	0.04–0.06
bright	40	100	0.06
polished sheet	100	200	0.05
Tungsten			
filament	550–1100	1000–2000	0.11–0.16
filament	2800	5000	0.39
filament, aged	40–3300	100–6000	0.03–0.35
polished	40–550	100–1000	0.04–0.08

* German silver is actually an alloy of copper, nickel and zinc.

APPENDIX H Total Emissivity Data I. Metals (continued)

Surface	°C	°F	ε
Zinc			
pure polished	40–250	100–500	0.02–0.03
oxidized at 750 °F	400	750	0.11
galvanized, gray	40	100	0.28
galvanized, fairly bright	40	100	0.23
dull	40–250	100–500	0.21

Total Emissivity Data II. Nonmetals

Surface	°C	°F	ε
Asbestos			
board	40	100	0.96
cement	40	100	0.96
paper	40	100	0.93–0.95
slate	40	100	0.97
Brick			
red, rough	40	100	0.93
silica	1000	1800	0.80–0.85
fireclay	1000	1800	0.75
ordinary refractory	1100	2000	0.59
magnesite refractory	1000	1800	0.38
white refractory	1100	2000	0.29
gray, glazed	1100	2000	0.75
Carbon			
filament	1050–1420	1900–2600	0.53
lampsoot	40	100	0.95
Clay			
fired	100	200	0.91
Concrete			
rough	40	100	0.94
Corundum			
emery rough	100	200	0.86
Glass			
smooth	40	100	0.94
quartz glass (2 mm)	250–550	500–1000	0.96–0.66
pyrex	250–550	500–1000	0.94–0.75
Gypsum	40	100	0.80–0.90
Ice			
smooth	0	32	0.97
rough crystals	0	32	0.99
hoarfrost	−18	0	0.99

APPENDIX H Total Emissivity Data II. Nonmetals (continued)

Surface	°C	°F	ε
Limestone	40–250	100–500	0.95–0.83
Marble			
light gray, polished	40	100	0.93
white	40	100	0.95
Mica	40	100	0.75
Paints			
aluminum, various ages and compositions	100	200	0.27–0.62
black gloss	40	100	0.90
black lacquer	40	100	0.80–0.93
white paint	40	100	0.89–0.97
white lacquer	40	100	0.80–0.95
various oil paints	40	100	0.92–0.96
red lead	100	200	0.93
Paper			
white	40	100	0.95
writing paper	40	100	0.98
any color	40	100	0.92–0.94
roofing	40	100	0.91
Plaster			
lime, rough	40–250	100–500	0.92
Porcelain			
glazed	40	100	0.93
Quartz	40–550	100–1000	0.89–0.58
Rubber			
hard	40	100	0.94
soft, gray rough	40	100	0.86
Sandstone	40–250	100–500	0.83–0.90
Snow	(–12)–(–6)	10–20	0.82
Water			
0.1 mm or more thick	40	100	0.96
Wood			
oak, planed	40	100	0.90
walnut, sanded	40	100	0.83
spruce, sanded	40	100	0.82
beech	40	100	0.94
planed	40	100	0.78
various	40	100	0.80–0.90
sawdust	40	100	0.75

†From *Radiation Heat Transfer*, Revised Edition, by E. M. Sparrow and R. D. Cess. Copyright © 1970 by Wadsworth Publishing Company, Inc. Reprinted by permission of the publisher, Brooks/Cole Publishing Company, Monterey, California.

Index

†